Advanced Studies of Flexible Robotic Manipulators

Modeling, Design, Control and Applications

SERIES IN INTELLIGENT CONTROL AND INTELLIGENT AUTOMATION

Editor-in-Charge: Fei-Yue Wang
(*University of Arizona*)

Vol. 1: Reliable Plan Selection by Intelligent Machines
(*J E McInroy, J C Musto, and G N Saridis*)

Vol. 2: Design of Intelligent Control Systems Based on Hierachical Stochastic Automata (*P Lima and G N Saridis*)

Vol. 3: Intelligent Task Planning Using Fuzzy Petri Nets
(*T Cao and A C Sanderson*)

Vol. 6: Modeling, Simulation, and Control of Flexible Manufacturing Systems: A Petri Net Approach (*M Zhou and K Venkatesh*)

Vol. 7: Intelligent Control: Principles, Techniques, and Applications (*Z-X Cai*)

Vol. 10: Autonomous Rock Excavation: Intelligent Control Techniques and Experimentation (*X Shi, P J A Lever and F Y Wang*)

Vol. 11: Multisensor Fusion: A Minimal Representation Framework
(*R Joshi and A C Sanderson*)

Vol. 12: Entropy in Control Engineering
(*George N Saridis*)

Forthcoming volumes:

Vol. 5: Computational Foundations for Intelligent Systems (*S J Yakowitz*)

Vol. 8: Advanced Topics in Computer Vision and Pattern Recognition
(*E Sung, D Mital, E K Teoh, H Wang, and Z Li*)

Vol. 9: Petri Nets for Supervisory Control of Discrete Event Systems: A Structural Approach (*A Giua and F DiCesare*)

Advanced Studies of Flexible Robotic Manipulators

Modeling, Design, Control and Applications

Fei-Yue Wang
Yanqing Gao

Chinese Academy of Sciences
University of Arizona

Vol. 4

Series in Intelligent Control and Intelligent Automation

World Scientific
New Jersey • London • Singapore • Hong Kong

Published by

World Scientific Publishing Co. Pte. Ltd.
5 Toh Tuck Link, Singapore 596224
USA office: Suite 202, 1060 Main Street, River Edge, NJ 07661
UK office: 57 Shelton Street, Covent Garden, London WC2H 9HE

British Library Cataloguing-in-Publication Data
A catalogue record for this book is available from the British Library.

ADVANCED STUDIES OF FLEXIBLE ROBOTIC MANIPULATORS
Modeling, Design, Control, and Applications
Series in Intelligent Control and Intelligent Automation — Vol. 4

ISBN 981-238-390-5

Printed in Singapore.

Contents

Preface

The research interest in flexible manipulators, i.e., lightweight and large dimension robotic manipulators, has increased significantly in recent years. Major advantages of flexible manipulators include, but not limited to, small mass, fast motion, and large force to mass ratio, which are reflected directly in the reduced energy consumption, increased productivity, and enhanced payload capacity. Flexible manipulators have important applications in space exploration, manufacturing automation, construction, mining, hazardous operations, and many other areas.

However, flexible robotic manipulators also impose various challenges in research in comparison to rigid robotic manipulators, ranging from system design, structural optimization, construction, to modeling, sensing, and control. Although significant progresses have been made in many aspects over the last one and half decades, many issues are not resolved yet, and simple, effective, and reliable controls of flexible manipulators still remain an open quest. Clearly, further efforts and results in this area would contribute significantly to robotics, and in particular, automation, as well as its application and education in general control engineering. To accelerate this process, this book presents the state of art of advanced studies in design, modeling, control and applications of flexible manipulators, from the leading experts in this important research area.

The book starts the modeling and control of flexible robotic manipulators, from chapters 1 to 7. It begins with Arteaga and Siciliano's chapter on modeling, nonlinear control and observer of flexible-link manipulators, where motion equations are derived with a combined Lagrange-assumed mode approach. The resulting model shows several similarities with that of a rigid manipulator, thus allowing important properties to be derived for designing controllers and observers. A nonlinear control scheme based on robust control techniques is proposed in order to improve the damping of the system. Since typically link coordinate rates cannot be measured, a nonlinear observer is presented which provides estimates of both joint and link coordinate rates while keeping stability of the system. Chapter 2 presents energy-based robust control strategies for the control of flexible link robots without using the dynamics of the systems explicitly. The energy-based controllers are independent of system parameters and thus possess stability robustness to system parametric uncertainties. Through the evaluation of vibrations of the links, direct control of link deflection is possible. Simulation results are provided to show the effectiveness of the presented approach. Chapter 3 addresses issues related to coordinating two robot manipulators to handle flexible materials, an application that has a wide range of applications in the manufacturing industry. In this case, the two robot manipulators have to follow complicated trajectories to maintain a minimum interaction force with the flexible beam. These trajectories are very complicated and not suitable for real time systems. Three approximation methods of the optimal trajectories and a compliant control scheme are introduced in this chapter. The first one uses a piece-wise linear approximation of the optimal trajectories, while the second one uses adaptive piece-wise linear approximation. The third method applies a continuous approximation of the optimal trajectories using an ellipsoid. Finally, a compliant motion scheme is proposed to reduce the interaction forces and moments in the first method. The stability of the proposed system is investigated. Experimental results encourage the proposed schemes. Chapter 4 discusses modeling methodology and force control scheme of constrained flexible manipulators. It is concluded that for the modeling of constrained flexible manipulators: 1) the end-point of the flexible manipulator is constrained and the boundary condition is non-homogeneous; and 2) the exact model can be described as a distributed parameter system, which can be approximated with an finite-dimensional model for designing controllers. The result here indicates that in order to suppress the spillover instability caused by the residual modes neglected at the controller design phase, a robust controller should be constructed. In chapter 5, the control of flexible link robots is examined. Modeling methods pertinent to control are briefly described and significant results from the control literature are discussed. Comparisons of some of these control techniques are conducted experimentally. The single flexible link and multi-link/multi-axis flexible link manipulators are both examined here. Chapter 6 presents sensor output feedback control laws for one-link flexible robot arms with rotational joints or prismatic joints driven by velocity-controlled motors. Specifically, issues related to mission function, strain feedback control, gain adaptive strain feedback control, and shear force feedback are discussed. Emphasis has been laid on the stability analysis of various sensor feedback closed-loop systems. It is shown that, in addition to being easily implemented, strain feedback and shear force feedback controls introduce damping for motion/vibration control, resulting in good control performance. It is also shown that simple gain adaptive strain feedback control can cope with tip load variations of the flexible arm in maintaining good vibration control performance. Chapter 7 addresses the issues related to the design of robust controllers using genetic algorithms (GA) for lightweight, one-link flexible manipulators working under dynamic environments and other uncertain influences. First, a design procedure based on improved GA is proposed to tune parameters in PID controllers to achieve mixed H_2/H_∞ optimal performances for flexible robotic manipulators. Graphic method, local search strategy, refusal strategy and renewal strategy are used to solve successfully constrains imposed in the design problem. Second, by selecting sensitivity weight functions properly using the GA method, a mixed sensitivity H_∞ controller is developed to ensure robustness of manipulator control systems for varying payloads and other modeling uncertainties. Numeric simulation has been conducted and the results have demonstrated the effectiveness of the proposed method.

The second part of the book, from chapters 8 to 10, focuses on various issues related to the optimal design of flexible manipulators. Chapter 8 analyzes the pole/zero locations of a linearly-tapered Euler-Bernoulli beam pinned at one end and free at the other end. Of particular interest is the location of zeros of the transfer function from torque applied at the pin to displacement of the free end. When tapered beams are used as the links of light-weight robots, the existence of non-minimum phase (right half plane) zeros complicates the robot control problem. Tapering the beam gives the robot designer an additional design parameter when establishing the flexible dynamics. The pole and zero locations are determined from a transfer matrix that is the exact solution for a uniform beam. The approximate results for a tapered model result from segmentation of the beam into segments of different but constant cross sections. The relative position of poles and zeros varies significantly as the rate of taper changes, which will have consequences on feedback stability and non-causal effects in inverse dynamics. Chapter 9 and 10 discuss the problem of optimum shape design of flexible manipulators using two different approaches. One approach employs an iterative method to solve the unconstrained analytical formulation of the optimal equations, while an optimization approach is developed that uses mathematical programming to solve the segmentized equation in order to accommodate various constraints on link design. The goal of each approach is to find the shape that maximizes the fundamental frequency, since this frequency is the governing factor for manipulator speeds. The requirement of multiple tip loads is formulated as a mini-max problem. Also, a multi-link design is implemented which utilizes single-link solutions. A sensitivity analysis of design parameters is conducted to reveal the robustness of optimum designs.

Finally, chapter 11 presents a comprehensive study of dynamic behaviors of flexible robotic links that may have more educational than research values, where dynamic models of flexible manipulators have been formulated using both the Euler-Bernoulli and Timoshenko beam theories. Based on a complete analysis of natural frequencies and modal shape functions, analytic expressions of step responses and general solutions of flexible manipulators have been obtained. Using the method of modal expansion, a finite dimensional nonlinear dynamic model has been derived. Explicit solutions of the asymptotic behavior of high order modal frequencies and vibration modes are given and verified with the numerical results. Both asymptotic analysis and numerical results indicate that the effect of shear deformation is significant for high frequency vibrations or shorter manipulators. Simulations indicate that the effects of both rotary inertia and shear deformation are significant and should be included in the high precision and high performance control design of flexible manipulators.

Both editors would like to thank our contributing authors for their outstanding research, hard work and great patience with our long and tedious editing process. We also like to thank our graduate students and associates at the Complex Systems and Intelligence Science Laboratory, Chinese Academy of Sciences, Beijing, and the Program for Advanced Research in Complex Systems, the University of Arizona, Tucson, Arizona, for their support and service. We thank Mr. Steven Patt of World Scientific Publishing Co. for his patience. This project is supported in part by the Outstanding Oversea Scientist Program, Outstanding Young Scientist Research Award, and a grant from the Chinese National Science Foundation, Chinese Academy of Sciences, and State Development and Planning Commission, China.

Fei-Yue Wang and Yanqing Gao
Beijing, China
Tucson, Arizona, USA

Contributing Authors

Omar Al-Jarrah
Department of Electrical and Computer Engineering
Ohio State University, USA

Marco A. Arteaga
Sección de Eléctrica, DEPFI
Universidad Nacional Autónoma de Méico, Mexico

Wayne J. Book
School of Mechanical Engineering
Georgia Institute of Technology, USA

Ling-Li Cui
Complex Systems and Intelligence Science Laboratory
Institute of Automation, Chinese Academy of Sciences, Beijing, China

Yanqing Gao
Systems and Industrial Engineering Department
University of Arizona, Tucson, Arizona 85721, USA
The Intelligent Control and Systems Engineering Center
Institute of Automation, Chinese Academy of Sciences, Beijing, China

Shuzhi Sam Ge
Department of Electrical and Computer Engineering
National University of Singapore
Singapore 117576

Douglas L. Girvin
School of Mechanical Engineering
Georgia Institute of Technology, USA

Zheng-Hua Luo,
Department of Control Engineering
Osaka University, Japan

Fumitoshi Matsuno
Department of Computational Intelligence and Systems Science
Tokyo Institute of Technology, Japan

Bruno Siciliano
Dipartimento di Informatica e Sistemistica
Università degli Studi di Napoli Federico II, Italy

Jeffery L. Russell
Systems and Industrial Engineering Department
University of Arizona, Tucson, Arizona 85721, USA

David Wang
Department of Electrical and Computer Engineering
University of Waterloo, Canada

Fei-Yue Wang
Complex Systems and Intelligence Science Laboratory
Institute of Automation, Chinese Academy of Sciences, China
Program for Advanced Research for Complex Systems
University of Arizona, Tucson, Arizona 85721, USA

Zhi-Quan Xiao
The Intelligent Control and Systems Engineering Center,
Institute of Automation, Chinese Academy of Science, Beijing, China

Keon-Young Yi
Department of Electrical and Computer Engineering
Ohio State University, USA

Yuan F. Zheng
Department of Electrical and Computer Engineering
Ohio State University, USA

Pixuan Zhou
Systems and Industrial Engineering Department
University of Arizona, Tucson, Arizona 85721, USA

Chapter 1

Flexible-link Manipulators: Modeling, Nonlinear Control and Observer

The interest in flexible robot manipulators has become greater in the latest years. In order to adequately exploit the advantages of this class of manipulators, accurate models and effective control schemes are necessary. This work collects a number of recent results on modeling, nonlinear control and observer for flexible–link manipulators. The equations of motion are derived on the basis of a combined Lagrange–assumed modes approach. The resulting model shows several similarities with that of a rigid manipulator, thus allowing important properties to be derived which are used to design controllers and observers. A nonlinear control scheme based on robust control techniques is proposed in order to improve the damping of the system. Since typically link coordinate rates cannot be measured, a nonlinear observer is presented which provides estimates of both joint and link coordinate rates while keeping stability of the system.

1.1 Introduction

Lightweight manipulators offer many challenges in comparison to rigid and bulky robot manipulators. Energy consumption is smaller, so that the payload–to–arm weight ratio can be increased as well as faster movements can be achieved. Due to their characteristics, this class of manipulators are specially suitable for a number of nonconventional robotic applications, including space missions. On the other hand, the study of link flexibility is enforced also for some kind of heavy manipulators such as large scale systems. In either case, it is no longer possible to assume that link deformation can be neglected. All these factors make the study of flexible robot manipulators quite interesting. The present work aims at presenting some of the latest results on modeling, nonlinear control and observer in this field.

The importance of having an accurate model that can adequately describe the dynamics of the manipulator is obvious. A common way of modeling a flexible robot manipulator consists in using a combined Lagrange–assumed modes approach, which allows deriving a dynamic model in closed form [Book (1984); De Luca and Siciliano (1991); Yuan *et al.* (1993); Canudas de Wit *et al.* (1996); Arteaga (1998)]. Just like in the case of the dynamic model of a rigid manipulator, which possesses many helpful properties [Ortega and Spong (1989); Nicosia and Tomei (1990); Canudas de Wit *et al.* (1990)], it is possible to compute a set of properties for the dynamic model of a flexible manipulator [Arteaga (1998)], whose knowledge facilitates the design of controllers and observers for this kind of system [De Luca and Siciliano (1993a); Lammerts *et al.* (1995); Arteaga (1996a); Arteaga (1996b); Arteaga (1996c)]. Perhaps the most well-known property of (rigid and flexible) manipulators is that referring to their passive structure. With the exception of those controllers based on inverse dynamics [Canudas de Wit *et al.* (1996); De Luca and Siciliano (1993a)], this property is usually employed to prove stability of several control schemes. However, there are many other properties which have been employed to design specific control laws [Ortega and Spong (1989); Nicosia and Tomei (1990); Canudas de Wit *et al.* (1990); De Luca and Siciliano (1993a); De Luca and Siciliano (1993b); De Luca and Panzieri (1994)].

Control of flexible robot manipulators shows the difficulty that there is not an independent control input for each degree of freedom. As in the case of rigid manipulators, there are mainly two goals to be achieved: point–to–point and tracking control. For the first case, some results are given in [De Luca and Siciliano (1993b); De Luca and Panzieri (1994)], where the regulation problem under gravity is studied. In [De Luca and Siciliano (1993b)] the case of no modal damping of the links is treated. By making some assumptions on the inertia matrix, it is possible to guarantee convergence of the link coordinates to certain constant values. In [De Luca and Panzieri (1994)] a solution is proposed for the case that the gravity vector is not perfectly known.

Because an arbitrary trajectory can only be assigned for the joint coordinates, the desired trajectory for the link coordinates must be computed in such a way that the control goal can be accomplished. In [De Luca and Siciliano (1993a); Lammerts *et al.* (1995)] this problem is addressed, and in particular in [Lammerts *et al.* (1995)] not only flexible links but also flexible joints are considered, but there is no guarantee that the computed desired trajectory remains bounded; when the model parameters are not well known, an adaptive algorithm can be used [Slotine and Li (1987)]. On the other hand, in [De Luca and Siciliano (1993a)] inverse control techniques are used [Canudas de Wit *et al.* (1996)], and it is shown that the computed desired trajectory remains bounded. In none of these works the problem of no damping is treated. In this work, the tracking control of flexible robot manipulators is studied [Arteaga (1996c); Arteaga and Siciliano (2000)]. A control law is proposed which is based on the passivity–based control approach with filtered reference velocity [Ortega and Spong (1989)]. It is proven that the desired trajectory for the link coordinates remains bounded. The no damping case is also treated and robust control techniques are used to increase the damping of the system [Dawson *et al.* (1991)].

A problem which deserves special attention regards the possible lack of measurement of link deflection rates, which typically requires the use of an observer. In addition, even though joint positions can be measured very accurately, tachometers (used to measured joint velocities) may not deliver reliable signals. That is why nonlinear observers are recommended to estimate joint speeds. In [Arteaga (1996a); Arteaga (1996b)] an observer for flexible robot manipulators is proposed. Although it is possible to measure link coordinates by using a strain gauge for each coordinate [Arteaga (1995)], the observer requires only a sensor for every flexible link. However, since it is designed independently of any control scheme, the stability of this observer together with the controller proposed in [Arteaga (1996c); Arteaga and Siciliano (2000)] and presented in this work can no longer be guaranteed. To overcome this difficulty, a new observer based on that given in [Nicosia and Tomei (1990)] is proposed [Arteaga (2000)]. In order to ensure enhancing of the damping of the system, some essential modifications are necessary.

The work is organized as follows: Section 1.2 briefly describes the kinematics of flexible robot manipulators and their dynamic modelling. Some of the most important properties of the model are listed. In Section 1.3, control of flexible manipulators is studied. By using robust control techniques, the damping of the system is increased. Since it is not always possible to measure link coordinate rates, a nonlinear observer is proposed in Section 1.4 in order to estimate them. Some simulation results are presented in Section 1.5, while Section 1.6 gives some concluding remarks.

1.2 Modeling

A common way of modeling flexible robot manipulators is using the so–called combined Lagrange–assumed modes approach [Book (1984); De Luca and Siciliano (1991); Yuan *et al.* (1993); Canudas de Wit *et al.* (1996); Arteaga (1998)]. In this case, it is necessary to describe the kinetic and potential energy of the system adequately. In order to compute them, it is advantageous to know the kinematics of the manipulator, which can be achieved by setting coordinate frames along the joint axes. In this section, the kinematics of flexible robot manipulators is briefly studied. By using Lagrange equations of motion, the dynamic model of this class of manipulators is derived in Section 1.2.2 and in Section 1.2.3 some of its most important properties are presented.

1.2.1 *Kinematics*

It is well known that the kinematics of a rigid robot manipulator can be described by employing the Denavit–Hartenberg representation [Denavit and Hartenberg (1955)]. The main idea is to use 4×4 transformation matrices which can be determined uniquely as a function of only 4 parameters. However, this procedure cannot be used directly to describe the kinematics of a flexible robot manipulator due to link deformation. In order to overcome this drawback, the procedure has been modified in [Book (1984); Book (1979)] by including some transformation matrices which take link elasticity into account. A description of the Denavit–Hartenberg representation for rigid manipulators is assumed to be known. Fig. 1.1 depicts a portion of the serial chain for a flexible robot manipulator. The case of revolute joints is considered.

Consider two coordinate frames i and j. Their mutual position and orientation can be expressed in terms of the homogeneous transformation matrix

$$ {}^j\boldsymbol{T}_i = \begin{bmatrix} {}^j\boldsymbol{R}_i & {}^j\boldsymbol{d}_i \\ \boldsymbol{0}^T & 1 \end{bmatrix} \tag{1.1} $$

where ${}^j\boldsymbol{R}_i$ is the 3×3 rotation matrix describing the orientation of the axes of frame i and ${}^j\boldsymbol{d}_i$ is the 3×1 vector describing the origin of frame i, both with respect to frame j; also, in (1.1) $\boldsymbol{0}$ denotes a 3×1 vector of null elements.

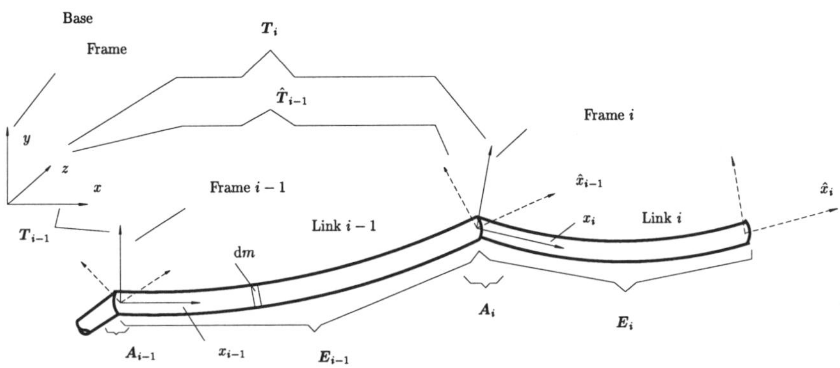

Fig. 1.1 Flexible manipulator serial chain.

The position of a point on link i with respect to frame i is given by

$$ {}^i\boldsymbol{p}_i = \begin{bmatrix} x_i \\ y_i \\ z_i \end{bmatrix}. \tag{1.2} $$

However, it is not possible to use a homogeneous transformation with a vector of the form (1.2), so that it is necessary to rewrite it as

$$ {}^i\boldsymbol{r}_i \triangleq \begin{bmatrix} {}^i\boldsymbol{p}_i \\ 1 \end{bmatrix}. \tag{1.3} $$

To express the position of this point in frame j, a homogeneous transformation is used, i.e.

$$ {}^j\boldsymbol{r}_i = {}^j\boldsymbol{T}_i {}^i\boldsymbol{r}_i. \tag{1.4} $$

In the case of the base frame one has

$$ {}^0\boldsymbol{r}_i \triangleq \boldsymbol{r}_i = {}^0\boldsymbol{T}_i {}^i\boldsymbol{r}_i \triangleq \boldsymbol{T}_i {}^i\boldsymbol{r}_i, \tag{1.5} $$

where the superscript 0 has been conveniently dropped.

In general, the homogeneous transformation of frame i with respect to the base frame can be characterized through the following composition of consecutive transformations:

$$ {}^0\boldsymbol{T}_i \triangleq \boldsymbol{T}_i = \boldsymbol{A}_1\boldsymbol{E}_1\boldsymbol{A}_2\boldsymbol{E}_2\ldots\boldsymbol{A}_{i-1}\boldsymbol{E}_{i-1}\boldsymbol{A}_i \triangleq \hat{\boldsymbol{T}}_{i-1}\boldsymbol{A}_i \tag{1.6} $$

$$ \hat{\boldsymbol{T}}_{i-1} \triangleq \boldsymbol{T}_{i-1}\boldsymbol{E}_{i-1} \tag{1.7} $$

$$ \boldsymbol{T}_1 = \boldsymbol{A}_1, \tag{1.8} $$

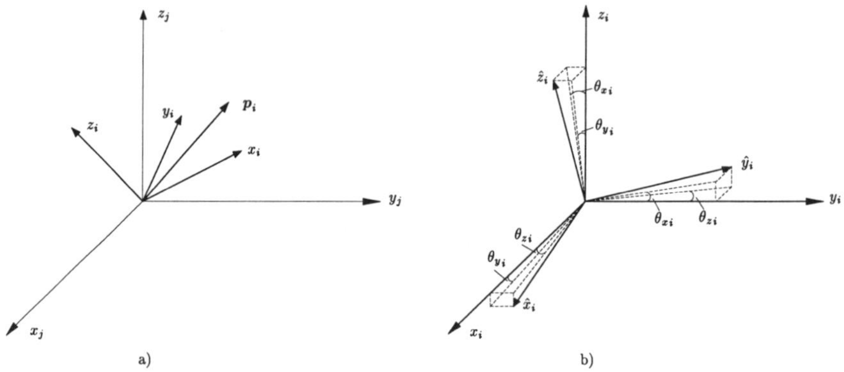

Fig. 1.2 a) Rotation of a coordinate frame; b) Rotation of a coordinate frame due to deformation of the flexible link.

where $\boldsymbol{A}_i$ is the standard homogeneous transformation matrix for joint i due to rigid motion and $\boldsymbol{E}_i$ is the homogeneous transformation matrix due to link i length and deflection. Notice that, even though the superscript is not explicitly indicated, each transformation matrix is referred to the frame determined by the preceding transformation.

The transformation matrix $\boldsymbol{A}_i$ can be computed just like in the case of rigid robot manipulators [Sciavicco and Siciliano (2000)]. On the other hand, the transformation matrix $\boldsymbol{E}_i$ deserves special attention. Firstly, consider the general form of a rotation matrix ${}^j\boldsymbol{R}_i$ between two coordinate frames of common origin (see Fig. 1.2 a)) [Sciavicco and Siciliano (2000)]:

$$ {}^j\boldsymbol{R}_i = \begin{bmatrix} \boldsymbol{x}_i^T\boldsymbol{x}_j & \boldsymbol{y}_i^T\boldsymbol{x}_j & \boldsymbol{z}_i^T\boldsymbol{x}_j \\ \boldsymbol{x}_i^T\boldsymbol{y}_j & \boldsymbol{y}_i^T\boldsymbol{y}_j & \boldsymbol{z}_i^T\boldsymbol{y}_j \\ \boldsymbol{x}_i^T\boldsymbol{z}_j & \boldsymbol{y}_i^T\boldsymbol{z}_j & \boldsymbol{z}_i^T\boldsymbol{z}_j \end{bmatrix} = \begin{bmatrix} \cos(\theta_{\boldsymbol{x}_i\boldsymbol{x}_j}) & \cos(\theta_{\boldsymbol{y}_i\boldsymbol{x}_j}) & \cos(\theta_{\boldsymbol{z}_i\boldsymbol{x}_j}) \\ \cos(\theta_{\boldsymbol{x}_i\boldsymbol{y}_j}) & \cos(\theta_{\boldsymbol{y}_i\boldsymbol{y}_j}) & \cos(\theta_{\boldsymbol{z}_i\boldsymbol{y}_j}) \\ \cos(\theta_{\boldsymbol{x}_i\boldsymbol{z}_j}) & \cos(\theta_{\boldsymbol{y}_i\boldsymbol{z}_j}) & \cos(\theta_{\boldsymbol{z}_i\boldsymbol{z}_j}) \end{bmatrix}, \tag{1.9} $$

where $\boldsymbol{x}$, $\boldsymbol{y}$, $\boldsymbol{z}$ denote the unit vectors of the respective axes. Then, the relationship between ${}^j\boldsymbol{p}_i$ and ${}^i\boldsymbol{p}_i$ is given by

$$ {}^j\boldsymbol{p}_i = {}^j\boldsymbol{R}_i{}^i\boldsymbol{p}_i. \tag{1.10} $$

From (1.9), it can easily be understood that the knowledge of the angles $\theta_{\boldsymbol{x}_i\boldsymbol{x}_j} \cdots \theta_{\boldsymbol{z}_i\boldsymbol{z}_j}$ is enough to compute ${}^j\boldsymbol{R}_i$. With this background, the matrices $\boldsymbol{E}_i$ can be determined as follows. Consider Fig. 1.1 again and assume that the x–axis of frame i is along the link. Assuming small link deformation [Book (1979); Meirovitch (1967); Meirovitch (1975)], $\boldsymbol{E}_i$ can be expressed as [Book (1979)]

$$\boldsymbol{E}_i = \begin{bmatrix} 1 & \cos(\pi/2+\theta_{z_i}) & \cos(\pi/2-\theta_{y_i}) & l_i+\delta_{x_i} \\ \cos(\pi/2-\theta_{z_i}) & 1 & \cos(\pi/2+\theta_{x_i}) & \delta_{y_i} \\ \cos(\pi/2+\theta_{y_i}) & \cos(\pi/2-\theta_{x_i}) & 1 & \delta_{z_i} \\ 0 & 0 & 0 & 1 \end{bmatrix}, \quad (1.11)$$

where θ_{x_i}, θ_{y_i}, θ_{z_i} are the angles of rotation, and δ_{x_i}, δ_{y_i}, δ_{z_i} represent link i deformation along x, y, z, respectively, being l_i the length of the link without deformation. The angles of rotation θ_{x_i}, θ_{y_i}, θ_{z_i} are depicted in Fig. 1.2 b). By taking into account the fact $\cos(\pi/2+\alpha) = -\sin(\alpha)$ and assuming small angles, so that $\sin(\alpha) \approx \alpha$ is valid, the matrix $\boldsymbol{E}_i$ can be approximated as

$$\boldsymbol{E}_i = \begin{bmatrix} 1 & -\theta_{z_i} & \theta_{y_i} & l_i+\delta_{x_i} \\ \theta_{z_i} & 1 & -\theta_{x_i} & \delta_{y_i} \\ -\theta_{y_i} & \theta_{x_i} & 1 & \delta_{z_i} \\ 0 & 0 & 0 & 1 \end{bmatrix}. \quad (1.12)$$

By using the homogeneous transformation matrices $\boldsymbol{A}_i$ and $\boldsymbol{E}_i$, the position of any point along the robot manipulator can uniquely be determined from Eqs. (1.5), (1.6) and (1.12).

1.2.2 *Dynamics*

In order to obtain a set of differential equations of motion to adequately describe the dynamics of a flexible–link manipulator, the Lagrange's approach can be used. A system with n generalized coordinates q_i must satisfy n differential equations of the form

$$\frac{\mathrm{d}}{\mathrm{dt}}\frac{\partial \mathcal{L}}{\partial \dot{q}_i} - \frac{\partial \mathcal{L}}{\partial q_i} + \frac{\partial \mathcal{D}}{\partial \dot{q}_i} = u_i \qquad i = 1, \ldots, n, \quad (1.13)$$

where $\mathcal{L}$ is the so called Lagrangian which is given by [Wellstead (1979)]

$$\mathcal{L} = \mathcal{T} - \mathcal{U}; \quad (1.14)$$

$\mathcal{T}$ represents the kinetic energy of the system and $\mathcal{U}$ the potential energy. Also, in (1.13) $\mathcal{D}$ is the Rayleigh's dissipation function which allows dissipative effects to be included, and u_i is the generalized force acting on q_i.

To compute the kinetic energy of the system, the manipulator kinematics can be described systematically as explained in the previous section.

The kinetic energy of link i link can be expressed as

$$\mathcal{T}_i = \int_{\text{link}_i} \mathrm{d}\mathcal{T}_i = \frac{1}{2}\int_{\text{link}_i} \mathrm{Tr}\left(\frac{\mathrm{d}\boldsymbol{r}_i}{\mathrm{d}t}\frac{\mathrm{d}\boldsymbol{r}_i^T}{\mathrm{d}t}\right)\mathrm{d}m, \tag{1.15}$$

which implies that the kinetic energy for the whole system is

$$\mathcal{T} = \sum_{i=1}^{n}\int_{\text{link}_i} \mathrm{d}\mathcal{T}_i = \frac{1}{2}\sum_{i=1}^{n}\int_{\text{link}_i} \mathrm{Tr}\left(\frac{\mathrm{d}\boldsymbol{r}_i}{\mathrm{d}t}\frac{\mathrm{d}\boldsymbol{r}_i^T}{\mathrm{d}t}\right)\mathrm{d}m; \tag{1.16}$$

$\mathrm{Tr}(\cdot)$ represents the trace operator of a square matrix. By accounting for (1.5), the kinetic energy (1.16) can be written in the form

$$\mathcal{T} = \frac{1}{2}\dot{\boldsymbol{q}}^T \boldsymbol{H}(\boldsymbol{q})\dot{\boldsymbol{q}}, \tag{1.17}$$

where

$$\begin{aligned}\boldsymbol{q}(t) &\triangleq [\theta_1(t)\cdots\theta_n(t)\ \ \delta_{11}(t)\cdots\delta_{1m_1}(t)\cdots\delta_{n1}(t)\cdots\delta_{nm_n}(t)]^T \\ &= [q_{10}(t)\cdots q_{n0}(t)\ \ q_{11}(t)\cdots q_{1m_1}(t)\cdots q_{n1}(t)\cdots q_{nm_n}(t)]^T\end{aligned} \tag{1.18}$$

is the vector of generalized coordinates which is formed by the rigid coordinates $\theta_1\ldots\theta_n$ $(q_{10}\cdots q_{n0})$ and the flexible coordinates $\delta_{11}\ldots\delta_{nm_n}$ $(q_{11}\cdots q_{nm_n})$, and

$$\boldsymbol{H}(\boldsymbol{q}) = \begin{bmatrix}\boldsymbol{H}_{\theta\theta}(\boldsymbol{q}) & \boldsymbol{H}_{\theta\delta}(\boldsymbol{q}) \\ \boldsymbol{H}_{\theta\delta}^T(\boldsymbol{q}) & \boldsymbol{H}_{\delta\delta}(\boldsymbol{q})\end{bmatrix} \tag{1.19}$$

is the inertia matrix. In particular, $\boldsymbol{H}_{\theta\theta}(\boldsymbol{q})$ is associated to the rigid coordinates, $\boldsymbol{H}_{\theta\delta}(\boldsymbol{q})$ takes into account the relationship between the flexible and rigid coordinates, and $\boldsymbol{H}_{\delta\delta}(\boldsymbol{q})$ is associated to the flexible coordinates.

In order to find an analytical form for the inertia matrix, it is necessary to describe the deflection and torsion of each link as a function of the link coordinates. These can be expressed according to the so–called assumed modes method, i.e. [De Luca and Siciliano (1991); Yuan *et al.* (1993);

Meirovitch (1967); Meirovitch (1975)]

$$\delta_{x_i} = \sum_{j=1}^{m_i} \phi_{x_{ij}} \delta_{ij} \qquad \theta_{x_i} = \sum_{j=1}^{m_i} \theta_{x_{ij}} \delta_{ij} \tag{1.20}$$

$$\delta_{y_i} = \sum_{j=1}^{m_i} \phi_{y_{ij}} \delta_{ij} \qquad \theta_{y_i} = \sum_{j=1}^{m_i} \theta_{y_{ij}} \delta_{ij} \tag{1.21}$$

$$\delta_{z_i} = \sum_{j=1}^{m_i} \phi_{z_{ij}} \delta_{ij} \qquad \theta_{z_i} = \sum_{j=1}^{m_i} \theta_{z_{ij}} \delta_{ij}, \tag{1.22}$$

where $\phi_{x_{ij}}$, $\phi_{y_{ij}}$, $\phi_{z_{ij}}$ ($\theta_{x_{ij}}$, $\theta_{y_{ij}}$, $\theta_{z_{ij}}$) are the spatial mode shapes used to model the deflection (torsion) of link i, being m_i the number of link coordinates.

From (1.16), the elements of $\boldsymbol{H}_{\theta\theta}(\boldsymbol{q})$ can be computed as

$$h_{\alpha 0 h 0} = \sum_{i=\max\{\alpha,h\}}^{n} \mathrm{Tr}\left(\left(\hat{\boldsymbol{T}}_{\alpha-1}\boldsymbol{U}_{\alpha}{}^{\alpha}\tilde{\boldsymbol{T}}_i\right)\boldsymbol{F}_i\left(\hat{\boldsymbol{T}}_{h-1}\boldsymbol{U}_{h}{}^{h}\tilde{\boldsymbol{T}}_i\right)^T\right) \tag{1.23}$$

with

$$\begin{aligned}
{}^{h}\boldsymbol{T}_i &\triangleq \boldsymbol{A}_{h+1}\boldsymbol{E}_{h+1}\boldsymbol{A}_{h+2}\boldsymbol{E}_{h+2}\cdots\boldsymbol{A}_{i-1}\boldsymbol{E}_{i-1}\boldsymbol{A}_i && (1.24)\\
{}^{h}\tilde{\boldsymbol{T}}_i &\triangleq \boldsymbol{E}_h{}^{h}\boldsymbol{T}_i && (1.25)\\
\boldsymbol{U}_h &\triangleq \frac{\partial \boldsymbol{A}_h}{\partial q_{h0}} && (1.26)\\
\boldsymbol{F}_i &\triangleq \boldsymbol{C}_i + \sum_{j=1}^{m_i}\delta_{ij}\left((\boldsymbol{C}_{ij}+\boldsymbol{C}_{ij}^T) + \sum_{k=1}^{m_i}\delta_{ik}\boldsymbol{C}_{ikj}\right) = \boldsymbol{F}_i^T && (1.27)\\
\boldsymbol{C}_i &\triangleq \int_{\mathrm{link}_i} [x_i\ y_i\ z_i\ 1]^T[x_i\ y_i\ z_i\ 1]\mathrm{d}m && (1.28)\\
\boldsymbol{C}_{ij} &\triangleq \int_{\mathrm{link}_i} [x_i\ y_i\ z_i\ 1]^T[\phi_{xij}\ \phi_{yij}\ \phi_{zij}\ 0]\mathrm{d}m && (1.29)\\
\boldsymbol{C}_{ikj} &\triangleq \int_{\mathrm{link}_i} [\phi_{xik}\ \phi_{yik}\ \phi_{zik}\ 0]^T[\phi_{xij}\ \phi_{yij}\ \phi_{zij}\ 0]\mathrm{d}m = \boldsymbol{C}_{ijk}^T, && (1.30)
\end{aligned}$$

the elements of $\boldsymbol{H}_{\theta\delta}(\boldsymbol{q})$ can be computed as

$$h_{h0\alpha\beta} = \gamma_{h\alpha} + \sum_{i=\max\{h,\alpha+1\}}^{n} \mathrm{Tr}\left(\left(\hat{\boldsymbol{T}}_{h-1}\boldsymbol{U}_h{}^{h}\tilde{\boldsymbol{T}}_i\right)\boldsymbol{F}_i\left(\boldsymbol{T}_{\alpha}\boldsymbol{N}_{\alpha\beta}{}^{\alpha}\boldsymbol{T}_i\right)^T\right) \tag{1.31}$$

with

$$\gamma_{h\alpha} = \begin{cases} 0 & \text{if } h > \alpha \\ \operatorname{Tr}\left(\left(\hat{\boldsymbol{T}}_{h-1}\boldsymbol{U}_h{}^h\tilde{\boldsymbol{T}}_\alpha\right)\boldsymbol{D}_{\alpha\beta}\boldsymbol{T}_\alpha^T\right) & \text{if } h \leq \alpha \end{cases}$$

$$\boldsymbol{N}_{\alpha\beta} \triangleq \begin{bmatrix} 0 & -\theta_{z\alpha\beta} & \theta_{y\alpha\beta} & \phi_{x\alpha\beta} \\ \theta_{z\alpha\beta} & 0 & -\theta_{x\alpha\beta} & \phi_{y\alpha\beta} \\ -\theta_{y\alpha\beta} & \theta_{x\alpha\beta} & 0 & \phi_{z\alpha\beta} \\ 0 & 0 & 0 & 0 \end{bmatrix} \tag{1.32}$$

$$\boldsymbol{D}_{\alpha\beta} \triangleq \boldsymbol{C}_{\alpha\beta} + \sum_{k=1}^{m_\alpha} \delta_{\alpha k}\boldsymbol{C}_{\alpha k\beta}, \tag{1.33}$$

and the elements of $\boldsymbol{H}_{\delta\delta}(\boldsymbol{q})$ can be computed as

$$h_{hk\alpha\beta} = \eta_{h\alpha} + \sum_{i=\max\{h,\alpha\}+1}^{n} \operatorname{Tr}\left(\left(\boldsymbol{T}_h\boldsymbol{N}_{hk}{}^h\boldsymbol{T}_i\right)\boldsymbol{F}_i\left(\boldsymbol{T}_\alpha\boldsymbol{N}_{\alpha\beta}{}^\alpha\boldsymbol{T}_i\right)^T\right) \tag{1.34}$$

with

$$\eta_{h\alpha} = \begin{cases} \operatorname{Tr}\left(\boldsymbol{T}_h\boldsymbol{C}_{hk\beta}\boldsymbol{T}_h^T\right) & \text{if } h = \alpha \\ \operatorname{Tr}\left(\left(\boldsymbol{T}_h\boldsymbol{N}_{hk}{}^h\mathbf{T}_\alpha\right)\boldsymbol{D}_{\alpha\beta}\boldsymbol{T}_\alpha^T\right) & \text{if } h < \alpha \\ \operatorname{Tr}\left(\left(\boldsymbol{T}_\alpha\mathbf{N}_{\alpha\beta}{}^\alpha\boldsymbol{T}_h\right)\boldsymbol{D}_{hk}\boldsymbol{T}_h^T\right) & \text{if } h > \alpha. \end{cases}$$

Notice that $\boldsymbol{H}_{\theta\theta}(\boldsymbol{q})$ and $\boldsymbol{H}_{\delta\delta}(\boldsymbol{q})$ are symmetric, so that it is only necessary to compute the terms for which $h \geq \alpha$.

The next step is to compute the potential energy of the system. In a flexible–link manipulator there are two sources of potential energy: link gravity and link elasticity.

The differential element of gravity potential energy of link i is given by

$$d\mathcal{U}_{gi} = -\boldsymbol{g}_0^T\boldsymbol{T}_i{}^i\boldsymbol{r}_i dm \tag{1.35}$$

where

$$\boldsymbol{g}_0 = [g_x \; g_y \; g_z \; 0]^T \tag{1.36}$$

is the gravity vector expressed in the base frame. The total gravitational energy is

$$\mathcal{U}_g = -\boldsymbol{g}_0^T \sum_{i=1}^{n} \boldsymbol{T}_i \boldsymbol{h}_i \tag{1.37}$$

with

$$\boldsymbol{h}_i \triangleq M_i \boldsymbol{l}_i + \sum_{k=1}^{m_i} \delta_{ik} \boldsymbol{s}_{ik} \tag{1.38}$$

$$\boldsymbol{l}_i = [l_{xi}\ l_{yi}\ l_{zi}\ 1]^T \tag{1.39}$$

$$\boldsymbol{s}_{ik} = \int_{\text{link}_i} [\phi_{xik}\ \phi_{yik}\ \phi_{zik}\ 0]^T \mathrm{d}m, \tag{1.40}$$

where $\boldsymbol{l}_i$ is the vector from joint i to the center of gravity when link i is undeformed and M_i is the total mass of the link.

The strain potential energy associated to the deformation of link i is given by [Yuan *et al.* (1993)]

$$\mathcal{U}_{ei} = \frac{1}{2} \int_{\text{link}_i} \left(EI_y \left(\frac{\partial^2 \delta_{y_i}}{\partial x_i^2} \right)^2 + EI_z \left(\frac{\partial^2 \delta_{z_i}}{\partial x_i^2} \right)^2 + E_G J_x \left(\frac{\partial \theta_{x_i}}{\partial x_i} \right)^2 \right) \mathrm{d}x_i, \tag{1.41}$$

where E is Young's modulus of elasticity, I_y (I_z) is the area moment of inertia of the link about an axis parallel to y (z) through the center of mass of the cross section, E_G is the shear modulus, and J_x is the polar area moment of inertia of the link about the center of mass. The integration in (1.41) is carried out along x–axis. Notice that in (1.41) the compression in the x direction has been assumed to be negligible. In view of (1.20)–(1.22), Eq. (1.41) can be rewritten as

$$\mathcal{U}_{ei} = \frac{1}{2} \sum_{j=1}^{m_i} \sum_{k=1}^{m_i} \delta_{ij} \delta_{ik} (k_{y_{ijk}} + k_{z_{ijk}} + k_{x_{ijk}}), \tag{1.42}$$

where $k_{x_{ijk}}$, $k_{y_{ijk}}$, $k_{z_{ijk}}$ are the stiffness coefficients given by

$$k_{x_{ijk}} = \int_{\text{link}_i} E_G J_x \frac{\mathrm{d}\theta_{x_{ij}}}{\mathrm{d}x_i} \frac{\mathrm{d}\theta_{x_{ik}}}{\mathrm{d}x_i} \mathrm{d}x_i \tag{1.43}$$

$$k_{y_{ijk}} = \int_{\text{link}_i} EI_y \frac{\mathrm{d}^2 \phi_{y_{ij}}}{\mathrm{d}x_i^2} \frac{\mathrm{d}^2 \phi_{y_{ik}}}{\mathrm{d}x_i^2} \mathrm{d}x_i \tag{1.44}$$

$$k_{z_{ijk}} = \int_{\text{link}_i} EI_z \frac{\mathrm{d}^2 \phi_{z_{ij}}}{\mathrm{d}x_i^2} \frac{\mathrm{d}^2 \phi_{z_{ik}}}{\mathrm{d}x_i^2} \mathrm{d}x_i. \tag{1.45}$$

The total elastic energy is

$$\mathcal{U}_e = \frac{1}{2}\sum_{i=1}^{n}\sum_{j=1}^{m_i}\sum_{k=1}^{m_i}\delta_{ij}\delta_{ik}k_{ijk} \tag{1.46}$$

with

$$k_{ijk} \triangleq k_{y_{ijk}} + k_{z_{ijk}} + k_{x_{ijk}} = k_{ikj} \tag{1.47}$$

or in matrix form:

$$\mathcal{U}_e = \frac{1}{2}\boldsymbol{\delta}^T\boldsymbol{K}\boldsymbol{\delta} = \frac{1}{2}\boldsymbol{q}^T\begin{bmatrix}\boldsymbol{O} & \boldsymbol{O}\\ \boldsymbol{O} & \boldsymbol{K}\end{bmatrix}\boldsymbol{q} \triangleq \frac{1}{2}\boldsymbol{q}^T\boldsymbol{K}_e\boldsymbol{q}, \tag{1.48}$$

where

$$\boldsymbol{\delta} \triangleq \begin{bmatrix}\delta_{11} & \cdots & \delta_{1m_1} & \cdots & \delta_{n1} & \cdots & \delta_{nm_n}\end{bmatrix}^T \tag{1.49}$$

is the vector of flexible coordinates, and

$$\boldsymbol{K} = \begin{bmatrix} k_{111} & \cdots & k_{11m_1} & \cdots & & \cdots & 0 \\ \vdots & \ddots & \vdots & & & & \vdots \\ k_{1m_11} & \cdots & k_{1m_1m_1} & & & & \\ \vdots & & & \ddots & & & \vdots \\ & & & & k_{n11} & \cdots & k_{n1m_n} \\ \vdots & & & & \vdots & \ddots & \vdots \\ 0 & \cdots & & \cdots & k_{nm_n1} & \cdots & k_{nm_nm_n} \end{bmatrix} \tag{1.50}$$

is called the stiffness matrix.

By taking (1.14), (1.17), (1.37) and (1.48) into account, the Lagrangian can be written as

$$\mathcal{L} = \frac{1}{2}\left(\dot{\boldsymbol{q}}^T\boldsymbol{H}(\boldsymbol{q})\dot{\boldsymbol{q}} - \boldsymbol{q}^T\boldsymbol{K}_e\boldsymbol{q}\right) + \boldsymbol{g}_0^T\sum_{i=1}^{n}\boldsymbol{T}_i\boldsymbol{h}_i. \tag{1.51}$$

In order to model link modal damping and joint viscous friction, the Rayleigh's dissipation function can be employed with the use of a matrix $\boldsymbol{D}$ so that [Meirovitch (1967)]:

$$\mathcal{D} = \frac{1}{2}\dot{\boldsymbol{q}}^T\boldsymbol{D}\dot{\boldsymbol{q}}\,. \tag{1.52}$$

Expressing (1.13) in matrix form yields

$$\frac{\mathrm{d}}{\mathrm{dt}}\left(\frac{\partial \mathcal{L}}{\partial \dot{\boldsymbol{q}}}\right)^T - \left(\frac{\partial \mathcal{L}}{\partial \boldsymbol{q}}\right)^T + \left(\frac{\partial \mathcal{D}}{\partial \dot{\boldsymbol{q}}}\right)^T = \boldsymbol{u} \tag{1.53}$$

with

$$\boldsymbol{u} \triangleq \begin{bmatrix} \boldsymbol{\tau} \\ \boldsymbol{0} \end{bmatrix}, \tag{1.54}$$

where $\boldsymbol{\tau}$ is the $n \times 1$ vector of the joint torques, and $\boldsymbol{0}$ denotes an $m \times 1$ vector of null elements ($m = m_1 + \cdots + m_n$) accounting for the fact that no generalized force acts on the flexible coordinates $\boldsymbol{\delta}$ as long as clamped boundary conditions at the joint side are assumed.

Then, substituting (1.52) into (1.53) and accounting for (1.51) leads to

$$\boldsymbol{H}(\boldsymbol{q})\ddot{\boldsymbol{q}} + \boldsymbol{h}_c(\boldsymbol{q}, \dot{\boldsymbol{q}}) + \boldsymbol{K}_e\boldsymbol{q} + \boldsymbol{D}\dot{\boldsymbol{q}} + \boldsymbol{g}(\boldsymbol{q}) = \boldsymbol{u}, \tag{1.55}$$

where

$$\boldsymbol{h}_c(\boldsymbol{q}, \dot{\boldsymbol{q}}) \triangleq \boldsymbol{C}(\boldsymbol{q}, \dot{\boldsymbol{q}})\dot{\boldsymbol{q}} \tag{1.56}$$

is the vector of Coriolis and centrifugal forces with

$$c_{rs\alpha\beta} = \sum_{i=1}^{n}\sum_{j=0}^{m_i} c_{ij\alpha\beta rs}\dot{q}_{ij} \tag{1.57}$$

$$c_{ij\alpha\beta rs} \triangleq \frac{1}{2}\left(\frac{\partial h_{rs\alpha\beta}}{\partial q_{ij}} + \frac{\partial h_{rsij}}{\partial q_{\alpha\beta}} - \frac{\partial h_{ij\alpha\beta}}{\partial q_{rs}}\right) \tag{1.58}$$

$$c_{ij\alpha\beta rs} = c_{\alpha\beta ijrs} \tag{1.59}$$

$$g_{rs} = \begin{cases} -\boldsymbol{g}_0^T \displaystyle\sum_{i=r}^{n} \frac{\partial \boldsymbol{T}_i}{\partial q_{rs}} \boldsymbol{h}_i & \text{if } s = 0 \\ -\boldsymbol{g}_0^T \left(\displaystyle\sum_{i=r+1}^{n} \frac{\partial \boldsymbol{T}_i}{\partial q_{rs}} \boldsymbol{h}_i + \boldsymbol{T}_r \boldsymbol{s}_{rs}\right) & \text{if } s \neq 0. \end{cases} \tag{1.60}$$

1.2.3 *Model Properties*

In this section some properties of model (1.55) are presented. Many of them are rather physical properties while other arise from the procedure used to derive the dynamic model of the manipulator. Several of these properties are similar to those of rigid manipulators. As a matter of fact, the properties presented in the following apply to rigid manipulators just by letting the link deformation be zero.

Hereafter, the Euclidean norm for vectors is used, i.e.

$$\|q\| \triangleq \sqrt{\sum_{i=1}^{n} q_i^2}. \tag{1.61}$$

The norm of a matrix A is the corresponding induced norm

$$\|A\| \triangleq \sqrt{\lambda_{\max}(A^T A)}, \tag{1.62}$$

where $\lambda_{\max}(\cdot)$ ($\lambda_{\min}(\cdot)$) denotes the largest (smallest) eigenvalue of a matrix. Since all norms in $\Re^n$ are equivalent [Desoer and Vidyasagar (1975)], the results presented in this section are valid for any norm in $\Re^n$.

A well-known property of the dynamic model of a robot manipulator is the following one.

Property 1.1 The inertia matrix $H(q)$ is symmetric positive definite.

Proof: It can be seen directly from (1.19), (1.23), (1.31) and (1.34) that $H(q)$ is symmetric. Since the kinetic energy of any mechanical system can be zero if and only if the system is in a steady state, and otherwise it is always greater than zero, it follows from (1.17) that $H(q)$ is positive definite.

△

The next property is very important. It is related to the passive structure of robot manipulators and it is frequently used in the proof of many control schemes. It gives a relationship between the inertia matrix $H(q)$ and the matrix $C(q, \dot{q})$ employed to compute the vector of Coriolis and centrifugal torques.

Property 1.2 The matrix $N(q, \dot{q}) \triangleq \dot{H}(q) - 2C(q, \dot{q})$ is skew symmetric.

Proof: Every element of $\dot{H}(q)$ satisfies

$$\dot{h}_{rs\alpha\beta} = \sum_{i=1}^{n} \sum_{j=0}^{m_i} \frac{\partial h_{rs\alpha\beta}}{\partial q_{ij}} \dot{q}_{ij}. \tag{1.63}$$

By taking (1.57) and (1.58) into account, the elements of $N(q, \dot{q})$ can be computed as

$$\begin{aligned}
n_{rs\alpha\beta} &\triangleq \dot{h}_{rs\alpha\beta} - 2c_{rs\alpha\beta} \\
&= \sum_{i=1}^{n}\sum_{j=0}^{m_i}\left(\frac{\partial h_{rs\alpha\beta}}{\partial q_{ij}} - \left(\frac{\partial h_{rs\alpha\beta}}{\partial q_{ij}} + \frac{\partial h_{rsij}}{\partial q_{\alpha\beta}} - \frac{\partial h_{ij\alpha\beta}}{\partial q_{rs}}\right)\right)\dot{q}_{ij} \\
&= \sum_{i=1}^{n}\sum_{j=0}^{m_i}\left(\frac{\partial h_{ij\alpha\beta}}{\partial q_{rs}} - \frac{\partial h_{rsij}}{\partial q_{\alpha\beta}}\right)\dot{q}_{ij}. \qquad (1.64)
\end{aligned}$$

Since $h_{rs\alpha\beta} = h_{\alpha\beta rs}$, the property holds true. △

Note that Property 1.2 has been proven using the definition of $\boldsymbol{C}(\boldsymbol{q},\dot{\boldsymbol{q}})$ which is in terms of the Christoffel symbols. Since there are many possible definitions for $\boldsymbol{C}(\boldsymbol{q},\dot{\boldsymbol{q}})$, it is worth pointing out that

$$\dot{\boldsymbol{q}}^T(\dot{\boldsymbol{H}}(\boldsymbol{q}) - 2\boldsymbol{C}(\boldsymbol{q},\dot{\boldsymbol{q}}))\dot{\boldsymbol{q}} = 0 \qquad (1.65)$$

is always true no matter what definition of $\boldsymbol{C}(\boldsymbol{q},\dot{\boldsymbol{q}})$ is used [Ortega and Spong (1989)]. To show this, rewrite (1.53) and (1.55) as

$$\frac{\mathrm{d}}{\mathrm{dt}}\left(\frac{\partial \mathcal{L}}{\partial \dot{\boldsymbol{q}}}\right)^T - \left(\frac{\partial \mathcal{L}}{\partial \boldsymbol{q}}\right)^T = \boldsymbol{\psi} = \boldsymbol{H}(\boldsymbol{q})\ddot{\boldsymbol{q}} + \boldsymbol{C}(\boldsymbol{q},\dot{\boldsymbol{q}})\dot{\boldsymbol{q}} + \boldsymbol{K}_e\boldsymbol{q} + \boldsymbol{g}(\boldsymbol{q}) \qquad (1.66)$$

with

$$\boldsymbol{\psi} \triangleq \boldsymbol{u} - \boldsymbol{D}\dot{\boldsymbol{q}}. \qquad (1.67)$$

The Hamiltonian of the system is given by [Ortega and Spong (1989); Greenwood (1977)]

$$\mathcal{H} = \boldsymbol{\pi}^T\dot{\boldsymbol{q}} - \mathcal{L}, \qquad (1.68)$$

where the generalized momentum $\boldsymbol{\pi}$ is defined as

$$\boldsymbol{\pi} = \left(\frac{\partial \mathcal{L}}{\partial \dot{\boldsymbol{q}}}\right)^T. \qquad (1.69)$$

On the other hand, by using (1.51), (1.68) and (1.69), the Hamiltonian can be expressed as the sum of the kinetic and the potential energy of the system, i.e.

$$\mathcal{H} = \frac{1}{2}\dot{\boldsymbol{q}}^T\boldsymbol{H}(\boldsymbol{q})\dot{\boldsymbol{q}} + \frac{1}{2}\boldsymbol{q}^T\boldsymbol{K}_e\boldsymbol{q} + \mathcal{U}_g = \mathcal{T} + \mathcal{U}, \qquad (1.70)$$

while the Hamilton's equations are given by

$$\dot{q}_i = \frac{\partial \mathcal{H}}{\partial p_i} \tag{1.71}$$

$$\dot{\pi}_i = -\frac{\partial \mathcal{H}}{\partial q_i} + \psi_i \qquad i = 1, \ldots, n+m. \tag{1.72}$$

By employing (1.71) and (1.72), the derivative of $\mathcal{H}$ can be computed as

$$\frac{\mathrm{d}\mathcal{H}}{\mathrm{d}t} = \sum_{i=1}^{n+m} \frac{\partial \mathcal{H}}{\partial q_i}\dot{q}_i + \sum_{i=1}^{n+m} \frac{\partial \mathcal{H}}{\partial \pi_i}\dot{\pi}_i = \dot{\boldsymbol{q}}^T \boldsymbol{\psi}. \tag{1.73}$$

Eqs. (1.66) and (1.70) can be used as well to obtain $\mathrm{d}\mathcal{H}/\mathrm{d}t$, i.e.

$$\begin{aligned}\frac{\mathrm{d}\mathcal{H}}{\mathrm{d}t} &= \dot{\boldsymbol{q}}^T \boldsymbol{H}(\boldsymbol{q})\ddot{\boldsymbol{q}} + \frac{1}{2}\dot{\boldsymbol{q}}^T \dot{\boldsymbol{H}}(\boldsymbol{q})\dot{\boldsymbol{q}} + \dot{\boldsymbol{q}}^T \boldsymbol{K}_e \boldsymbol{q} + \dot{\boldsymbol{q}}^T \left(\frac{\partial \mathcal{U}_g}{\partial \boldsymbol{q}}\right)^T \\ &= \dot{\boldsymbol{q}}^T \boldsymbol{\psi} + \frac{1}{2}\dot{\boldsymbol{q}}^T \left(\dot{\boldsymbol{H}}(\boldsymbol{q}) - 2\boldsymbol{C}(\boldsymbol{q}, \dot{\boldsymbol{q}})\right)\dot{\boldsymbol{q}}. \end{aligned} \tag{1.74}$$

By comparing (1.73) and (1.74), one can conclude that (1.65) is valid for any possible choice of $\boldsymbol{C}(\boldsymbol{q}, \dot{\boldsymbol{q}})$.

The following property holds for the vector $\boldsymbol{h}_c(\boldsymbol{q}, \dot{\boldsymbol{q}})$ of Coriolis and centrifugal torques.

Property 1.3 The vector $\boldsymbol{h}_c(\boldsymbol{q}, \dot{\boldsymbol{q}})$ of Coriolis and centrifugal torques satisfies the equalities:

$$\boldsymbol{h}_c(\boldsymbol{q}, \boldsymbol{x}, \boldsymbol{y}) = \boldsymbol{C}(\boldsymbol{q}, \mathrm{x})\boldsymbol{y} = \boldsymbol{C}(\boldsymbol{q}, \boldsymbol{y})\boldsymbol{x} = \boldsymbol{h}_c(\boldsymbol{q}, \boldsymbol{y}, \boldsymbol{x}) \quad \forall \boldsymbol{x}, \boldsymbol{y} \in \Re^{n+m}. \tag{1.75}$$

Proof: The element rs of vector $\boldsymbol{h}_c(\boldsymbol{q}, \boldsymbol{x}, \boldsymbol{y})$ can be expressed as (see (1.56) and (1.57))

$$\begin{aligned} h_{c_{rs}}(\boldsymbol{q}, \boldsymbol{x}, \boldsymbol{y}) &= \sum_{\alpha=1}^{n}\sum_{\beta=0}^{m_\alpha}\left(\sum_{i=1}^{n}\sum_{j=0}^{m_i} c_{ij\alpha\beta rs} x_{ij}\right) y_{\alpha\beta} \\ &= \sum_{i=1}^{n}\sum_{j=0}^{m_i}\left(\sum_{\alpha=1}^{n}\sum_{\beta=0}^{m_\alpha} c_{ij\alpha\beta rs} y_{\alpha\beta}\right) x_{ij} \\ &= \sum_{i=1}^{n}\sum_{j=0}^{m_i}\left(\sum_{\alpha=1}^{n}\sum_{\beta=0}^{m_\alpha} c_{\alpha\beta ijrs} y_{\alpha\beta}\right) x_{ij} = h_{c_{rs}}(\boldsymbol{q}, \boldsymbol{y}, \boldsymbol{x}). \end{aligned} \tag{1.76}$$

△

Due to the orthogonality of the modes shapes of a flexible–link manipulator, the following property can be obtained.

Property 1.4 The stiffness matrix $\boldsymbol{K}$ is diagonal and positive definite and satisfies

$$\lambda_{\min}(\boldsymbol{K})\|\boldsymbol{x}\|^2 \leq \boldsymbol{x}^T\boldsymbol{K}\boldsymbol{x} \leq \lambda_{\max}(\boldsymbol{K})\|\boldsymbol{x}\|^2 \qquad \forall \boldsymbol{x} \in \Re^m. \tag{1.77}$$

Proof: To prove the positive definiteness of the stiffness matrix, the definition of its elements can be used (see (1.43)–(1.45) and (1.47)). Since the mode shapes are orthogonal [Meirovitch (1967)], the result of the integrals must be zero if $j \neq k$ and positive otherwise. Eq. (1.77) follows from the fact that $\boldsymbol{K}$ is positive definite.

△

Regarding the matrix $\boldsymbol{D}$ of link modal damping and joint viscous friction, the following property can be established.

Property 1.5 The matrix $\boldsymbol{D}$ is diagonal positive semidefinite and satisfies

$$\lambda_{\min}(\boldsymbol{D})\|\boldsymbol{x}\|^2 \leq \boldsymbol{x}^T\boldsymbol{D}\boldsymbol{x} \leq \lambda_{\max}(\boldsymbol{D})\|\boldsymbol{x}\|^2 \qquad \forall \boldsymbol{x} \in \Re^{n+m}. \tag{1.78}$$

Proof: $\boldsymbol{D}$ is positive semidefinite because it is defined on the basis of the Rayleigh's dissipation function (see (1.52)). Assuming it to be diagonal is actually a special but very important and common case of the definition of the Rayleigh's dissipation function [Meirovitch (1967)].

△

Finding bounds on the norms of the matrices of model (1.55) plays an important role in the control of robot manipulators because such bounds are helpful for design of many control schemes. Norm bounds are especially advantageous when Lyapunov theory is used. As a matter of fact, for any mechanical system, the vectors $\boldsymbol{q}$ and $\dot{\boldsymbol{q}}$ are bounded. Taking only link deformation into account and in view of the small deformation assumption, the potential energy due to elasticity cannot be infinite, i.e. $U_e < \infty$ [De Luca and Siciliano (1993b); De Luca and Panzieri (1994)], so that it is possible to find a bound for the vector of link coordinates $\boldsymbol{\delta}$.

Property 1.6 The norm of $\boldsymbol{\delta}$ is bounded by

$$\|\boldsymbol{\delta}\| \leq \sqrt{\frac{2U_{e,\max}}{\lambda_{\min}(\boldsymbol{K})}} \triangleq \bar{\delta}, \tag{1.79}$$

where $U_{e,\max}$ is the maximum of the link strain potential energy.

Proof: In view of the assumption of small link deformation, there must be a maximum link strain potential energy. Directly from (1.48) it is

$$\boldsymbol{\delta}^T \boldsymbol{K} \boldsymbol{\delta} \leq 2U_{e,\max} < \infty, \tag{1.80}$$

from which Property 1.6 follows by taking Property 1.4 into account.

△

Notice that Property 1.6 means that $\boldsymbol{\delta}$ belongs to a set Δ whose elements are bounded. For simplicity, every vector $\boldsymbol{q} \in \Re^n \times \Delta \subset \Re^{n+m}$ will be assumed to belong to a set $\mathcal{Q}^{n+m}$.

The next four properties are related to the inertia matrix and can easily be derived from Property 1.1.

Property 1.7 The inertia matrix $\boldsymbol{H}(\boldsymbol{q})$ satisfies

$$\lambda_{\min}(\boldsymbol{H}(\boldsymbol{q}))\|\boldsymbol{y}\|^2 \leq \boldsymbol{y}^T \boldsymbol{H}(\boldsymbol{q})\boldsymbol{y} \leq \lambda_{\max}(\boldsymbol{H}(\mathbf{q}))\|\boldsymbol{y}\|^2 \quad \forall \boldsymbol{y} \in \Re^{n+m}. \tag{1.81}$$

Proof: Since $\boldsymbol{H}(\boldsymbol{q})$ is positive definite, each vector $\boldsymbol{y}$ in $\Re^{n+m}$ can be expressed in terms of an orthonormal basis $(\boldsymbol{y}_1, \ldots, \boldsymbol{y}_{n+m})$ as

$$\boldsymbol{y} = \sum_{i=1}^{n+m} c_i \boldsymbol{y}_i, \tag{1.82}$$

implying that

$$\boldsymbol{y}^T \boldsymbol{H}(\boldsymbol{q})\boldsymbol{y} = c_1^2 \lambda_1(\boldsymbol{H}(\boldsymbol{q})) + \cdots + c_{n+m}^2 \lambda_{n+m}(\boldsymbol{H}(\boldsymbol{q})) \tag{1.83}$$

$$\boldsymbol{y}^T \boldsymbol{y} = \|\boldsymbol{y}\|^2 = c_1^2 + \cdots + c_{n+m}^2, \tag{1.84}$$

from which (1.81) follows. △

Property 1.8 The matrix $\boldsymbol{H}^{-1}(\boldsymbol{q})$ exists and satisfies

$$\lambda_{\max}^{-1}(\boldsymbol{H}(\boldsymbol{q}))\|\boldsymbol{y}\|^2 \leq \mathbf{y}^T \boldsymbol{H}^{-1}(\boldsymbol{q})\boldsymbol{y} \leq \lambda_{\min}^{-1}(\boldsymbol{H}(\boldsymbol{q}))\|\boldsymbol{y}\|^2 \; Consaid \forall \boldsymbol{y} \in \Re^{n+m} \tag{1.85}$$

Proof: This property follows directly from Property 1.7. △

Property 1.9 The inertia matrix satisfies

$$\lambda_h \leq \|\boldsymbol{H}(\boldsymbol{q})\| \leq \lambda_H < \infty. \tag{1.86}$$

Proof: Since the vector of generalized coordinates is bounded, i.e. $\|\boldsymbol{q}\| < \infty$, it is easy to see from (1.81) that

$$\lambda_h = \min_{\boldsymbol{q}\in\mathcal{Q}^{n+m}} \lambda_{\min}(\boldsymbol{H}(\boldsymbol{q})) \tag{1.87}$$

$$\lambda_H = \max_{\boldsymbol{q}\in\mathcal{Q}^{n+m}} \lambda_{\max}(\boldsymbol{H}(\boldsymbol{q})). \tag{1.88}$$

△

Property 1.10 The inverse of the inertia matrix satisfies

$$\sigma_h \le \|\boldsymbol{H}^{-1}(\boldsymbol{q})\| \le \sigma_H < \infty. \tag{1.89}$$

Proof: The proof is the same as in Property 1.9 with

$$\sigma_h = \min_{\boldsymbol{q}\in\mathcal{Q}^{n+m}} \lambda_{\max}^{-1}(\boldsymbol{H}(\boldsymbol{q})) \tag{1.90}$$

$$\sigma_H = \max_{\boldsymbol{q}\in\mathcal{Q}^{n+m}} \lambda_{\min}^{-1}(\boldsymbol{H}(\boldsymbol{q})). \tag{1.91}$$

△

It is easy to recognize that Properties 1.7 to 1.10 are closely related. Of course, by taking only Property 1.7 into account, it is not difficult to develop the other three properties. These properties are very important because many Lyapunov functions employed to prove the stability of a control approach make use of the inertia matrix and its boundedness properties.

Since $\boldsymbol{q}$ is bounded, it is possible to find the following bound for the matrix $\boldsymbol{C}(\boldsymbol{q},\dot{\boldsymbol{q}})$.

Property 1.11 The matrix $\boldsymbol{C}(\boldsymbol{q},\dot{\boldsymbol{q}})$ satisfies

$$\|\boldsymbol{C}(\boldsymbol{q},\dot{\boldsymbol{q}})\| \le k_c\|\dot{\boldsymbol{q}}\|. \tag{1.92}$$

Proof: From (1.57), it can be seen that matrix $\boldsymbol{C}(\boldsymbol{q},\dot{\boldsymbol{q}})$ can be written as

$$\boldsymbol{C}(\boldsymbol{q},\dot{\boldsymbol{q}}) = \frac{1}{2}\sum_{i=1}^{n}\sum_{j=0}^{m_i} \boldsymbol{C}_{ij}(\boldsymbol{q})\dot{q}_{ij}, \tag{1.93}$$

so that each element of matrix $\boldsymbol{C}_{ij}(\boldsymbol{q})$ is given by

$$\frac{\partial h_{rs\alpha\beta}}{\partial q_{ij}} + \frac{\partial h_{rsij}}{\partial q_{\alpha\beta}} - \frac{\partial h_{ij\alpha\beta}}{\partial q_{rs}}.$$

Computing the norm of $\boldsymbol{C}(\boldsymbol{q}, \dot{\boldsymbol{q}})$ leads to

$$\begin{aligned} \|\boldsymbol{C}(\boldsymbol{q}, \dot{\boldsymbol{q}})\| &= \frac{1}{2}\left\|\sum_{i=1}^{n}\sum_{j=0}^{m_i} \boldsymbol{C}_{ij}(\boldsymbol{q})\dot{q}_{ij}\right\| \\ &\le \frac{1}{2}\sum_{i=1}^{n}\sum_{j=0}^{m_i} \|\boldsymbol{C}_{ij}(\boldsymbol{q})\dot{q}_{ij}\| \\ &= \frac{1}{2}\sum_{i=1}^{n}\sum_{j=0}^{m_i} \|\boldsymbol{C}_{ij}(\boldsymbol{q})\|\ |\dot{q}_{ij}| \\ &\le \frac{1}{2}\sum_{i=1}^{n}\sum_{j=0}^{m_i} \|\boldsymbol{C}_{ij}(\boldsymbol{q})\|\ \|\dot{\boldsymbol{q}}\|. \end{aligned} \tag{1.94}$$

With

$$k_c \triangleq \frac{1}{2} \max_{\boldsymbol{q}\in\mathcal{Q}^{n+m}} \sum_{i=1}^{n}\sum_{j=0}^{m_i} \|\boldsymbol{C}_{ij}(\boldsymbol{q})\|, \tag{1.95}$$

Property 1.11 follows. △

It is worth noticing that Property 1.11 applies to every vector $\boldsymbol{y} \in \Re^{n+m}$.

Because the vector of gravitational torques $\boldsymbol{g}(\boldsymbol{q})$ is only a function of $\boldsymbol{q}$, one can find a bound related to it as well.

Property 1.12 The vector of gravitational torques $\boldsymbol{g}(\boldsymbol{q})$ is bounded by a constant $\sigma_g > 0$, i.e.

$$\|\boldsymbol{g}(\boldsymbol{q})\| \le \sigma_g. \tag{1.96}$$

Proof: Since

$$\|\boldsymbol{g}(\boldsymbol{q})\| = \sqrt{\sum_{r=1}^{n}\sum_{s=0}^{m_r} g_{rs}^2}, \tag{1.97}$$

it should be proven that each term g_{rs} is bounded. By recalling that $\boldsymbol{q}$ is bounded and taking (1.60) into account, it can easily be seen that each g_{rs} is bounded, so that (1.96) follows. △

1.3 Nonlinear Control

Perhaps the most relevant issue in controlling flexible robot manipulators resides in the fact that there are fewer inputs than degrees of freedom, thus making impossible to choose a desired trajectory for each generalized coordinate. An obvious solution to this problem consists in renouncing to choose an arbitrary trajectory for the link coordinates and select only one for the joint coordinates. By doing so, a new difficulty might arise: since the desired trajectory for the link coordinates is computed to achieve some control goal for the joint coordinates, it is necessary to guarantee, at least, that it remains bounded. In Section 1.3.1, some stability theorems related to the boundedness of the state of a system are presented. Although they are mainly used to design robust controllers, it will be shown in Sections 1.3.2 and 1.3.3 that they can be employed to design controllers for flexible robot manipulators as well. Section 1.3.2 presents a controller based on the well–known approach with filtered reference velocity at the basis of passivity. The damping problem is specifically treated in Section 1.3.3, where a solution to increase damping is proposed.

1.3.1 *State Boundedness*

Consider a dynamical continuous system

$$\dot{\boldsymbol{x}} = \boldsymbol{f}(\boldsymbol{x}, t) \qquad \boldsymbol{x}(t_0) = \boldsymbol{x}_0 \tag{1.98}$$

where $\boldsymbol{f}(\cdot) : \Re^n \times \Re \rightarrow \Re^n$ is known and $\boldsymbol{x}$ is an $n \times 1$ vector.

Definition 1.1 [Leitmann (1981)] Given a solution $\boldsymbol{x}(\cdot) : [t_0, t_1] \rightarrow \Re^n$ of (1.98), we say that it is *uniformly bounded* (UB) if there is a positive constant $d(\boldsymbol{x}_0) < \infty$, possibly dependent on $\boldsymbol{x}_0$ but not on t_0, such that

$$\|\boldsymbol{x}(t)\| \leq d(\boldsymbol{x}_0) \qquad \forall t \in [t_0, t_1]. \tag{1.99}$$

$\triangle$

Definition 1.2 [Leitmann (1981)] Given a solution $\boldsymbol{x}(\cdot) : [t_0, \infty) \rightarrow \Re^n$ of (1.98), we say that it it *uniformly ultimately bounded* (UUB) with respect to the set S if there is a non–negative constant $T(\boldsymbol{x}_0, S) < \infty$, possibly dependent on $\boldsymbol{x}_0$ and S but not on t_0, such that

$$\boldsymbol{x}(t) \in S \qquad \forall t \geq t_0 + T(\boldsymbol{x}_0, S). \tag{1.100}$$

$\triangle$

Notice that if the constants $d(\boldsymbol{x}_0)$ in Definition 1.1 and $T(\boldsymbol{x}_0, S)$ in Definition 1.2 are independent of $\boldsymbol{x}_0$, then we say that the solution is *globally uniformly bounded* (GUB) or *globally uniformly ultimately bounded* (GUUB). The next two theorems are related to the stability of a system of form (1.98).

Theorem 1.1 *[Dawson* et al. *(1991)] Let a continuous system be described by (1.98) and let $V(\boldsymbol{x},t)$ be an associated Lyapunov function with the following properties:*

$$\lambda_1\|\boldsymbol{x}(t)\|^2 \le V(\boldsymbol{x},t) \le \lambda_2\|\boldsymbol{x}(t)\|^2 \tag{1.101}$$
$$\dot{V}(\boldsymbol{x},t) \le -\lambda_3\|\boldsymbol{x}(t)\|^2 + \epsilon e^{-\beta t} \tag{1.102}$$

$\forall(\boldsymbol{x},t) \in \Re^n \times \Re$, where $\lambda_1, \lambda_2, \lambda_3$ and ϵ are positive scalar constants. If $\beta = 0$ in (1.102), then the state $\boldsymbol{x}$ is GUUB in the sense that

$$\|\boldsymbol{x}(t)\| \le \left(\frac{\lambda_2}{\lambda_1}\|\boldsymbol{x}_0\|^2 e^{-\lambda t} + \frac{\epsilon}{\lambda_1\lambda}(1-e^{-\lambda t})\right)^{1/2}, \tag{1.103}$$

where λ_1, λ_2, λ_3 are defined in (1.101) and (1.102) and $\lambda = \lambda_3/\lambda_2$. If $\beta > 0$ in (1.102), then the state $\boldsymbol{x}$ is globally exponentially stable *(GES) in the sense that*

$$\|\boldsymbol{x}(t)\| \le \begin{cases} \left(\dfrac{\lambda_2}{\lambda_1}\|\boldsymbol{x}_0\|^2 e^{-\lambda t} + \dfrac{\epsilon}{\lambda_1} t e^{-\lambda t}\right)^{1/2} & \text{if } \beta = \lambda \\ \left(\dfrac{\lambda_2}{\lambda_1}\|\boldsymbol{x}_0\|^2 e^{-\lambda t} + \dfrac{\epsilon}{\lambda_1(\lambda-\beta)}(e^{-\beta t} - e^{-\lambda t})\right)^{1/2} & \text{if } \beta \ne \lambda \end{cases} \tag{1.104}$$

where λ_1, λ_2, λ_3 are defined in (1.101) and (1.102) and $\lambda = \lambda_3/\lambda_2$. △

Theorem 1.2 *[Dawson* et al. *(1991)] Let a continuous system be described by (1.98) and let $V(\boldsymbol{x},t)$ be an associated Lyapunov function with the following properties:*

$$\lambda_1\|\boldsymbol{x}(t)\|^2 \le V(\boldsymbol{x},t) \le \lambda_2\|\boldsymbol{x}(t)\|^2 \tag{1.105}$$
$$\dot{V}(\boldsymbol{x},t) \le -\lambda_3\|\boldsymbol{x}(t)\|^2 + \|\boldsymbol{x}(t)\|\sigma e^{-\gamma t} \tag{1.106}$$

$\forall(\boldsymbol{x},t) \in \Re^n \times \Re$, where λ_1, λ_2, λ_3 and σ are positive scalar constants. If $\gamma = 0$ in (1.106), then the state $\boldsymbol{x}$ is GUUB in the sense that

$$\|\boldsymbol{x}(t)\| \le \frac{1}{\sqrt{\lambda_1}}\left(\sqrt{\lambda_2}\|\boldsymbol{x}_0\|e^{-\lambda t/2} + \frac{\zeta}{\lambda}\left(1-e^{-\lambda t/2}\right)\right), \tag{1.107}$$

where λ_1, λ_2, λ_3 *are defined in (1.105) and (1.106),* $\zeta = \sigma/\sqrt{\lambda_1}$ *and* $\lambda = \lambda_3/\lambda_2$. *If* $\gamma > 0$, *then the state* $\boldsymbol{x}$ *is GES in the sense that*

$$\|\boldsymbol{x}(t)\| \leq \begin{cases} \frac{1}{\sqrt{\lambda_1}}\left(\sqrt{\lambda_2}\|\boldsymbol{x}_0\|e^{-\lambda t/2} + \frac{\zeta}{2}te^{-\lambda t/2}\right) & \text{if } \lambda = 2\gamma \\ \frac{1}{\sqrt{\lambda_1}}\left(\sqrt{\lambda_2}\|\boldsymbol{x}_0\|e^{-\lambda t/2} + \frac{\zeta}{\lambda - 2\gamma}\left(e^{-\gamma t} - e^{-\lambda t/2}\right)\right) & \text{if } \lambda \neq 2\gamma \end{cases} \tag{1.108}$$

where λ_1, λ_2, λ_3 *are defined in (1.105) and (1.106),* $\zeta = \sigma/\sqrt{\lambda_1}$ *and* $\lambda = \lambda_3/\lambda_2$. $\triangle$

1.3.2 *Tracking Control*

In this section, the tracking control problem of flexible manipulators is studied. Consider model (1.55) again:

$$\boldsymbol{H}(\boldsymbol{q})\ddot{\boldsymbol{q}} + \boldsymbol{C}(\boldsymbol{q}, \dot{\boldsymbol{q}})\dot{\boldsymbol{q}} + \boldsymbol{K}_e\boldsymbol{q} + \boldsymbol{D}\dot{\boldsymbol{q}} + \boldsymbol{g}(\boldsymbol{q}) = \boldsymbol{u} \tag{1.109}$$

with $\boldsymbol{u}$ as in (1.54).

Given a bounded continuous desired trajectory $\boldsymbol{q}_d$, with bounded velocity $\dot{\boldsymbol{q}}_d$ and acceleration $\ddot{\boldsymbol{q}}_d$, the tracking errors $\tilde{\boldsymbol{q}}$ and $\dot{\tilde{\boldsymbol{q}}}$ can be defined as

$$\tilde{\boldsymbol{q}} \triangleq \boldsymbol{q} - \boldsymbol{q}_d \tag{1.110}$$

$$\dot{\tilde{\boldsymbol{q}}} \triangleq \dot{\boldsymbol{q}} - \dot{\boldsymbol{q}}_d. \tag{1.111}$$

Before the controller can be introduced, the following definitions are necessary:

$$\dot{\boldsymbol{q}}_r \triangleq \dot{\boldsymbol{q}}_d - \boldsymbol{\Lambda}\tilde{\boldsymbol{q}} \triangleq \begin{bmatrix} \dot{\boldsymbol{\theta}}_r \\ \dot{\boldsymbol{\delta}}_r \end{bmatrix} = \begin{bmatrix} \dot{\boldsymbol{\theta}}_d - \boldsymbol{\Lambda}_\theta \tilde{\boldsymbol{\theta}} \\ \dot{\boldsymbol{\delta}}_d - \boldsymbol{\Lambda}_\delta \tilde{\boldsymbol{\delta}} \end{bmatrix} \tag{1.112}$$

$$\boldsymbol{s} \triangleq \dot{\boldsymbol{q}} - \dot{\boldsymbol{q}}_r = \dot{\tilde{\boldsymbol{q}}} + \boldsymbol{\Lambda}\tilde{\boldsymbol{q}} \triangleq \begin{bmatrix} \boldsymbol{s}_\theta \\ \boldsymbol{s}_\delta \end{bmatrix} \tag{1.113}$$

$$\boldsymbol{\Lambda} \triangleq \begin{bmatrix} \boldsymbol{\Lambda}_\theta & \boldsymbol{O} \\ \boldsymbol{O} & \boldsymbol{\Lambda}_\delta \end{bmatrix} \tag{1.114}$$

$$\boldsymbol{K}_p \triangleq \begin{bmatrix} \boldsymbol{K}_{p\theta} & \boldsymbol{O} \\ \boldsymbol{O} & \boldsymbol{K}_{p\delta} \end{bmatrix} \tag{1.115}$$

$$\boldsymbol{D} \triangleq \begin{bmatrix} \boldsymbol{D}_\theta & \boldsymbol{O} \\ \boldsymbol{O} & \boldsymbol{D}_\delta \end{bmatrix} \tag{1.116}$$

$$\boldsymbol{K}_{pD} \triangleq \boldsymbol{K}_p + \boldsymbol{D} = \begin{bmatrix} \boldsymbol{K}_{p\theta} + \boldsymbol{D}_\theta & \boldsymbol{O} \\ \boldsymbol{O} & \boldsymbol{K}_{p\delta} + \boldsymbol{D}_\delta \end{bmatrix} \triangleq \begin{bmatrix} \boldsymbol{K}_{pD\theta} & \boldsymbol{O} \\ \boldsymbol{O} & \boldsymbol{K}_{pD\delta} \end{bmatrix} \tag{1.117}$$

$$\boldsymbol{C}(\boldsymbol{q}, \dot{\boldsymbol{q}}) \triangleq \begin{bmatrix} \boldsymbol{C}_{\theta\theta}(\boldsymbol{q}, \dot{\boldsymbol{q}}) & \boldsymbol{C}_{\theta\delta}(\boldsymbol{q}, \dot{\boldsymbol{q}}) \\ \boldsymbol{C}_{\delta\theta}(\boldsymbol{q}, \dot{\boldsymbol{q}}) & \boldsymbol{C}_{\delta\delta}(\boldsymbol{q}, \dot{\boldsymbol{q}}) \end{bmatrix} \tag{1.118}$$

$$\boldsymbol{g}(\boldsymbol{q}) \triangleq \begin{bmatrix} \boldsymbol{g}_\theta \\ \boldsymbol{g}_\delta \end{bmatrix}, \tag{1.119}$$

where $\boldsymbol{\Lambda}$ and $\boldsymbol{K}_p$ are diagonal and positive definite.

The proposed controller is given by

$$\boldsymbol{\tau} = \boldsymbol{H}_{\theta\theta}\ddot{\boldsymbol{\theta}}_r + \boldsymbol{H}_{\theta\delta}\ddot{\boldsymbol{\delta}}_r + \boldsymbol{C}_{\theta\theta}\dot{\boldsymbol{\theta}}_r + \boldsymbol{C}_{\theta\delta}\dot{\boldsymbol{\delta}}_r + \boldsymbol{D}_\theta\dot{\boldsymbol{\theta}}_r + \boldsymbol{g}_\theta - \boldsymbol{K}_{p\theta}\boldsymbol{s}_\theta, \tag{1.120}$$

while the desired trajectory $\boldsymbol{\delta}_d$ is computed from

$$\ddot{\boldsymbol{\delta}}_d = \boldsymbol{\Lambda}_\delta \dot{\tilde{\boldsymbol{\delta}}} - \boldsymbol{H}_{\delta\delta}^{-1}(\boldsymbol{C}_{\delta\delta}\dot{\boldsymbol{\delta}}_r + \boldsymbol{D}_\delta\dot{\boldsymbol{\delta}}_r - \boldsymbol{K}_{p\delta}\boldsymbol{s}_\delta + \boldsymbol{K}\boldsymbol{\delta}_d + \boldsymbol{H}_{\theta\delta}^T\ddot{\boldsymbol{\theta}}_r + \boldsymbol{C}_{\delta\theta}\dot{\boldsymbol{\theta}}_r + \boldsymbol{g}_\delta) \tag{1.121}$$

with initial condition

$$\boldsymbol{\delta}_d(0) = \dot{\boldsymbol{\delta}}_d(0) = \boldsymbol{0}. \tag{1.122}$$

It is not difficult to realize that Eqs. (1.120) and (1.121) are equivalent to

$$\boldsymbol{u} = \boldsymbol{H}(\boldsymbol{q})\ddot{\boldsymbol{q}}_r + \boldsymbol{C}(\boldsymbol{q}, \dot{\boldsymbol{q}})\dot{\boldsymbol{q}}_r + \boldsymbol{K}_e\boldsymbol{q}_d + \boldsymbol{D}\dot{\boldsymbol{q}}_r + \boldsymbol{g}(\boldsymbol{q}) - \boldsymbol{K}_p\boldsymbol{s}.$$

Notice that in contrast with the control given in [Lammerts *et al.* (1995)], the term $\boldsymbol{K}_e$ is not multiplied by $\boldsymbol{q}_r$ but by $\boldsymbol{q}_d$. It will be proven that $\boldsymbol{\delta}_d$ and $\dot{\boldsymbol{\delta}}_d$ computed from (1.121)–(1.122) are bounded.

By substituting (1.123) in (1.109), the error dynamics can be expressed as

$$\boldsymbol{H}(\boldsymbol{q})\dot{\boldsymbol{s}} = -\left(\boldsymbol{C}(\boldsymbol{q},\dot{\boldsymbol{q}})\boldsymbol{s} + \boldsymbol{K}_e\tilde{\boldsymbol{q}} + \boldsymbol{K}_{pD}\boldsymbol{s}\right). \tag{1.123}$$

In order to simplify the stability discussion, the state $\boldsymbol{x}$ is introduced:

$$\boldsymbol{x} \triangleq \begin{bmatrix} \tilde{\boldsymbol{q}} \\ \dot{\tilde{\boldsymbol{q}}} \end{bmatrix}. \tag{1.124}$$

By taking advantage of the fact that for any mechanical system the velocity vector $\dot{\boldsymbol{q}}$ is bounded, the following theorem establishes the boundedness of $\boldsymbol{\delta}_d$ and $\dot{\boldsymbol{\delta}}_d$ and the stability of $\boldsymbol{x}$.

Theorem 1.3 *Given a bounded continuous desired trajectory $\boldsymbol{\theta}_d$ with bounded velocity and acceleration, if $\|\dot{\boldsymbol{q}}\| \leq v_m$, where v_m is a positive scalar constant, then the desired trajectory $\boldsymbol{\delta}_d$ and $\dot{\boldsymbol{\delta}}_d$ given by (1.121) and (1.122) remains bounded. In addition, by using the input vector (1.120), the equilibrium point $\boldsymbol{x} = \boldsymbol{0}$ of (1.123) is globally asymptotically stable.*

Proof:

a) In order to prove the asymptotic stability of $\boldsymbol{x}$, consider the Lyapunov function $V(\boldsymbol{x},t) = V(\boldsymbol{x})$:

$$V(\boldsymbol{x}) = \frac{1}{2}\boldsymbol{s}^T\boldsymbol{H}(\boldsymbol{q})\boldsymbol{s} + \tilde{\boldsymbol{q}}^T\left(\boldsymbol{\Lambda}\boldsymbol{K}_{pD} + \frac{1}{2}\boldsymbol{K}_e\right)\tilde{\boldsymbol{q}}.$$

Its derivative along (1.123) is

$$\begin{aligned}\dot{V}(\boldsymbol{x}) &= \frac{1}{2}\boldsymbol{s}^T\dot{\boldsymbol{H}}(\boldsymbol{q})\boldsymbol{s} + \dot{\tilde{\boldsymbol{q}}}^T\boldsymbol{\Lambda}\boldsymbol{K}_{pD}\tilde{\boldsymbol{q}} + \tilde{\boldsymbol{q}}^T\boldsymbol{\Lambda}\boldsymbol{K}_{pD}\dot{\tilde{\boldsymbol{q}}} + \dot{\tilde{\boldsymbol{q}}}^T\boldsymbol{K}_e\tilde{\boldsymbol{q}} \\ &\quad -\boldsymbol{s}^T(\boldsymbol{C}(\boldsymbol{q},\dot{\boldsymbol{q}})\boldsymbol{s} + \boldsymbol{K}_e\tilde{\boldsymbol{q}} + \boldsymbol{K}_{pD}\boldsymbol{s}) \\ &= -\boldsymbol{s}^T\boldsymbol{K}_{pD}\boldsymbol{s} + \dot{\tilde{\boldsymbol{q}}}^T\boldsymbol{\Lambda}\boldsymbol{K}_{pD}\tilde{\boldsymbol{q}} + \tilde{\boldsymbol{q}}^T\boldsymbol{\Lambda}\boldsymbol{K}_{pD}\dot{\tilde{\boldsymbol{q}}} - \boldsymbol{s}^T\boldsymbol{K}_e\tilde{\boldsymbol{q}} + \dot{\tilde{\boldsymbol{q}}}^T\boldsymbol{K}_e\tilde{\boldsymbol{q}} \\ &= -(\dot{\tilde{\boldsymbol{q}}}^T + \tilde{\boldsymbol{q}}^T\boldsymbol{\Lambda})\boldsymbol{K}_{pD}(\dot{\tilde{\boldsymbol{q}}} + \boldsymbol{\Lambda}\tilde{\boldsymbol{q}}) + \dot{\tilde{\boldsymbol{q}}}^T\boldsymbol{\Lambda}\boldsymbol{K}_{pD}\tilde{\boldsymbol{q}} + \tilde{\boldsymbol{q}}^T\boldsymbol{\Lambda}\boldsymbol{K}_{pD}\dot{\tilde{\boldsymbol{q}}} \\ &\quad -(\dot{\tilde{\boldsymbol{q}}}^T + \tilde{\boldsymbol{q}}^T\boldsymbol{\Lambda})\boldsymbol{K}_e\tilde{\boldsymbol{q}} + \dot{\tilde{\boldsymbol{q}}}^T\boldsymbol{K}_e\tilde{\boldsymbol{q}} \\ &= -\dot{\tilde{\boldsymbol{q}}}^T\boldsymbol{K}_{pD}\dot{\tilde{\boldsymbol{q}}} - \tilde{\boldsymbol{q}}^T(\boldsymbol{\Lambda}\boldsymbol{K}_{pD}\boldsymbol{\Lambda} + \boldsymbol{\Lambda}\boldsymbol{K}_e)\tilde{\boldsymbol{q}} \\ &= -\boldsymbol{x}^T\boldsymbol{Q}\boldsymbol{x} \\ &\leq 0\end{aligned} \tag{1.125}$$

with

$$\boldsymbol{Q} \triangleq \begin{bmatrix} \boldsymbol{\Lambda}\boldsymbol{K}_{pD}\boldsymbol{\Lambda} + \boldsymbol{\Lambda}\boldsymbol{K}_e & \boldsymbol{O} \\ \boldsymbol{O} & \boldsymbol{K}_{pD} \end{bmatrix}.$$

To compute (1.125), Property 1.2 has been used. Because $\boldsymbol{Q} > \boldsymbol{O}$, $\dot{V}(\boldsymbol{x})$ is 0 if and only if $\boldsymbol{x} = \boldsymbol{0}$, which implies that the equilibrium point is asymptotically stable ([Vidyasagar (1978)], p. 154).

b) To prove the boundedness of $\boldsymbol{\delta}_d$ and $\dot{\boldsymbol{\delta}}_d$, consider the following notation:

$$\boldsymbol{s}_0 \triangleq \dot{\boldsymbol{\delta}}_d + \boldsymbol{\Lambda}_\delta \boldsymbol{\delta}_d \tag{1.126}$$

$$\boldsymbol{x}_\delta \triangleq \begin{bmatrix} \boldsymbol{\delta}_d \\ \dot{\boldsymbol{\delta}}_d \end{bmatrix} \tag{1.127}$$

and simplify (1.121) to get

$$\boldsymbol{H}_{\delta\delta}\dot{\boldsymbol{s}}_0 = -(\boldsymbol{C}_{\delta\delta}\boldsymbol{s}_0 + \boldsymbol{K}_{pD\delta}\boldsymbol{s}_0 + \boldsymbol{K}\boldsymbol{\delta}_d + \boldsymbol{f}_r) \tag{1.128}$$

$$\begin{aligned} \boldsymbol{f}_r \triangleq\ & \boldsymbol{H}_{\theta\delta}^T\ddot{\boldsymbol{\theta}}_r + \boldsymbol{C}_{\delta\theta}\dot{\boldsymbol{\theta}}_r + \boldsymbol{g}_\delta - \boldsymbol{H}_{\delta\delta}\boldsymbol{\Lambda}_\delta\dot{\boldsymbol{\delta}} - \boldsymbol{C}_{\delta\delta}\boldsymbol{\Lambda}_\delta\boldsymbol{\delta} \\ & -\boldsymbol{D}_\delta\boldsymbol{\Lambda}_\delta\boldsymbol{\delta} - \boldsymbol{K}_{p\delta}(\dot{\boldsymbol{\delta}} + \boldsymbol{\Lambda}_\delta\boldsymbol{\delta}). \end{aligned} \tag{1.129}$$

Since $||\boldsymbol{q}|| < \infty$ and $||\dot{\boldsymbol{q}}|| \leq v_m$, the vector $\boldsymbol{f}_r$ is bounded by a positive constant $f_{r,\max}$, i.e. $||\boldsymbol{f}_r|| \leq f_{r,\max}$ (see Properties 1.9 to 1.12).

Consider the Lyapunov function $V_\delta(\boldsymbol{x}_\delta, t) = V_\delta(\boldsymbol{x}_\delta)$:

$$V_\delta(\boldsymbol{x}_\delta) = \frac{1}{2}\boldsymbol{s}_0^T\boldsymbol{H}_{\delta\delta}\boldsymbol{s}_0 + \boldsymbol{\delta}_d^T\left(\boldsymbol{\Lambda}_\delta\boldsymbol{K}_{pD\delta} + \frac{1}{2}\boldsymbol{K}\right)\boldsymbol{\delta}_d = \frac{1}{2}\boldsymbol{x}_\delta^T\boldsymbol{M}\boldsymbol{x}_\delta, \tag{1.130}$$

with

$$\boldsymbol{M} \triangleq \begin{bmatrix} 2\boldsymbol{\Lambda}_\delta\boldsymbol{K}_{pD\delta} + \boldsymbol{K} + \boldsymbol{\Lambda}_\delta\boldsymbol{H}_{\delta\delta}\boldsymbol{\Lambda}_\delta & \boldsymbol{\Lambda}_\delta\boldsymbol{H}_{\delta\delta} \\ \boldsymbol{H}_{\delta\delta}\boldsymbol{\Lambda}_\delta & \boldsymbol{H}_{\delta\delta} \end{bmatrix}.$$

Since $\boldsymbol{M} > \boldsymbol{O}$, there exist two constants λ_1 and λ_2 such that

$$\lambda_1||\boldsymbol{x}_\delta||^2 \leq V_\delta(\boldsymbol{x}_\delta) \leq \lambda_2||\boldsymbol{x}_\delta||^2 \tag{1.131}$$

$$\lambda_1 \triangleq \frac{1}{2}\min_{\boldsymbol{q}\in Q^{n+m}} \lambda_{\min}(\boldsymbol{M}) \tag{1.132}$$

$$\lambda_2 \triangleq \frac{1}{2}\max_{\boldsymbol{q}\in Q^{n+m}} \lambda_{\max}(\boldsymbol{M}). \tag{1.133}$$

The derivative of (1.130) is

$$\begin{aligned} \dot{V}_\delta(\boldsymbol{x}_\delta) =\ & \frac{1}{2}\boldsymbol{s}_0^T\dot{\boldsymbol{H}}_{\delta\delta}\boldsymbol{s}_0 + \dot{\boldsymbol{\delta}}_d^T\boldsymbol{\Lambda}_\delta\boldsymbol{K}_{pD\delta}\boldsymbol{\delta}_d + \boldsymbol{\delta}_d^T\boldsymbol{\Lambda}_\delta\boldsymbol{K}_{pD\delta}\dot{\boldsymbol{\delta}}_d \\ & + \dot{\boldsymbol{\delta}}_d^T\boldsymbol{K}\boldsymbol{\delta}_d + \boldsymbol{s}_0^T\boldsymbol{H}_{\delta\delta}\dot{\boldsymbol{s}}_0. \end{aligned} \tag{1.134}$$

Evaluating (1.134) along (1.128) and taking Property 1.2 into account yields

$$\begin{aligned}
\dot{V}_\delta(\boldsymbol{x}_\delta) &= \frac{1}{2}\boldsymbol{s}_0^T\dot{\boldsymbol{H}}_{\delta\delta}\boldsymbol{s}_0 + \dot{\boldsymbol{\delta}}_d^T\boldsymbol{\Lambda}_\delta\boldsymbol{K}_{pD\delta}\boldsymbol{\delta}_d + \boldsymbol{\delta}_d^T\boldsymbol{\Lambda}_\delta\boldsymbol{K}_{pD\delta}\dot{\boldsymbol{\delta}}_d + \dot{\boldsymbol{\delta}}_d^T\boldsymbol{K}\boldsymbol{\delta}_d \\
&\quad -\boldsymbol{s}_0^T(\boldsymbol{C}_{\delta\delta}\boldsymbol{s}_0 + \boldsymbol{K}_{pD\delta}\boldsymbol{s}_0 + \boldsymbol{K}\boldsymbol{\delta}_d + \boldsymbol{f}_r) \\
&= -\boldsymbol{s}_0^T\boldsymbol{K}_{pD\delta}\boldsymbol{s}_0 + \dot{\boldsymbol{\delta}}_d^T\boldsymbol{\Lambda}_\delta\boldsymbol{K}_{pD\delta}\boldsymbol{\delta}_d + \boldsymbol{\delta}_d^T\boldsymbol{\Lambda}_\delta\boldsymbol{K}_{pD\delta}\dot{\boldsymbol{\delta}}_d - \boldsymbol{s}_0^T\boldsymbol{K}\boldsymbol{\delta}_d \\
&\quad +\dot{\boldsymbol{\delta}}_d^T\boldsymbol{K}\boldsymbol{\delta}_d - \boldsymbol{s}_0^T\boldsymbol{f}_r \\
&= -(\dot{\boldsymbol{\delta}}_d^T + \boldsymbol{\delta}_d^T\boldsymbol{\Lambda}_\delta)\boldsymbol{K}_{pD\delta}(\dot{\boldsymbol{\delta}}_d + \boldsymbol{\Lambda}_\delta\boldsymbol{\delta}_d) + \dot{\boldsymbol{\delta}}_d^T\boldsymbol{\Lambda}_\delta\boldsymbol{K}_{pD\delta}\boldsymbol{\delta}_d + \boldsymbol{\delta}_d^T\boldsymbol{\Lambda}_\delta\boldsymbol{K}_{pD\delta}\dot{\boldsymbol{\delta}}_d \\
&\quad -(\dot{\boldsymbol{\delta}}_d^T + \boldsymbol{\delta}_d^T\boldsymbol{\Lambda}_\delta)\boldsymbol{K}\boldsymbol{\delta}_d + \dot{\boldsymbol{\delta}}_d^T\boldsymbol{K}\boldsymbol{\delta}_d - \boldsymbol{s}_0^T\boldsymbol{f}_r \\
&= -\dot{\boldsymbol{\delta}}_d^T\boldsymbol{K}_{pD\delta}\dot{\boldsymbol{\delta}}_d - \boldsymbol{\delta}_d^T(\boldsymbol{\Lambda}_\delta\boldsymbol{K}_{pD\delta}\boldsymbol{\Lambda}_\delta + \boldsymbol{\Lambda}_\delta\boldsymbol{K})\boldsymbol{\delta}_d - \boldsymbol{s}_0^T\boldsymbol{f}_r \\
&= -\boldsymbol{x}_\delta^T\boldsymbol{P}\boldsymbol{x}_\delta - \boldsymbol{s}_0^T\boldsymbol{f}_r \\
&\leq -\lambda_3\|\boldsymbol{x}_\delta\|^2 - \boldsymbol{s}_0^T\boldsymbol{f}_r
\end{aligned} \tag{1.135}$$

with

$$\boldsymbol{P} \triangleq \begin{bmatrix} \boldsymbol{\Lambda}_\delta\boldsymbol{K}_{pD\delta}\boldsymbol{\Lambda}_\delta + \boldsymbol{\Lambda}_\delta\boldsymbol{K} & \boldsymbol{O} \\ \boldsymbol{O} & \boldsymbol{K}_{pD\delta} \end{bmatrix}$$

and

$$\begin{aligned}
\lambda_3 &\triangleq \lambda_{\min}(\boldsymbol{P}) \\
&= \min\{\lambda_{\min}(\boldsymbol{K}_{pD\delta}), \lambda_{\min}(\boldsymbol{\Lambda}_\delta)^2\lambda_{\min}(\boldsymbol{K}_{pD\delta}) + \lambda_{\min}(\boldsymbol{\Lambda}_\delta)\lambda_{\min}(\boldsymbol{K})\}.
\end{aligned} \tag{1.136}$$

Since

$$\begin{aligned}
\|\boldsymbol{s}_0\| = \|\dot{\boldsymbol{\delta}}_d + \boldsymbol{\Lambda}_\delta\boldsymbol{\delta}_d\| &\leq \|\dot{\boldsymbol{\delta}}_d\| + \lambda_{\max}(\boldsymbol{\Lambda}_\delta)\|\boldsymbol{\delta}_d\| \\
&\leq (1 + \lambda_{\max}(\boldsymbol{\Lambda}_\delta))\|\boldsymbol{x}_\delta\|,
\end{aligned} \tag{1.137}$$

Eq. (1.135) can be rewritten as

$$\begin{aligned}
\dot{V}_\delta(\boldsymbol{x}_\delta) &\leq -\lambda_3\|\boldsymbol{x}_\delta\|^2 + (1 + \lambda_{\max}(\boldsymbol{\Lambda}_\delta))f_{r,\max}\|\boldsymbol{x}_\delta\| \\
&\triangleq -\lambda_3\|\boldsymbol{x}_\delta\|^2 + \sigma\|\boldsymbol{x}_\delta\|.
\end{aligned} \tag{1.138}$$

By applying Theorem 1.2 with $\gamma = 0$, it can be proven that $\|\boldsymbol{x}_\delta\|$ is bounded. △

Now, suppose that the desired trajectory $\boldsymbol{\theta}_d$ is constant and assume that $\boldsymbol{x} = \boldsymbol{0}$. In this case (1.121) becomes

$$\boldsymbol{H}_{\delta\delta}\ddot{\boldsymbol{\delta}}_d + \boldsymbol{C}_{\delta\delta}\dot{\boldsymbol{\delta}}_d + \boldsymbol{D}_\delta\dot{\boldsymbol{\delta}}_d + \boldsymbol{K}\boldsymbol{\delta}_d + \boldsymbol{g}_\delta = \boldsymbol{0}.$$

Assume that the vector $\boldsymbol{g}_\delta$ is only a function of $\boldsymbol{\theta}$ [De Luca and Siciliano (1993b); De Luca and Panzieri (1994)], i.e. $\boldsymbol{g}_\delta(\boldsymbol{q}) = \boldsymbol{g}_\delta(\boldsymbol{\theta})$. Since $\boldsymbol{\theta}_d$ is constant and $\boldsymbol{\theta}_d \equiv \boldsymbol{\theta}$, the vector $\boldsymbol{g}_\delta$ is constant as well, so that a new variable $\boldsymbol{y}$ can be defined:

$$\boldsymbol{y} \triangleq \boldsymbol{\delta}_d + \boldsymbol{K}^{-1}\boldsymbol{g}_\delta, \tag{1.139}$$

and (1.139) can be rewritten as

$$\boldsymbol{H}_{\delta\delta}\ddot{\boldsymbol{y}} + \boldsymbol{C}_{\delta\delta}\dot{\boldsymbol{y}} + \boldsymbol{D}_\delta\dot{\boldsymbol{y}} + \boldsymbol{K}\boldsymbol{y} = \boldsymbol{0}. \tag{1.140}$$

Consider the Lyapunov function

$$V_y(\boldsymbol{y}, \dot{\boldsymbol{y}}) = \frac{1}{2}\boldsymbol{y}^T\boldsymbol{K}\boldsymbol{y} + \frac{1}{2}\dot{\boldsymbol{y}}^T\boldsymbol{H}_{\delta\delta}\dot{\boldsymbol{y}}. \tag{1.141}$$

Its derivative along (1.140) is

$$\begin{aligned} \dot{V}_y &= \dot{\boldsymbol{y}}^T\boldsymbol{K}\boldsymbol{y} + \frac{1}{2}\dot{\boldsymbol{y}}^T\dot{\boldsymbol{H}}_{\delta\delta}\dot{\boldsymbol{y}} + \dot{\boldsymbol{y}}^T\boldsymbol{H}_{\delta\delta}\ddot{\boldsymbol{y}} \\ &= \dot{\boldsymbol{y}}^T\boldsymbol{K}\boldsymbol{y} + \frac{1}{2}\dot{\boldsymbol{y}}^T\dot{\boldsymbol{H}}_{\delta\delta}\dot{\boldsymbol{y}} - \dot{\boldsymbol{y}}^T(\boldsymbol{C}_{\delta\delta}\dot{\boldsymbol{y}} + \boldsymbol{D}_\delta\dot{\boldsymbol{y}} + \boldsymbol{K}\boldsymbol{y}) \\ &= -\dot{\boldsymbol{y}}^T\boldsymbol{D}_\delta\dot{\boldsymbol{y}} \leq 0. \end{aligned} \tag{1.142}$$

Property 1.2 has been used to compute (1.142). Assuming that $\boldsymbol{D}_\delta > \boldsymbol{O}$, $\dot{V}_y$ is 0 if and only if $\dot{\boldsymbol{y}}$ is $\boldsymbol{0}$. By applying the invariant set theorem, it can be proven that the equilibrium point $\dot{\boldsymbol{y}} = \boldsymbol{y} = \boldsymbol{0}$ of (1.140) is asymptotically stable [De Luca and Siciliano (1993a)]. When $\boldsymbol{y} = \boldsymbol{0}$, $\boldsymbol{\delta}_d$ becomes $-\boldsymbol{K}^{-1}\boldsymbol{g}_\delta$, which is the same result as in [De Luca and Siciliano (1993b); De Luca and Panzieri (1994)].

1.3.3 *No Damping*

In the preceding section, a nonlinear controller for the tracking problem of flexible robot manipulators was introduced. By studying the case when the desired trajectory for the joint coordinates is constant and the error vector $\boldsymbol{x} = \boldsymbol{0}$, Eq. (1.139) was obtained. Assuming that $\boldsymbol{D}_\delta > \boldsymbol{O}$, it was proven that $\boldsymbol{\delta}_d \to -\boldsymbol{K}^{-1}\boldsymbol{g}_\delta$. Although this is the case for any mechanical system, if the elements of $\boldsymbol{D}_\delta$ are small, it may last before possible oscillations disappear. Therefore, it is desirable that the controller increases the damping of the system.

Suppose that $\boldsymbol{D}_\delta = \boldsymbol{O}$. Since (1.139) belongs to the controller rather than to the manipulator equations of motion, it is always possible to add, arbitrarily, a term $\boldsymbol{D}_\triangle\dot{\boldsymbol{\delta}}_d$, with $\boldsymbol{D}_\triangle$ diagonal and positive definite. If $\boldsymbol{x} \to \boldsymbol{0}$ were still true, then the desired trajectory of the link coordinates would be damped and as a direct consequence of this, the real trajectory $\boldsymbol{\delta}$ too. Of course, by doing so, the stability analysis of Section 1.3.2 is no longer valid. However, if the term $\boldsymbol{D}_\triangle\dot{\boldsymbol{\delta}}_d$ is treated as a well–known perturbation, some robust control techniques [Dawson *et al.* (1991); Leitmann (1981); Corless and Leitmann (1981); Qu and Dawson (1991); Spong (1992); Yaz (1993)] can be used to guarantee that $\boldsymbol{x} \to \boldsymbol{0}$.

Consider the following equation to compute the desired trajectory $\boldsymbol{\delta}_d$ and $\dot{\boldsymbol{\delta}}_d$:

$$\begin{aligned}\ddot{\boldsymbol{\delta}}_d = \boldsymbol{\Lambda}_\delta\dot{\tilde{\boldsymbol{\delta}}} - \boldsymbol{H}_{\delta\delta}^{-1}(&\boldsymbol{C}_{\delta\delta}\dot{\boldsymbol{\delta}}_r + \boldsymbol{D}_\delta\dot{\boldsymbol{\delta}}_r - \boldsymbol{K}_{p\delta}\boldsymbol{s}_\delta + \boldsymbol{K}\boldsymbol{\delta}_d \\ &+ \boldsymbol{H}_{\theta\delta}^T\ddot{\boldsymbol{\theta}}_r + \boldsymbol{C}_{\delta\theta}\dot{\boldsymbol{\theta}}_r + \boldsymbol{g}_\delta + \boldsymbol{D}_\triangle\dot{\boldsymbol{\delta}}_d + \boldsymbol{f})\end{aligned} \tag{1.143}$$

with initial condition

$$\boldsymbol{\delta}_d(0) = \dot{\boldsymbol{\delta}}_d(0) = \boldsymbol{0} \tag{1.144}$$

and

$$f_{ij} \triangleq -\dot{\delta}_{dij}\frac{(\dot{\delta}_{dij}s_{\delta ij})d_{ij}}{\|\dot{\delta}_{dij}s_{\delta ij}\| + \epsilon_{ij}e^{-\beta_{ij}t}} \qquad i = 1, \ldots, n \quad j = 1, \ldots, m_i, \tag{1.145}$$

where $s_{\delta ij}$ is an element of $\boldsymbol{s}_\delta$, $\dot{\delta}_{dij}$ is an element of $\dot{\boldsymbol{\delta}}_d$, f_{ij} is an element of $\boldsymbol{f}$ and d_{ij} is an element of $\boldsymbol{D}_\triangle$. Control (1.120) together with (1.143) can be expressed as

$$\boldsymbol{u} = \boldsymbol{H}(\boldsymbol{q})\ddot{\boldsymbol{q}}_r + \boldsymbol{C}(\boldsymbol{q}, \dot{\boldsymbol{q}})\dot{\boldsymbol{q}}_r + \boldsymbol{K}_e\boldsymbol{q}_d + \boldsymbol{D}\dot{\boldsymbol{q}}_r + \boldsymbol{g}(\boldsymbol{q}) - \boldsymbol{K}_p\boldsymbol{s} + \boldsymbol{f}_1, \tag{1.146}$$

with

$$\boldsymbol{f}_1 \triangleq \begin{bmatrix} \boldsymbol{0} \\ \boldsymbol{D}_\triangle\dot{\boldsymbol{\delta}}_d + \boldsymbol{f}\end{bmatrix}. \tag{1.147}$$

By taking (1.146) into account, the error dynamics becomes

$$\boldsymbol{H}(\boldsymbol{q})\dot{\boldsymbol{s}} = -(\boldsymbol{C}(\boldsymbol{q}, \dot{\boldsymbol{q}})\boldsymbol{s} + \boldsymbol{K}_e\tilde{\boldsymbol{q}} + \boldsymbol{K}_{pD}\boldsymbol{s}) + \boldsymbol{f}_1. \tag{1.148}$$

Theorem 1.4 establishes the stability of the state $\boldsymbol{x}$ of (1.148).

Theorem 1.4 *Given a bounded continuous desired trajectory $\boldsymbol{\theta}_d$ with bounded velocity and acceleration, if $\|\dot{\boldsymbol{q}}\| \le v_m$, where v_m is a positive scalar* *constant, then the desired trajectory* $\boldsymbol{\delta}_d$ *and* $\dot{\boldsymbol{\delta}}_d$ *given by (1.143) and (1.144) remains bounded if* $\lambda_3 \triangleq \lambda_{\min}(\boldsymbol{P}) - (1 + \lambda_{\max}(\boldsymbol{\Lambda}_\delta))\sqrt{m}\,\lambda_{\max}(\boldsymbol{D}_\Delta) > 0$, *with*

$$\boldsymbol{P} \triangleq \begin{bmatrix} \boldsymbol{\Lambda}_\delta \boldsymbol{K}_{pD\delta} \boldsymbol{\Lambda}_\delta + \boldsymbol{\Lambda}_\delta \boldsymbol{K} & \boldsymbol{O} \\ \boldsymbol{O} & \boldsymbol{K}_{pD\delta} + \boldsymbol{D}_\Delta \end{bmatrix}$$

and

$$\begin{aligned}\lambda_{\min}(\boldsymbol{P}) = \min\{&\lambda_{\min}(\boldsymbol{K}_{pD\delta}) + \lambda_{\min}(\boldsymbol{D}_\Delta), \\ &\lambda_{\min}(\boldsymbol{\Lambda}_\delta)^2 \lambda_{\min}(\boldsymbol{K}_{pD\delta}) + \lambda_{\min}(\boldsymbol{\Lambda}_\delta)\lambda_{\min}(\boldsymbol{K})\}.\end{aligned} \tag{1.149}$$

In addition, by using the input vector (1.120) the state $\boldsymbol{x}$ *of (1.148) is GES in the sense of Theorem 1.1.*

Proof: The proof is similar at all to that of Theorem 1.3.

a) For the stability analysis of $\boldsymbol{x}$, the Lyapunov function (1.125) is used, which can be rewritten as

$$V(\boldsymbol{x}) = \frac{1}{2}\boldsymbol{x}^T \boldsymbol{N} \boldsymbol{x} \tag{1.150}$$

with

$$\boldsymbol{N} \triangleq \begin{bmatrix} 2\boldsymbol{\Lambda}\boldsymbol{K}_{pD} + \boldsymbol{K}_e + \boldsymbol{\Lambda}\boldsymbol{H}(\boldsymbol{q})\boldsymbol{\Lambda} & \boldsymbol{\Lambda}\boldsymbol{H}(\boldsymbol{q}) \\ \boldsymbol{H}(\boldsymbol{q})\boldsymbol{\Lambda} & \boldsymbol{H}(\boldsymbol{q}) \end{bmatrix}.$$

Notice that $\boldsymbol{N} > \boldsymbol{O}$, implying that

$$\hat{\lambda}_1 \|\boldsymbol{x}\|^2 \le V(\boldsymbol{x}) \le \hat{\lambda}_2 \|\boldsymbol{x}\|^2 \tag{1.151}$$

$$\hat{\lambda}_1 \triangleq \frac{1}{2} \min_{\boldsymbol{q} \in \mathcal{Q}^{n+m}} \lambda_{\min}(\boldsymbol{N}) \tag{1.152}$$

$$\hat{\lambda}_2 \triangleq \frac{1}{2} \max_{\boldsymbol{q} \in \mathcal{Q}^{n+m}} \lambda_{\max}(\boldsymbol{N}). \tag{1.153}$$

The derivative $\dot{V}(\boldsymbol{x})$ can be obtained from (1.125) and (1.148), using Property 1.2, i.e.

$$\begin{aligned}\dot{V}(\boldsymbol{x}) &= -\boldsymbol{x}^T\boldsymbol{Q}\boldsymbol{x} + \boldsymbol{s}_\delta^T(\boldsymbol{D}_\Delta\dot{\boldsymbol{\delta}}_d + \boldsymbol{f}) \\ &\leq -\lambda_{\min}(\boldsymbol{Q})\|\boldsymbol{x}\|^2 + \sum_{i=1}^{n}\sum_{j=1}^{m_i}\left(\dot{\delta}_{dij}s_{\delta ij}d_{ij} - d_{ij}\frac{\|\dot{\delta}_{dij}s_{\delta ij}\|^2}{\|\dot{\delta}_{dij}s_{\delta ij}\| + \epsilon_{ij}e^{-\beta_{ij}t}}\right) \\ &\leq -\lambda_{\min}(\boldsymbol{Q})\|\boldsymbol{x}\|^2 + \sum_{i=1}^{n}\sum_{j=1}^{m_i}\left(\|\dot{\delta}_{dij}s_{\delta ij}\|d_{ij} - d_{ij}\frac{\|\dot{\delta}_{dij}s_{\delta ij}\|^2}{\|\dot{\delta}_{dij}s_{\delta ij}\| + \epsilon_{ij}e^{-\beta_{ij}t}}\right) \\ &= -\lambda_{\min}(\boldsymbol{Q})\|\boldsymbol{x}\|^2 + \sum_{i=1}^{n}\sum_{j=1}^{m_i}\left(d_{ij}\frac{\|\dot{\delta}_{dij}s_{\delta ij}\|\epsilon_{ij}e^{-\beta_{ij}t}}{\|\dot{\delta}_{dij}s_{\delta ij}\| + \epsilon_{ij}e^{-\beta_{ij}t}}\right) \\ &\leq -\lambda_{\min}(\boldsymbol{Q})\|\boldsymbol{x}\|^2 + \sum_{i=1}^{n}\sum_{j=1}^{m_i}(d_{ij}\epsilon_{ij}e^{-\beta_{ij}t}) \\ &= -\hat{\lambda}_3\|\boldsymbol{x}\|^2 + \epsilon e^{-\beta t}\end{aligned} \tag{1.154}$$

with

$$\hat{\lambda}_3 \triangleq \lambda_{\min}(\boldsymbol{Q}) \tag{1.155}$$

$$\epsilon \triangleq \sum_{i=1}^{n}\sum_{j=1}^{m_i} d_{ij}\epsilon_{ij} \tag{1.156}$$

$$\beta \triangleq \min\{\beta_{ij}, i = 1, \ldots, n, j = 1, \ldots, m_i\}. \tag{1.157}$$

The proof concludes by applying Theorem 1.1 with $\hat{\lambda}_1$, $\hat{\lambda}_2$, $\hat{\lambda}_3$.

b) To prove the boundedness of $\boldsymbol{\delta}_d$ and $\dot{\boldsymbol{\delta}}_d$, the variables $\boldsymbol{x}_\delta$ and $\boldsymbol{s}_0$ are used again (see (1.126) and (1.127)), such that the dynamics of $\boldsymbol{\delta}_d$ and $\dot{\boldsymbol{\delta}}_d$ can be expressed as

$$\boldsymbol{H}_{\delta\delta}\dot{\boldsymbol{s}}_0 = -(\boldsymbol{C}_{\delta\delta}\boldsymbol{s}_0 + \boldsymbol{K}_{pD\delta}\boldsymbol{s}_0 + \boldsymbol{K}\boldsymbol{\delta}_d + \boldsymbol{f}_r + \boldsymbol{f} + \boldsymbol{D}_\Delta\dot{\boldsymbol{\delta}}_d). \tag{1.158}$$

The term $\boldsymbol{f}_r$ is the same as in Theorem 1.3 (see (1.129)). Note that $\|\boldsymbol{f}_r\| \leq f_{r,\max}$. The following Lyapunov function is employed:

$$V_\delta(\boldsymbol{x}_\delta) = \frac{1}{2}\boldsymbol{s}_0^T\boldsymbol{H}_{\delta\delta}\boldsymbol{s}_0 + \frac{1}{2}\boldsymbol{\delta}_d^T\left(2\boldsymbol{\Lambda}_\delta\boldsymbol{K}_{pD\delta} + \boldsymbol{K} + \boldsymbol{\Lambda}_\delta\boldsymbol{D}_\Delta\right)\boldsymbol{\delta}_d = \frac{1}{2}\boldsymbol{x}_\delta^T\boldsymbol{M}\boldsymbol{x}_\delta \tag{1.159}$$

with

$$\boldsymbol{M} \triangleq \begin{bmatrix} 2\boldsymbol{\Lambda}_\delta\boldsymbol{K}_{pD\delta} + \boldsymbol{K} + \boldsymbol{\Lambda}_\delta\boldsymbol{H}_{\delta\delta}\boldsymbol{\Lambda}_\delta + \boldsymbol{\Lambda}_\delta\boldsymbol{D}_\Delta & \boldsymbol{\Lambda}_\delta\boldsymbol{H}_{\delta\delta} \\ \boldsymbol{H}_{\delta\delta}\boldsymbol{\Lambda}_\delta & \boldsymbol{H}_{\delta\delta}\end{bmatrix}.$$

Since $\boldsymbol{M} > \boldsymbol{O}$, there exist two constants λ_1 and λ_2 such that

$$\lambda_1 \|\boldsymbol{x}_\delta\|^2 \leq V_\delta(\boldsymbol{x}_\delta) \leq \lambda_2 \|\boldsymbol{x}_\delta\|^2 \tag{1.160}$$

$$\lambda_1 \triangleq \frac{1}{2} \min_{\boldsymbol{q} \in \mathcal{Q}^{n+m}} \lambda_{\min}(\boldsymbol{M}) \tag{1.161}$$

$$\lambda_2 \triangleq \frac{1}{2} \max_{\boldsymbol{q} \in \mathcal{Q}^{n+m}} \lambda_{\max}(\boldsymbol{M}). \tag{1.162}$$

It is not difficult to obtain $\dot{V}_\delta(\boldsymbol{x}_\delta)$ from (1.135) and (1.158), using Property 1.2, i.e.

$$\begin{aligned} \dot{V}_\delta(\boldsymbol{x}_\delta) &= -\dot{\boldsymbol{\delta}}_d^T \boldsymbol{K}_{pD\delta} \dot{\boldsymbol{\delta}}_d - \boldsymbol{\delta}_d^T (\boldsymbol{\Lambda}_\delta \boldsymbol{K}_{pD\delta} \boldsymbol{\Lambda}_\delta + \boldsymbol{\Lambda}_\delta \boldsymbol{K}) \boldsymbol{\delta}_d \\ &\quad + \boldsymbol{\delta}_d^T \boldsymbol{\Lambda}_\delta \boldsymbol{D}_\Delta \dot{\boldsymbol{\delta}}_d - \boldsymbol{s}_0^T (\boldsymbol{f}_r + \boldsymbol{f} + \boldsymbol{D}_\Delta \dot{\boldsymbol{\delta}}_d) \\ &\leq -\lambda_{\min}(\boldsymbol{P}) \|\boldsymbol{x}_\delta\|^2 - \boldsymbol{s}_0^T (\boldsymbol{f}_r + \boldsymbol{f}), \end{aligned} \tag{1.163}$$

where $\lambda_{\min}(\boldsymbol{P})$ is given by (1.149). By noting that $\boldsymbol{f}$ can be written as

$$\boldsymbol{f} = -\boldsymbol{D}_\Delta \mathrm{diag}\{\dot{\delta}_{d11}, \ldots, \dot{\delta}_{dnm_n}\} \hat{\boldsymbol{f}} \tag{1.164}$$

with

$$\hat{\boldsymbol{f}} \triangleq [\hat{f}_{11} \ \ldots \ \hat{f}_{nm_n}]^T \tag{1.165}$$

$$\hat{f}_{ij} \triangleq \frac{\dot{\delta}_{dij} s_{\delta ij}}{\|\dot{\delta}_{dij} s_{\delta ij}\| + \epsilon_{ij} e^{-\beta_{ij} t}}, \tag{1.166}$$

and since $\|\hat{f}_{ij}\| < 1$, $\forall i = 1, \ldots, n, \ j = 1, \ldots, m_i$, it can be seen from (1.164) that

$$\|\boldsymbol{f}\| \leq \lambda_{\max}(\boldsymbol{D}_\Delta) \sqrt{m} \ \|\dot{\boldsymbol{\delta}}_d\|, \tag{1.167}$$

such that (see (1.137))

$$\|\boldsymbol{s}_0^T(\boldsymbol{f}_r + \boldsymbol{f})\| \leq (1 + \lambda_{\max}(\boldsymbol{\Lambda}_\delta)) \|\boldsymbol{x}_\delta\| (f_{r,\max} + \lambda_{\max}(\boldsymbol{D}_\Delta)\sqrt{m} \|\boldsymbol{x}_\delta\|). \tag{1.168}$$

Eq. (1.163) then becomes

$$\begin{aligned} \dot{V}_\delta(\boldsymbol{x}_\delta) &\leq -\lambda_{\min}(\boldsymbol{P}) \|\boldsymbol{x}_\delta\|^2 + (1 + \lambda_{\max}(\boldsymbol{\Lambda}_\delta)) \|\boldsymbol{x}_\delta\| (f_{r,\max} \\ &\quad + \lambda_{\max}(\boldsymbol{D}_\Delta) \sqrt{m} \ \|\boldsymbol{x}_\delta\|) \\ &= -(\lambda_{\min}(\boldsymbol{P}) - (1 + \lambda_{\max}(\boldsymbol{\Lambda}_\delta)) \lambda_{\max}(\boldsymbol{D}_\Delta) \sqrt{m}) \|\boldsymbol{x}_\delta\|^2 \\ &\quad + (1 + \lambda_{\max}(\boldsymbol{\Lambda}_\delta)) f_{r,\max} \|\boldsymbol{x}_\delta\| \\ &\triangleq -\lambda_3 \|\boldsymbol{x}_\delta\|^2 + \sigma \|\boldsymbol{x}_\delta\|. \end{aligned} \tag{1.169}$$

By using λ_1, λ_2 given in (1.161), (1.162) and taking into account that $\lambda_3 > 0$ (by assumption), Theorem 1.2 can be applied with $\gamma = 0$ to prove the boundedness of $||\boldsymbol{x}_\delta||$. $\triangle$

The control approach of Theorem 1.4 is similar to that given in [Dawson *et al.* (1991)], although here it is used to increase damping, rather than to consider possible uncertainties in the manipulator model. Notice that it is always possible to get $\lambda_3 > 0$ just by letting the elements of $\boldsymbol{K}_{p\delta}$ be large enough.

In order to show that the damping of the system becomes greater, assume that $\boldsymbol{x}$ is negligible and that $\boldsymbol{D}_\delta = \boldsymbol{O}$. In this case, the vector $\boldsymbol{D}_\triangle \dot{\boldsymbol{\delta}}_d + \boldsymbol{f}$ represents the damping of the system. By factorizing every element of this vector as

$$\dot{\delta}_{dij} d_{ij} \left(1 - \frac{\dot{\delta}_{dij} s_{\delta ij}}{||\dot{\delta}_{dij} s_{\delta ij}|| + \epsilon_{ij} e^{-\beta_{ij} t}}\right),$$

it can be seen that $\boldsymbol{\delta}_d$ is damped because the second factor belongs to the open set $(0, 2)$. This becomes clearer if one considers again the case when $\boldsymbol{\theta}_d$ is constant. The dynamics of $\boldsymbol{\delta}_d$ is described by

$$\boldsymbol{H}_{\delta\delta} \ddot{\boldsymbol{\delta}}_d + \boldsymbol{C}_{\delta\delta} \dot{\boldsymbol{\delta}}_d + \boldsymbol{D}_\triangle \dot{\boldsymbol{\delta}}_d + \boldsymbol{K} \boldsymbol{\delta}_d + \boldsymbol{g}_\delta + \boldsymbol{f} = \boldsymbol{0}. \tag{1.170}$$

If $\boldsymbol{g}_\delta(\boldsymbol{q}) = \boldsymbol{g}_\delta(\boldsymbol{\theta})$, the vector $\boldsymbol{y}$ (see (1.139)) can be employed, so that (1.170) becomes

$$\boldsymbol{H}_{\delta\delta} \ddot{\boldsymbol{y}} + \boldsymbol{C}_{\delta\delta} \dot{\boldsymbol{y}} + \boldsymbol{D}_\triangle \dot{\boldsymbol{y}} + \boldsymbol{K} \boldsymbol{y} + \boldsymbol{f}(\dot{\boldsymbol{y}}) = \boldsymbol{0}. \tag{1.171}$$

Using the Lyapunov function (1.141) leads to (see (1.142))

$$\begin{aligned} \dot{V}_y &= -\dot{\boldsymbol{y}}^T \boldsymbol{D}_\triangle \dot{\boldsymbol{y}} - \dot{\boldsymbol{y}}^T \boldsymbol{f}(\dot{\boldsymbol{y}}) \qquad (1.172) \\ &= -\sum_{i=1}^{n} \sum_{j=1}^{m_i} \left(d_{ij} \dot{y}_{ij}^2 - d_{ij} \dot{y}_{ij}^2 \frac{\dot{y}_{ij} s_{\delta ij}}{||\dot{y}_{ij} s_{\delta ij}|| + \epsilon_{ij} e^{-\beta_{ij} t}} \right) \\ &= -\sum_{i=1}^{n} \sum_{j=1}^{m_i} d_{ij} \dot{y}_{ij}^2 \left(1 - \frac{\dot{y}_{ij} s_{\delta ij}}{||\dot{y}_{ij} s_{\delta ij}|| + \epsilon_{ij} e^{-\beta_{ij} t}}\right) \\ &\leq 0. \end{aligned}$$

Because $\dot{V}_y$ is 0 if and only if $\dot{\boldsymbol{y}} = \boldsymbol{0}$, the asymptotic stability of the equilibrium point $\boldsymbol{y}$ and $\dot{\boldsymbol{y}}$ of (1.171) can be proven by using the same argument as before. Note that $\boldsymbol{\delta}_d$ becomes $-\boldsymbol{K}^{-1} \boldsymbol{g}_\delta$ as well and that this would not be true if $\boldsymbol{D}_\triangle = \boldsymbol{O}$.

1.4 Control with Nonlinear Observer

In the previous section, the tracking control problem of flexible robot manipulators was studied. It was shown how to get a bounded desired trajectory for the link coordinates $(\boldsymbol{\delta}_d, \dot{\boldsymbol{\delta}}_d)$ and how to increase the damping of the system by resorting to a robust control technique. However, it was assumed that link and joint coordinate rates were available. Although one can use tachometers to measure joint speeds, it is not always possible to measure link coordinate rates. In this section, the controller given in Section 1.3 is slightly modified and a nonlinear observer is proposed to estimate link and joint coordinate rates.

1.4.1 *Nonlinear Observer*

In this section, the tracking control problem with a nonlinear observer is studied. Consider model (1.55) again:

$$\boldsymbol{H}(\boldsymbol{q})\ddot{\boldsymbol{q}} + \boldsymbol{C}(\boldsymbol{q}, \dot{\boldsymbol{q}})\dot{\boldsymbol{q}} + \boldsymbol{K}_e\boldsymbol{q} + \boldsymbol{D}\dot{\boldsymbol{q}} + \boldsymbol{g}(\boldsymbol{q}) = \boldsymbol{u} \tag{1.173}$$

and the tracking errors $\tilde{\boldsymbol{q}}$ and $\dot{\tilde{\boldsymbol{q}}}$ given in Section 1.3.2, together with definitions (1.112) to (1.119). In addition, define:

$$\boldsymbol{z} \triangleq \boldsymbol{q} - \hat{\boldsymbol{q}} = \begin{bmatrix} \boldsymbol{\theta} - \hat{\boldsymbol{\theta}} \\ \boldsymbol{\delta} - \hat{\boldsymbol{\delta}} \end{bmatrix} = \begin{bmatrix} \boldsymbol{z}_\theta \\ \boldsymbol{z}_\delta \end{bmatrix} \tag{1.174}$$

$$\ddot{\hat{\boldsymbol{q}}}_r = \begin{bmatrix} \ddot{\hat{\boldsymbol{\theta}}}_r \\ \ddot{\hat{\boldsymbol{\delta}}}_r \end{bmatrix} \triangleq \ddot{\boldsymbol{q}}_d - \boldsymbol{\Lambda}(\dot{\hat{\boldsymbol{q}}} - \dot{\boldsymbol{q}}_d) = \ddot{\boldsymbol{q}}_r + \boldsymbol{\Lambda}\dot{\boldsymbol{z}} \tag{1.175}$$

$$= \begin{bmatrix} \ddot{\boldsymbol{\theta}}_d - \boldsymbol{\Lambda}_\theta(\dot{\hat{\boldsymbol{\theta}}} - \dot{\boldsymbol{\theta}}_d) \\ \ddot{\boldsymbol{\delta}}_d - \boldsymbol{\Lambda}_\delta(\dot{\hat{\boldsymbol{\delta}}} - \dot{\boldsymbol{\delta}}_d) \end{bmatrix} = \begin{bmatrix} \ddot{\boldsymbol{\theta}}_r + \boldsymbol{\Lambda}_\theta \dot{\boldsymbol{z}}_\theta \\ \ddot{\boldsymbol{\delta}}_r + \boldsymbol{\Lambda}_\delta \dot{\boldsymbol{z}}_\delta \end{bmatrix}$$

$$\hat{\boldsymbol{s}} \triangleq \dot{\hat{\boldsymbol{q}}} - \dot{\boldsymbol{q}}_r = \boldsymbol{s} - \dot{\boldsymbol{z}} = \begin{bmatrix} \boldsymbol{s}_\theta - \dot{\boldsymbol{z}}_\theta \\ \boldsymbol{s}_\delta - \dot{\boldsymbol{z}}_\delta \end{bmatrix} \triangleq \begin{bmatrix} \hat{\boldsymbol{s}}_\theta \\ \hat{\boldsymbol{s}}_\delta \end{bmatrix} = \begin{bmatrix} \dot{\hat{\boldsymbol{\theta}}} - \dot{\boldsymbol{\theta}}_r \\ \dot{\hat{\boldsymbol{\delta}}} - \dot{\boldsymbol{\delta}}_r \end{bmatrix} \tag{1.176}$$

$$\boldsymbol{s}_x \triangleq \dot{\hat{\boldsymbol{\delta}}} - \dot{\boldsymbol{\delta}}_d + \boldsymbol{\Lambda}_\delta(\hat{\boldsymbol{\delta}} - \boldsymbol{\delta}_d) \tag{1.177}$$

$$\boldsymbol{K}_q \triangleq \begin{bmatrix} \boldsymbol{K}_{q\theta} & \boldsymbol{O} \\ \boldsymbol{O} & \boldsymbol{K}_{q\delta} \end{bmatrix} \tag{1.178}$$

$$\boldsymbol{K}_{qe} \triangleq \boldsymbol{K}_q + \boldsymbol{K}_e = \begin{bmatrix} \boldsymbol{K}_{q\theta} & \boldsymbol{O} \\ \boldsymbol{O} & \boldsymbol{K}_{q\delta} + \boldsymbol{K} \end{bmatrix} \triangleq \begin{bmatrix} \boldsymbol{K}_{q\theta} & \boldsymbol{O} \\ \boldsymbol{O} & \boldsymbol{K}_{qe\delta} \end{bmatrix}, \tag{1.179}$$

where $\hat{(\cdot)}$ denotes the estimate of $(\cdot)$ and z and $\dot{z}$ are the observation errors. In view of Property 1.11, there exists a constant $k_{c\delta\delta}$ such that

$$\|C_{\delta\delta}(q,y)\| \le k_{c\delta\delta}\|y\| \qquad \forall q \in \mathcal{Q}^{n+m},\ \forall y \in \Re^{n+m}. \tag{1.180}$$

The proposed controller is given by

$$\tau = H_{\theta\theta}\ddot{\hat{\theta}}_r + H_{\theta\delta}\ddot{\hat{\delta}}_r + C_{\theta\theta}(q,\dot{\hat{q}})\dot{\theta}_r + C_{\theta\delta}(q,\dot{\hat{q}})\dot{\delta}_r + D_\theta\dot{\theta}_r + g_\theta - K_{q\theta}\tilde{\theta} - K_{p\theta}\hat{s}_\theta, \tag{1.181}$$

while the desired trajectory for the link coordinates is computed from

$$\begin{aligned}\ddot{\delta}_d = \Lambda_\delta(\dot{\hat{\delta}} - \dot{\delta}_d) - H_{\delta\delta}^{-1}\Big(&H_{\theta\delta}^T\ddot{\hat{\theta}}_r + C_{\delta\theta}(q,\dot{\hat{q}})\dot{\theta}_r + C_{\delta\delta}(q,\dot{\hat{q}})\dot{\delta}_r \\ &+K\delta_d + D_\delta\dot{\delta}_r + g_\delta - K_{q\delta}\tilde{\delta} - K_{p\delta}\hat{s}_\delta + D_\Delta\dot{\delta}_d + f\Big),\end{aligned} \tag{1.182}$$

with initial condition

$$\delta_d(0) = \dot{\delta}_d(0) = \mathbf{0} \tag{1.183}$$

and

$$f_{ij} \triangleq -\dot{\delta}_{dij}\frac{(\dot{\delta}_{dij}s_{xij})d_{ij}}{\|\dot{\delta}_{dij}s_{xij}\| + \epsilon_{ij}e^{-\beta_{ij}t}} \qquad i = 1,\dots,n,\ j = 1,\dots,m_i, \tag{1.184}$$

where s_{xij} is an element of the vector s_x. The inclusion of the positive definite matrix K_q should be noticed. As pointed out before, f helps increase the damping of the system. Eqs. (1.181) and (1.182) are equivalent to

$$u = H(q)\ddot{\hat{q}}_r + C(q,\dot{\hat{q}})\dot{q}_r + K_e q_d + D\dot{q}_r + g(q) - K_q\tilde{q} - K_p\hat{s} + f_1 \tag{1.185}$$

with

$$f_1 \triangleq \begin{bmatrix} \mathbf{0} \\ D_\Delta\dot{\delta}_d + f \end{bmatrix}. \tag{1.186}$$

It is not difficult to show that control (1.185) can be rewritten as

$$\begin{aligned}u = {}& H(q)\ddot{q}_r + C(q,\dot{q})\dot{q}_r + K_e q_d + D\dot{q}_r + g(q) - K_q\tilde{q} \\ &-K_p s + f_1 + H(q)\Lambda\dot{z} - C(q,\dot{z})\dot{q}_r + K_p\dot{z}.\end{aligned} \tag{1.187}$$

By substituting (1.187) into (1.173), the error dynamics is given by

$$\begin{aligned}H(q)\dot{s} = {}& -\left(C(q,\dot{q})s + K_{pD}s + K_{qe}\tilde{q}\right) + f_1 + H(q)\Lambda\dot{z} \\ &-C(q,\dot{q})\dot{z} + C(q,\dot{z})s + K_p\dot{z}.\end{aligned} \tag{1.188}$$

In the following, an observer based on that given in [Nicosia and Tomei (1990)] is presented, although some essential modifications are necessary in order to guarantee the stability of the whole system and the improvement of link damping [Arteaga (2000)]. Consider the following definitions:

$$\dot{\boldsymbol{q}}_o \triangleq \dot{\hat{\boldsymbol{q}}} - \boldsymbol{\Lambda z} \tag{1.189}$$

$$\boldsymbol{r} \triangleq \dot{\boldsymbol{q}} - \dot{\boldsymbol{q}}_o = \dot{\boldsymbol{q}} - \dot{\hat{\boldsymbol{q}}} + \boldsymbol{\Lambda z} = \dot{\boldsymbol{z}} + \boldsymbol{\Lambda z}. \tag{1.190}$$

By using an estimate of $\boldsymbol{q}_o$, the proposed observer is given by

$$\dot{\hat{\boldsymbol{q}}} = \dot{\hat{\boldsymbol{q}}}_o + \boldsymbol{\Lambda z} + k_d \boldsymbol{z} \tag{1.191}$$

$$\boldsymbol{H}(\boldsymbol{q})\ddot{\hat{\boldsymbol{q}}}_o + \boldsymbol{C}(\boldsymbol{q}, \dot{\hat{\boldsymbol{q}}})\dot{\boldsymbol{q}}_o + \boldsymbol{K}_e\hat{\boldsymbol{q}} + \boldsymbol{D}\dot{\boldsymbol{q}}_o + \boldsymbol{g}(\boldsymbol{q}) = \boldsymbol{u} + \boldsymbol{K}_v\boldsymbol{z} + \boldsymbol{f}_1 \tag{1.192}$$

where $\boldsymbol{K}_v$ is diagonal and positive definite and $k_d > 0$. Since one has

$$\ddot{\hat{\boldsymbol{q}}} = \ddot{\hat{\boldsymbol{q}}}_o + \boldsymbol{\Lambda}\dot{\boldsymbol{z}} + k_d\dot{\boldsymbol{z}}, \tag{1.193}$$

Eqs. (1.191) and (1.192) are equivalent to

$$\begin{aligned}\boldsymbol{H}(\boldsymbol{q})\ddot{\boldsymbol{q}}_o + \boldsymbol{C}(\boldsymbol{q}, \dot{\boldsymbol{q}})\dot{\boldsymbol{q}}_o + \boldsymbol{K}_e\hat{\boldsymbol{q}} + \boldsymbol{D}\dot{\boldsymbol{q}}_o + \boldsymbol{g}(\boldsymbol{q}) &= \boldsymbol{u} + \boldsymbol{K}_v\boldsymbol{z} + \boldsymbol{f}_1 \\ &\quad + k_d\boldsymbol{H}(\boldsymbol{q})\dot{\boldsymbol{z}} + \boldsymbol{C}(\boldsymbol{q}, \dot{\boldsymbol{z}})\dot{\boldsymbol{q}}_o.\end{aligned} \tag{1.194}$$

Subtracting (1.194) from (1.173) yields

$$\boldsymbol{H}(\boldsymbol{q})\dot{\boldsymbol{r}} + \boldsymbol{C}(\boldsymbol{q}, \dot{\boldsymbol{q}})\boldsymbol{r} + \boldsymbol{K}_e\boldsymbol{z} + \boldsymbol{D}\boldsymbol{r} = -\boldsymbol{K}_v\boldsymbol{z} - \boldsymbol{f}_1 - k_d\boldsymbol{H}(\boldsymbol{q})\dot{\boldsymbol{z}} - \boldsymbol{C}(\boldsymbol{q}, \dot{\boldsymbol{z}})\dot{\boldsymbol{q}}_o. \tag{1.195}$$

Due to Property 1.3, the following equality is valid:

$$\boldsymbol{C}(\boldsymbol{q}, \dot{\boldsymbol{z}})\dot{\boldsymbol{q}}_o = \boldsymbol{C}(\boldsymbol{q}, \dot{\boldsymbol{z}})(\dot{\boldsymbol{q}} - \boldsymbol{r}) = \boldsymbol{C}(\boldsymbol{q}, \dot{\boldsymbol{q}})\dot{\boldsymbol{z}} - \boldsymbol{C}(\boldsymbol{q}, \dot{\boldsymbol{z}})\boldsymbol{r}, \tag{1.196}$$

so that (1.195) can be rewritten as

$$\boldsymbol{H}(\boldsymbol{q})\dot{\boldsymbol{r}} = -\left(\boldsymbol{C}(\boldsymbol{q}, \dot{\boldsymbol{q}})\boldsymbol{r} + \boldsymbol{K}_{ve}\boldsymbol{z} + \boldsymbol{D}\boldsymbol{r} + k_d\boldsymbol{H}(\boldsymbol{q})\dot{\boldsymbol{z}}\right) + \boldsymbol{C}(\boldsymbol{q}, \dot{\boldsymbol{z}})\boldsymbol{r} - \boldsymbol{C}(\boldsymbol{q}, \dot{\boldsymbol{q}})\dot{\boldsymbol{z}} - \boldsymbol{f}_1 \tag{1.197}$$

with $\boldsymbol{K}_{ve} \triangleq \boldsymbol{K}_v + \boldsymbol{K}_e$.

In order to simplify the stability discussion, the following definition is introduced:

$$\boldsymbol{x} \triangleq \begin{bmatrix} \tilde{\boldsymbol{q}} \\ \dot{\tilde{\boldsymbol{q}}} \\ \boldsymbol{z} \\ \dot{\boldsymbol{z}} \end{bmatrix}. \tag{1.198}$$

By taking advantage again of the fact that $\dot{\boldsymbol{q}}$ is bounded, Theorem 1.5 establishes that the state $\boldsymbol{x}$ of (1.188) and (1.197) is *exponentially stable* (ES). In order to prove it, Theorems 1.1 and 1.2 can be employed. Although they are helpful to prove whether the state of a system is GUUB or GES, they can be used as well to analyze if it is UUB or ES. To show this, suppose that there exists a ball $B_r = \{\boldsymbol{x} : \|\boldsymbol{x}\| \leq b_r\}$ so that conditions (1.101)–(1.102) of Theorem 1.1 or (1.105)–(1.106) of Theorem 1.2 are satisfied. If it is possible to find a region of attraction $S = \{\boldsymbol{x} : \|\boldsymbol{x}\| \leq b_s\}$ with $0 < b_s \leq b_r$, so that $\|\boldsymbol{x}\|$ never abandons the ball B_r, then it can be proven that $\|\boldsymbol{x}\|$ is UUB or ES.

Theorem 1.5 *Given a bounded continuous desired trajectory $\boldsymbol{\theta}_d$ with bounded velocity and acceleration, if $\|\dot{\boldsymbol{q}}\| \leq v_m$ where v_m is a positive scalar constant, then the state $\boldsymbol{x}$ of (1.188) and (1.197) is ES in the sense of Theorem 1.1 as long as the following inequalities are satisfied:*

$$k_d > \frac{9}{4}\frac{(\lambda_M\lambda_H + k_c v_m + \lambda_P)^2}{\lambda_{pd}\lambda_h} + \frac{k_c v_m - \lambda_d}{\lambda_h} \tag{1.199}$$

$$\lambda_q > \frac{9}{4}\frac{\lambda_M(\lambda_M\lambda_H + k_c v_m + \lambda_P)}{\lambda_m(\lambda_d + k_d\lambda_h - k_c v_m)} - \lambda_m\lambda_{pd} - \lambda_{\min}(\boldsymbol{K}_e) \tag{1.200}$$

$$\lambda_v > \frac{9}{4}\frac{(k_d\lambda_M\lambda_H + \lambda_M k_c v_m)^2}{\lambda_m(\lambda_d + k_d\lambda_h - k_c v_m)} - \lambda_m\lambda_d - \lambda_{\min}(\boldsymbol{K}_e) \tag{1.201}$$

$$\varphi^2 z_m^2 > \frac{\epsilon e^{-1}}{\lambda_1\lambda} \quad \textit{if } \beta = \lambda \tag{1.202}$$

$$\varphi^2 z_m^2 > \frac{\epsilon}{\lambda_1(\lambda-\beta)}\left(e^{-\beta\left(\frac{\ln(\lambda/\beta)}{\lambda-\beta}\right)} - e^{-\lambda\left(\frac{\ln(\lambda/\beta)}{\lambda-\beta}\right)}\right) \quad \textit{if } \beta \neq \lambda \tag{1.203}$$

with

$$\lambda_m \triangleq \lambda_{\min}(\boldsymbol{\Lambda}) \quad \lambda_M \triangleq \lambda_{\max}(\boldsymbol{\Lambda}) \quad \lambda_{pd} \triangleq \lambda_{\min}(\boldsymbol{K}_{pD})$$
$$\lambda_d \triangleq \lambda_{\min}(\boldsymbol{D}) \quad \lambda_{qe} \triangleq \lambda_{\min}(\boldsymbol{K}_{qe}) \; \lambda_q \triangleq \lambda_{\min}(\boldsymbol{K}_q)$$
$$\lambda_P \triangleq \lambda_{\max}(\boldsymbol{K}_p) \; \lambda_{ve} \triangleq \lambda_{\min}(\boldsymbol{K}_{ve}) \; \lambda_v \triangleq \lambda_{\min}(\boldsymbol{K}_v)$$

and λ_h, λ_H are defined in Property 1.9. A region of attraction is given by

$$S = \left\{\boldsymbol{x} \in \Re^{4(n+m)} : \; \|\boldsymbol{x}\| \leq \sqrt{\frac{\lambda_1}{\lambda_2}}\left(\varphi^2 z_m^2 - \frac{\epsilon e^{-1}}{\lambda_1\lambda}\right)^{\frac{1}{2}}\right\} \tag{1.204}$$

if $\beta = \lambda$, and by

$$S = \Big\{\boldsymbol{x} \in \Re^{4(n+m)} : \tag{1.205}$$

$$\|\boldsymbol{x}\| \le \sqrt{\frac{\lambda_1}{\lambda_2}} \left(\varphi^2 z_m^2 - \frac{\epsilon}{\lambda_1(\lambda - \beta)} \left(e^{-\beta\left(\frac{\ln(\lambda/\beta)}{\lambda-\beta}\right)} - e^{-\lambda\left(\frac{\ln(\lambda/\beta)}{\lambda-\beta}\right)} \right) \right)^{\frac{1}{2}} \Bigg\}$$

if $\beta \neq \lambda$, *where*

$$\lambda_1 \triangleq \frac{1}{2} \min_{\boldsymbol{q} \in \mathcal{Q}^{n+m}} \lambda_{\min}(\boldsymbol{N}) \tag{1.206}$$

$$\lambda_2 \triangleq \frac{1}{2} \max_{\boldsymbol{q} \in \mathcal{Q}^{n+m}} \lambda_{\max}(\boldsymbol{N}) \tag{1.207}$$

$$\lambda_3 \triangleq \min \lambda_{\min}(\boldsymbol{Q}) \qquad 0 \le \|\dot{\boldsymbol{z}}\| \le \varphi z_m, \ 0 < \varphi < 1 \tag{1.208}$$

$$\lambda \triangleq \frac{\lambda_3}{\lambda_2} \tag{1.209}$$

$$\epsilon \triangleq \sum_{i=1}^{n} \sum_{j=1}^{m_i} d_{ij}\epsilon_{ij} \tag{1.210}$$

$$\beta \triangleq \min\{\beta_{ij},\ i = 1, \ldots, n;\ j = 1, \ldots, m_i\} \tag{1.211}$$

$$z_m \triangleq \min \left\{ \begin{array}{c} \dfrac{\lambda_m^2\lambda_{pd}+\lambda_m\lambda_{qe}}{\lambda_M^2 k_c}, \dfrac{\lambda_{pd}}{k_c}, \dfrac{\lambda_m^2\lambda_d+\lambda_m\lambda_{ve}}{\lambda_M^2 k_c}, \dfrac{\lambda_d+k_d\lambda_h-k_c v_m}{k_c}, \\[2ex] \dfrac{(\lambda_m^2\lambda_{pd}+\lambda_m\lambda_{qe})(\lambda_d+k_d\lambda_h-k_c v_m)-\frac{9}{4}\lambda_M(\lambda_M\lambda_H+k_c v_m+\lambda_P)}{k_c(\lambda_m^2\lambda_{pd}+\lambda_m\lambda_{qe}+\lambda_M^2(\lambda_d+k_d\lambda_h-k_c v_m))}, \\[2ex] \dfrac{\lambda_{pd}(\lambda_d+k_d\lambda_h-k_c v_m)-\frac{9}{4}(\lambda_M\lambda_H+k_c v_m+\lambda_P)^2}{k_c(\lambda_{pd}+\lambda_d+k_d\lambda_h-k_c v_m)}, \\[2ex] \left(\dfrac{\lambda_m^2\lambda_{pd}^2+\lambda_m\lambda_{qe}\lambda_{pd}}{8\lambda_M^2 k_c^2}+\left(\dfrac{2\lambda_M^2\lambda_{pd}+\lambda_m\lambda_{qe}}{16\lambda_M^2 k_c}\right)^2\right)^{\frac{1}{2}} - \dfrac{2\lambda_M^2\lambda_{pd}+\lambda_m\lambda_{qe}}{16\lambda_M^2 k_c}, \\[2ex] \left(\dfrac{(\lambda_m\lambda_{ve}+\lambda_m^2\lambda_d)(\lambda_d+k_d\lambda_h-k_c v_m)-\frac{9}{4}(k_d\lambda_M\lambda_H+\lambda_M k_c v_m)^2}{8\lambda_M^2 k_c^2}\right. \\[2ex] \left. +\left(\dfrac{\lambda_m\lambda_{ve}+\lambda_m^2\lambda_d+\lambda_M^2(\lambda_d+k_d\lambda_h-k_c v_m)+9\lambda_M^2(k_d\lambda_H+k_c v_m)}{16k_c\lambda_M^2}\right)^2\right)^{\frac{1}{2}} \\[2ex] -\dfrac{\lambda_m\lambda_{ve}+\lambda_m^2\lambda_d+\lambda_M^2(\lambda_d+k_d\lambda_h-k_c v_m)+9\lambda_M^2(k_d\lambda_H+k_c v_m)}{16k_c\lambda_M^2} \end{array} \right\} \tag{1.212}$$

$$
\boldsymbol{Q} \triangleq \begin{bmatrix}
\begin{array}{c} \boldsymbol{\Lambda K}_{pD}\boldsymbol{\Lambda} \\ +\boldsymbol{\Lambda K}_{qe} \\ -\frac{1}{2}\boldsymbol{\Lambda C}(\boldsymbol{q},\dot{\boldsymbol{z}})\boldsymbol{\Lambda} \\ -\frac{1}{2}\boldsymbol{\Lambda C}^T(\boldsymbol{q},\dot{\boldsymbol{z}})\boldsymbol{\Lambda} \end{array} &
\begin{array}{c} -\frac{1}{2}\boldsymbol{\Lambda C}^T(\boldsymbol{q},\dot{\boldsymbol{z}}) \\ -\frac{1}{2}\boldsymbol{\Lambda C}(\boldsymbol{q},\dot{\boldsymbol{z}}) \end{array} &
\boldsymbol{O} &
\begin{array}{c} -\frac{1}{2}\boldsymbol{\Lambda H}(\boldsymbol{q})\boldsymbol{\Lambda} \\ -\frac{1}{2}\boldsymbol{\Lambda K}_p \\ +\frac{1}{2}\boldsymbol{\Lambda C}(\boldsymbol{q},\dot{\boldsymbol{q}}) \end{array} \\[2ex]
\begin{array}{c} -\frac{1}{2}\boldsymbol{C}(\boldsymbol{q},\dot{\boldsymbol{z}})\boldsymbol{\Lambda} \\ -\frac{1}{2}\boldsymbol{C}^T(\boldsymbol{q},\dot{\boldsymbol{z}})\boldsymbol{\Lambda} \end{array} &
\begin{array}{c} \boldsymbol{K}_{pD} \\ -\frac{1}{2}\boldsymbol{C}(\boldsymbol{q},\dot{\boldsymbol{z}}) \\ -\frac{1}{2}\boldsymbol{C}^T(\boldsymbol{q},\dot{\boldsymbol{z}}) \end{array} &
\boldsymbol{O} &
\begin{array}{c} -\frac{1}{2}\boldsymbol{H}(\boldsymbol{q})\boldsymbol{\Lambda} \\ -\frac{1}{2}\boldsymbol{K}_p \\ +\frac{1}{2}\boldsymbol{C}(\boldsymbol{q},\dot{\boldsymbol{q}}) \end{array} \\[2ex]
\boldsymbol{O} & \boldsymbol{O} &
\begin{array}{c} \boldsymbol{\Lambda K}_{ve} \\ +\boldsymbol{\Lambda D \Lambda} \\ -\frac{1}{2}\boldsymbol{\Lambda C}(\boldsymbol{q},\dot{\boldsymbol{z}})\boldsymbol{\Lambda} \\ -\frac{1}{2}\boldsymbol{\Lambda C}^T(\boldsymbol{q},\dot{\boldsymbol{z}})\boldsymbol{\Lambda} \end{array} &
\begin{array}{c} \frac{1}{2}k_d\boldsymbol{\Lambda H}(\boldsymbol{q}) \\ -\frac{1}{2}\boldsymbol{\Lambda C}^T(\boldsymbol{q},\dot{\boldsymbol{z}}) \\ -\frac{1}{2}\boldsymbol{\Lambda C}(\boldsymbol{q},\dot{\boldsymbol{z}}) \\ +\frac{1}{2}\boldsymbol{\Lambda C}(\boldsymbol{q},\dot{\boldsymbol{q}}) \end{array} \\[2ex]
\begin{array}{c} -\frac{1}{2}\boldsymbol{\Lambda H}(\boldsymbol{q})\boldsymbol{\Lambda} \\ -\frac{1}{2}\boldsymbol{K}_p\boldsymbol{\Lambda} \\ +\frac{1}{2}\boldsymbol{C}^T(\boldsymbol{q},\dot{\boldsymbol{q}})\boldsymbol{\Lambda} \end{array} &
\begin{array}{c} -\frac{1}{2}\boldsymbol{\Lambda H}(\boldsymbol{q}) \\ -\frac{1}{2}\boldsymbol{K}_p \\ +\frac{1}{2}\boldsymbol{C}^T(\boldsymbol{q},\dot{\boldsymbol{q}}) \end{array} &
\begin{array}{c} \frac{1}{2}k_d\boldsymbol{H}(\boldsymbol{q})\boldsymbol{\Lambda} \\ -\frac{1}{2}\boldsymbol{C}(\boldsymbol{q},\dot{\boldsymbol{z}})\boldsymbol{\Lambda} \\ -\frac{1}{2}\boldsymbol{C}^T(\boldsymbol{q},\dot{\boldsymbol{z}})\boldsymbol{\Lambda} \\ +\frac{1}{2}\boldsymbol{C}^T(\boldsymbol{q},\dot{\boldsymbol{q}})\boldsymbol{\Lambda} \end{array} &
\begin{array}{c} \boldsymbol{D} \\ +k_d\boldsymbol{H}(\boldsymbol{q}) \\ -\frac{1}{2}\boldsymbol{C}(\boldsymbol{q},\dot{\boldsymbol{z}}) \\ -\frac{1}{2}\boldsymbol{C}^T(\boldsymbol{q},\dot{\boldsymbol{z}}) \\ +\frac{1}{2}\boldsymbol{C}(\boldsymbol{q},\dot{\boldsymbol{q}}) \\ +\frac{1}{2}\boldsymbol{C}^T(\boldsymbol{q},\dot{\boldsymbol{q}}) \end{array}
\end{bmatrix} \tag{1.213}
$$

$$
\boldsymbol{N} \triangleq \begin{bmatrix}
\begin{array}{c} 2\boldsymbol{\Lambda K}_{pD} + \boldsymbol{K}_{qe} \\ +\boldsymbol{\Lambda H}(\boldsymbol{q})\boldsymbol{\Lambda} \end{array} & \boldsymbol{\Lambda H}(\boldsymbol{q}) & \boldsymbol{O} & \boldsymbol{O} \\[2ex]
\boldsymbol{H}(\boldsymbol{q})\boldsymbol{\Lambda} & \boldsymbol{H}(\boldsymbol{q}) & \boldsymbol{O} & \boldsymbol{O} \\[2ex]
\boldsymbol{O} & \boldsymbol{O} & \begin{array}{c} 2\boldsymbol{\Lambda D} + \boldsymbol{K}_{ve} \\ +\boldsymbol{\Lambda H}(\boldsymbol{q})\boldsymbol{\Lambda} \end{array} & \boldsymbol{\Lambda H}(\boldsymbol{q}) \\[2ex]
\boldsymbol{O} & \boldsymbol{O} & \boldsymbol{H}(\boldsymbol{q})\boldsymbol{\Lambda} & \boldsymbol{H}(\boldsymbol{q})
\end{bmatrix}. \tag{1.214}
$$

In addition, the desired trajectory for the link coordinates $\boldsymbol{\delta}_d$ and $\dot{\boldsymbol{\delta}}_d$ given by (1.182) and (1.183) remains bounded if

$$
\hat{\lambda}_3 \triangleq \lambda_{\min}(\boldsymbol{P}) - (1+\lambda_{\max}(\boldsymbol{\Lambda}_\delta))(\lambda_{\max}(\boldsymbol{D}_\Delta)\sqrt{m} + k_{c\delta\delta}\varphi z_m(1+\lambda_{\max}(\boldsymbol{\Lambda}_\delta)) > 0 \tag{1.215}
$$

with

$$\boldsymbol{P} \triangleq \begin{bmatrix} \boldsymbol{\Lambda}_\delta \boldsymbol{K}_{qe\delta} + \boldsymbol{\Lambda}_\delta \boldsymbol{K}_{pD\delta} \boldsymbol{\Lambda}_\delta & \boldsymbol{O} \\ \boldsymbol{O} & \boldsymbol{D}_\Delta + \boldsymbol{K}_{pD\delta} \end{bmatrix} \tag{1.216}$$

$$\begin{aligned} \lambda_{\min}(\boldsymbol{P}) = \min\{ & \lambda_{\min}(\boldsymbol{\Lambda}_\delta)\lambda_{\min}(\boldsymbol{K}_{qe\delta}) + \lambda_{\min}^2(\boldsymbol{\Lambda}_\delta)\lambda_{\min}(\boldsymbol{K}_{pD\delta}), \\ & \lambda_{\min}(\boldsymbol{K}_{pD\delta}) + \lambda_{\min}(\boldsymbol{D}_\Delta)\}. \end{aligned} \tag{1.217}$$

Proof:

a) Firstly, the stability of the whole system will be proven. Consider the following Lyapunov function:

$$\begin{aligned} V(\boldsymbol{x},t) \triangleq V(\boldsymbol{x}) &= \frac{1}{2}\boldsymbol{s}^T \boldsymbol{H}(\boldsymbol{q})\boldsymbol{s} + \frac{1}{2}\tilde{\boldsymbol{q}}^T \left(2\boldsymbol{\Lambda}\boldsymbol{K}_{pD} + \boldsymbol{K}_{qe}\right)\tilde{\boldsymbol{q}} \\ &\quad + \frac{1}{2}\boldsymbol{r}^T \boldsymbol{H}(\boldsymbol{q})\boldsymbol{r} + \frac{1}{2}\boldsymbol{z}^T \left(2\boldsymbol{\Lambda}\boldsymbol{D} + \boldsymbol{K}_{ve}\right)\boldsymbol{z} \\ &= \frac{1}{2}\boldsymbol{x}^T \boldsymbol{N}\boldsymbol{x}, \end{aligned} \tag{1.218}$$

whose derivative is given by

$$\begin{aligned} \dot{V}(\boldsymbol{x}) = & \frac{1}{2}\boldsymbol{s}^T \dot{\boldsymbol{H}}(\boldsymbol{q})\boldsymbol{s} + \tilde{\boldsymbol{q}}^T \left(2\boldsymbol{\Lambda}\boldsymbol{K}_{pD} + \boldsymbol{K}_{qe}\right)\dot{\tilde{\boldsymbol{q}}} + \frac{1}{2}\boldsymbol{r}^T \dot{\boldsymbol{H}}(\boldsymbol{q})\boldsymbol{r} \\ & + \boldsymbol{z}^T \left(2\boldsymbol{\Lambda}\boldsymbol{D} + \boldsymbol{K}_{ve}\right)\dot{\boldsymbol{z}} + \boldsymbol{s}^T \boldsymbol{H}(\boldsymbol{q})\dot{\boldsymbol{s}} + \boldsymbol{r}^T \boldsymbol{H}(\boldsymbol{q})\dot{\boldsymbol{r}}. \end{aligned} \tag{1.219}$$

Substituting (1.188) and (1.197) into (1.219) leads to

$$\begin{aligned} \dot{V}(\boldsymbol{x}) = & \frac{1}{2}\boldsymbol{s}^T \dot{\boldsymbol{H}}(\boldsymbol{q})\boldsymbol{s} + \tilde{\boldsymbol{q}}^T \left(2\boldsymbol{\Lambda}\boldsymbol{K}_{pD} + \boldsymbol{K}_{qe}\right)\dot{\tilde{\boldsymbol{q}}} + \frac{1}{2}\boldsymbol{r}^T \dot{\boldsymbol{H}}(\boldsymbol{q})\boldsymbol{r} \\ & + \boldsymbol{z}^T \left(2\boldsymbol{\Lambda}\boldsymbol{D} + \boldsymbol{K}_{ve}\right)\dot{\boldsymbol{z}} \\ & - \boldsymbol{s}^T \left(\boldsymbol{C}(\boldsymbol{q},\dot{\boldsymbol{q}})\boldsymbol{s} + \boldsymbol{K}_{pD}\boldsymbol{s} + \boldsymbol{K}_{qe}\tilde{\boldsymbol{q}}\right) \\ & + \boldsymbol{s}^T \left(\boldsymbol{f}_1 + \boldsymbol{H}(\boldsymbol{q})\boldsymbol{\Lambda}\dot{\boldsymbol{z}} - \boldsymbol{C}(\boldsymbol{q},\dot{\boldsymbol{q}})\dot{\boldsymbol{z}} + \boldsymbol{C}(\boldsymbol{q},\dot{\boldsymbol{z}})\boldsymbol{s} + \boldsymbol{K}_p\dot{\boldsymbol{z}}\right) \\ & - \boldsymbol{r}^T \left(\boldsymbol{C}(\boldsymbol{q},\dot{\boldsymbol{q}})\boldsymbol{r} + \boldsymbol{K}_{ve}\boldsymbol{z} + \boldsymbol{D}\boldsymbol{r} + k_d\boldsymbol{H}(\boldsymbol{q})\dot{\boldsymbol{z}}\right) \\ & + \boldsymbol{r}^T \left(\boldsymbol{C}(\boldsymbol{q},\dot{\boldsymbol{z}})\boldsymbol{r} - \boldsymbol{C}(\boldsymbol{q},\dot{\boldsymbol{q}})\dot{\boldsymbol{z}} - \boldsymbol{f}_1\right). \end{aligned} \tag{1.220}$$

Since

$$\begin{aligned} -\boldsymbol{s}^T \boldsymbol{K}_{pD}\boldsymbol{s} - \boldsymbol{s}^T \boldsymbol{K}_{qe}\tilde{\boldsymbol{q}} &= -(\dot{\tilde{\boldsymbol{q}}}^T + \tilde{\boldsymbol{q}}^T\boldsymbol{\Lambda})\boldsymbol{K}_{pD}(\dot{\tilde{\boldsymbol{q}}} + \boldsymbol{\Lambda}\tilde{\boldsymbol{q}}) \\ &\quad -(\dot{\tilde{\boldsymbol{q}}}^T + \tilde{\boldsymbol{q}}^T\boldsymbol{\Lambda})\boldsymbol{K}_{qe}\tilde{\boldsymbol{q}} \\ &= -\dot{\tilde{\boldsymbol{q}}}^T\boldsymbol{K}_{pD}\dot{\tilde{\boldsymbol{q}}} - \tilde{\boldsymbol{q}}^T\boldsymbol{\Lambda}\boldsymbol{K}_{pD}\dot{\tilde{\boldsymbol{q}}} - \dot{\tilde{\boldsymbol{q}}}^T\boldsymbol{K}_{pD}\boldsymbol{\Lambda}\tilde{\boldsymbol{q}} \\ &\quad -\tilde{\boldsymbol{q}}^T\boldsymbol{\Lambda}\boldsymbol{K}_{pD}\boldsymbol{\Lambda}\tilde{\boldsymbol{q}} - \dot{\tilde{\boldsymbol{q}}}^T\boldsymbol{K}_{qe}\tilde{\boldsymbol{q}} - \tilde{\boldsymbol{q}}^T\boldsymbol{\Lambda}\boldsymbol{K}_{qe}\tilde{\boldsymbol{q}} \end{aligned} \tag{1.221}$$

and

$$
\begin{aligned}
-\boldsymbol{r}^T\boldsymbol{D}\boldsymbol{r} - \boldsymbol{r}^T\boldsymbol{K}_{ve}\boldsymbol{z} &= -(\dot{\boldsymbol{z}}^T + \boldsymbol{z}^T\boldsymbol{\Lambda})\boldsymbol{D}(\dot{\boldsymbol{z}} + \boldsymbol{\Lambda}\boldsymbol{z}) \\
&\quad -(\dot{\boldsymbol{z}}^T + \boldsymbol{z}^T\boldsymbol{\Lambda})\boldsymbol{K}_{ve}\boldsymbol{z} \\
&= -\dot{\boldsymbol{z}}^T\boldsymbol{D}\dot{\boldsymbol{z}} - \boldsymbol{z}^T\boldsymbol{\Lambda}\boldsymbol{D}\dot{\boldsymbol{z}} - \dot{\boldsymbol{z}}^T\boldsymbol{D}\boldsymbol{\Lambda}\boldsymbol{z} \\
&\quad -\boldsymbol{z}^T\boldsymbol{\Lambda}\boldsymbol{D}\boldsymbol{\Lambda}\boldsymbol{z} - \dot{\boldsymbol{z}}^T\boldsymbol{K}_{ve}\boldsymbol{z} - \boldsymbol{z}^T\boldsymbol{\Lambda}\boldsymbol{K}_{ve}\boldsymbol{z},
\end{aligned}
\tag{1.222}
$$

by taking Property 1.2 into account, one gets from (1.220)–(1.222)

$$
\begin{aligned}
\dot{V}(\boldsymbol{x}) &= -\tilde{\boldsymbol{q}}^T(\boldsymbol{\Lambda}\boldsymbol{K}_{pD}\boldsymbol{\Lambda} + \boldsymbol{\Lambda}\boldsymbol{K}_{qe})\tilde{\boldsymbol{q}} - \dot{\tilde{\boldsymbol{q}}}^T\boldsymbol{K}_{pD}\dot{\tilde{\boldsymbol{q}}} \\
&\quad -\boldsymbol{z}^T(\boldsymbol{\Lambda}\boldsymbol{D}\boldsymbol{\Lambda} + \boldsymbol{\Lambda}\boldsymbol{K}_{ve})\boldsymbol{z} - \dot{\boldsymbol{z}}^T(\boldsymbol{D} + k_d\boldsymbol{H}(\boldsymbol{q}))\dot{\boldsymbol{z}} \\
&\quad +\boldsymbol{s}^T(\boldsymbol{H}(\boldsymbol{q})\boldsymbol{\Lambda}\dot{\boldsymbol{z}} - \boldsymbol{C}(\boldsymbol{q},\dot{\boldsymbol{q}})\dot{\boldsymbol{z}} + \boldsymbol{C}(\boldsymbol{q},\dot{\boldsymbol{z}})\boldsymbol{s} + \boldsymbol{K}_p\dot{\boldsymbol{z}}) \\
&\quad -k_d\boldsymbol{z}^T\boldsymbol{\Lambda}\boldsymbol{H}(\boldsymbol{q})\dot{\boldsymbol{z}} + \boldsymbol{r}^T(\boldsymbol{C}(\boldsymbol{q},\dot{\boldsymbol{z}})\boldsymbol{r} - \boldsymbol{C}(\boldsymbol{q},\dot{\boldsymbol{q}})\dot{\boldsymbol{z}}) + (\boldsymbol{s}^T - \boldsymbol{r}^T)\boldsymbol{f}_1 \\
&= -\tilde{\boldsymbol{q}}^T(\boldsymbol{\Lambda}\boldsymbol{K}_{pD}\boldsymbol{\Lambda} + \boldsymbol{\Lambda}\boldsymbol{K}_{qe} - \boldsymbol{\Lambda}\boldsymbol{C}(\boldsymbol{q},\dot{\boldsymbol{z}})\boldsymbol{\Lambda})\tilde{\boldsymbol{q}} - \dot{\tilde{\boldsymbol{q}}}^T(\boldsymbol{K}_{pD} - \boldsymbol{C}(\boldsymbol{q},\dot{\boldsymbol{z}}))\dot{\tilde{\boldsymbol{q}}} \\
&\quad -\boldsymbol{z}^T(\boldsymbol{\Lambda}\boldsymbol{D}\boldsymbol{\Lambda} + \boldsymbol{\Lambda}\boldsymbol{K}_{ve} - \boldsymbol{\Lambda}\boldsymbol{C}(\boldsymbol{q},\dot{\boldsymbol{z}})\boldsymbol{\Lambda})\boldsymbol{z} \\
&\quad -\dot{\boldsymbol{z}}^T(\boldsymbol{D} + k_d\boldsymbol{H}(\boldsymbol{q}) - \boldsymbol{C}(\boldsymbol{q},\dot{\boldsymbol{z}}) + \boldsymbol{C}(\boldsymbol{q},\dot{\boldsymbol{q}}))\dot{\boldsymbol{z}} \\
&\quad +\tilde{\boldsymbol{q}}^T(\boldsymbol{\Lambda}\boldsymbol{C}^T(\boldsymbol{q},\dot{\boldsymbol{z}}) + \boldsymbol{\Lambda}\boldsymbol{C}(\boldsymbol{q},\dot{\boldsymbol{z}}))\dot{\tilde{\boldsymbol{q}}} \\
&\quad +\boldsymbol{z}^T(-k_d\boldsymbol{\Lambda}\boldsymbol{H}(\boldsymbol{q}) + \boldsymbol{\Lambda}\boldsymbol{C}^T(\boldsymbol{q},\dot{\boldsymbol{z}}) + \boldsymbol{\Lambda}\boldsymbol{C}(\boldsymbol{q},\dot{\boldsymbol{z}}) - \boldsymbol{\Lambda}\boldsymbol{C}(\boldsymbol{q},\dot{\boldsymbol{q}}))\dot{\boldsymbol{z}} \\
&\quad +\tilde{\boldsymbol{q}}^T(\boldsymbol{\Lambda}\boldsymbol{H}(\boldsymbol{q})\boldsymbol{\Lambda} - \boldsymbol{\Lambda}\boldsymbol{C}(\boldsymbol{q},\dot{\boldsymbol{q}}) + \boldsymbol{\Lambda}\boldsymbol{K}_p)\dot{\boldsymbol{z}} \\
&\quad +\dot{\tilde{\boldsymbol{q}}}^T(\boldsymbol{H}(\boldsymbol{q})\boldsymbol{\Lambda} - \boldsymbol{C}(\boldsymbol{q},\dot{\boldsymbol{q}}) + \boldsymbol{K}_p)\dot{\boldsymbol{z}} + \boldsymbol{s}_x^T(\boldsymbol{D}_\Delta\dot{\boldsymbol{\delta}}_d + \boldsymbol{f}) \\
&= -\boldsymbol{x}^T\boldsymbol{Q}\boldsymbol{x} + \boldsymbol{s}_x^T(\boldsymbol{D}_\Delta\dot{\boldsymbol{\delta}}_d + \boldsymbol{f}).
\end{aligned}
\tag{1.223}
$$

Finally, from (1.184) it is

$$
\begin{aligned}
\dot{V}(\boldsymbol{x}) &= -\boldsymbol{x}^T\boldsymbol{Q}\boldsymbol{x} \\
&\quad +\sum_{i=1}^{n}\sum_{j=1}^{m_i}\left(\dot{\delta}_{dij}s_{xij}d_{ij} - d_{ij}\frac{\|\dot{\delta}_{dij}s_{xij}\|^2}{\|\dot{\delta}_{dij}s_{xij}\| + \epsilon_{ij}e^{-\beta_{ij}t}}\right) \\
&\leq -\boldsymbol{x}^T\boldsymbol{Q}\boldsymbol{x} + \sum_{i=1}^{n}\sum_{j=1}^{m_i}\left(\|\dot{\delta}_{dij}s_{xij}\|d_{ij} - d_{ij}\frac{\|\dot{\delta}_{dij}s_{xij}\|^2}{\|\dot{\delta}_{dij}s_{xij}\| + \epsilon_{ij}e^{-\beta_{ij}t}}\right) \\
&= -\boldsymbol{x}^T\boldsymbol{Q}\boldsymbol{x} + \sum_{i=1}^{n}\sum_{j=1}^{m_i}\left(d_{ij}\frac{\|\dot{\delta}_{dij}s_{xij}\|}{\|\dot{\delta}_{dij}s_{xij}\| + \epsilon_{ij}e^{-\beta_{ij}t}}\epsilon_{ij}e^{-\beta_{ij}t}\right) \\
&\leq -\boldsymbol{x}^T\boldsymbol{Q}\boldsymbol{x} + \sum_{i=1}^{n}\sum_{j=1}^{m_i} d_{ij}\epsilon_{ij}e^{-\beta_{ij}t} \\
&\leq -\boldsymbol{x}^T\boldsymbol{Q}\boldsymbol{x} + \epsilon e^{-\beta t}.
\end{aligned}
\tag{1.224}
$$

The stability of the system can be proven by using Theorem 1.1 with λ_3 given by (1.208) and assuming $\boldsymbol{Q}$ to be positive definite. It can be shown (see Section 1.4.2), that conditions (1.199) to (1.201) together with the definition of λ_3 guarantee that $\boldsymbol{Q} > \boldsymbol{O}$. Since

$$\|\boldsymbol{x}\|^2 \leq \varphi^2 z_m^2 \Rightarrow \|\dot{\boldsymbol{z}}\|^2 \leq \varphi^2 z_m^2, \tag{1.225}$$

a proper region of attraction should be selected in such a way that (1.225) is always true for $t \geq 0$. If $\beta = \lambda$ and $\boldsymbol{Q} > \boldsymbol{O}$, the following must hold:

$$\|\boldsymbol{x}\|^2 \leq \frac{\lambda_2}{\lambda_1}\|\boldsymbol{x}_0\|^2 e^{-\lambda t} + \frac{\epsilon}{\lambda_1} t e^{-\lambda t}. \tag{1.226}$$

In view of (1.225) and (1.226), it should be satisfied that

$$\frac{\lambda_2}{\lambda_1}\|\boldsymbol{x}_0\|^2 e^{-\lambda t} + \frac{\epsilon}{\lambda_1} t e^{-\lambda t} \leq \varphi^2 z_m^2. \tag{1.227}$$

Since the maximum of the left–hand side of (1.227) is a function of $\|\boldsymbol{x}_0\|$, it is easier, although more conservative, to compute the maxima of both terms separately, so that one has

$$\frac{\lambda_2}{\lambda_1}\|\boldsymbol{x}_0\|^2 + \frac{\epsilon}{\lambda_1 \lambda} e^{-1} \leq \varphi^2 z_m^2. \tag{1.228}$$

The region of attraction (1.204) comes from (1.228). If $\beta \neq \lambda$, it should be true that

$$\|\boldsymbol{x}\|^2 \leq \frac{\lambda_2}{\lambda_1}\|\boldsymbol{x}_0\|^2 e^{-\lambda t} + \frac{\epsilon}{\lambda_1(\lambda - \beta)}(e^{-\beta t} - e^{-\lambda t}). \tag{1.229}$$

From (1.225) it is

$$\frac{\lambda_2}{\lambda_1}\|\boldsymbol{x}_0\|^2 e^{-\lambda t} + \frac{\epsilon}{\lambda_1(\lambda - \beta)}(e^{-\beta t} - e^{-\lambda t}) \leq \varphi^2 z_m^2. \tag{1.230}$$

By calculating the maxima of both terms on the right–hand side of (1.230) as before, the region of attraction (1.205) can be derived.

b) It must be proven that the desired trajectory given by (1.182) and (1.183) remains bounded. In view of part *a)* of the proof, a constant $\hat{v}_m$ exists such that $\|\dot{\hat{\boldsymbol{q}}}\| < \hat{v}_m$. By using again

$$\boldsymbol{s}_0 = \dot{\boldsymbol{\delta}}_d + \boldsymbol{\Lambda}_\delta \boldsymbol{\delta}_d \tag{1.231}$$

$$\boldsymbol{x}_\delta = \begin{bmatrix} \boldsymbol{\delta}_d \\ \dot{\boldsymbol{\delta}}_d \end{bmatrix}, \tag{1.232}$$

Eq. (1.182) can be written as

$$\begin{aligned} \boldsymbol{H}_{\delta\delta}\dot{\boldsymbol{s}}_0 = &-(\boldsymbol{C}_{\delta\delta}(\boldsymbol{q},\dot{\boldsymbol{q}})\boldsymbol{s}_0 + \boldsymbol{K}_{pD\delta}\boldsymbol{s}_0 + \boldsymbol{K}_{qe\delta}\boldsymbol{\delta}_d + \boldsymbol{D}_{\triangle}\dot{\boldsymbol{\delta}}_d + \boldsymbol{f}_r \\ &- \boldsymbol{C}_{\delta\delta}(\boldsymbol{q},\dot{\boldsymbol{z}})\boldsymbol{s}_0 + \boldsymbol{f}), \end{aligned} \tag{1.233}$$

with

$$\begin{aligned} \boldsymbol{f}_r \triangleq \boldsymbol{H}_{\theta\delta}^T\ddot{\hat{\boldsymbol{\theta}}}_r + \boldsymbol{C}_{\delta\theta}(\boldsymbol{q},\dot{\hat{\boldsymbol{q}}})\dot{\boldsymbol{\theta}}_r + \boldsymbol{g}_\delta - \boldsymbol{H}_{\delta\delta}\boldsymbol{\Lambda}_\delta\dot{\hat{\boldsymbol{\delta}}} - \boldsymbol{C}_{\delta\delta}(\boldsymbol{q},\dot{\boldsymbol{q}})\boldsymbol{\Lambda}_\delta\boldsymbol{\delta} \\ -\boldsymbol{D}_\delta\boldsymbol{\Lambda}_\delta\boldsymbol{\delta} - \boldsymbol{K}_{q\delta}\boldsymbol{\delta} - \boldsymbol{K}_{p\delta}\dot{\hat{\boldsymbol{\delta}}} - \boldsymbol{K}_{p\delta}\boldsymbol{\Lambda}_\delta\boldsymbol{\delta} + \boldsymbol{C}_{\delta\delta}(\boldsymbol{q},\dot{\boldsymbol{z}})\boldsymbol{\Lambda}_\delta\boldsymbol{\delta}. \end{aligned} \tag{1.234}$$

To compute (1.233) and (1.234), Property 1.3 has been used in the form:

$$\begin{aligned} \begin{bmatrix} \boldsymbol{C}_{\theta\theta}(\boldsymbol{q},\dot{\hat{\boldsymbol{q}}}) & \boldsymbol{C}_{\theta\delta}(\boldsymbol{q},\dot{\hat{\boldsymbol{q}}}) \\ \boldsymbol{C}_{\delta\theta}(\boldsymbol{q},\dot{\hat{\boldsymbol{q}}}) & \boldsymbol{C}_{\delta\delta}(\boldsymbol{q},\dot{\hat{\boldsymbol{q}}}) \end{bmatrix} \begin{bmatrix} \dot{\boldsymbol{\theta}}_r \\ \dot{\boldsymbol{\delta}}_r \end{bmatrix} = &\begin{bmatrix} \boldsymbol{C}_{\theta\theta}(\boldsymbol{q},\dot{\boldsymbol{q}}) & \boldsymbol{C}_{\theta\delta}(\boldsymbol{q},\dot{\boldsymbol{q}}) \\ \boldsymbol{C}_{\delta\theta}(\boldsymbol{q},\dot{\boldsymbol{q}}) & \boldsymbol{C}_{\delta\delta}(\boldsymbol{q},\dot{\boldsymbol{q}}) \end{bmatrix} \begin{bmatrix} \dot{\boldsymbol{\theta}}_r \\ \dot{\boldsymbol{\delta}}_r \end{bmatrix} \\ &- \begin{bmatrix} \boldsymbol{C}_{\theta\theta}(\boldsymbol{q},\dot{\boldsymbol{z}}) & \boldsymbol{C}_{\theta\delta}(\boldsymbol{q},\dot{\boldsymbol{z}}) \\ \boldsymbol{C}_{\delta\theta}(\boldsymbol{q},\dot{\boldsymbol{z}}) & \boldsymbol{C}_{\delta\delta}(\boldsymbol{q},\dot{\boldsymbol{z}}) \end{bmatrix} \begin{bmatrix} \dot{\boldsymbol{\theta}}_r \\ \dot{\boldsymbol{\delta}}_r \end{bmatrix}. \end{aligned} \tag{1.235}$$

Since $||\boldsymbol{q}|| < \infty$, $||\dot{\boldsymbol{q}}|| \leq v_m$ and $||\dot{\hat{\boldsymbol{q}}}|| \leq \hat{v}_m$, $\boldsymbol{f}_r$ is only a function of bounded variables, which implies that a positive constant $f_{r,\max}$ must exist such that (see Properties 1.9 to 1.12)

$$||\boldsymbol{f}_r|| \leq f_{r,\max}. \tag{1.236}$$

A Lyapunov function for system (1.233) is

$$\begin{aligned} V_\delta(\boldsymbol{x}_\delta, t) \triangleq V_\delta(\boldsymbol{x}_\delta) &= \frac{1}{2}\boldsymbol{s}_0^T\boldsymbol{H}_{\delta\delta}\boldsymbol{s}_0 + \frac{1}{2}\boldsymbol{\delta}_d^T(2\boldsymbol{\Lambda}_\delta\boldsymbol{K}_{pD\delta} + \boldsymbol{K}_{qe\delta} \\ &\quad + \boldsymbol{\Lambda}_\delta\boldsymbol{D}_\triangle)\boldsymbol{\delta}_d \\ &= \frac{1}{2}\boldsymbol{x}_\delta^T \begin{bmatrix} \boldsymbol{\Lambda}_\delta\boldsymbol{H}_{\delta\delta}\boldsymbol{\Lambda}_\delta + 2\boldsymbol{\Lambda}_\delta\boldsymbol{K}_{pD\delta} + \boldsymbol{K}_{qe\delta} + \boldsymbol{\Lambda}_\delta\boldsymbol{D}_\triangle & \boldsymbol{\Lambda}_\delta\boldsymbol{H}_{\delta\delta} \\ \boldsymbol{H}_{\delta\delta}\boldsymbol{\Lambda}_\delta & \boldsymbol{H}_{\delta\delta} \end{bmatrix} \boldsymbol{x}_\delta \\ &\triangleq \frac{1}{2}\boldsymbol{x}_\delta^T\boldsymbol{M}\boldsymbol{x}_\delta. \end{aligned} \tag{1.237}$$

In view of Property 1.9, one has

$$\hat{\lambda}_1||\boldsymbol{x}_\delta||^2 \leq V_\delta(\boldsymbol{x}_\delta) \leq \hat{\lambda}_2||\boldsymbol{x}_\delta||^2 \tag{1.238}$$

$$\hat{\lambda}_1 \triangleq \frac{1}{2}\min_{\boldsymbol{q}\in\mathcal{Q}^{n+m}}(\lambda_{\min}(\boldsymbol{M})) \tag{1.239}$$

$$\hat{\lambda}_2 \triangleq \frac{1}{2}\max_{\boldsymbol{q}\in\mathcal{Q}^{n+m}}(\lambda_{\max}(\boldsymbol{M})). \tag{1.240}$$

The derivative of (1.237) is

$$\dot{V}_\delta(\boldsymbol{x}_\delta) = \frac{1}{2}\boldsymbol{s}_0^T\dot{\boldsymbol{H}}_{\delta\delta}\boldsymbol{s}_0 + \boldsymbol{\delta}_d^T(2\boldsymbol{\Lambda}_\delta\boldsymbol{K}_{pD\delta} + \boldsymbol{K}_{qe\delta} + \boldsymbol{\Lambda}_\delta\boldsymbol{D}_\Delta)\dot{\boldsymbol{\delta}}_d + \boldsymbol{s}_0^T\boldsymbol{H}_{\delta\delta}\dot{\boldsymbol{s}}_0. \tag{1.241}$$

Substituting (1.233) into (1.241) and taking Property 1.2 into account leads to

$$\begin{aligned}
\dot{V}_\delta(\boldsymbol{x}_\delta) &= \frac{1}{2}\boldsymbol{s}_0^T\dot{\boldsymbol{H}}_{\delta\delta}\boldsymbol{s}_0 + \boldsymbol{\delta}_d^T(2\boldsymbol{\Lambda}_\delta\boldsymbol{K}_{pD\delta} + \boldsymbol{K}_{qe\delta} + \boldsymbol{\Lambda}_\delta\boldsymbol{D}_\Delta)\dot{\boldsymbol{\delta}}_d \\
&\quad -\boldsymbol{s}_0^T\left(\boldsymbol{C}_{\delta\delta}(\boldsymbol{q},\dot{\boldsymbol{q}})\boldsymbol{s}_0 + \boldsymbol{K}_{qe\delta}\boldsymbol{\delta}_d + \boldsymbol{K}_{pD\delta}\boldsymbol{s}_0 + \boldsymbol{D}_\Delta\dot{\boldsymbol{\delta}}_d\right) \\
&\quad -\boldsymbol{s}_0^T\left(\boldsymbol{f}_r + \boldsymbol{f} - \boldsymbol{C}_{\delta\delta}(\boldsymbol{q},\dot{\boldsymbol{z}})\boldsymbol{s}_0\right) \\
&= -\boldsymbol{\delta}_d^T\boldsymbol{\Lambda}_\delta\boldsymbol{K}_{qe\delta}\boldsymbol{\delta}_d - \dot{\boldsymbol{\delta}}_d^T\boldsymbol{D}_\Delta\dot{\boldsymbol{\delta}}_d - \dot{\boldsymbol{\delta}}_d^T\boldsymbol{K}_{pD\delta}\dot{\boldsymbol{\delta}}_d \\
&\quad -\dot{\boldsymbol{\delta}}_d^T\boldsymbol{K}_{pD\delta}\boldsymbol{\Lambda}_\delta\boldsymbol{\delta}_d - \boldsymbol{\delta}_d^T\boldsymbol{\Lambda}_\delta\boldsymbol{K}_{pD\delta}\dot{\boldsymbol{\delta}}_d - \boldsymbol{\delta}_d^T\boldsymbol{\Lambda}_\delta\boldsymbol{K}_{pD\delta}\boldsymbol{\Lambda}_\delta\boldsymbol{\delta}_d \\
&\quad +\dot{\boldsymbol{\delta}}_d^T\boldsymbol{K}_{pD\delta}\boldsymbol{\Lambda}_\delta\boldsymbol{\delta}_d + \boldsymbol{\delta}_d^T\boldsymbol{\Lambda}_\delta\boldsymbol{K}_{pD\delta}\dot{\boldsymbol{\delta}}_d - \boldsymbol{s}_0^T\left(\boldsymbol{f}_r + \boldsymbol{f} - \boldsymbol{C}_{\delta\delta}(\boldsymbol{q},\dot{\boldsymbol{z}})\boldsymbol{s}_0\right) \\
&= -\boldsymbol{\delta}_d^T(\boldsymbol{\Lambda}_\delta\boldsymbol{K}_{qe\delta} + \boldsymbol{\Lambda}_\delta\boldsymbol{K}_{pD\delta}\boldsymbol{\Lambda}_\delta)\boldsymbol{\delta}_d - \dot{\boldsymbol{\delta}}_d^T(\boldsymbol{D}_\Delta + \boldsymbol{K}_{pD\delta})\dot{\boldsymbol{\delta}}_d \\
&\quad -\boldsymbol{s}_0^T\left(\boldsymbol{f}_r + \boldsymbol{f} - \boldsymbol{C}_{\delta\delta}(\boldsymbol{q},\dot{\boldsymbol{z}})\boldsymbol{s}_0\right) \\
&= -\boldsymbol{x}_\delta^T\boldsymbol{P}\boldsymbol{x}_\delta - \boldsymbol{s}_0^T\left(\boldsymbol{f}_r + \boldsymbol{f} - \boldsymbol{C}_{\delta\delta}(\boldsymbol{q},\dot{\boldsymbol{z}})\boldsymbol{s}_0\right),
\end{aligned} \tag{1.242}$$

where $\boldsymbol{P}$ is given by (1.216). Notice that

$$||\boldsymbol{s}_0|| = ||\dot{\boldsymbol{\delta}}_d + \boldsymbol{\Lambda}_\delta\boldsymbol{\delta}_d|| \leq ||\dot{\boldsymbol{\delta}}_d|| + \lambda_{\max}(\boldsymbol{\Lambda}_\delta)||\boldsymbol{\delta}_d|| \leq (1 + \lambda_{\max}(\boldsymbol{\Lambda}_\delta))||\boldsymbol{x}_\delta||. \tag{1.243}$$

Rewrite $\boldsymbol{f}$ as

$$\boldsymbol{f} = -\boldsymbol{D}_\Delta \mathrm{diag}\{\dot{\delta}_{d11}, \ldots, \dot{\delta}_{dnm_n}\}\begin{bmatrix} \dfrac{\dot{\delta}_{d11}s_{x11}}{||\dot{\delta}_{d11}s_{x11}|| + \epsilon_{11}e^{-\beta_{11}t}} \\ \vdots \\ \dfrac{\dot{\delta}_{dnm_n}s_{xnm_n}}{||\dot{\delta}_{dnm_n}s_{xnm_n}|| + \epsilon_{nm_n}e^{-\beta_{nm_n}t}} \end{bmatrix}, \tag{1.244}$$

to show that

$$||\boldsymbol{f}|| \leq \lambda_{\max}(\boldsymbol{D}_\Delta)\sqrt{m}\;||\boldsymbol{x}_d||. \tag{1.245}$$

To find a bound for $\boldsymbol{C}_{\delta\delta}(\boldsymbol{q},\dot{\boldsymbol{z}})$, consider (1.180). From part *a*) of the proof, it is

$$||\boldsymbol{C}_{\delta\delta}(\boldsymbol{q},\dot{\boldsymbol{z}})|| \leq k_{c\delta\delta}\varphi z_m. \tag{1.246}$$

By taking Eqs. (1.243)–(1.246) into account, (1.242) can be written as

$$\begin{aligned}\dot{V}_\delta(\boldsymbol{x}_\delta) \le & -\lambda_{\min}(\boldsymbol{P})\|\boldsymbol{x}_\delta\|^2 + f_{r,\max}(1+\lambda_{\max}(\boldsymbol{\Lambda}_\delta))\|\boldsymbol{x}_\delta\| \\ & +(1+\lambda_{\max}(\boldsymbol{\Lambda}_\delta))\left(\lambda_{\max}(\boldsymbol{D}_\Delta)\sqrt{m} + k_{c\delta\delta}\varphi z_m(1+\lambda_{\max}(\boldsymbol{\Lambda}_\delta))\right)\|\boldsymbol{x}_\delta\|^2 \\ \triangleq & -\hat{\lambda}_3\|\boldsymbol{x}_\delta\|^2 + \sigma\|\boldsymbol{x}_\delta\|.\end{aligned} \quad (1.247)$$

Since $\hat{\lambda}_3 > 0$ (by assumption), the boundedness of $\|\boldsymbol{x}_\delta\|$ can be proven using Theorem 1.2 with $\gamma = 0$. $\triangle$

In the case that the term $\boldsymbol{f}$ is not to be used, i.e. $\boldsymbol{D}_\Delta = \boldsymbol{O}$, $\boldsymbol{\delta}_d$ can be computed from

$$\begin{aligned}\ddot{\boldsymbol{\delta}}_d = \boldsymbol{\Lambda}_\delta(\dot{\hat{\boldsymbol{\delta}}} - \dot{\boldsymbol{\delta}}_d) - \boldsymbol{H}_{\delta\delta}^{-1}\big(&\boldsymbol{H}_{\theta\delta}^T\ddot{\boldsymbol{\theta}}_r + \boldsymbol{C}_{\delta\theta}(\boldsymbol{q},\dot{\hat{\boldsymbol{q}}})\dot{\boldsymbol{\theta}}_r + \boldsymbol{C}_{\delta\delta}(\boldsymbol{q},\dot{\hat{\boldsymbol{q}}})\dot{\boldsymbol{\delta}}_r \\ & + \boldsymbol{K}\boldsymbol{\delta}_d + \boldsymbol{D}_\delta\dot{\boldsymbol{\delta}}_r + \boldsymbol{g}_\delta - \boldsymbol{K}_{q\delta}\tilde{\boldsymbol{\delta}} - \boldsymbol{K}_{p\delta}\hat{\boldsymbol{s}}_\delta\big)\end{aligned} \quad (1.248)$$

with initial condition

$$\boldsymbol{\delta}_d(0) = \dot{\boldsymbol{\delta}}_d(0) = \boldsymbol{0}. \quad (1.249)$$

Eqs. (1.188) and (1.197) then become

$$\begin{aligned}\boldsymbol{H}(\boldsymbol{q})\dot{\boldsymbol{s}} = & -(\boldsymbol{C}(\boldsymbol{q},\dot{\boldsymbol{q}})\boldsymbol{s} + \boldsymbol{K}_{pD}\boldsymbol{s} + \boldsymbol{K}_{qe}\tilde{\boldsymbol{q}}) + \boldsymbol{H}(\boldsymbol{q})\boldsymbol{\Lambda}\dot{\boldsymbol{z}} - \boldsymbol{C}(\boldsymbol{q},\dot{\boldsymbol{q}})\dot{\boldsymbol{z}} \\ & + \boldsymbol{C}(\boldsymbol{q},\dot{\boldsymbol{z}})\boldsymbol{s} + \boldsymbol{K}_p\dot{\boldsymbol{z}} \quad (1.250)\\ \boldsymbol{H}(\boldsymbol{q})\dot{\boldsymbol{r}} = & -(\boldsymbol{C}(\boldsymbol{q},\dot{\boldsymbol{q}})\boldsymbol{r} + \boldsymbol{K}_{ve}\boldsymbol{z} + \boldsymbol{D}\boldsymbol{r} + k_d\boldsymbol{H}(\boldsymbol{q})\dot{\boldsymbol{z}}) + \boldsymbol{C}(\boldsymbol{q},\dot{\boldsymbol{z}})\boldsymbol{r} \quad (1.251)\\ & - \boldsymbol{C}(\boldsymbol{q},\dot{\boldsymbol{q}})\dot{\boldsymbol{z}}.\end{aligned}$$

The next Corollary establishes the stability of system (1.250) and (1.251) when $\boldsymbol{D}_\Delta = \boldsymbol{O}$.

Corollary 1.1

Given a bounded continuous desired trajectory $\boldsymbol{\theta}_d$ with bounded velocity and acceleration, if $\|\dot{\boldsymbol{q}}\| \le v_m$ where v_m is a positive scalar constant, then the equilibrium point $\boldsymbol{x} = \boldsymbol{0}$ of (1.250) and (1.251) is asymptotically stable if conditions (1.199)–(1.201) are satisfied. A region of attraction is given by

$$S = \left\{\boldsymbol{x} \in \Re^{4(n+m)} : \|\boldsymbol{x}\| \le \sqrt{\frac{\lambda_1}{\lambda_2}}\varphi z_m\right\}, \quad (1.252)$$

where λ_1, λ_2, φ and z_m are the same as in Theorem 1.5. In addition, the desired trajectory $\boldsymbol{\delta}_d$ and $\dot{\boldsymbol{\delta}}_d$ given by (1.248) and (1.249) remains bounded *if*

$$\hat{\lambda}_3 \triangleq \lambda_{\min}(\boldsymbol{P}) - k_{c\delta\delta}\varphi z_m(1 + \lambda_{\max}(\boldsymbol{\Lambda}_\delta))^2 > 0 \tag{1.253}$$

with

$$\boldsymbol{P} \triangleq \begin{bmatrix} \boldsymbol{\Lambda}_\delta \boldsymbol{K}_{qe\delta} + \boldsymbol{\Lambda}_\delta \boldsymbol{K}_{pD\delta}\boldsymbol{\Lambda}_\delta & \boldsymbol{O} \\ \boldsymbol{O} & \boldsymbol{K}_{pD\delta} \end{bmatrix} \tag{1.254}$$

$$\lambda_{\min}(\boldsymbol{P}) = \min\{\lambda_{\min}(\boldsymbol{\Lambda}_\delta)\lambda_{\min}(\boldsymbol{K}_{qe\delta}) + \lambda^2_{\min}(\boldsymbol{\Lambda}_\delta)\lambda_{\min}(\boldsymbol{K}_{pD\delta}), \lambda_{\min}(\boldsymbol{K}_{pD\delta})\}. \tag{1.255}$$

Proof: The proof is similar to that of Theorem 1.5 just by letting $\boldsymbol{D}_\Delta = \boldsymbol{O}$.

a) Firstly, the stability of the whole system will be proven. Consider again the Lyapunov function (1.218). From (1.224) one has

$$\dot{V}(\boldsymbol{x}) = -\boldsymbol{x}^T \boldsymbol{Q} \boldsymbol{x}. \tag{1.256}$$

Since

$$\lambda_1 ||\boldsymbol{x}||^2 \leq V(\boldsymbol{x}, t) \leq \lambda_2 ||\boldsymbol{x}||^2, \tag{1.257}$$

$V(\boldsymbol{x}, t)$ is a decrescent function. On the other hand, since

$$\dot{V}(\boldsymbol{x}) \leq -\lambda_3 ||\boldsymbol{x}||^2 \tag{1.258}$$

with λ_3 as in (1.208), system (1.250) and (1.251) is asymptotically stable [Vidyasagar (1978)]. As before, the region of attraction (1.252) can be computed by taking (1.257) and the definition of λ_3 into account.

b) To prove the boundedness of the desired trajectory given by (1.248) and (1.249), the existence of a constant $\hat{v}_m$ such that $||\dot{\boldsymbol{q}}|| < \hat{v}_m$ is necessary. Note that $\hat{v}_m$ must exist in view of part *a*) of the proof. By employing (1.231)–(1.233) and the Lyapunov function (1.237), and setting $\boldsymbol{f}$ to $\mathbf{0}$, one gets (see (1.247))

$$\dot{V}_\delta(\boldsymbol{x}_\delta) \leq -\hat{\lambda}_3 ||\boldsymbol{x}_\delta||^2 + \sigma ||\boldsymbol{x}_\delta||. \tag{1.259}$$

The proof is accomplished by using Theorem 1.2 with $\gamma = 0$. △

As discussed in Section 1.3.3, $\boldsymbol{f}$ helps increase the damping of the system since every element of $\boldsymbol{D}_\Delta \dot{\boldsymbol{\delta}}_d + \boldsymbol{f}$ can be expressed as

$$d_{ij}\dot{\delta}_{ij}\left(1 - \frac{\dot{\delta}_{dij}s_{xij}}{\|\dot{\delta}_{dij}s_{xij}\| + \epsilon_{ij}e^{-\beta_{ij}t}}\right) \qquad i = 1,\dots,n,\ j = 1,\dots,m_i,$$

so that $\boldsymbol{\delta}_d$ is damped. When the tracking error becomes negligible, $\boldsymbol{\delta}$ becomes damped as well. Note that this is true even if $\boldsymbol{\theta}_d$ is not a constant vector. It is not difficult to show that if $\boldsymbol{\theta}_d$ is constant and $\boldsymbol{g}_\delta(\boldsymbol{q}) = \boldsymbol{g}_\delta(\boldsymbol{\theta})$, then $\boldsymbol{\delta}_d$ becomes $-\boldsymbol{K}^{-1}\boldsymbol{g}_\delta$.

It should be noticed that both Theorem 1.5 and Corollary 1.1 assume that $\hat{\lambda}_3 > 0$. From its definition, it is obvious that it is always possible to achieve this goal by letting φ be small enough and $\lambda_{\min}(\boldsymbol{K}_p)$ large enough. Nevertheless, it is not obvious whether one can enlarge the region of attraction arbitrarily. To show that this is actually possible, let $\lambda_{\min}(\boldsymbol{K}_{p\delta})$ be large enough and choose k_d, $\boldsymbol{K}_q$, $\boldsymbol{K}_v$ to satisfy conditions (1.199)–(1.201). Note that $\boldsymbol{\Lambda}$ can be selected freely. From (1.212) it can be seen that

$$z_m = \frac{\lambda_{pd}}{k_c} \tag{1.260}$$

with a proper choice of k_d, λ_q and λ_v. Although some of these constants appear not only in the numerator but also in the denominator, it should be pointed out that they appear in the numerator as a product and in the denominator as a summation, so that a parameter selection can always be found so that the assumption (1.260) is valid. Without loss of generality, it can be assumed that

$$\lambda_{\min}(\boldsymbol{K}_p) = \lambda_{\min}(\boldsymbol{K}_{p\delta}) \tag{1.261}$$

$$z_m = \frac{\lambda_{\min}(\boldsymbol{K}_{p\delta})}{k_c}. \tag{1.262}$$

Also without loss of generality, from (1.217) it can be assumed that

$$\lambda_{\min}(\boldsymbol{P}) = \lambda_{\min}(\boldsymbol{K}_{p\delta}) + \lambda_{\min}(\boldsymbol{D}_\Delta), \tag{1.263}$$

since $\lambda_{\min}(\boldsymbol{K}_{qe\delta})$ can be tuned arbitrarily large. Taking into account

Eqs. (1.261)–(1.263) together with (1.215) yields

$$\begin{aligned}\hat{\lambda}_3 &= \lambda_{\min}(\boldsymbol{K}_{p\delta}) + \lambda_{\min}(\boldsymbol{D}_\Delta) \\ &\quad -(1+\lambda_{\max}(\boldsymbol{\Lambda}_\delta))\left(\lambda_{\max}(\boldsymbol{D}_\Delta)\sqrt{m} + k_{c\delta\delta}\varphi z_m(1+\lambda_{\max}(\boldsymbol{\Lambda}_\delta))\right) \\ &= \lambda_{\min}(\boldsymbol{K}_{p\delta}) + \lambda_{\min}(\boldsymbol{D}_\Delta) \\ &\quad -(1+\lambda_{\max}(\boldsymbol{\Lambda}_\delta))\left(\lambda_{\max}(\boldsymbol{D}_\Delta)\sqrt{m} + \frac{\lambda_{\min}(\boldsymbol{K}_{p\delta})}{k_c}\varphi k_{c\delta\delta}(1+\lambda_{\max}(\boldsymbol{\Lambda}_\delta))\right),\end{aligned} \tag{1.264}$$

which implies that

$$\lambda_{\min}(\boldsymbol{K}_{p\delta})\left(1-\frac{k_{c\delta\delta}}{k_c}\varphi(1+\lambda_{\max}(\boldsymbol{\Lambda}_\delta))^2\right) \tag{1.265}$$

$$-\lambda_{\max}(\boldsymbol{D}_\Delta)\sqrt{m}(1+\lambda_{\max}(\boldsymbol{\Lambda}_\delta)) + \lambda_{\min}(\boldsymbol{D}_\Delta) > 0 \tag{1.266}$$

must be satisfied, or in other form

$$\lambda_{\min}(\boldsymbol{K}_{p\delta}) > \frac{\lambda_{\max}(\boldsymbol{D}_\Delta)\sqrt{m}(1+\lambda_{\max}(\boldsymbol{\Lambda}_\delta)) - \lambda_{\min}(\boldsymbol{D}_\Delta)}{\left(1-\frac{k_{c\delta\delta}}{k_c}\varphi(1+\lambda_{\max}(\boldsymbol{\Lambda}_\delta))^2\right)} \tag{1.267}$$

$$\frac{k_c}{k_{c\delta\delta}(1+\lambda_{\max}(\boldsymbol{\Lambda}_\delta))^2} > \varphi. \tag{1.268}$$

Since condition (1.268) can always be accomplished, it is possible to choose $\lambda_{\min}(\boldsymbol{K}_{p\delta})$ as large as wished and, as a direct consequence, z_m and the regions of attraction (1.204) and (1.205). The same conclusion can be derived if $\boldsymbol{D}_\Delta = \boldsymbol{O}$. Conditions (1.202) and (1.203) can also be satisfied by letting either ϵ be small enough or z_m be large enough.

1.4.2 *Analysis of Matrix Q*

In this section, the conditions under which the matrix $\boldsymbol{Q}$ (used in the proofs of Section 1.4.1) is positive definite will be studied. A direct analysis is too difficult. Consider again Eqs. (1.223) and (1.224) to get

$$\begin{aligned}
\dot{V}(\boldsymbol{x}) \leq & -\tilde{\boldsymbol{q}}^T(\boldsymbol{\Lambda}\boldsymbol{K}_{pD}\boldsymbol{\Lambda} + \boldsymbol{\Lambda}\boldsymbol{K}_{qe} - \boldsymbol{\Lambda}\boldsymbol{C}(\boldsymbol{q},\dot{\boldsymbol{z}})\boldsymbol{\Lambda})\tilde{\boldsymbol{q}} \\
& -\dot{\tilde{\boldsymbol{q}}}^T(\boldsymbol{K}_{pD} - \boldsymbol{C}(\boldsymbol{q},\dot{\boldsymbol{z}}))\dot{\tilde{\boldsymbol{q}}} \\
& -\boldsymbol{z}^T(\boldsymbol{\Lambda}\boldsymbol{D}\boldsymbol{\Lambda} + \boldsymbol{\Lambda}\boldsymbol{K}_{ve} - \boldsymbol{\Lambda}\boldsymbol{C}(\boldsymbol{q},\dot{\boldsymbol{z}})\boldsymbol{\Lambda})\boldsymbol{z} \\
& -\dot{\boldsymbol{z}}^T(\boldsymbol{D} + k_d\boldsymbol{H}(\boldsymbol{q}) - \boldsymbol{C}(\boldsymbol{q},\dot{\boldsymbol{z}}) + \boldsymbol{C}(\boldsymbol{q},\dot{\boldsymbol{q}}))\dot{\boldsymbol{z}} \\
& +\tilde{\boldsymbol{q}}^T(\boldsymbol{\Lambda}\boldsymbol{C}^T(\boldsymbol{q},\dot{\boldsymbol{z}}) + \boldsymbol{\Lambda}\boldsymbol{C}(\boldsymbol{q},\dot{\boldsymbol{z}}))\dot{\tilde{\boldsymbol{q}}} \\
& +\boldsymbol{z}^T(-k_d\boldsymbol{\Lambda}\boldsymbol{H}(\boldsymbol{q}) + \boldsymbol{\Lambda}\boldsymbol{C}^T(\boldsymbol{q},\dot{\boldsymbol{z}}) + \boldsymbol{\Lambda}\boldsymbol{C}(\boldsymbol{q},\dot{\boldsymbol{z}}) - \boldsymbol{\Lambda}\boldsymbol{C}(\boldsymbol{q},\dot{\boldsymbol{q}}))\dot{\boldsymbol{z}} \\
& +\tilde{\boldsymbol{q}}^T(\boldsymbol{\Lambda}\boldsymbol{H}(\boldsymbol{q})\boldsymbol{\Lambda} - \boldsymbol{\Lambda}\boldsymbol{C}(\boldsymbol{q},\dot{\boldsymbol{q}}) + \boldsymbol{\Lambda}\boldsymbol{K}_p)\dot{\boldsymbol{z}} \\
& +\dot{\tilde{\boldsymbol{q}}}^T(\boldsymbol{H}(\boldsymbol{q})\boldsymbol{\Lambda} - \boldsymbol{C}(\boldsymbol{q},\dot{\boldsymbol{q}}) + \boldsymbol{K}_p)\dot{\boldsymbol{z}} + \epsilon e^{-\beta t}
\end{aligned} \tag{1.269}$$

and take norm bounds into account, so that (1.269) can be written as

$$\begin{aligned}
\dot{V}(\boldsymbol{x}) \leq & -\|\tilde{\boldsymbol{q}}\|^2(\lambda_m^2\lambda_{pd} + \lambda_m\lambda_{qe} - \lambda_M^2 k_c\|\dot{\boldsymbol{z}}\|) \\
& -\|\dot{\tilde{\boldsymbol{q}}}\|^2(\lambda_{pd} - k_c\|\dot{\boldsymbol{z}}\|) \\
& -\|\boldsymbol{z}\|^2(\lambda_m\lambda_{ve} + \lambda_m^2\lambda_d - \lambda_M^2 k_c\|\dot{\boldsymbol{z}}\|) \\
& -\|\dot{\boldsymbol{z}}\|^2(\lambda_d + k_d\lambda_h - k_c\|\dot{\boldsymbol{z}}\| - k_c v_m) \\
& +2\lambda_M k_c\|\dot{\boldsymbol{z}}\|\,\|\tilde{\boldsymbol{q}}\|\,\|\dot{\tilde{\boldsymbol{q}}}\| \\
& +\|\boldsymbol{z}\|\,\|\dot{\boldsymbol{z}}\|(k_d\lambda_M\lambda_H + \lambda_M k_c v_m + 2\lambda_M k_c\|\dot{\boldsymbol{z}}\|) \\
& +\|\tilde{\boldsymbol{q}}\|\,\|\dot{\boldsymbol{z}}\|(\lambda_M^2\lambda_H + \lambda_M k_c v_m + \lambda_M\lambda_P) \\
& +\|\dot{\tilde{\boldsymbol{q}}}\|\,\|\dot{\boldsymbol{z}}\|(\lambda_M\lambda_H + k_c v_m + \lambda_P) + \epsilon e^{-\beta t}.
\end{aligned} \tag{1.270}$$

Define

$$\hat{\boldsymbol{x}} \triangleq \begin{bmatrix} \|\tilde{\boldsymbol{q}}\| \\ \|\dot{\tilde{\boldsymbol{q}}}\| \\ \|\boldsymbol{z}\| \\ \|\dot{\boldsymbol{z}}\| \end{bmatrix} \tag{1.271}$$

$$\hat{Q} \triangleq \begin{bmatrix} \begin{matrix}\lambda_m^2\lambda_{pd}\\ +\lambda_m\lambda_{qe}\\ -\lambda_M^2 k_c\|\dot{z}\|\end{matrix} & -\lambda_M k_c\|\dot{z}\| & 0 & \begin{matrix}-\frac{1}{2}(\lambda_M^2\lambda_H\\ +\lambda_M k_c v_m\\ +\lambda_M\lambda_P)\end{matrix} \\ -\lambda_M k_c\|\dot{z}\| & \begin{matrix}\lambda_{pd}\\ -k_c\|\dot{z}\|\end{matrix} & 0 & \begin{matrix}-\frac{1}{2}(\lambda_M\lambda_H\\ +k_c v_m\\ +\lambda_P)\end{matrix} \\ 0 & 0 & \begin{matrix}\lambda_m\lambda_{ve}\\ +\lambda_m^2\lambda_d\\ -\lambda_M^2 k_c\|\dot{z}\|\end{matrix} & \begin{matrix}-\frac{1}{2}(k_d\lambda_M\lambda_H\\ +\lambda_M k_c v_m)\\ -\lambda_M k_c\|\dot{z}\|\end{matrix} \\ \begin{matrix}-\frac{1}{2}(\lambda_M^2\lambda_H\\ +\lambda_M k_c v_m\\ +\lambda_M\lambda_P)\end{matrix} & \begin{matrix}-\frac{1}{2}(\lambda_M\lambda_H\\ +k_c v_m\\ +\lambda_P)\end{matrix} & \begin{matrix}-\frac{1}{2}(k_d\lambda_M\lambda_H\\ +\lambda_M k_c v_m)\\ -\lambda_M k_c\|\dot{z}\|\end{matrix} & \begin{matrix}\lambda_d\\ +k_d\lambda_h\\ -k_c v_m\\ -k_c\|\dot{z}\|\end{matrix} \end{bmatrix} \tag{1.272}$$

and rewrite (1.270) as

$$\dot{V}(\boldsymbol{x}) \leq -\hat{\boldsymbol{x}}^T\hat{\boldsymbol{Q}}\hat{\boldsymbol{x}} + \epsilon e^{-\beta t}. \tag{1.273}$$

From (1.273) it can be seen that $\boldsymbol{Q} > \boldsymbol{O}$ as long as $\hat{\boldsymbol{Q}} > \boldsymbol{O}$. However, it is still too hard to say whether or not $\hat{\boldsymbol{Q}} > \boldsymbol{O}$. On the other hand, one can express $\hat{\boldsymbol{Q}}$ as

$$\hat{Q} = \begin{bmatrix} \frac{1}{3}(\lambda_m^2\lambda_{pd} + \lambda_m\lambda_{qe} - \lambda_M^2 k_c\|\dot{z}\|) & -\lambda_M k_c\|\dot{z}\| & 0 & 0 \\ -\lambda_M k_c\|\dot{z}\| & \frac{1}{3}(\lambda_{pd} - k_c\|\dot{z}\|) & 0 & 0 \\ 0 & 0 & 0 & 0 \\ 0 & 0 & 0 & 0 \end{bmatrix} \tag{1.274}$$

$$+ \begin{bmatrix} \frac{1}{3}(\lambda_m^2\lambda_{pd} + \lambda_m\lambda_{qe} - \lambda_M^2 k_c\|\dot{z}\|) & 0 & 0 & 0 \\ 0 & 0 & 0 & 0 \\ 0 & 0 & \frac{1}{3}(\lambda_m\lambda_{ve} + \lambda_m^2\lambda_d - \lambda_M^2 k_c\|\dot{z}\|) & 0 \\ 0 & 0 & 0 & 0 \end{bmatrix}$$

$$+\begin{bmatrix} \frac{1}{3}(\lambda_m^2\lambda_{pd}+\lambda_m\lambda_{qe}-\lambda_M^2k_c\|\dot{z}\|) & 0 & 0 & -\frac{1}{2}(\lambda_M^2\lambda_H+\lambda_Mk_cv_m+\lambda_M\lambda_P) \\ 0 & 0 & 0 & 0 \\ 0 & 0 & 0 & 0 \\ -\frac{1}{2}(\lambda_M^2\lambda_H+\lambda_Mk_cv_m+\lambda_M\lambda_P) & 0 & 0 & \frac{1}{3}(\lambda_d+k_d\lambda_h-k_cv_m-k_c\|\dot{z}\|) \end{bmatrix}$$

$$+\begin{bmatrix} 0 & 0 & 0 & 0 \\ 0 & \frac{1}{3}(\lambda_{pd}-k_c\|\dot{z}\|) & 0 & 0 \\ 0 & 0 & \frac{1}{3}(\lambda_m\lambda_{ve}+\lambda_m^2\lambda_d-\lambda_M^2k_c\|\dot{z}\|) & 0 \\ 0 & 0 & 0 & 0 \end{bmatrix}$$

$$+\begin{bmatrix} 0 & 0 & 0 & 0 \\ 0 & \frac{1}{3}(\lambda_{pd}-k_c\|\dot{z}\|) & 0 & -\frac{1}{2}(\lambda_M\lambda_H+k_cv_m+\lambda_P) \\ 0 & 0 & 0 & 0 \\ 0 & -\frac{1}{2}(\lambda_M\lambda_H+k_cv_m+\lambda_P) & 0 & \frac{1}{3}(\lambda_d+k_d\lambda_h-k_cv_m-k_c\|\dot{z}\|) \end{bmatrix}$$

$$+\begin{bmatrix} 0 & 0 & 0 & 0 \\ 0 & 0 & 0 & 0 \\ 0 & 0 & \frac{1}{3}(\lambda_m\lambda_{ve}+\lambda_m^2\lambda_d-\lambda_M^2k_c\|\dot{z}\|) & -\frac{1}{2}(k_d\lambda_M\lambda_H+\lambda_Mk_cv_m)-\lambda_Mk_c\|\dot{z}\| \\ 0 & 0 & -\frac{1}{2}(k_d\lambda_M\lambda_H+\lambda_Mk_cv_m)-\lambda_Mk_c\|\dot{z}\| & \frac{1}{3}(\lambda_d+k_d\lambda_h-k_cv_m-k_c\|\dot{z}\|) \end{bmatrix}$$

and define the matrices

$$\boldsymbol{Q}_1 \triangleq \begin{bmatrix} \frac{1}{3}(\lambda_m^2\lambda_{pd}+\lambda_m\lambda_{qe}-\lambda_M^2k_c\|\dot{z}\|) & -\lambda_Mk_c\|\dot{z}\| \\ -\lambda_Mk_c\|\dot{z}\| & \frac{1}{3}(\lambda_{pd}-k_c\|\dot{z}\|) \end{bmatrix} \quad (1.275)$$

$$\boldsymbol{Q}_2 \triangleq \begin{bmatrix} \frac{1}{3}(\lambda_m^2\lambda_{pd}+\lambda_m\lambda_{qe}-\lambda_M^2k_c\|\dot{z}\|) & 0 \\ 0 & \frac{1}{3}(\lambda_m\lambda_{ve}+\lambda_m^2\lambda_d-\lambda_M^2k_c\|\dot{z}\|) \end{bmatrix} \quad (1.276)$$

$$\boldsymbol{Q}_3 \triangleq \begin{bmatrix} \frac{1}{3}(\lambda_m^2\lambda_{pd}+\lambda_m\lambda_{qe}-\lambda_M^2k_c\|\dot{z}\|) & -\frac{1}{2}(\lambda_M^2\lambda_H+\lambda_Mk_cv_m+\lambda_M\lambda_P) \\ -\frac{1}{2}(\lambda_M^2\lambda_H+\lambda_Mk_cv_m+\lambda_M\lambda_P) & \frac{1}{3}(\lambda_d+k_d\lambda_h-k_cv_m-k_c\|\dot{z}\|) \end{bmatrix} \quad (1.277)$$

(1.278)

$$\boldsymbol{Q}_4 \triangleq \begin{bmatrix} \frac{1}{3}(\lambda_{pd} - k_c\|\dot{\boldsymbol{z}}\|) & 0 \\ 0 & \frac{1}{3}(\lambda_m\lambda_{ve} + \lambda_m^2\lambda_d - \lambda_M^2 k_c\|\dot{\boldsymbol{z}}\|) \end{bmatrix} \tag{1.279}$$

$$\boldsymbol{Q}_5 \triangleq \begin{bmatrix} \frac{1}{3}(\lambda_{pd} - k_c\|\dot{\boldsymbol{z}}\|) \\ -\frac{1}{2}(\lambda_M\lambda_H + k_c v_m + \lambda_P) \end{bmatrix} \tag{1.280}$$

$$\begin{matrix} -\frac{1}{2}(\lambda_M\lambda_H + k_c v_m + \lambda_P) \\ \frac{1}{3}(\lambda_d + k_d\lambda_h - k_c v_m - k_c\|\dot{\boldsymbol{z}}\|) \end{matrix} \Bigg] \tag{1.281}$$

$$\boldsymbol{Q}_6 \triangleq \begin{bmatrix} \frac{1}{3}(\lambda_m\lambda_{ve} + \lambda_m^2\lambda_d & -\frac{1}{2}(k_d\lambda_M\lambda_H + \lambda_M k_c v_m) \\ -\lambda_M^2 k_c\|\dot{\boldsymbol{z}}\|) & -\lambda_M k_c\|\dot{\boldsymbol{z}}\| \\ & \\ -\frac{1}{2}(k_d\lambda_M\lambda_H + \lambda_M k_c v_m) & \frac{1}{3}(\lambda_d + k_d\lambda_h - k_c v_m \\ -\lambda_M k_c\|\dot{\boldsymbol{z}}\| & -k_c\|\dot{\boldsymbol{z}}\|) \end{bmatrix} \tag{1.282}$$

Of course, if every matrix $\boldsymbol{Q}_i$ is positive definite, so is $\hat{\boldsymbol{Q}}$. This implies that every term $\boldsymbol{Q}_{i_{11}}$ and $\boldsymbol{Q}_{i_{22}}$ together with $\det(\boldsymbol{Q}_i)$ must be positive.

Regarding $\boldsymbol{Q}_1$, one has

$$\frac{\lambda_m^2\lambda_{pd} + \lambda_m\lambda_{qe}}{\lambda_M^2 k_c} > \|\dot{\boldsymbol{z}}\| \tag{1.283}$$

$$\frac{\lambda_{pd}}{k_c} > \|\dot{\boldsymbol{z}}\| \tag{1.284}$$

$$\left(\frac{\lambda_m^2\lambda_{pd}^2 + \lambda_m\lambda_{qe}\lambda_{pd}}{8\lambda_M^2 k_c^2} + \left(\frac{2\lambda_M^2\lambda_{pd} + \lambda_m\lambda_{qe}}{16\lambda_M^2 k_c}\right)^2\right)^{\frac{1}{2}} - \frac{2\lambda_M^2\lambda_{pd} + \lambda_m\lambda_{qe}}{16\lambda_M^2 k_c} > \|\dot{\boldsymbol{z}}\|. \tag{1.285}$$

As for $\boldsymbol{Q}_2$, (1.283) must be satisfied together with

$$\frac{\lambda_m^2\lambda_d + \lambda_m\lambda_{ve}}{\lambda_M^2 k_c} > \|\dot{\boldsymbol{z}}\|. \tag{1.286}$$

As for $\boldsymbol{Q}_3$, one has condition (1.283) and

$$\lambda_d + k_d\lambda_h - k_c v_m > 0 \tag{1.287}$$

$$\frac{\lambda_d + k_d\lambda_h - k_c v_m}{k_c} > \|\dot{\boldsymbol{z}}\| \tag{1.288}$$

$$\begin{aligned}(\lambda_m^2\lambda_{pd} + \lambda_m\lambda_{qe})(\lambda_d + k_d\lambda_h - k_c v_m)&\\ - \frac{9}{4}(\lambda_M^2\lambda_H + \lambda_M k_c v_m + \lambda_M\lambda_P) > 0&\end{aligned} \tag{1.289}$$

$$\begin{aligned}&\frac{(\lambda_m^2\lambda_{pd} + \lambda_m\lambda_{qe})(\lambda_d+k_d\lambda_h-k_c v_m)-\frac{9}{4}\lambda_M(\lambda_M\lambda_H+k_c v_m+\lambda_P)}{k_c(\lambda_m^2\lambda_{pd}+\lambda_m\lambda_{qe}+\lambda_M^2(\lambda_d+k_d\lambda_h-k_c v_m))}\\ &> \|\dot{\boldsymbol{z}}\|.\end{aligned} \tag{1.290}$$

Note that $\boldsymbol{Q}_4$ is positive definite if conditions (1.284) and (1.286) are satisfied.

As for $\boldsymbol{Q}_5$, one gets (1.284), (1.287), (1.288) together with

$$\lambda_{pd}(\lambda_d + k_d\lambda_h - k_c v_m) - \frac{9}{4}(\lambda_M\lambda_H + k_c v_m + \lambda_P)^2 > 0 \tag{1.291}$$

$$\frac{\lambda_{pd}(\lambda_d + k_d\lambda_h - k_c v_m) - \frac{9}{4}(\lambda_M\lambda_H + k_c v_m + \lambda_P)^2}{k_c(\lambda_{pd} + \lambda_d + k_d\lambda_h - k_c v_m)} > \|\dot{\boldsymbol{z}}\|. \tag{1.292}$$

Finally, as for $\boldsymbol{Q}_6$, one has (1.286)–(1.288) and

$$\begin{aligned}(\lambda_m\lambda_{ve} + \lambda_m^2\lambda_d)(\lambda_d + k_d\lambda_h - k_c v_m)&\\ - \frac{9}{4}(k_d\lambda_M\lambda_H + \lambda_M k_c v_m)^2 > 0&\end{aligned} \tag{1.293}$$

$$\begin{aligned}&\left(\frac{(\lambda_m\lambda_{ve}+\lambda_m^2\lambda_d)(\lambda_d+k_d\lambda_h-k_c v_m)-\frac{9}{4}(k_d\lambda_M\lambda_H+\lambda_M k_c v_m)^2}{8\lambda_M^2 k_c^2}\right.\\ &\left. + \left(\frac{\lambda_m\lambda_{ve}+\lambda_m^2\lambda_d+\lambda_M^2(\lambda_d+k_d\lambda_h-k_c v_m)+9\lambda_M^2(k_d\lambda_H+k_c v_m)}{16k_c\lambda_M^2}\right)^2\right)^{\frac{1}{2}}\\ &-\frac{\lambda_m\lambda_{ve}+\lambda_m^2\lambda_d+\lambda_M^2(\lambda_d+k_d\lambda_h-k_c v_m)+9\lambda_M^2(k_d\lambda_H+k_c v_m)}{16k_c\lambda_M^2}\\ &> \|\dot{\boldsymbol{z}}\|.\end{aligned} \tag{1.294}$$

By sorting Eqs. (1.283) to (1.294), one obtains conditions (1.199)–(1.201) and (1.212).

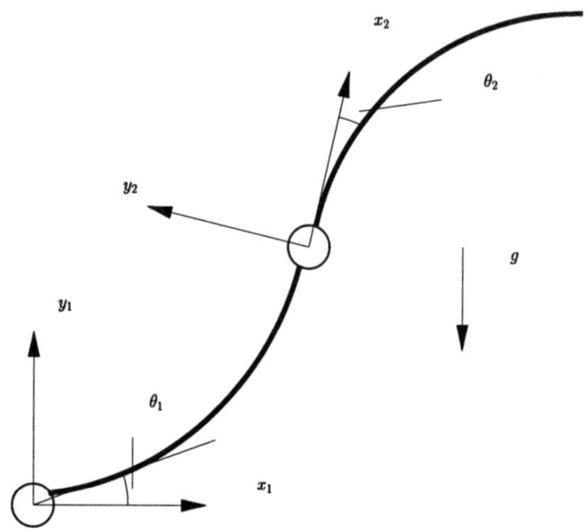

Fig. 1.3 Planar two-link flexible robot arm.

1.5 Simulation Results

In order to test the controllers given in Sections 1.3 and 1.4, the planar two-link flexible robot arm modeled in [De Luca and Siciliano (1991)] (Fig. 1.3), and modified in [De Luca and Siciliano (1993b)] to take gravity into account, has been used. A complete description of the arm equations of motion together with the parameters employed in the simulations can be found in [De Luca and Siciliano (1991); De Luca and Siciliano (1993b)].

1.5.1 *Control without Observer*

The first simulations are aimed at testing the controllers given in Section 1.3. The arm is assumed to be initially in its vertical equilibrium configuration, i.e.

$$\boldsymbol{\theta} = \begin{bmatrix}-90 \; 0\end{bmatrix}^T \; [°]$$

and

$$\boldsymbol{\delta} = \begin{bmatrix}0 \; 0 \; 0 \; 0\end{bmatrix}^T \; [\mathrm{m}].$$

As for the control goal, the final desired joint angles

$$\boldsymbol{\theta}_d = \begin{bmatrix}-45 \; 45\end{bmatrix}^T \; [°]$$

should be reached at a final time $t_f = 7.5$[s]. Since control (1.120) has been designed for time–varying trajectories, a linear trajectory interpolation with a fifth–order polynomial is used for the time $0 \leq t \leq 7.5$[s]:

$$\boldsymbol{\theta}_d = \begin{bmatrix} -90 + 1.0667t^3 - 0.2133t^4 + 0.0114t^5 \\ 1.0667t^3 - 0.2133t^4 + 0.0114t^5 \end{bmatrix} [^\circ].$$

The matrices $\boldsymbol{K}_p$ and $\boldsymbol{\Lambda}$ are

$$\boldsymbol{K}_p = \text{diag}\{10, 10, 0.5, 0.5, 0.5, 0.5\}$$

and

$$\boldsymbol{\Lambda} = \text{diag}\{1, 1, 1, 1, 1, 1\}.$$

In the first simulation, the desired trajectory $\boldsymbol{\delta}_d$ is computed from (1.121). The results are depicted in Fig. 1.4 in terms of the joint torques, the joint coordinate errors and the link coordinates with their desired values. As foreseen, the tracking errors converge to zero while $\boldsymbol{\delta}_d$ and $\dot{\boldsymbol{\delta}}_d$ remain bounded. The final value for $\boldsymbol{\delta}_d$ is

$$\boldsymbol{\delta}_d = \text{diag}\{-0.1832, -0.0048, -0.0078, -0.000106\}.$$

In the second simulation, it is assumed that $\boldsymbol{D}_\delta \equiv \boldsymbol{O}$. The desired trajectory $\boldsymbol{\delta}_d$ is computed from (1.143), while the elements of $\boldsymbol{D}_\Delta$ take on the values of $\boldsymbol{D}_\delta$ given in [De Luca and Siciliano (1991); De Luca and Siciliano (1993b)]. Also, with reference to (1.145), it is $\epsilon_{11} = 0.1$, $\epsilon_{12} = 0.0001$, $\epsilon_{21} = 0.0001$, $\epsilon_{22} = 0.00001$ and $\beta_{11} = \beta_{12} = \beta_{21} = \beta_{22} = 0.001$. The results are depicted in Fig. 1.5. It is worth observing how the overall behavior remains satisfactory; this is because the desired trajectories are damped and control (1.120) guarantees that the tracking error converges to zero. As expected, the final vector $\boldsymbol{\delta}_d$ is the same as before. Notice that a bad choice of the parameters ϵ_{ij} and β_{ij} may cause chattering behavior. More simulation results can be found in [Arteaga (1996c)].

1.5.2 *Control with Observer*

In order to test the observers of Section 1.4, the same control goal as in Section 1.5.1 has to be achieved. The initial conditions for the joint and link coordinates of the arm are

$$\boldsymbol{\theta} = \begin{bmatrix} -90 \; 0 \end{bmatrix}^T [^\circ]$$

and

$$\boldsymbol{\delta} = \left[0\ 0\ 0.05\ 0.02\right]^T \text{[m]},$$

while the velocity vector initial condition is $\mathbf{0}$. The gains of the controller and the observer have been selected to be:

$$\begin{aligned}
\boldsymbol{K}_p &= \text{diag}\{5, 5, 0.25, 0.25, 0.25, 0.25\}\\
\boldsymbol{\Lambda} &= \text{diag}\{0.75, 0.75, 0.75, 0.75, 0.75, 0.75\}\\
\boldsymbol{K}_q &= \text{diag}\{1.5, 1.5, 1, 1, 1, 1\}\\
\boldsymbol{K}_v &= \text{diag}\{50, 50, 50, 50, 50, 50\}\\
k_d &= 10.
\end{aligned}$$

In the first simulation, the desired trajectory $\boldsymbol{\delta}_d$ is computed from (1.182) by setting $\boldsymbol{D}_\Delta$ to $\boldsymbol{O}$. The results for the tracking errors are depicted in Fig. 1.6. Again, they converge to zero while $\boldsymbol{\delta}_d$ and $\dot{\boldsymbol{\delta}}_d$ remain bounded. The final value for $\boldsymbol{\delta}_d$ is the same as in Section 1.5.1. The observation errors are depicted in Fig. 1.7. It can be seen that they converge to zero.

In the second simulation, $\boldsymbol{D}_\delta$ is set to $\boldsymbol{O}$ and $\boldsymbol{D}_\Delta$ takes on the old values of $\boldsymbol{D}_\delta$. The desired trajectory $\boldsymbol{\delta}_d$ is computed from (1.182). Also, with reference to (1.184), it is $\epsilon_{11} = 0.1$, $\epsilon_{12} = 0.01$, $\epsilon_{21} = 0.01$, $\epsilon_{22} = 0.05$ and $\beta_{11} = \beta_{12} = \beta_{21} = \beta_{22} = 0.001$. The results for the tracking errors and input torques are shown in Fig. 1.8, while those for the observation errors are shown in Fig. 1.9. As expected, the final vector $\boldsymbol{\delta}_d$ is the same as before. More simulation results can be found in [Arteaga (2000)].

1.5.3 *Control with a Reduced-Order Model*

In this section, the controllers given in Section 1.3 are used again with the purpose to test their sensitivity to the number of modes taken into account. In particular, it is assumed that only one link coordinate for each link is included in the controller. It is not troublesome to get this reduced–order model from the original one, since it can be obtained by just letting the high–order terms be zero. On the other hand, the simulated model includes two link coordinates for each link as ever before, but no gravitational term is considered so as to obtain somewhat less damped motions. The initial values of the generalized coordinates are

$$\boldsymbol{\theta} = \left[0\ 0\right]^T [°]$$

and

$$\boldsymbol{\delta} = \left[0\ 0\ 0\ 0\right]^T [\text{m}].$$

As for the control goal, the final desired joint angles

$$\boldsymbol{\theta}_d = \left[45\ 45\right]^T [^\circ]$$

should be reached. In order to excite the unmodelled high frequencies, no trajectory for the joint coordinates is assigned. The matrices $\boldsymbol{K}_p$ and $\boldsymbol{\Lambda}$ are

$$\boldsymbol{K}_p = \text{diag}\{10, 10, 0.5, 0.5\}$$

and

$$\boldsymbol{\Lambda} = \text{diag}\{1, 1, 1, 1\}.$$

In the first simulation, the desired trajectory $\boldsymbol{\delta}_d$ is computed from (1.121). The results are depicted in Fig. 1.10. It can be observed that the tracking errors converge to zero. Notice that there is no desired trajectory for δ_{12} nor for δ_{22} and that those for δ_{11} and δ_{21} remain bounded.

The next simulation is aimed at testing the robust term used to increase the damping of the system. As a matter of fact, any mechanical system must be damped, so it makes more sense to show whether a poor damping can be improved. To do this, only the damping of the modeled dynamics is set to zero while that for the unmodeled dynamics is the same as before. The elements of $\boldsymbol{D}_\Delta$ are assumed to be the ones of the original system, while $\epsilon_{11} = 0.1$, $\epsilon_{21} = 0.0001$ and $\beta_{11} = \beta_{21} = 0.001$. The results are depicted in Fig. 1.11. It is worth observing how the overall behavior remains satisfactory. Note that δ_{11} and δ_{21} are damped because δ_{d11} and δ_{d21} are damped as well.

It is interesting to carry out the same simulation while letting $\boldsymbol{D}_\Delta$ be $\boldsymbol{O}$. As can be appreciated in Fig. 1.12, δ_{11} and δ_{21} are no longer damped, which shows the efficacy of the proposed solution to increase the damping of the system. Nonetheless, the tracking errors become zero in all the cases, as it could have been foreseen.

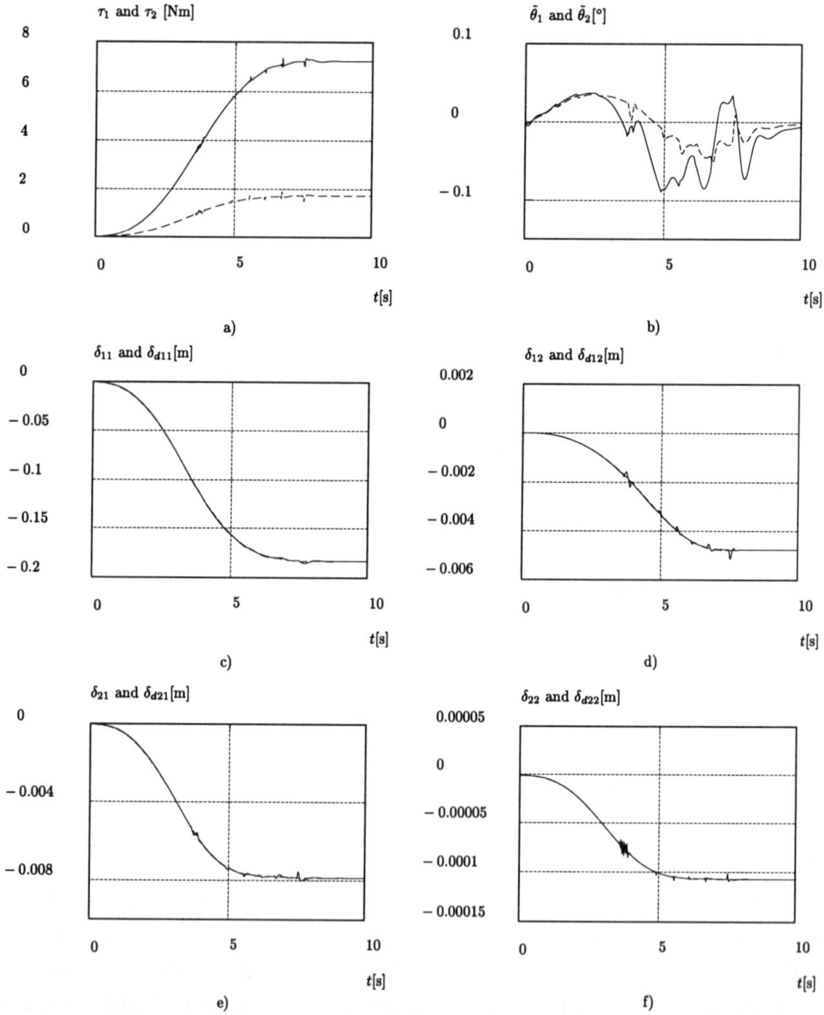

Fig. 1.4 Simulation results for a planar two-link flexible robot arm: Damping case. a) Input torques τ_1 (—) and τ_2 (- - -); b) Joint errors $\tilde{\theta}_1$ (—) and $\tilde{\theta}_2$ (- - -); c) Link coordinate δ_{11} (—) and its desired value δ_{d11} (- - -); d) Link coordinate δ_{12} (—) and its desired value δ_{d12} (- - -); e) Link coordinate δ_{21} (—) and its desired value δ_{d21} (- - -); f) Link coordinate δ_{22} (—) and its desired value δ_{d22} (- - -).

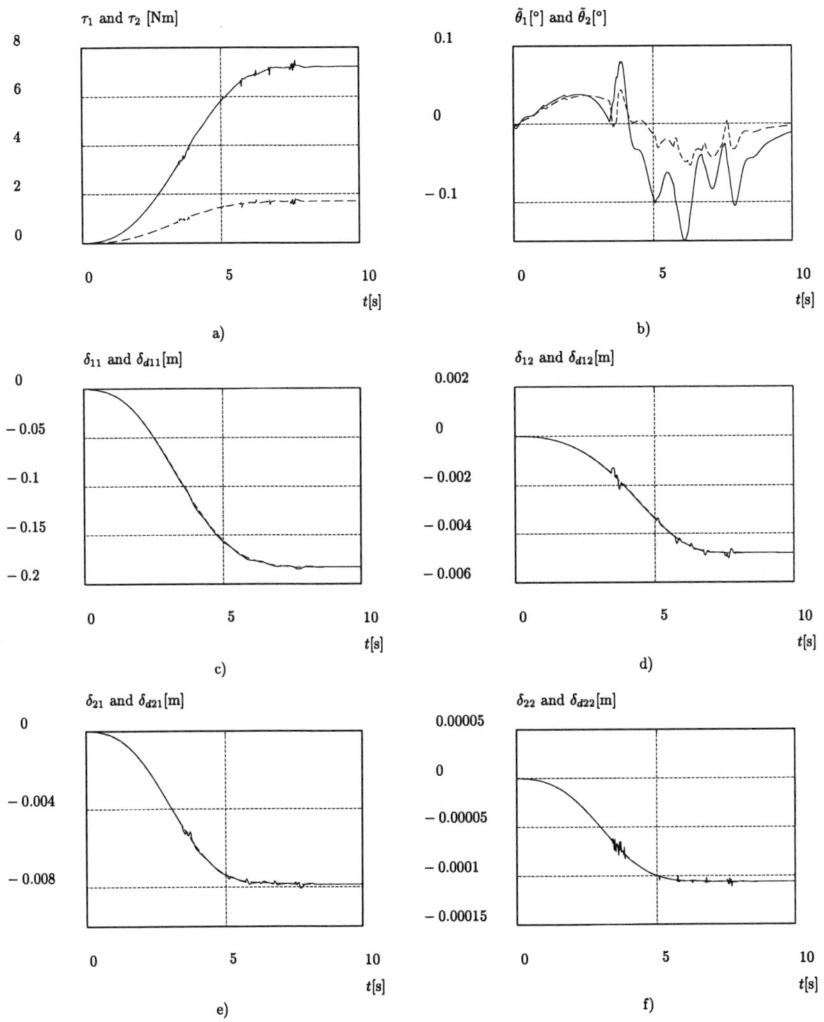

Fig. 1.5 Simulation results for a planar two-link flexible robot arm: No damping case. a) Input torques τ_1 (—) and τ_2 (- - -); b) Joint errors $\tilde{\theta}_1$ (—) and $\tilde{\theta}_2$ (- - -); c) Link coordinate δ_{11} (—) and its desired value δ_{d11} (- - -); d) Link coordinate δ_{12} (—) and its desired value δ_{d12} (- - -); e) Link coordinate δ_{21} (—) and its desired value δ_{d21} (- - -); f) Link coordinate δ_{22} (—) and its desired value δ_{d22} (- - -).

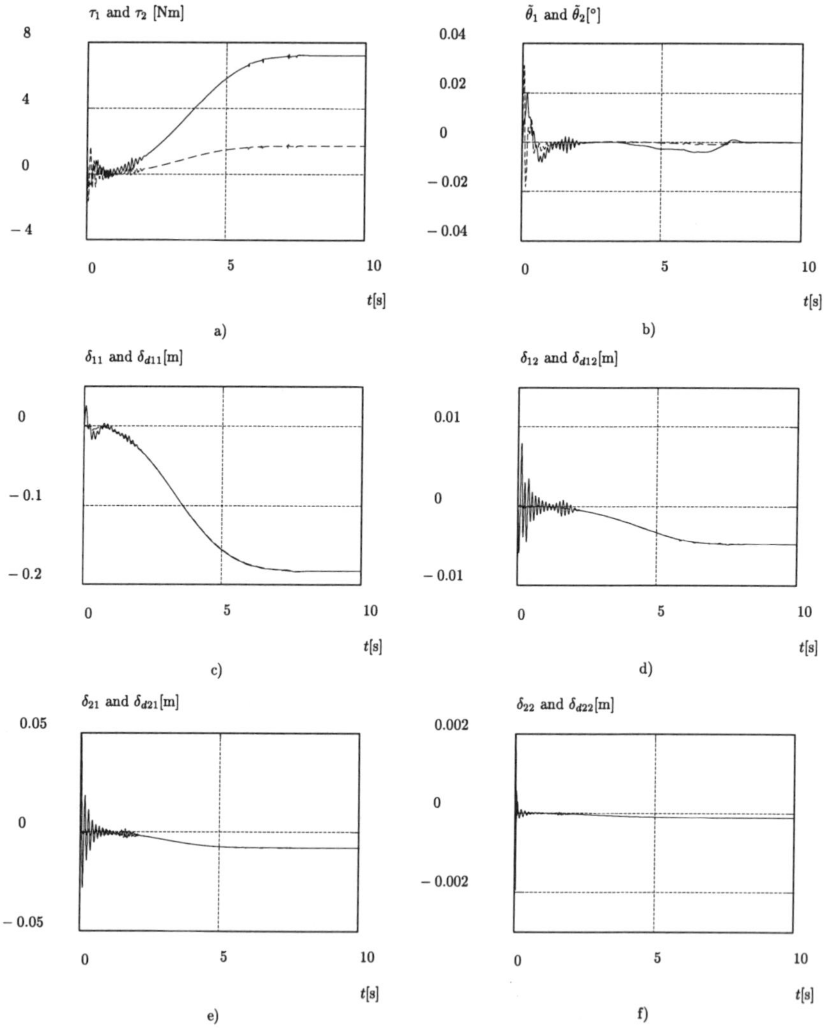

Fig. 1.6 Simulation results for a planar two-link flexible robot arm: Damping case with observer. a) Input torques τ_1 (—) and τ_2 (- - -); b) Joint errors $\tilde{\theta}_1$ (—) and $\tilde{\theta}_2$ (- - -); c) Link coordinate δ_{11} (—) and its desired value δ_{d11} (- - -); d) Link coordinate δ_{12} (—) and its desired value δ_{d12} (- - -); e) Link coordinate δ_{21} (—) and its desired value δ_{d21} (- - -); f) Link coordinate δ_{22} (—) and its desired value δ_{d22} (- - -).

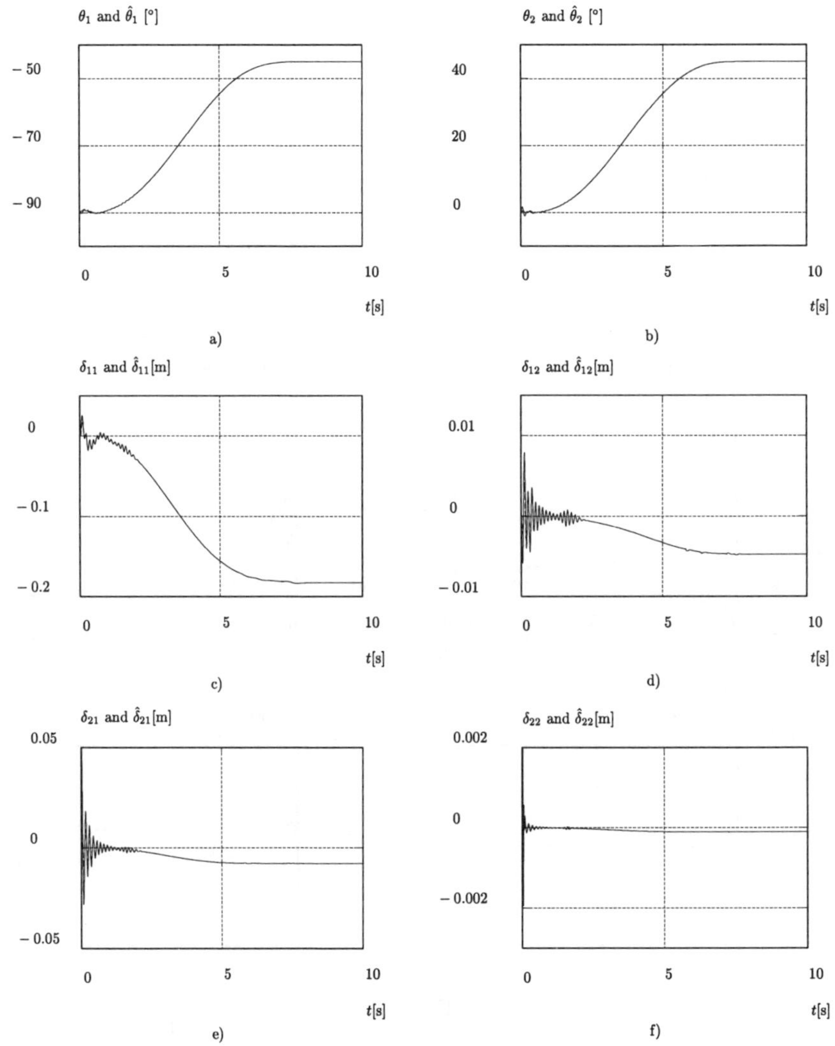

Fig. 1.7 Simulation results for a planar two-link flexible robot arm: Damping case with observer. a) Joint coordinate θ_1 (—) and its estimate $\hat{\theta}_1$ (- - -); b) Joint coordinate θ_2 (—) and its estimate $\hat{\theta}_2$ (- - -); c) Link coordinate δ_{11} (—) and its estimate $\hat{\delta}_{11}$ (- - -); d) Link coordinate δ_{12} (—) and its estimate $\hat{\delta}_{12}$ (- - -); e) Link coordinate δ_{21} (—) and its estimate $\hat{\delta}_{21}$ (- - -); f) Link coordinate δ_{22} (—) and its estimate $\hat{\delta}_{22}$ (- - -);

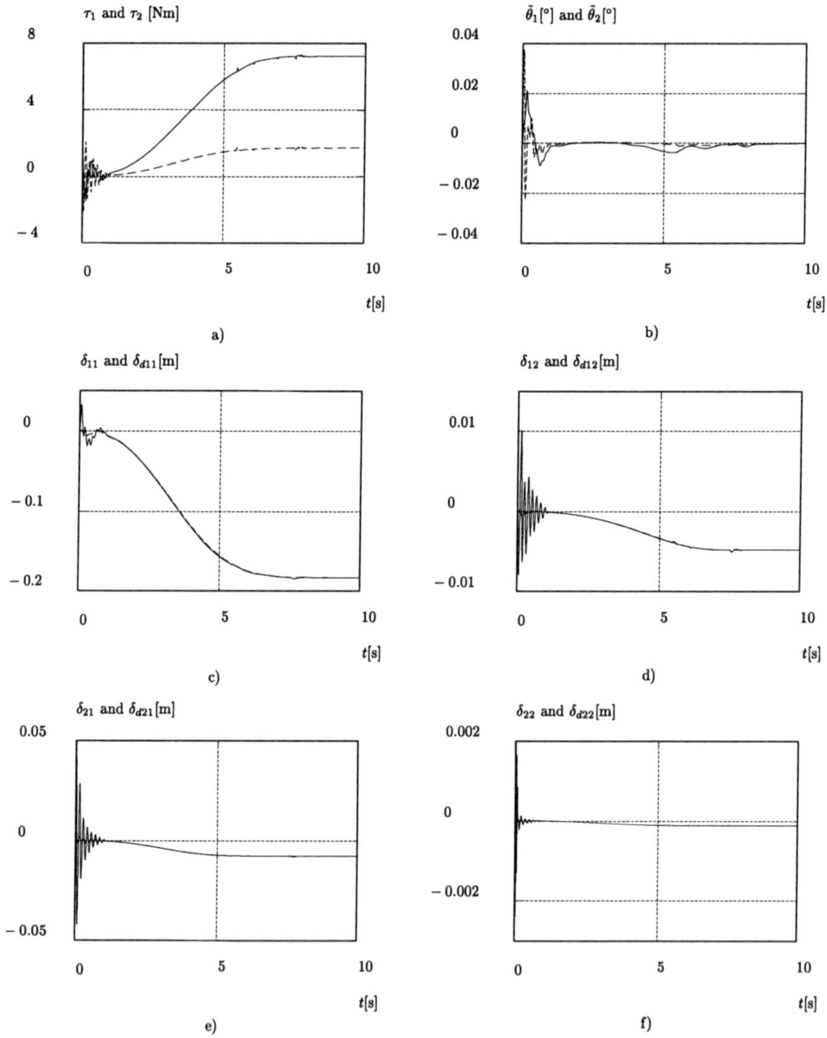

Fig. 1.8 Simulation results for a planar two-link flexible robot arm: No damping case with observer. a) Input torques τ_1 (—) and τ_2 (- - -); b) Joint errors $\tilde{\theta}_1$ (—) and $\tilde{\theta}_2$ (- - -); c) Link coordinate δ_{11} (—) and its desired value δ_{d11} (- - -); d) Link coordinate δ_{12} (—) and its desired value δ_{d12} (- - -); e) Link coordinate δ_{21} (—) and its desired value δ_{d21} (- - -); f) Link coordinate δ_{22} (—) and its desired value δ_{d22} (- - -).

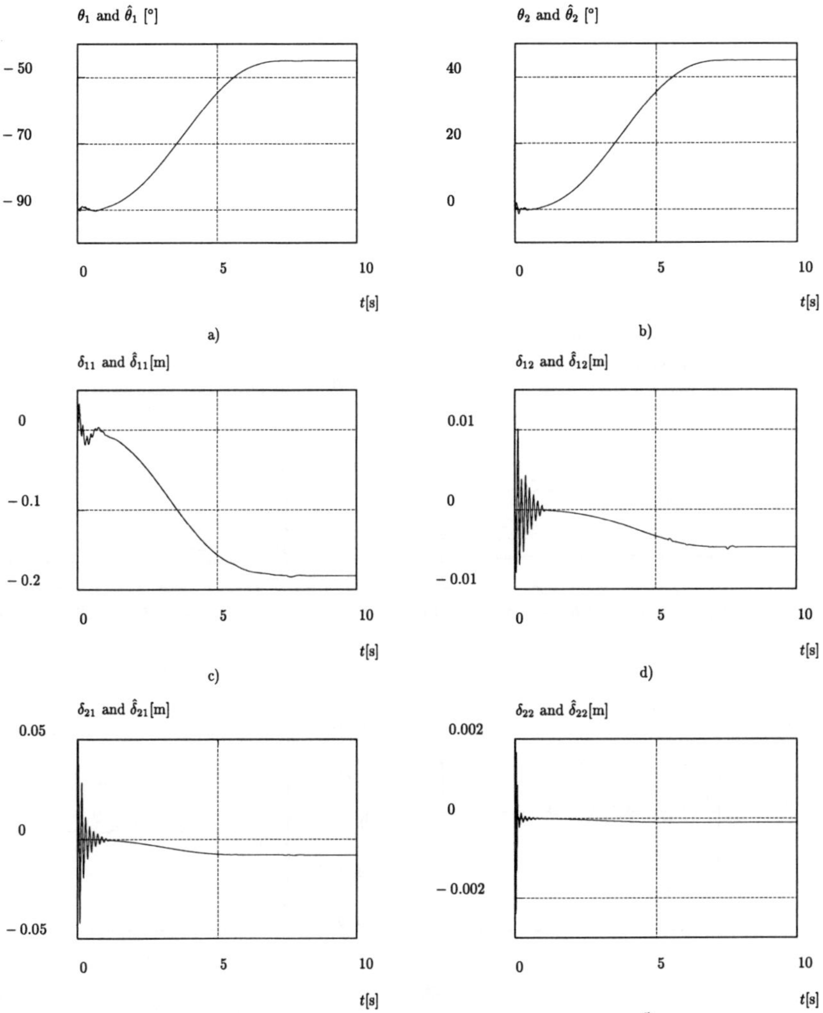

Fig. 1.9 Simulation results for a planar two-link flexible robot arm: No damping case with observer. a) Joint coordinate θ_1 (—–) and its estimate $\hat{\theta}_1$ (- - -); b) Joint coordinate θ_2 (—–) and its estimate $\hat{\theta}_2$ (- - -); c) Link coordinate δ_{11} (—–) and its estimate $\hat{\delta}_{11}$ (- - -); d) Link coordinate δ_{12} (—–) and its estimate $\hat{\delta}_{12}$ (- - -); e) Link coordinate δ_{21} (—–) and its estimate $\hat{\delta}_{21}$ (- - -); f) Link coordinate δ_{22} (—–) and its estimate $\hat{\delta}_{22}$ (- - -);

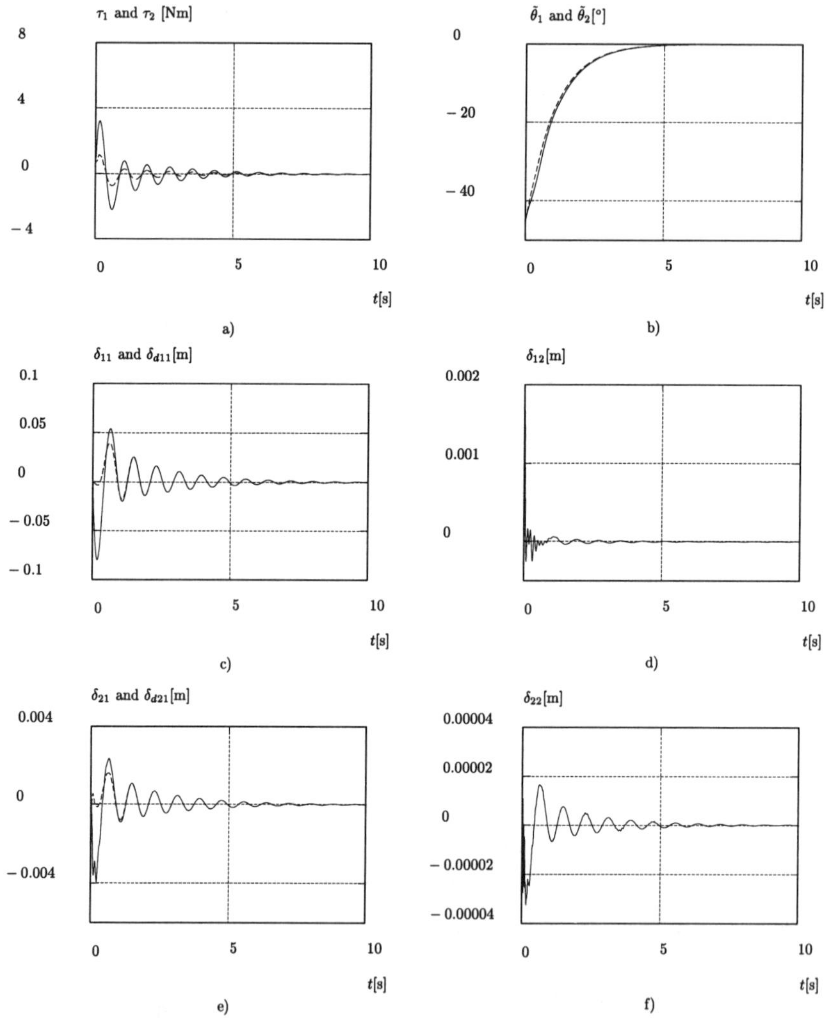

Fig. 1.10 Simulation results for a planar two-link flexible robot arm with reduced-order model: Damping case. a) Input torques τ_1 (—) and τ_2 (- - -); b) Joint errors $\tilde{\theta}_1$ (—) and $\tilde{\theta}_2$ (- - -); c) Link coordinate δ_{11} (—) and its desired value δ_{d11} (- - -); d) Link coordinate δ_{12}; e) Link coordinate δ_{21} (—) and its desired value δ_{d21} (- - -); f) Link coordinate δ_{22}.

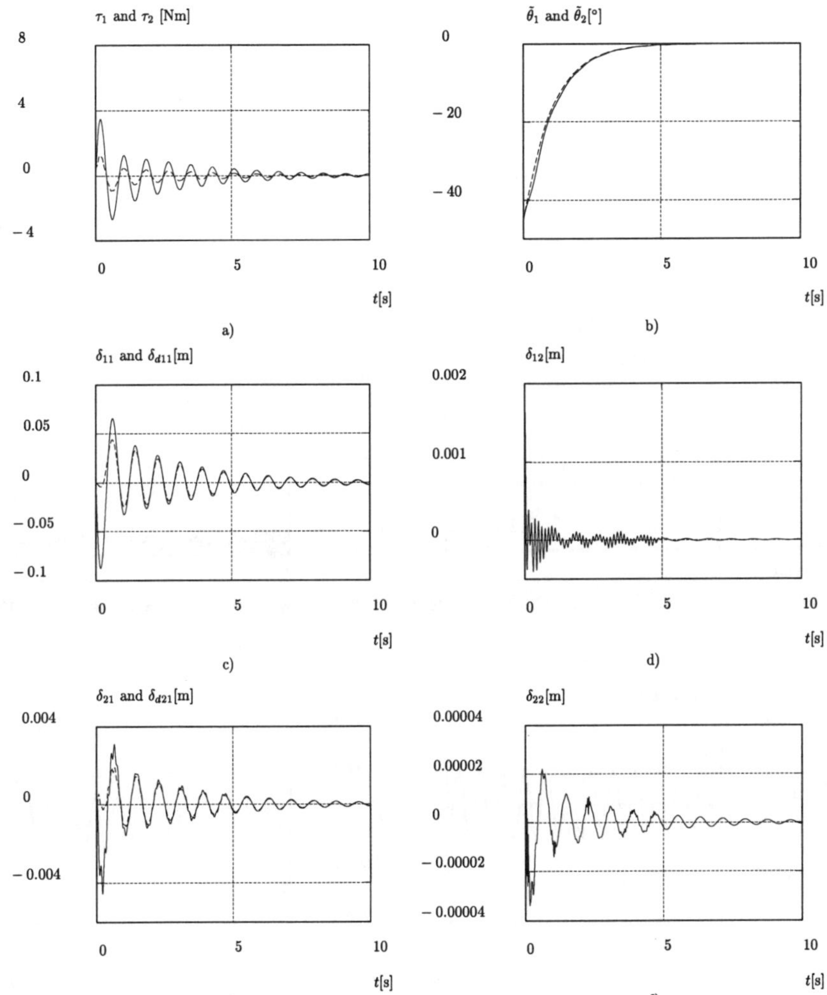

Fig. 1.11 Simulation results for a planar two-link flexible robot arm with reduced-order model: No damping case with desired trajectories δ_{d11} and δ_{d21} damped. a) Input torques τ_1 (—) and τ_2 (- - -); b) Joint errors $\tilde{\theta}_1$ (—) and $\tilde{\theta}_2$ (- - -); c) Link coordinate δ_{11} (—) and its desired value δ_{d11} (- - -); d) Link coordinate δ_{12}; e) Link coordinate δ_{21} (—) and its desired value δ_{d21} (- - -); f) Link coordinate δ_{22}.

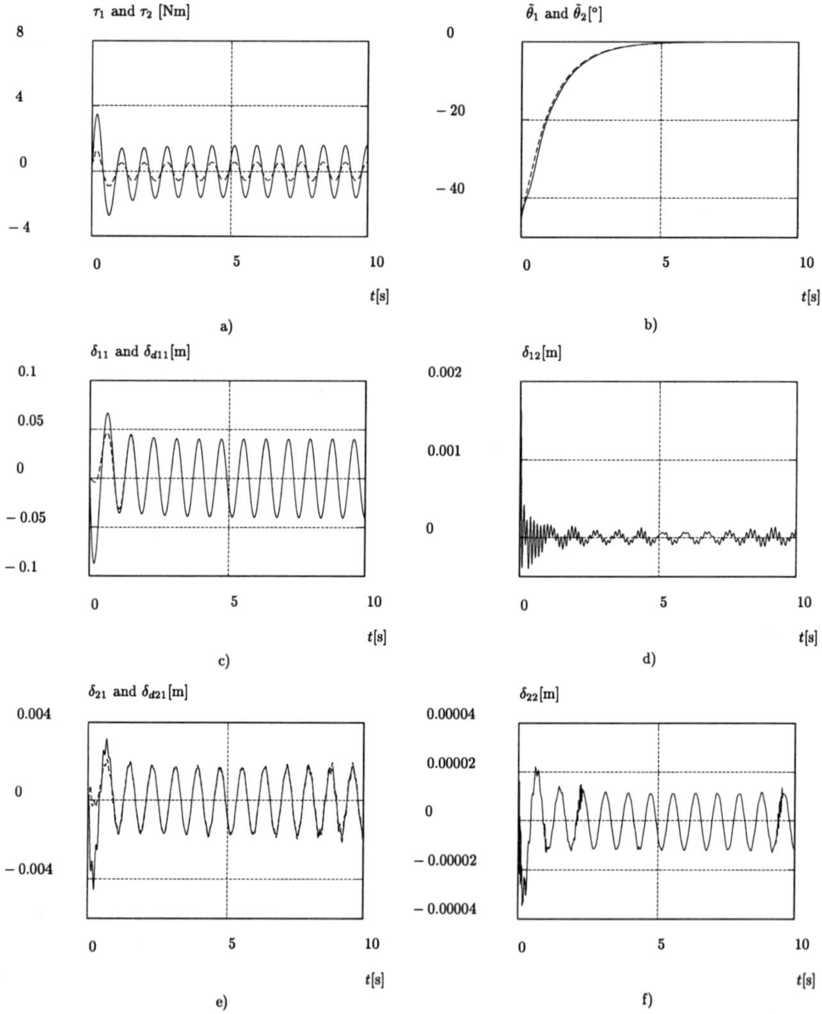

Fig. 1.12 Simulation results for a planar two-link flexible robot arm with reduced-order model: No damping case with desired trajectories δ_{d11} and δ_{d21} undamped. a) Input torques τ_1 (——) and τ_2 (- - -); b) Joint errors $\tilde{\theta}_1$ (——) and $\tilde{\theta}_2$ (- - -); c) Link coordinate δ_{11} (——) and its desired value δ_{d11} (- - -); d) Link coordinate δ_{12}; e) Link coordinate δ_{21} (——) and its desired value δ_{d21} (- - -); f) Link coordinate δ_{22}.

1.6 Conclusion

The modeling and control problem of flexible robot manipulators has been studied in the present work. By using Lagrange's equations of motion, a closed–form dynamic model has been obtained where link deflection has been described in terms of assumed modes. For control design purposes, several important properties of the manipulator model have been given and proven. Some of them are physical properties whereas others do arise from the method used to derive the model.

The tracking control problem has been studied. Because a flexible robot has fewer inputs than degrees of freedom, a desired trajectory for the flexible coordinates cannot be selected arbitrarily and has to be computed on–line to accomplish the tracking control goal for the rigid coordinates. It has been proven that this trajectory remains bounded. In order to increase the damping of the system, the equations used for the desired trajectory of the link coordinates have been modified. To ensure global stability of the system, robust control techniques have been employed.

Since the possible lack of measurements of link deflection rates is a drawback of the proposed control scheme, a nonlinear observer has been designed which can still guarantee global stability of the system and boundedness of the desired link coordinate trajectory. It has been shown that the region of attraction for the observer can be enlarged arbitrarily.

In order to test the controller, with and without observer, several simulations have been carried out. It has been shown that the proposed solution does actually increase the damping of the system. To investigate how the controller works in the presence of unmodelled dynamics, some simulations have been accomplished with a reduced–order model yielding satisfactory results.

Acknowledgments

This work is based on research sponsored by DAAD (German Academic Exchange Service) and ASI (Italian Space Agency).

Bibliography

Arteaga, M. A. (1995). Position Determination for Flexible Robot Arms. Research Report (in German) No. 9/95, Department of Measurement and Control,

University of Duisburg, Germany.

Arteaga, M. A. (1996a). A nonlinear observer for flexible robot arms, *Proc. of 4th IEEE Mediterranean Symposium on New Directions in Control and Automation*, Maleme, Krete, Greece, pp. 144–149.

Arteaga, M. A. (1996b). Nonlinear Observers for Flexible Robot Arms, Research Report (in German) No. 4/96, Department of Measurement and Control, University of Duisburg, Germany.

Arteaga, M. A. (1996c). On the Control of Flexible Robot Arms, Research Report (in German) No. 14/96, Department of Measurement and Control, University of Duisburg, Germany.

Arteaga, M.A. (1998). On the properties of a dynamic model of flexible robot manipulators, *ASME Journal of Dynamic Systems, Measurement, and Control*, **120**, pp. 8–14.

Arteaga, M. A. (2000). Tracking control of flexible robot arms with a nonlinear observer, *Automatica*, **36**, pp. 1329–1337.

Arteaga, M. A. and Siciliano, B. (2000). On tracking control of flexible robot arms, *IEEE Transactions on Automatic Control*, **45**, pp. 520–527.

Book, W. J. (1979). Analysis of massless elastic chains with servo controlled joints, *ASME Journal of Dynamic Systems, Measurement, and Control*, **101**, pp. 187–192.

Book, W. J. (1984). Recursive Lagrangian dynamics of flexible manipulator arms, *International Journal of Robotics Research*, **3**(3), pp. 87–101.

Canudas de Wit, C., Åström, K. J. and Fixot, N. (1990). Computed torque control via a nonlinear observer, *International Journal of Adaptive Control and Signal Processing*, **4**, pp. 443–452.

Canudas de Wit, C., Siciliano, B. and Bastin, G. (Eds.) (1996). *Theory of Robot Control*, Springer–Verlag, London.

Corless, M. J. and Leitmann, G. (1981). Continuous state feedback guaranteeing uniform ultimate boundedness for uncertain dynamic systems, *IEEE Transactions on Automatic Control*, **26**, pp. 1139–1144.

Dawson, D. M., Qu, Z. and Lim, S. (1991). Re–thinking the robust control of robot manipulators, *Proc. 30th IEEE Conference on Decision and Control*, Brighton, England, pp. 1043–1045.

De Luca, A. and Panzieri, S. (1994). An iterative scheme for learning gravity compensation in flexible robot arms, *Automatica*, **30**, pp. 993–1002.

De Luca, A. and Siciliano, B. (1991). Closed–form dynamic model of planar multilink lightweight robots, *IEEE Transactions on Systems, Man and Cybernetics*, **21**, pp. 826–839.

De Luca, A. and Siciliano, B. (1993a). Inversion–based nonlinear control of robot arms with flexible links, *AIAA Journal of Guidance, Control, and Dynamics*, **16**, pp. 1169–1176.

De Luca, A. and Siciliano, B. (1993b). Regulation of flexible arms under gravity, *IEEE Transactions on Robotics and Automation*, **9**, pp. 463–467.

Denavit, J. and Hartenberg, R. S. (1955). A kinematic notation for lower pair mechanisms based on matrices, *ASME Journal of Applied Mechanics*, **22**, pp. 215–221.

Desoer, C. A. and Vidyasagar, M. (1975). *Feedback Systems: Input–Output Properties*, Academic Press, New York.

Greenwood, D. T. (1977). *Classical Dynamics*, Prentice–Hall, Englewood Cliffs, N.J.

Lammerts, I. M. M., Veldpaus, F. E., Van de Molengraft, M. J. G. and Kok, J. J. (1995). Adaptive computed reference computed torque control of flexible robots, *ASME Journal of Dynamic Systems, Measurement, and Control*, **117** pp. 31–36.

Leitmann, G. (1981). On the efficacy of nonlinear control in uncertain linear systems, *ASME Journal of Dynamic Systems, Measurement, and Control*, **102**, pp. 95–102.

Meirovitch, L. (1967). *Analytical Methods in Vibrations*, The Macmillan Company, New York.

Meirovitch, L. (1975). *Elements of Vibration Analysis*, McGraw–Hill, New York.

Nicosia, S. and Tomei, P. (1990). Robot control by using only joint position measurements, *IEEE Transactions on Automatic Control*, **35**, pp. 1058–1061.

Ortega, R. and Spong, M. W. (1989). Adaptive motion control of rigid robots: a tutorial, *Automatica*, **25**, pp. 877–888.

Qu, Z. and Dawson, D. M. (1991). Continuous state feedback control guaranteeing exponential stability for uncertain dynamical systems, *Proc. 30th IEEE Conference on Decision and Control*, Brighton, England, pp. 2636–2638.

Sciavicco, L. and Siciliano, B. (2000), *Modelling and Control of Robot Manipulators*, 2nd Ed., Springer–Verlag, London.

Slotine, J. J. E. and Li, W. (1987). On the adaptive control of robot manipulators, *International Journal of Robotics Research*, **6**(3), pp. 49–59.

Spong, M. W. (1992). On the robust control of robot manipulators, *IEEE Transactions on Automatic Control*, **37**, pp. 1782–1786.

Vidyasagar, M. (1978). *Nonlinear Systems Analysis*, Prentice–Hall, Englewood Cliffs, N.J.

Wellstead, P. E. (1979). *Introduction to Physical System Modelling*, Academic Press, London.

Yaz, E. (1993). Comments on "On the robust control of robot manipulators", *IEEE Transactions on Automatic Control*, **38**, pp. 511–512.

Yuan, B. S., Book, W. J. and Huggins, J. D. (1993). Dynamics of flexible manipulator arms: alternative derivation, verification, and characteristics for control, *ASME Journal of Dynamic Systems, Measurement, and Control*, **115**, pp. 394–404.

Chapter 2

Energy-based Robust Control of Flexible Link Robots

In this chapter, energy-based robust control strategies are presented for the control of flexible link robots without using the dynamics of the systems explicitly. Firstly, the dynamics of a single link flexible robot are introduced. Then, robust none-model-based control is presented for the system without employing the dynamics of the system explicitly. Finally, the results for the single link case is extended to the multi-link case. The energy-based controllers are independent of system parameters and thus possess stability robustness to system parametric uncertainties. Through the evaluation of vibrations of the links, direct control of link deflection is possible. Simulation results are provided to show the effectiveness of the presented approach.

2.1 Introduction

Conventional rigid-link robots have been widely used in industrial production. However, to obtain high accuracy at the end-point position control of these robots, the weight to payload ratio of the robots must be high, and the operation speed is normally quite low. At the same time, large power supply and thus considerable energy consumption are necessary to operate these large-weight robots. These drawbacks greatly limit the application of these robots to the tasks where high speed, high accuracy and low energy support are required. A flexible robot, made of slender links which are considerably more rigid in compression than in flexure, is free of these drawbacks (Krishnan, 1988; Book, 1990; Korolov and Chen, 1988; Kotnik *et al.*, 1988; Rattan and Feliu, 1992). The light-weight flexible link robot needs lower power consumption for high speed operation, which makes it more feasible than the conventional rigid-link robots in many applications, such as in space applications. These advantages greatly motivate the research in modelling and control of flexible robots. When such a robot is carrying a large payload and moving at a high speed, the effect of link flexibility will be non-neglectable, and the traditional collocated control, e.g., the PD feedback is no longer sufficient for fast and accurate positioning. The system, described by partial differential equations (PDEs), is actually a distributed-parameter system of infinite dimensions (Spector and Flashner, 1990; Luo, 1993; Shifman, 1993). Its nonminimum phase behavior from the base input to the end-point output makes it very difficult to achieve high level performance and robustness simultaneously. The approaches established for control of rigid-link manipulators are theoretically no longer applicable to flexible ones. The commonly used modelling methods, such as Assumed Modes Method (AMM) and the Finite Element Method (FEM), allow us to obtain approximated finite dimensional models of flexible-link robots without losing much system information (Kanoh *et al.*, 1986; Hastings and Book, 1987; Wang and Vidyasagar, 1989; Bayo, 1987). With such a finite dimensional model in hand, various control strategies are then applicable (Siciliano and Book, 1988; R.H. Cannon and Schmitz, 1984; Zuo and Wang, 1992; Tzes and Yurkovich, 1993). However, it is noted that the truncation in system dimensionality, though normally provides good approximations to the original system, also have some problems including (i) control and observation spillovers due to the ignored unmodelled dynamics; (ii) high order of controller for the large number of flexible modes/elements included in the model to achieve high accuracy of performance; and (iii) difficulty in implementing the controller from the engineering point of view since full states measurements/observers are often required. For truncated-model-based controllers, extra efforts are needed to overcome these problems in practice. In an attempt to overcome these shortcomings, an alternative method has been developed for the control of flexible robot system in recent years. In the method, dynamic analysis and controller design are carried out directly based on the original PDE dynamics, named "PDEs-based approach". This approach, since it avoids the undesirable model truncation, is more attractive to practical applications. For the PDEs-based approach, some control design examples can be found in (Luo, 1993; Shifman, 1993; Luo *et al.*, 1995; Yigit, 1994). In general, this approach needs more complicated derivations than the truncated-model-based methods, because the PDEs of the original flexible robot systems are quite complicated, especially for the multi-link flexible robots, which is also the reason that the four references above only considered the single flexible link case.

In this chapter, energy based robust controller design is presented for both single- and multi-link flexible robots without invoking the undesirable model truncation. Rather than using the dynamics of the links, the main stability results are obtained by using the total energy and the energy-work relationship of the whole systems. As the controllers are designed without invoking the dynamics of the systems, they are model free and thus robust to uncertainties in the systems. In addition, the non-model-based nature also makes the controller free of the problems associated with the truncated-model-based methods such as the control/observation spillovers. Through the evaluation of vibration of the links, direct control of link deflection is possible, and leads to improved control performance. As the controllers are designed based on signals that are easy to measure, they can be easily implemented in practice.

The rest part of the paper is organized as follows. Section 2.2 presents the dynamics of a single link flexible robot and the some properties of the systems. In Section 2.3, energy based robust control is presented for the system without employing the dynamics of the system explicitly. Finally, the results for the single link case is extended to the multi-link case in Section 2.4. Section 2.5 conclude the Chapter.

2.2 Dynamics of a Single-link Flexible Robot

Consider a flexible beam which is clamped at its base on the rotor of a motor, and loaded by a point mass at its tip, as shown in Figure 2.1. The flexible manipulator is rotating in the horizontal plane, thus, gravitational loading does not feature in this situation. Frame XOY is the fixed base frame and frame X_1OY_1 is the local frame rotating with the hub. The relevant system parameters and variables are defined as follows: L denotes the length of the beam; EI the uniform flexural rigidity of the beam; M_t the point mass tip payload; I_h the total moment of inertia with respect to the rotation joint axis of the beam; τ the control torque; $\theta(t)$ the joint angle; and $y(x,t)$ the elastic deflection measured from the undeformed beam.

Let $p(x,t) = x\theta(t) + y(x,t)$ represent the position of a point on the flexible beam, consistent with notation used elsewhere in the literature (Krishnan, 1988; Sakawa *et al.*, 1985). The total kinetic energy E_k can be described by

$$E_k = \frac{1}{2}I_h\dot{\theta}^2 + \frac{\rho}{2}\int_0^L \dot{p}^2(x,t)dx + \frac{1}{2}M_t\dot{p}^2(L,t) \tag{2.1}$$

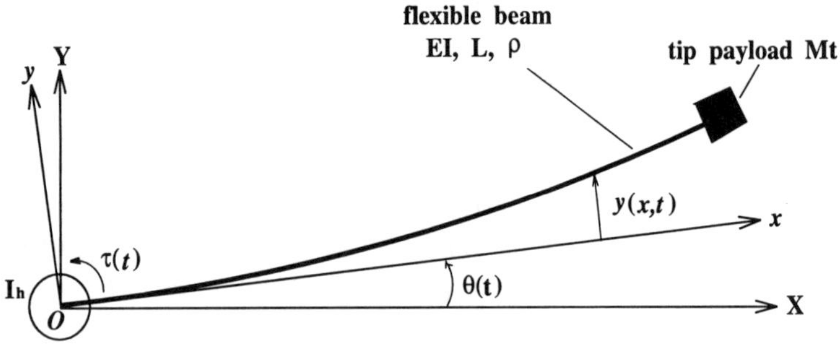

Fig. 2.1 Single-link flexible robot

and the total potential energy E_p is given by

$$E_p = \frac{EI}{2}\int_0^L [y''(x,t)]^2 dx \tag{2.2}$$

where, in the usual manner, the dots and primes denote the derivatives with respect to time and space, respectively.

Using the extended Hamilton's Principle:

$$\int_{t_o}^{t_f} \delta(E_k - E_p + \tau\theta)dt = 0 \tag{2.3}$$

and substituting (2.1) and (2.2) into (2.3), we obtain the system dynamics described by the following Partial Differential Equations (PDE):

$$(I_h + \frac{1}{3}\rho L^3)\ddot{\theta}(t) + \rho\int_0^L x\ddot{y}(x,t)dx + M_t L[L\ddot{\theta}(t) + \ddot{y}(L,t)] = \tau \tag{2.4}$$

$$\rho[x\ddot{\theta}(t) + \ddot{y}(x,t)] = -EIy''''(x,t) \tag{2.5}$$

The corresponding boundary conditions are given by

$$y(0,t) = 0 \tag{2.6}$$

$$y'(0,t) = 0 \tag{2.7}$$

$$y''(L,t) = 0 \tag{2.8}$$

$$EIy'''(L,t) = M_t[L\ddot{\theta}(t) + \ddot{y}(L,t)] \tag{2.9}$$

Equations (2.6)-(2.7) hold because frame xOy is selected such that the axis Ox is tangent to the beam at the base. The third boundary condition (2.8) comes directly from the zero value of the bending moment at the tip, and the fourth one (2.9) is actually the motion equation of the tip payload M_t.

For the system described above, the total work done by external inputs is given by

$$W = \int_0^t \tau(t)\dot{\theta}(t)dt \tag{2.10}$$

From the energy-work relationship, i.e., the increment of the system energy is equal to the work done by external inputs, we have the following equation

$$[E_k(t) + E_p(t)] - [E_k(0) + E_p(0)] = \int_0^t \dot{\theta}(t)\tau(t)dt \tag{2.11}$$

where $E_k(t)$ and $E_p(t)$ are the total kinetic energy and total potential energy of system at time t, $E_k(0)$ and $E_p(0)$ are constants representing the kinetic and potential energies at time 0. Therefore, by taking time derivatives of both sides of equation (2.11), we arrive at

$$\dot{E}_k(t) + \dot{E}_p(t) = \dot{\theta}(t)\tau(t) \tag{2.12}$$

which will be used in the controller design in the next section.

2.3 Controller Design

In this section, energy-based robust control (ERC) is investigated for the single link flexible robot described in Section 2.2. The proposed controllers include (i) stable ERC, (ii) asymptotic stable ERC, and (iii) adaptive ERC.

2.3.1 Stable Energy Based Robust Control

It is well-known that the pure joint PD controller is able to stabilize the system, however, the vibration of the flexible beam cannot be effectively controlled. We hereby introduce the nonlinear strain feedback to improve the performance of the pure joint PD controller given by

$$\tau = -k_p[\theta - \theta_f] - k_v\dot{\theta} - k_f y''(x_s, t)\int_0^t \dot{\theta}(\sigma)y''(x_s, \sigma)d\sigma \tag{2.13}$$

where k_p, k_v are positive constants and $k_f \geq 0$, and $0 \leq x_s < L$ which gives the location of the strain gauge on the beam. Obviously, $k_f = 0$ leads to the pure joint PD controller.

Theorem 2.1 *The closed-loop system described by (2.4), (2.5) and (2.13) is stable (Ge* et al., *1998*c*).*

Proof. Consider the following Lyapunov function candidate

$$V(t) = E_k + E_p + \frac{1}{2}k_p[\theta - \theta_f]^2 + \frac{1}{2}k_f\left[\int_0^t \dot{\theta}(\sigma)y''(x_s,\sigma)d\sigma\right]^2 \tag{2.14}$$

where E_k and E_p are the total kinetic energy and the total potential energy of the system. Obviously, $V(t)$ is positive definite.

The time derivative of $V(t)$ is

$$\dot{V}(t) = k_p\dot{\theta}(t)[\theta - \theta_f] + k_f\dot{\theta}y''(x_s,t)\int_0^t \dot{\theta}(\sigma)y''(x_s,\sigma)d\sigma + \dot{\theta}\tau \tag{2.15}$$

in which we have used

$$\dot{E}_k + \dot{E}_p = \dot{\theta}\tau \tag{2.16}$$

Now, substituting controller (2.13) into (2.15), we have

$$\dot{V}(t) = -k_v\dot{\theta}^2(t)$$

which is negative semi-definite. Thus, the closed-loop system is stable. □

Remark 2.1 *As the proof of stability shown involves the system energy-work relationship (2.11), the resulting controller is named as "energy-based". In addition, it can be seen from the proof that, when we use (2.11) to prove the stability, there is no need to know the dynamic equations of the system, and thus the controller is actually "non-model-based".*

Remark 2.2 *There is no model truncation in the stability proof. This implies the aforementioned problems associated with truncated-model-based controllers are essentially avoided. The controller (2.13) is independent of system parameters and thus possesses stability robustness to system parameters uncertainties. As a matter of fact, the closed-loop system is stable as long as* $k_p, k_v > 0$ *and* $k_f \geq 0$.

Remark 2.3 *In (2.13), only the measurements of joint angle* θ*, joint velocity* $\dot{\theta}$*, and strain (the bending moment) signal at* $x = x_s$*, i.e.,* $y''(x_s,t)$ *are needed. The integral type of the* k_f *item avoids measurements of high order signals, such as* $\ddot{\theta}$ *and* $\dot{y}''(x_s,t)$ *in the resulting controller. Therefore, the controller is very easy to implement from the engineering point of view. Moreover, from the proof, it is easy to see that any possible measurement* *noises existing in $y''(x_s, t)$ feedback will not affect the closed-loop stability. This is a very desirable feature in practical applications.*

It is noted that the joint PD controller is a special case of controller (2.13) by setting k_f to be zero, which corresponds to the case when there is no strain feedback is available. Accordingly, the Lyapunov function candidate $V(t)$ in (2.14) is reduced to

$$V(t) = E_k + E_p + \frac{1}{2}k_p[\theta - \theta_f]^2$$

Although the joint PD controller will not destabilize the system, the system performance, as will be shown in simulations later, is not satisfactory, because the flexible vibration of the link cannot be effectively suppressed. The introduction of the k_f term into $\tau(t)$ allows us to explicitly evaluate the vibration in the Lyapunov function. Subsequently, the controller (2.13) is expected to provide direct control effort on vibration suppression, and improve the system performance.

In the literature, tip acceleration feedback are also frequently used to improve the performance of the flexible manipulator systems (Kotnik *et al.*, 1988; Khorrami and Jain, 1993). If the tip acceleration feedback is available, the following controller can be proposed based on the idea of Theorem 2.1 as

$$\tau = -k_p[\theta(t) - \theta_f] - k_v\dot{\theta}(t) - k_f a_{tip}(t)\int_0^t \dot{\theta}(\sigma)a_{tip}(\sigma)d\sigma$$

where a_{tip} is the tip acceleration which can be measured by an accelerometer attached at the tip of the flexible beam. The closed-loop system stability can be guaranteed by using the following Lyapunov function:

$$V(t) = E_k + E_p + \frac{1}{2}k_p[\theta(t) - \theta_f]^2 + \frac{1}{2}k_f\left[\int_0^t \dot{\theta}(\sigma)a_{tip}(\sigma)d\sigma\right]^2$$

In summary, depending on the actual instrumentation of the system, a general stable controller can be obtained given by Lemma 2.1.

Lemma 2.1 *The closed-loop system consisting of the plant described by (2.4)-(2.5) and the following controller*

$$\tau = -k_p[\theta - \theta_f] - k_v\dot{\theta} - k_f f(t)\int_0^t \dot{\theta}(\sigma)f(\sigma)d\sigma \tag{2.17}$$

is stable, where, depending on the actual instrumentation, $f(t)$ can be strain or bending moment $y''(x,t)$, acceleration $a(x,t)$, deflection $y(x,t)$, rotation

$y'(\bullet,t)$ or shear force $y'''(\bullet,t)$, or any kind combination of these feedbacks, $\forall x \in [0,L]$.

When multiple feedback signals are available, the following general control follows

$$\tau = -k_p[\theta(t) - \theta_f] - k_v\dot{\theta}(t) - \sum_{i=1}^{n} k_{fi} f(x_{si}, t) \int_0^t \dot{\theta}(\sigma) f(x_{si}, \sigma) d\sigma$$

where $k_{fi} \geq 0$, $i = 1, 2, \cdots, n$, and x_{si} is the location of the ith appropriate measurement on the flexible beam. The stability of the closed-loop system can be guaranteed by choosing the following Lyapunov function (Ge *et al.*, 1998*c*):

$$V(t) = E_k + E_p + \frac{1}{2}k_p[\theta(t) - \theta_f]^2 + \frac{1}{2}\sum_{i=1}^{N} k_{fi} \left[\int_0^t \dot{\theta}(\sigma) f(x_{si}, \sigma) d\sigma\right]^2$$

and following the same proving procedure as shown earlier.

The only restriction on choosing $f(t)$ is that it must be zero when the flexible robot is static and undergoes no deformation. This restriction is reasonable, since $f(t)$ is introduced to evaluate the elastic vibration of the flexible link.

Remark 2.4 *Note that in Theorem 2.1 and Lemma 2.1, we can only claim the stability of closed-loop systems. To prove the asymptotic stability is difficult due to the infinite dimensionality of the system. As a result, the well known LaSalle's Theorem (Khalil, 1992) is not applicable here. This means the robot may stop before reaching the final position and thus the regulation will fail. Asymptotic control of an Euler-Bernoulli beam has been achieved in (Shifman, 1993), but it was assumed that there was no hub inertia, i.e., $I_h = 0$, which is not realizable in practice.*

However, we shall show in the following that practically the flexible robot can only possibly stop at the final position $\theta = \theta_f$ without vibrating. Consider the general form controller (2.17). Assume that the joint stops at a position $\theta \equiv \alpha$ (hence $\dot{\theta} \equiv 0$) with $\alpha \neq \theta_f$, thus there is no energy input to the system since $\dot{\theta} \equiv 0$. Due to the existence of internal structural damping in a flexible link in practice (structural damping is neglected in the proof of Theorem 2.1 and Lemma 2.1), the flexible robot must tend to stop vibrating and finally be static at the undeformed position. Consequently, the first term in (2.17) is a nonzero constant, the middle term is zero, and the last term approaches zero (note that $f(t)$ is selected such that it is zero when the link is not deformed, as stated in Lemma 2.1). Therefore, the controller *in (2.17) approaches a nonzero constant and thus $\theta \equiv \alpha$ cannot hold. The only possibility is that the flexible link is at the final position $\theta \equiv \theta_f$ without vibrating, which implies the tip regulation is achieved.*

Next, let us show the effectiveness of the proposed controllers through simulation studies. In the simulations, the system parameters are given such values: $L = 1.0\ m$, $EI = 2.0\ Nm^2$, $\rho = 0.1\ Kg/m$, $I_h = 0.5\ Kgm^2$ and $M_t = 0.05\ Kg$. The final joint position is $\theta_f = 0.5\ rad$. The flexible link robot is modeled by a 4-mode AMM model and simulated using an adaptive step size numerical algorithm documented in (Ge *et al.*, 1998*a*).

We shall first give the simulation results of the following joint PD controller

$$\tau_{PD} = -k_p(\theta - \theta_f) - k_v\dot{\theta}$$

The feedback gains k_p and k_v are determined using the method in (Yigit, 1994). Assuming the beam is rigid instead of flexible, i.e., $y(x,t) \equiv 0$, and substituting τ_{PD} into (2.4) yields the closed-loop joint error equation

$$(I_h + \frac{1}{3}\rho L^3 + M_t L^2)\ddot{e}(t) + k_v\dot{e}(t) + k_p e = 0$$

where $e = \theta - \theta_f$. It should be noted that $\ddot{e} = \ddot{\theta}$ and $\dot{e} = \dot{\theta}$ since the final joint position θ_f is a constant. For this $2nd$-order system, if critical damping is assumed ($\xi = 1$) and the natural frequency $\omega_n = 2.5$, we have

$$k_p = (I_h + \frac{1}{3}\rho L^3 + M_t L^2)\omega_n^2$$
$$k_v = 2(I_h + \frac{1}{3}\rho L^3 + M_t L^2)\omega_n$$

and the following PD controller:

$$\tau_{PD1} = -3.65(\theta - \theta_f) - 2.92\dot{\theta}$$

For a faster joint motion, $\omega_n = 5.0$, we have

$$\tau_{PD2} = -14.58(\theta - \theta_f) - 5.83\dot{\theta}$$

For the controller (2.13), k_p and k_v are selected to be the same as in τ_{PD2} for easy comparison. The strain gauge can be attached at $0 \leq x_s < L$ on the beam (note $y''(L,t) = 0$ since the tip payload is assumed to be a concentrated mass). Considering the fact that usually the base strain $y''(0,t)$ is comparatively large, it is used here for feedback control. A comparatively large sensor output will increase the signal to noise ratio and benefit the measurement. The controller is given by

$$\tau = \tau_N = -14.58(\theta - \theta_f) - 5.83\dot{\theta} - k_f y''(0,t) \int_0^t \dot{\theta}(\sigma) y''(0,\sigma) d\sigma$$

Three simulations are carried out for $k_f = 200$, 500 and 1000 and are shown by dashed, solid and dash-dotted lines, respectively. Joint trajectories, tip motions, control torques and the base strain feedback are shown in Fig. 2.2 – Fig. 2.5. The corresponding results of τ_{PD2} are also plotted in these figures by dotted lines for easy comparison.

Comparing with the results of the pure joint PD controller τ_{PD2} (the dotted lines), it can be seen that the system performance has been improved because of the introduction of the k_f item in the sense of smaller tip deflection and less oscillations in the tip motion $p(L,t)$, and simultaneously without reducing the converging speed of joint motion $\theta(t)$.

It was found that the performance is very good when $k_f = 500.0$ shown by the solid lines in comparison with that when $k_f = 200.0$ shown by the dashed lines. The joint motion (Fig. 2.2) is retained to be fast, and the tip of the flexible beam (what we care about most in tip regulation) moves to the final position along a smooth trajectory very fast with only negligible overshoot (Fig. 2.3). When a large gain $k_f = 1000.0$ was used, the system performance (dash-dotted lines) shows no much differences from that of $k_f = 500.0$'s except that the overshoot of $p(L,t)$ becomes visible though it is still very small. The strain feedbacks at the base of the flexible beam, $y''(0,t)$, are given in Fig. 2.5. The strain feedback is introduced to representing the vibration of the flexible beam. From Fig. 2.5, it can be seen that the base bending strain has also been effectively suppressed.

2.3.2 Asymptotic Stable ERC

In this subsection, we shall present a modified controller of (2.17). The modification allows us, *without including any damping terms*, to (i) prove the closed-loop stability in the absence of system dynamics using the same energy relationship (2.11), and (ii) achieve asymptotic convergence of the joint motion and an arbitrarily finite number of flexible modes. In addition, the only restriction on $f(t)$, that it should be zero when the link is not deformed (Lemma 2.1), can be removed. Thus, theoretically the $f(t)$ for the modified controller can be arbitrary any vibration feedback.

For simplicity, the modification is made directly based on the general

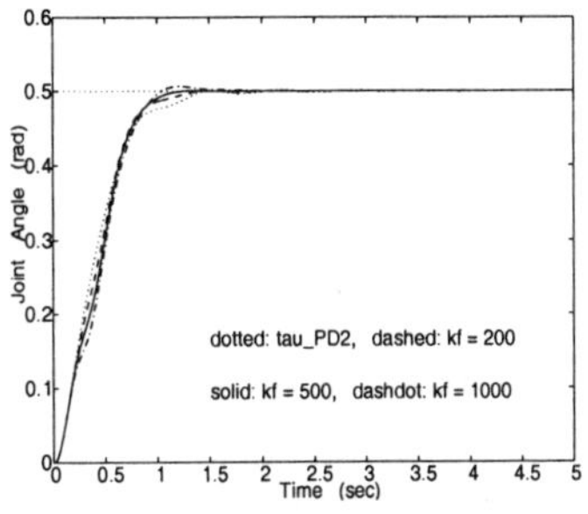

Fig. 2.2 Joint angle trajectories $\theta(t)$

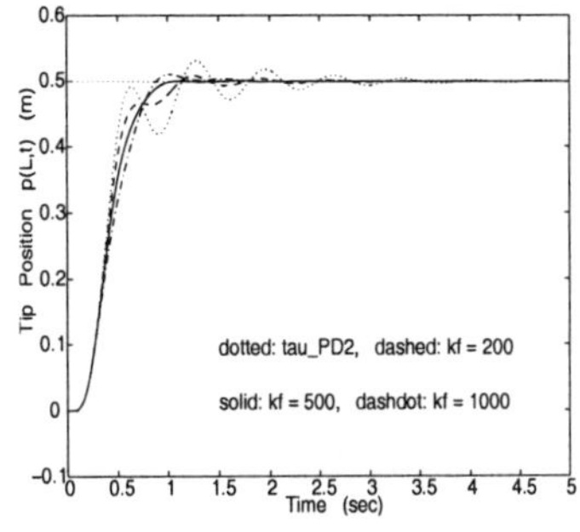

Fig. 2.3 Tip motion trajectories $p(L,t)$

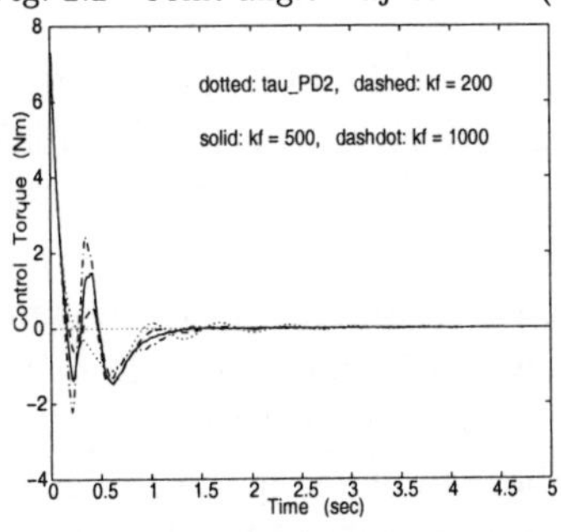

Fig. 2.4 Control signals $\tau(t)$

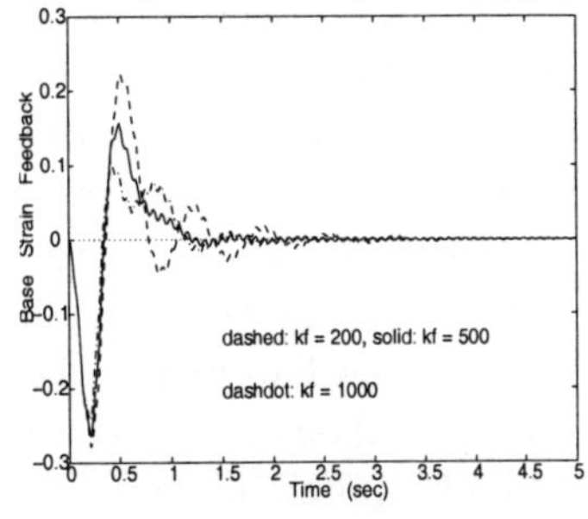

Fig. 2.5 Base strains $y''(0,t)$

controller (2.17). The modified controller is given by

$$\tau = -k_p[\theta - \theta_f] - k_v\dot{\theta} - k_f f(t)\text{sgn}(\dot{\theta}) \int_0^t |\dot{\theta}(\sigma)| f(\sigma) d\sigma \tag{2.18}$$

where $k_p,\ k_v, k_f$ are defined as before, and sgn($\bullet$) is defined by

$$\text{sgn}(\dot{\theta}) = \begin{cases} 1, & \dot{\theta} > 0 \\ 0, & \dot{\theta} = 0 \\ -1, & \dot{\theta} < 0 \end{cases} \tag{2.19}$$

The stability of the system is stated in the following two theorems.

Theorem 2.2 *The closed-loop system described by (2.4), (2.5) and (2.18) is stable (Ge* et al.*, 1997).*

Proof. Consider the following Lyapunov function candidate

$$V(t) = E_k + E_p + \frac{1}{2}k_p[\theta - \theta_f]^2 + \frac{1}{2}k_f \left[\int_0^t |\dot{\theta}(\sigma)| f(\sigma) d\sigma\right]^2 \tag{2.20}$$

Taking time derivative of (2.20), and invoking the energy relationship (2.11), $\dot{V}(t) = -k_v\dot{\theta}^2 \leq 0$ directly follows, thus the closed-loop system is stable. □

Obviously, if we replace $|\dot{\theta}(\sigma)|$ by $\dot{\theta}(\sigma)$ in the last term in (2.20), we arrive at the Laypunov function for controller (2.17). The energy relationship (2.11) again makes us avoid using any dynamics of the system. In this sense, we can similarly conclude that controller (2.18) is "non-model-based" in achieving closed-loop stability. The closed-loop stability holds for the original distributed parameter system, and the problems associated with the truncated-model-based methods mentioned at the beginning of this chapter are essentially avoided.

To prove the asymptotic stability of the original system is difficult, because the original system is of infinite dimensions, to which the LaSalle's Theorem (Khalil, 1992) is not applicable. In (Shifman, 1993), some results due to Hale (Hale, 1969) were used to establish asymptotically stable tracking control of an Euler-Bernoulli beam. However, an assumption was made that there is no hub inertia, i.e., $I_h = 0$, which is not realizable in practice. Here, without this assumption, we will prove that if the deflection of the beam is approximated by a finite number of flexible modes, then these modes, together with the joint motion, can be stabilized by the controller asymptotically .

Theorem 2.3 *The controller (2.18) asymptotically stabilizes the joint motion and an arbitrarily finite number of flexible modes (Ge* et al.*, 1997).*

Proof. See Appendix A. □

Remark 2.5 *The truncation of flexible modes is used only in the final stage of the proof. Without the truncation, i.e. for the infinite dimensional system, even though we can prove that the final position is the largest invariant set in the set of $\dot{V} = 0$, we cannot guarantee that the system motion will converge to the largest invariant set. However, as in the proof above, this can be assured for the truncated finite dimensional system by just directly applying the LaSalle's Theorem (Khalil, 1992). For invariance properties of some infinite dimensional systems, interest readers are referred to (Hale, 1969).*

Remark 2.6 *Although asymptotic stability is proved for only finite number of flexible modes, controller (2.18) is totally different from those truncated-model-based ones in the literature. Controller (2.18) does not make use of any flexible modes feedback. Therefore, (i) the problem of* *spillovers will not affect system's closed-loop stability, and (ii) the order of the controller is independent of the number of modes one would like to consider in stability proof, and the computing burden in realizing the controller can thus be always kept light.*

Remark 2.7 *Similarly as for controller (2.17), the $f(t)$ in (2.18) allows great freedom in feedback design, since it can be chosen depending on available sensor facilities. It should be noted that the proof of Theorem 2.3 does not require $f(t)$ to be zero when the flexible link is undeformed. Therefore, the $f(t)$ in (2.18) can be any vibration function theoretically.*

Computer simulation results are provided below. The system parameters are chosen the same as in the last section. The pure joint PD controller τ_{PD2} given in the last section is also simulated for comparison. Introducing base strain $y''(0,t)$ and tip acceleration $a_{tip}(t)$ feedbacks into τ_{PD2} leads to the following two controllers,

$$\tau_s = -14.6(\theta - \theta_f) - 5.8\dot{\theta} - 500y''(0,t)\text{sgn}(\dot{\theta}) \int_0^t |\dot{\theta}(s)| y''(0,s) ds$$

$$\tau_a = -14.6(\theta - \theta_f) - 5.8\dot{\theta} - 0.6a_{tip}(t)\text{sgn}(\dot{\theta}) \int_0^t |\dot{\theta}(s)| a_{tip}(s) ds$$

The tip deflections $y(L,t)$ and joint motions $\theta(t)$ of τ_{PD2}, τ_s and τ_a are given in Fig. 2.6 and Fig. 2.7. The joint PD controller, though gives quite fast joint motion, exhibits very large deflection. While τ_s and τ_a can effectively damped out the elastic vibration without slowing down the joint motion.

The tip trajectory $p(L,t)$ is plotted in Fig. 2.8. The performance of joint PD control has been greatly improved, in the sense that the tip trajectories corresponding to τ_s and τ_a converge quickly and there are only very little vibration and overshoot. These properties are very important to achieve accurate tip positioning. For completeness, the three control torques are also plotted in Fig. 2.9.

The energy-based control can also be extended to other types of flexible link robots. Asymptotically stable end-point regulation was presented in (Ge *et al.*, 1998*b*) for a flexible SCARA/Cartesian robot.

2.3.3 Adaptive ERC

Although the energy-based controller always guarantees closed-loop stability, it has been shown that the gain of the nonlinear control term plays an important role in obtaining satisfactory performance. It was found that

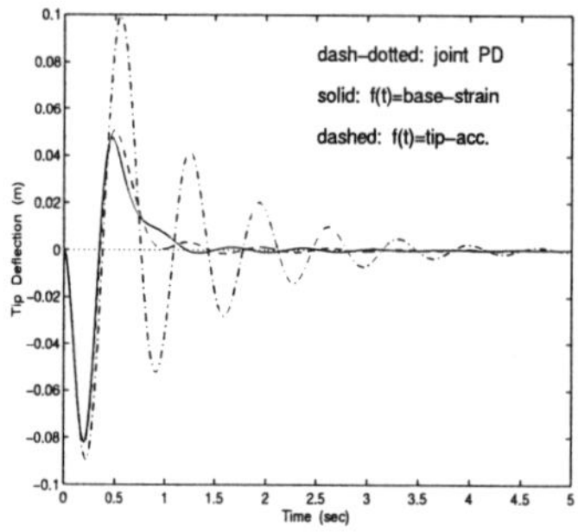

Fig. 2.6 Tip deflections $y(L,t)$

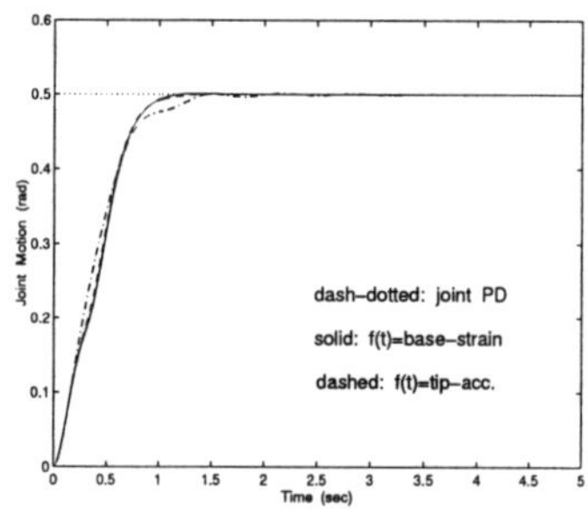

Fig. 2.7 Joint angles $\theta(t)$

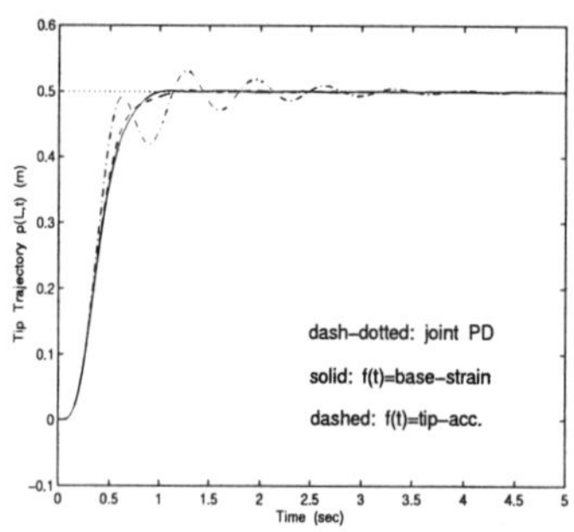

Fig. 2.8 Tip motion trajectories $p(L,t)$

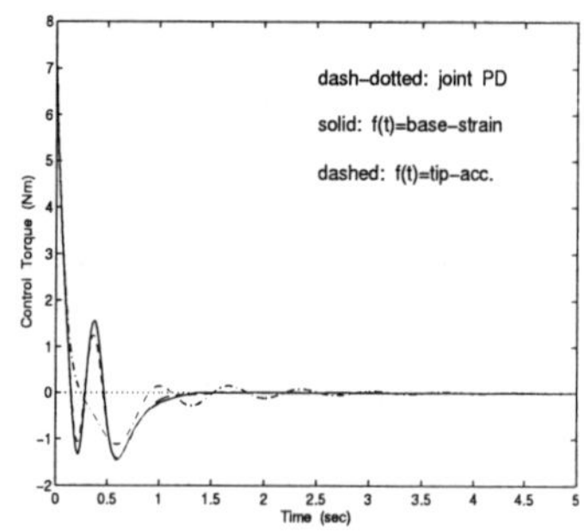

Fig. 2.9 Control torques

when the gain is too small, the control performance is very oscillatory. If the gain is too large, the control action may become not admissible. In addition, the existence of measurement noise also prohibit the use of high gain control. Unfortunately, it is not easy to find optimal gain of this robust term. Therefore, a trade-off has to be made in performance improvement and feedback gain selection by trial and error. An alternative way is to use adaptive techniques thus the control gain can be automatically tuned on-line.

Consider the following control law

$$\tau = -k_p(\theta - \theta_f) - k_v\dot{\theta} - k_s f(t, x_s)\text{sgn}(\dot{\theta}) \int_0^t |\dot{\theta}| f(s, x_s) ds \qquad (2.21)$$

where $k_s = Y_s^2$ and Y_s is adaptively tuned by

$$\dot{Y}_s(t) = \alpha Y_s |\dot{\theta}| f(t, x_s) \int_0^t |\dot{\theta}| f(s, x_s) ds - \sigma Y_s(t) \qquad (2.22)$$

with $x_s \in [0, L]$ being the location of the sensor on the ith link.

Now, we are ready to present the following theorem on the stability of the single-link flexible robot system.

Theorem 2.4 *The proposed control law (2.21) and the adaptation law (2.22) guarantee the stability of the closed-loop single link flexible robotic system.*

Proof. Consider the following Lyapunov function candidate

$$V = E_k + E_p + \frac{1}{2}k_p(\theta - \theta_f)^2 + \frac{1}{2}\alpha^{-1}Y_s^2 \tag{2.23}$$

By virtue of equation (2.12), the time derivative of V is given by

$$\dot{V} = \tau\dot{\theta} + k_p(\theta - \theta_f)\dot{\theta} + \alpha^{-1}Y_s\dot{Y}_s$$

Substituting the control law (2.21) and the adaptation law (2.22) into the above equation, we have

$$\begin{aligned} \dot{V} &= -k_v\dot{\theta}^2 - \alpha^{-1}\sigma Y_s^2 \\ &\leq 0 \end{aligned} \tag{2.24}$$

It follows that $0 \leq V(t) \leq V(0)$, $\forall t \geq 0$, hence $V(t) \in L_\infty$. Since $V(t)$ is a continuous function of Y_s, $V(t)$ is nonincreasing in t, which implies the boundedness of Y_s and, hence, the boundedness of k_s.

In addition, it also implies that the closed-loop system is energy dissipative and, hence, stable. □

Next, numerical simulations are provided to verify the presented adaptive ERC. For comparison, the pure joint PD controller τ_{PD2} given in Section 2.3.1 and the asymptotic stable ERC τ_s given in Section 2.3.2 are simulated again. Introducing base strain $y''(0,t)$ feedback into τ_{PD2} leads to the following ERC control law

$$\tau_{ad} = -14.6(\theta - \theta_f) - 5.8\dot{\theta} - k_s y''(0,t)\text{sgn}(\dot{\theta})\int_0^t |\dot{\theta}(s)|y''(0,s)ds \tag{2.25}$$

and the adaptive law for the control parameter k_s is

$$\begin{aligned} k_s &= Y_s \\ \dot{Y}_s(t) &= \alpha Y_s|\dot{\theta}|y''(0,t)\int_0^t |\dot{\theta}|y''(0,s)ds - \sigma Y_s(t) \end{aligned} \tag{2.26}$$

In the simulation, the system parameters are chosen the same as in the subsection 2.3.2. Other control parameters are selected as: $k_s(0) = 320$,

$\alpha = 20$ and $\sigma = 0.1$ for adaptive ERC and $k_f = 500$ for asymptotic stable ERC. The tip deflections $y(L,t)$ and joint motions $\theta(t)$ of τ_{PD2}, τ_s and τ_{ad} are given in Fig. 2.10 and Fig. 2.11. It can be seen that the joint PD controller exhibits very large deflection, while both τ_s and τ_{ad} can effectively damped out the elastic vibration without slowing down the joint motion. Fig. 2.12 shows the changes of k_s for the adaptive ERC controller. It can be seen that k_s are relatively high at the beginning, then with the residual vibrations become small, k_{si} also become small. This is meaningful in engineering practice as well. For completeness, the three control torques are also plotted in Fig. 2.13.

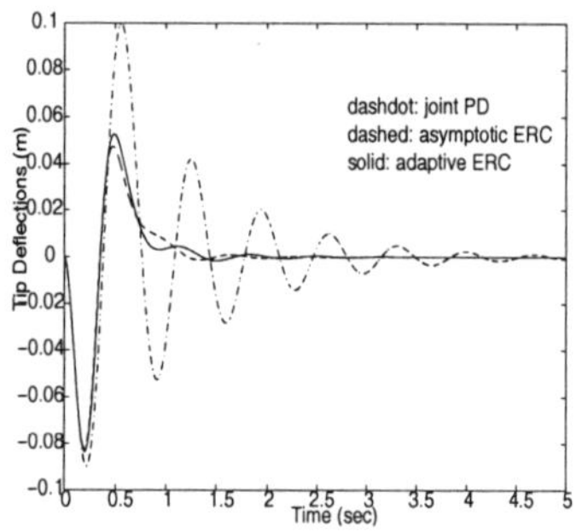

Fig. 2.10 Tip deflections $y(L,t)$

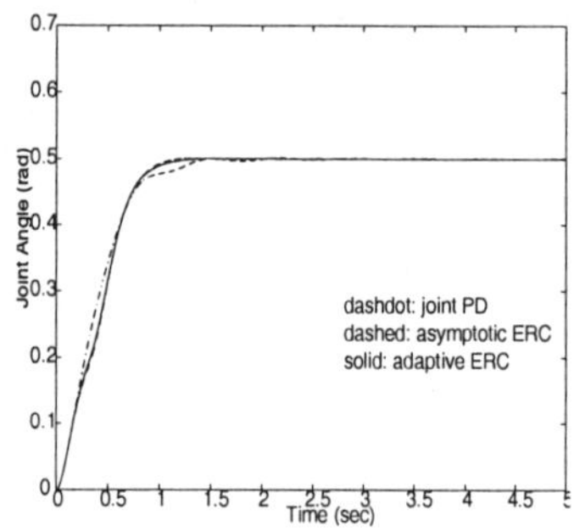

Fig. 2.11 Joint angles $\theta(t)$

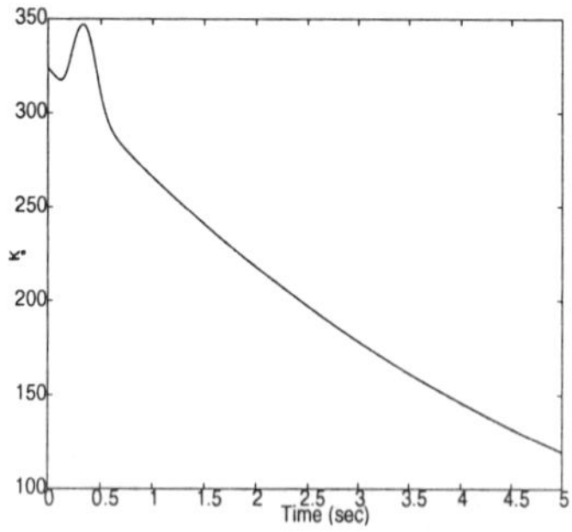

Fig. 2.12 Adaptively tuned k_s

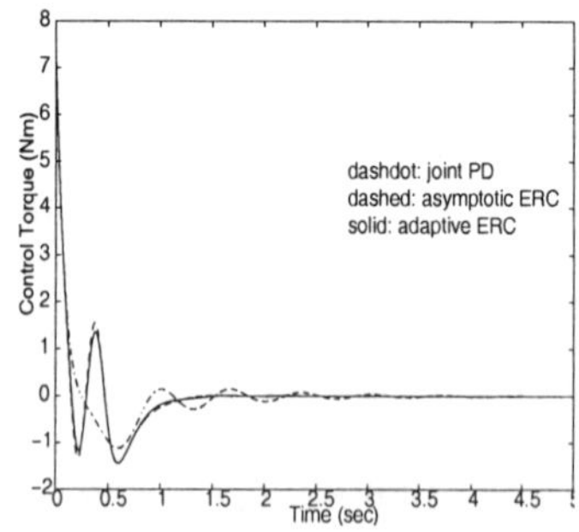

Fig. 2.13 Control torques

2.3.4 Genetic Algorithm Tuning of ERC

In subsection 2.3.3, adaptive ERC is presented for online tuning of the controller parameter. In this subsection, inspired by the natural search and selection processes leading to the survival of the fittest individuals, we shall presented the effectiveness of genetic algorithm (GA) in tunning the parameters for optimal performance.

Since the optimization of the fitness functions is the basic goal of GA, different fitness functions will produce, in general, different optimization results. The objective of the optimization here is to obtain better tip motion performance of the single-link flexible robot, i.e., to find a set of feedback gains with which the controller can drive the tip of the flexible beam to a pre-defined position as fast as possible and with minimal overshoot/oscillation.

If the final position is given by $p(L,t) = L\theta_f$, then the following are two appropriate choices of fitness functions:

$$ISE = \int_0^T [L\theta_f - p(L,t)]^2 \, dt \tag{2.27}$$

$$ITAE = \int_0^T t|L\theta_f - p(L,t)|dt \tag{2.28}$$

where *ISE* means the Integral of Squared Errors and *ITAE* is the Integral Time-multiplied Absolute value of Errors (Schultz and Rideout, 1961). Since overshoot of the tip trajectory is quite undesirable in tip position control of flexible robots, the following two Modified *ISE* and Modified *ITAE* fitness functions maybe more appropriate:

$$MISE = \int_0^T K\,[L\theta_f - p(L,t)]^2 \, dt \tag{2.29}$$

$$MITAE = \int_0^T Kt|L\theta_f - p(L,t)|dt \tag{2.30}$$

where the penalty weight K is defined as

$$K = \begin{cases} 1.0, & when \;\; |p(L,t)| \leq L\theta_f \;\; (no \;\; overshoot) \\ 10.0, & when \;\; |p(L,t)| > L\theta_f \;\; (overshoot) \end{cases} \tag{2.31}$$

i.e., a large weight has been put on the overshoot of the tip performance. For the four fitness functions (2.27)–(2.30), the integrals are taken within the evaluation time interval $[0,\ T]$. Obviously, the smaller the values of the four fitness functions, the better tip motion performances we can achieve.

It has been shown in Section 2.3.1 that the following simple tip position controller

$$\tau = -k_p[\theta(t) - \theta_f] - k_v\dot{\theta}(t) - k_f y''(x_s, t)\int_0^t \dot{\theta}(s)y''(x_s, s)ds \qquad (2.32)$$

can guarantee the system to be stable, where $k_p,\ k_v,\ k_f\ >\ 0$ are three controller parameters needed to be tuned. To further simplify the genetic optimization process, the set of tuning parameters ($k_p,\ k_v,\ k_f$) is reduced to ($\omega_n,\ k_f$), where ω_n is the natural frequency of the following $2nd$-order system

$$I_{eq}\ddot{e}(t) + k_v\dot{e}(t) + k_p e = 0 \qquad (2.33)$$

Since θ_f is constant, we have $\ddot{e} = \ddot{\theta}$ and $\dot{e} = \dot{\theta}$. For the $2nd$-order system (2.33), if critical damping ($\xi = 1$) is assumed for fastest overshoot free performance of θ, the feedback gains k_p and k_v are given by

$$\begin{cases} k_p = I_{eq}\cdot\omega_n^2 \\ k_v = 2I_{eq}\cdot\omega_n \end{cases} \qquad (2.34)$$

When I_{eq} is known, we shall consider only ω_n and k_f instead of $k_p,\ k_v$ and k_f to simplify the genetic optimization process. If I_{eq} is unknown, the parameters k_p, k_v and k_f can then be simultaneously tuned. As a matter of fact, it is very desirable to have less tuning parameters for actual engineering implementation. Although (2.34) actually puts an extra constraint on the optimization, it is shown that comparatively good joint motion can be obtained while the computational cost is reduced.

Genetic algorithm generates a sequence of populations by using a selection mechanism, and use crossover and mutation as search mechanism. It relies on crossover, a mechanism of probabilistic and useful exchange of information among solutions, to locate better solutions, see (Ge *et al.*, 1996*b*) for the detailed algorithms.

For the system "parametried" by $L = 1.0\ m$, $EI = 2.0\ Nm^2$, $\rho = 0.1\ Kg/m$, $I_h = 0.5\ Kgm^2$ and $M_t = 0.05\ Kg$, we reproduce the results of GA optimization (Ge *et al.*, 1996*b*) here to show (i) the effectiveness of GA tunning, and (ii) the elegant performance of the simple and effective controller. The evaluation time interval is set to be $0 \leq t \leq 2\ s$. From the engineering point of view, a convergence decision is needed to terminate the GA operations when the result is good enough. For this consideration, the following termination decisions are introduced: (i) if the considered fitness function does not change much (say less than $T_M = 0.005$) in a certain number of generations (say $G_N = 50$ generations), we assume that the optimization has converged and the process is terminated; and (ii) otherwise, the process shall be terminated at a the maximum number of generations (say $G_M = 500$). Four optimization experiments, corresponding to the four fitness functions (2.27)–(2.30), are carried out using the same set of initial population which is generated randomly. The terminating generations of the four fitness functions are 253 (*ISE*), 247 (*MISE*), 235 (*ITAE*) and 212 (*MITAE*). Figure 2.14 shows the feedback gains k_p and k_v (obtained from (2.34)) in accordance with the best individuals. It can be noted that k_p and k_v with respect to the fitness functions *ISE* and *MISE* (Figure 2.14 (b)) are very close to each other. The corresponding k_f's are plotted in Figure 2.15. The best values of the four fitness functions are shown in Figure 2.16. From Figure 2.16, one can observe that better performance (corresponding to smaller values of the fitness functions) will be obtained with the progression of the GA optimization processes. Tip motion trajectories, $p(L,t)$, are plotted in Figure 2.17 and Figure 2.18. It can be seen from these figures that tip motion performance is improved as the number of generations increases. Finally, the best tip performances (i.e., the results of the last generations) with respect to the four fitness functions are plotted together in Figure 2.19. The "best" control performance is unbelievably good, the tip settles down in less than 1 second, a control performance that other complicated controllers may not be able to achieve.

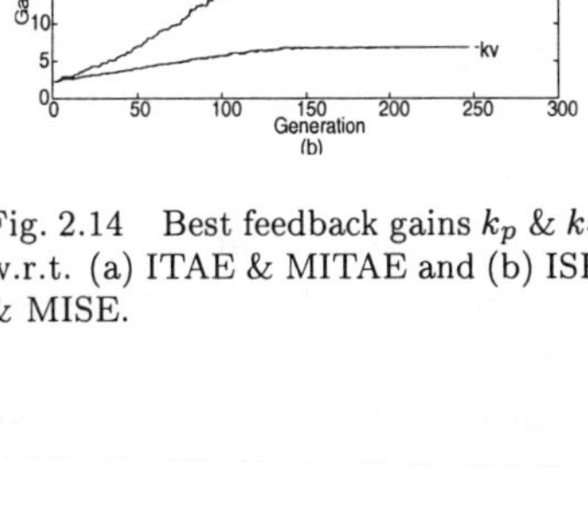

Fig. 2.14 Best feedback gains k_p & k_v w.r.t. (a) ITAE & MITAE and (b) ISE & MISE.

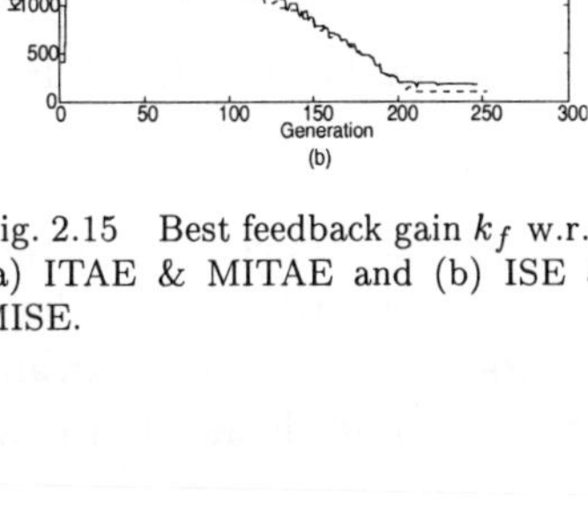

Fig. 2.15 Best feedback gain k_f w.r.t. (a) ITAE & MITAE and (b) ISE & MISE.

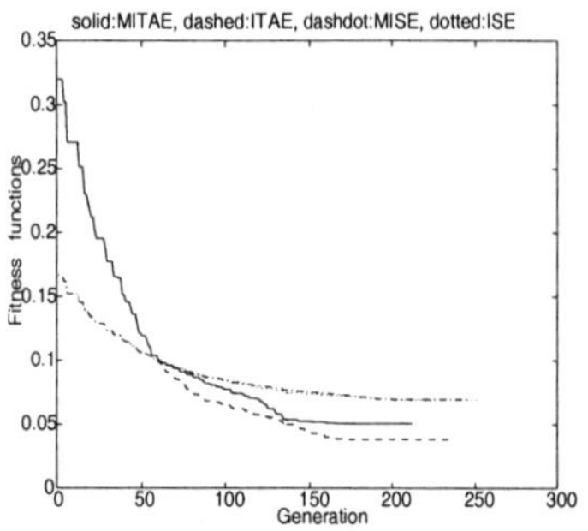

Fig. 2.16 Best (smallest) values of ITAE, MITAE, ISE & MISE.

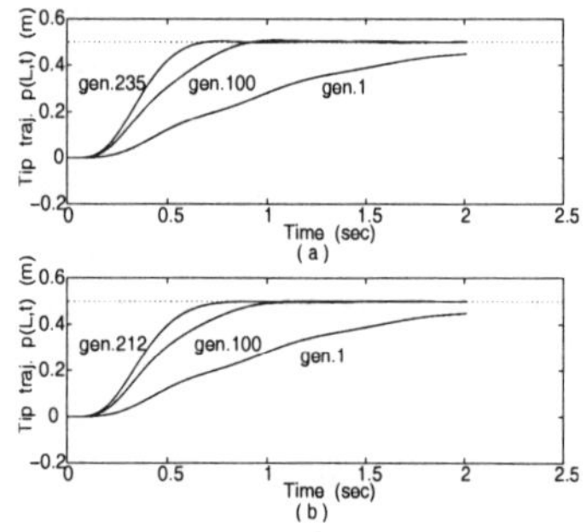

Fig. 2.17 Best tip performance of Generation 1, 100 and the final generation w.r.t (a) ITAE and (b) MITAE.

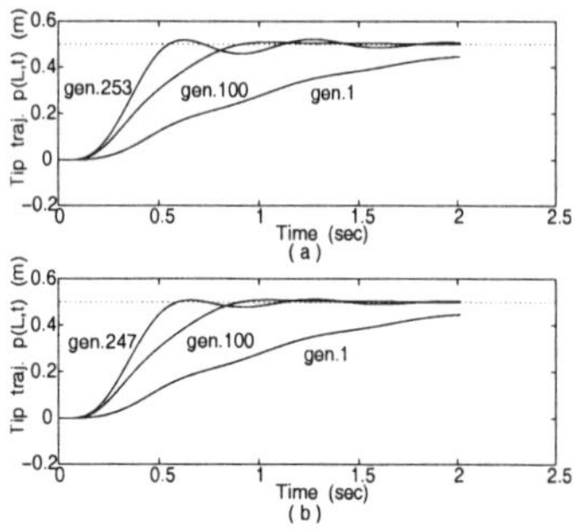

Fig. 2.18 Best tip performance of Generation 1, 100 and the final generation w.r.t (a) ISE and (b) MISE.

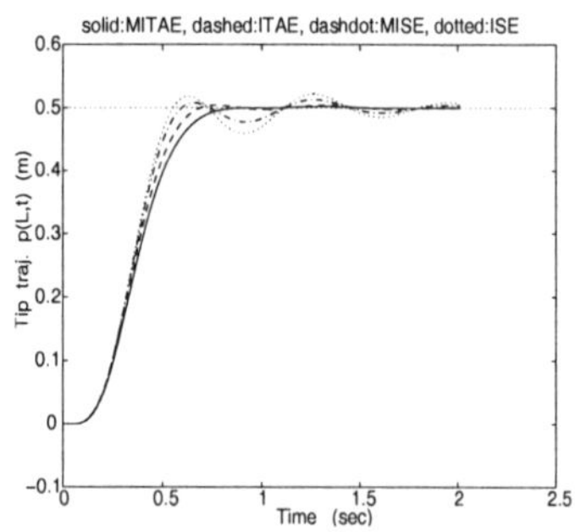

Fig. 2.19 Best tip performance w.r.t. ITAE, MITAE, ISE & MISE.

2.4 Multi-link Flexible Robots

Dynamic modelling and control of multi-link flexible robots are much more difficult and complex in general. If the truncated model of finite dimension is used for control design, the complexity and order of the corresponding controller increases dramatically as the order of the model increases. The extension from single link to multi-link is usually not that straightforward.

However, the approach presented based on the system energy-and-work relationship applied to systems of any number of links, and the results obtained in Section 2.3 can be easily extended to multi-link cases without the influence of gravity.

2.4.1 Dynamic Properties of Multi-link Flexible Robots

We shall consider the flexible robots (i) deployed in space, or (ii) moving in the horizontal plane. In both cases, the effect of gravity is ignored. The N links are connected using N motors. Motor 1 is fixed in position, which is at the origin of the fixed base frame $X_1O_1Y_1$. The remaining motors, each being supported by a roller, are movable on the platform. The free tip of the last link has a point mass payload attached.

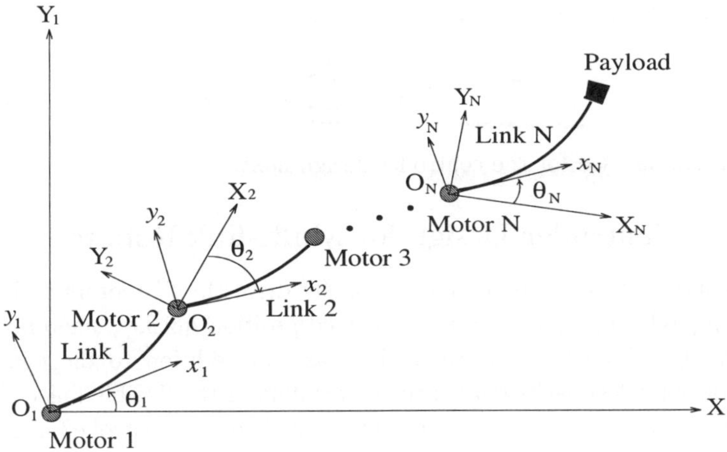

Fig. 2.20 Geometry of the multi-link flexible robot

For clarity, the geometry of the robot is shown in Fig. 2.20. There are in total $2N$ frames being used to describe the system, i.e., $X_iO_iY_i$ and $x_iO_iy_i$, $i = 1, 2, \cdots, N$. Frame $X_1O_1Y_1$, as stated above, is the fixed base frame. Other frames are all local reference frames attached to the corresponding motors, specifically axis O_iX_i ($i = 2, 3, \cdots, N$) is defined as the tangent to the end tip of link $i-1$, and axis O_ix_i ($i = 1, 2, \cdots, N$) is tangent to link i at its base. The angular position of the ith link is denoted by θ_i measured in frame $X_iO_iY_i$. θ_i is actually the angular difference between frames $x_iO_iy_i$

and $X_iO_iY_i$.

For the system described above, the total work done by external inputs is given by

$$W = \sum_{i=1}^{N} \int_0^t \tau_i(t)\dot{\theta}_i(t)dt \tag{2.35}$$

From the energy-work relationship, i.e., the increment of the system energy is equal to the work done by external inputs, we have the following equation

$$[E_k(t) + E_p(t)] - [E_k(0) + E_p(0)] = \sum_{i=1}^{N} \int_0^t \dot{\theta}_i(t)\tau_i(t)dt \tag{2.36}$$

where $E_k(t)$ and $E_p(t)$ are the total kinetic energy and total potential energy of system at time t, $E_k(0)$ and $E_p(0)$ are constants representing the kinetic and potential energies at time 0. Therefore, by taking time derivatives of both sides of equation (2.36), we arrive at

$$\dot{E}_k(t) + \dot{E}_p(t) = \sum_{i=1}^{N} \dot{\theta}_i(t)\tau_i(t) \tag{2.37}$$

which will be used for the controller design next.

2.4.2 Controller Design for Multi-link Robots

The control objective is to simultaneously drive the N motors such that the tip payload is moved to a pre-defined position quickly, smoothly and accurately. Similar to the single-link case, in addition to the joint PD control, some feedbacks related to the bending of the flexible links will be introduced into the controller, and thus provide direct control effort on the elastic vibrations.

Now, we are ready to present the following theorem on the stability of the multi-link flexible robot system.

Theorem 2.5 *The closed-loop multi-link flexible robot system is stable if the N control torques are given by*

$$\tau_i = -k_{pi}[\theta_i(t) - \theta_{fi}] - k_{vi}\dot{\theta}_i - k_i f_i(t) \int_0^t \dot{\theta}_i(s) f_i(s) ds \tag{2.38}$$
$$i = 1, 2, \cdots, N$$

where $k_{pi} > 0$, $k_{vi} > 0$, $k_i \geq 0$; θ_{fi} is the constant final position of the ith link; $f_i(t)$ is the variable related to the bending of the ith link, or any *combination of bending variables of the ith link. The condition on choosing $f_i(t)$ is that they must be zero when the ith link undergoes no deformation (Ge* et al., *1996*a*).*

Proof. Considering the Lyapunov function candidate

$$V = E_k + E_p + \frac{1}{2}\sum_{i=1}^{N} k_{pi}[\theta_i(t) - \theta_{fi}]^2 + \frac{1}{2}\sum_{i=1}^{N} k_i \left[\int_0^t \dot{\theta}_i(s) f_i(s) ds\right]^2 \tag{2.39}$$

where E_k and E_p are the total kinetic energy and total potential energy of the system.

Recalling (2.37), the time derivative of energy function V is given by

$$\dot{V} = \sum_{i=1}^{N} \tau_i \dot{\theta}_i(t) + \sum_{i=1}^{N} k_{pi}\dot{\theta}_i[\theta_i(t) - \theta_{fi}] + \sum_{i=1}^{N} k_i \dot{\theta}_i(t) f_i(t) \int_0^t \dot{\theta}_i(s) f_i(s) ds \tag{2.40}$$

Substituting (2.38) into (2.40) yields

$$\dot{V} = -\sum_{i=1}^{N} k_{vi}\dot{\theta}_i^2 \tag{2.41}$$

which is negative semi-definite. Hence, the closed-loop system is stable. □

The controller possesses similar advantages as those described in Remarks 2.1-2.3 for single link case. Furthermore, we have the following remarks to make:

Remark 2.8 *When the bending functions $f_i(t)$ are chosen in the local reference frame $x_iO_iy_i$, such as those afore-mentioned, the controllers presented in (2.38) are of decentralized type, which has the advantages of requiring low computing power, and giving ease of implementation and tolerance to failure, since the N controllers in (2.38) can be implemented in parallel.*

Remark 2.9 *In Theorem 2.5, we only claim the closed-loop stability of the system. To prove the asymptotic stability of the system is not easy due to the infinite dimensionality of the system. In the following, instead of giving rigorous proof, we shall show that practically the flexible robot can only possibly stop at $\theta_i = \theta_{fi}$ $(i = 1, 2, \cdots, N)$ without vibrating. Assuming* *that the N links stop at the position $\theta_i \equiv \alpha_i$ (hence $\dot{\theta}_i \equiv 0$ for $i = 1, 2, \cdots, N$) with $\alpha_i \neq \theta_{fi}$, thus there is no energy input to the system since $\dot{\theta}_i \equiv 0$. Due to the existence of internal structural damping in a flexible link in practice (structural damping is neglected in the proof of Theorem 2.5), the links must tend to stop vibrating and finally be static at the undeformed position. Consequently, the first term in τ_i is a nonzero constant, the middle term is zero, and the last term approaches zero (note that $f_i(t)$ are zero when the link i is not deformed). Therefore, τ_i approaches a nonzero constant and thus $\theta_i \equiv \alpha_i$ cannot hold. The only possibility is that the flexible link is at the final position $\theta_i \equiv \theta_{fi}$ without vibrating, which implies the tip regulation can be practically achieved.*

In the following, some numerical simulations are provided for a two-link flexible robot. The plant is simulated by a FEM model in which each link is divided into 4 elements with the same length.

The system parameters are given in Table 2.1, in which M_{t1} denotes the mass of the second motor, and M_{t2} is the payload attached to the end tip of the second Link.

Table 2.1 System Parameters

	Link 1	Link 2
Length	$L_1 = 1.0\ m$	$L_2 = 0.8\ m$
Flexural Rigidity	$EI_1 = 5.0\ Nm^2$	$EI_2 = 3.0\ Nm^2$
Linear Density	$\rho_1 = 0.1\ Kg/m$	$\rho_2 = 0.1\ Kg/m$
Hub Initial	$I_{h1} = 3.0\ Kgm^2$	$I_{h2} = 1.5\ Kgm^2$
Payload	$M_{t1} = 0.1\ Kg$	$M_{t2} = 0.05\ Kg$

The initial position and the final position of the robot are described in Fig. 2.21, i.e., the initial joint positions of the both links are all zeros, and the set point values of the two links are $\theta_{f1} = 20^o$ and $\theta_{f2} = 10^o$.

Similarly, a set of simulations of pure joint PD control are also carried out for comparison purpose. The joint PD controller is given by

$$\tau_{PDi} = -k_{pi}(\theta_i - \theta_{fi}) - k_{vi}\dot{\theta}_i, \quad i = 1, 2$$

From Theorem 2.5, any $k_{pi}, k_{vi} > 0$ will not destabilize the closed-loop system. Let $\omega_{n1} = 2.5$ and $\omega_{n2} = 3.0$, and the two joint PD controllers are given by (Ge *et al.*, 1996*a*)

$$\tau_{PD1} = -21.6(\theta_1 - \theta_{f1}) - 17.3\dot{\theta}_1 \quad (2.42)$$

$$\tau_{PD2} = -14.0(\theta_2 - \theta_{f2}) - 9.3\dot{\theta}_2 \quad (2.43)$$

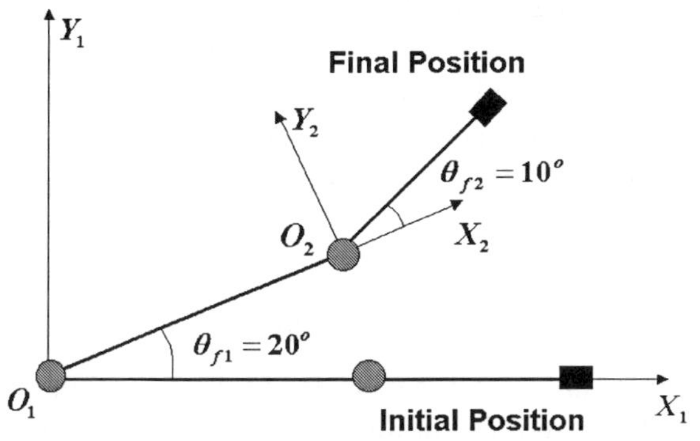

Fig. 2.21 Two-link flexible robot

Considering that strain gauges have been very widely used in control of flexible robots, we assume that they are available and base strain of each link is used for feedback in the following simulations. From (2.38), we have the following controllers:

$$\tau_1 = \tau_{PD1} - 1800 y_1^{''}(0,t) \int_0^t \dot{\theta}_1(s) y_1^{''}(0,s) ds \tag{2.44}$$

$$\tau_2 = \tau_{PD2} - 1800 y_2^{''}(0,t) \int_0^t \dot{\theta}_2(s) y_2^{''}(0,s) ds \tag{2.45}$$

where τ_{PD1} and τ_{PD2} are also given in (2.42) and (2.43). Let $k_1 = k_2 = 32000$, and $f_1(t) = y_1^{''}(0,t)$ and $f_2(t) = y_2^{''}(0,t)$ which are the base strains of link 1 and link 2, respectively. It should be noted that the values of k_1 and k_2 given above may not be optimal. The simulations here are only used to show the effectiveness of the ERC approach.

The tip deflections, joint motions and tip trajectories in this case are shown in Fig. 2.22–Fig. 2.24. The results of ERC, compared with that of joint PD control, exhibit the similar improvements to that in Case 1, i.e., the tip deflections are suppressed effectively, the joint motions are very smooth, and the tip trajectories are converging fast with only very little vibrations and overshoots. The base strain feedback signals $y_1^{''}(0,t)$ and $y_2^{''}(0,t)$ are shown in Fig. 2.25, in which the base strains of joint PD control are also plotted with dashed lines for comparison. Finally, the corresponding control efforts and joint velocities are given in Fig. 2.26 and Fig. 2.27 for completeness.

In the above, through simulations, we have shown the effectiveness of the controller in (2.38). It should be pointed out that the simulations presented here are not complete, and other kinds of feedbacks and/or other kinds of combinations of feedbacks can also be considered depending on the available sensor facilities, and controller (2.38) actually allows great freedom of feedback design.

2.4.3 Other Variations of ERC

Similar to the single-link case, the vibration feedback $f_i(t)$ in the above ERC is also required to be zero for an undeformed link i, such that asymptotic control can be achieved in practice (Remark 2.9). It has been shown that by replacing the joint velocity feedback by its absolute value, this restriction on vibration feedback can be removed in the improved version of single-link ERC. As a matter of fact, there also exists a similar improved version for the multi-link ERC. The improved controller is able to (i) guarantee the closed-loop stability; and (ii) achieve asymptotic convergence of the joints motion, and an arbitrarily finite number of flexible modes of each link. At the same time, there will be no restriction on choosing the vibration feedback. Moreover, the stability proof needs only the PDEs of a single-link flexible robot, and the PDEs of the multi-link flexible robot (which are very complicated) are not necessary.

A theorem similar to Theorem 2.5 on the closed-loop stability exists.

Theorem 2.6 *The closed-loop multi-link flexible robot system is stable if the* N *control torques are given by*

$$\tau_i = -k_{pi}[\theta_i(t) - \theta_{fi}] - k_{vi}\dot{\theta}_i - k_i \mathrm{sgn}(\dot{\theta}_i) f_i(t) \int_0^t |\dot{\theta}_i(s)| f_i(s) ds \quad (2.46)$$
$$i = 1, 2, \cdots, N,$$

where $k_{pi}, k_{vi} > 0$, $k_i \geq 0$, θ_{fi} *is the constant final position of the* i*th link,* $\dot{\theta}_i$ *is the time derivative of* θ_i, $f_i(t)$ *is the extra feedback introduced for link* i *and will be defined later, and the signum function is defined as in (2.19).*

Proof. See proof of Theorem 1 in (Ge *et al.*, 1997). □

Similar to Section 2.3, the control gain of the robust term in the energy-based controller for the multi-link flexible robots can be tuned adaptively on line instead of keeping it constant as in (Ge *et al.*, 1997). Thus, a general adaptive energy-based controller for the multi-link flexible robots is given

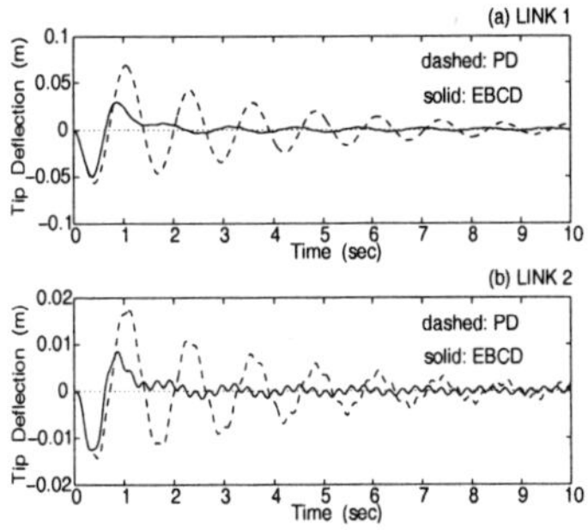

Fig. 2.22 Tip deflections: (a) $y_1(L_1, t)$ and (b) $y_2(L_2, t)$

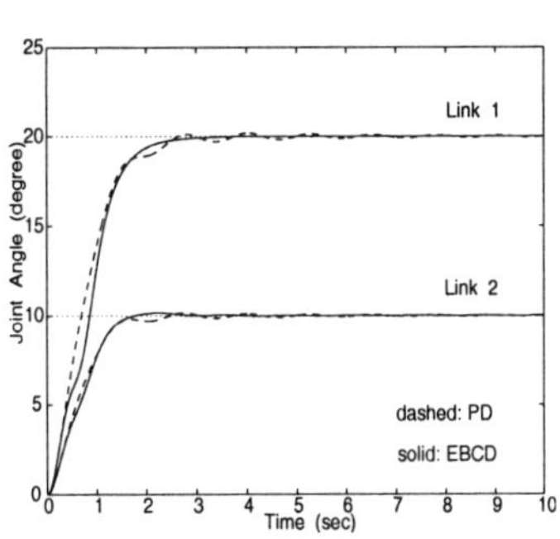

Fig. 2.23 Joint angles

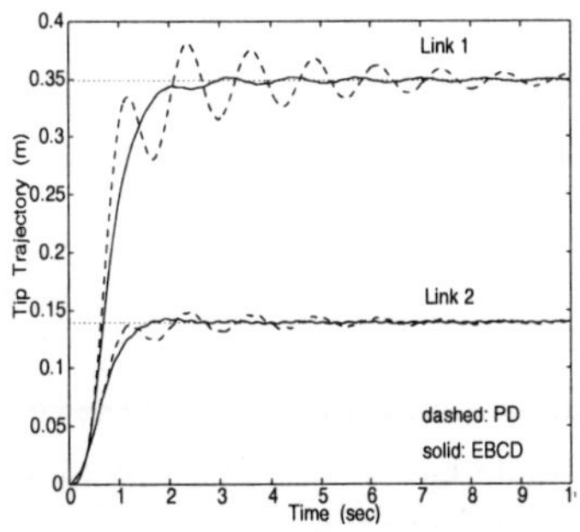

Fig. 2.24 Tip trajectories

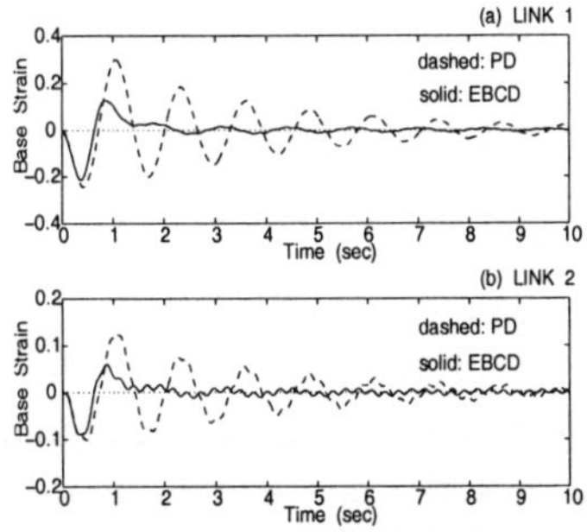

Fig. 2.25 Base strains: (a) $y_1^{''}(0, t)$ and (b) $y_2^{''}(0, t)$

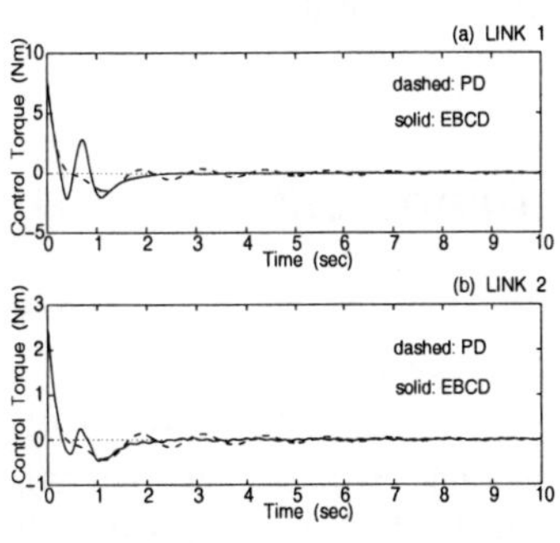

Fig. 2.26 Control efforts

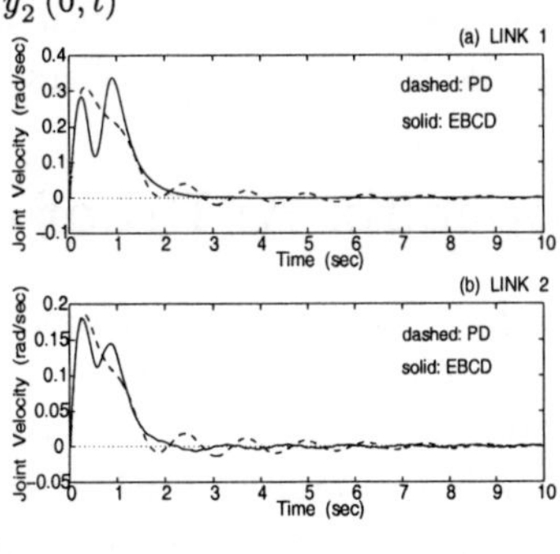

Fig. 2.27 Joint velocities: (a) $\dot{\theta}_1$ and (b) $\dot{\theta}_2$

by

$$\tau_i = -k_{pi}(\theta_i - \theta_{di}) - k_{di}\dot{\theta}_i - \sum_{j=1}^{m_i} k_{sij} f_i(t, x_{sij}) sgn(\dot{\theta}_i) \int_0^t |\dot{\theta}_i| f_i(s, x_{sij}) ds \qquad (2.47)$$

where $k_{sij} = Y_{sij}^2$ and Y_{sij} is adaptively tuned by

$$\dot{Y}_{sij}(t) = \alpha_{ij} Y_{sij} |\dot{\theta}_i| f_i(t, x_{sij}) \int_0^t |\dot{\theta}_i| f_i(s, x_{sij}) ds - \sigma_{ij} Y_{sij}(t) \tag{2.48}$$

with $j = 1, 2, \ldots, m_i$, $x_{sij} \in [0, L_j]$ being the location of the jth sensor on the ith link, and $i = 1, 2, \ldots, N$. The stability of the closed-loop system can be shown easily by choosing the following Lyapunov function candidate

$$V = E_k + E_p + \frac{1}{2}\sum_{i=1}^{N} k_{pi}(\theta_i - \theta_{di})^2 + \frac{1}{2}\sum_{i=1}^{N}\sum_{j=1}^{m_i} \alpha_{ij}{}^{-1} Y_{sij}^2 \tag{2.49}$$

The readers are referred to (Lee *et al.*, 2001) for details.

2.5 Conclusion

In this Chapter, energy-based control strategies has been presented for both single-link and multi-link flexible robots. Theoretical proofs have shown that closed-loop system is stable, and the energy-based controllers are independent of system parameters and hence possess stability robustness to parameter variations. Furthermore the controllers can be easily implemented because the signals used in the feedback control can be measured directly or be chosen as measurable ones. Numerical simulations have shown that system response converges fast and the residue vibrations are effectively suppressed using the controllers proposed.

Appendix A: Proof of Theorem 2.3

Proof. The proof of Theorem 2.3 can be done by following the same procedure in (Ge *et al.*, 1997). It is given here for completeness.

For the Laypunov function (2.20), let us consider the motion of the system in the largest invariant set in the set of $\dot{V} = 0$, i.e., when $\dot{V} \equiv 0$.

When $\dot{V} \equiv 0$, we have $\dot{\theta} \equiv 0$ and subsequently $\ddot{\theta} = 0$. Controller (2.18) becomes

$$\tau = -k_p[\theta - \theta_f]$$

which is constant. Equations (2.4) and (2.5) are reduced to

$$EIy''(0, t) - k_p[\theta - \theta_f] = 0 \tag{2.50}$$

$$\rho\ddot{y}(x, t) = -EIy''''(x, t) \tag{2.51}$$

The corresponding boundary conditions (2.6)-(2.9) become

$$y(0,t) = 0 \tag{2.52}$$
$$y'(0,t) = 0 \tag{2.53}$$
$$y''(L,t) = 0 \tag{2.54}$$
$$EIy'''(L,t) = M_t\ddot{y}(L,t) \tag{2.55}$$

The solution of (2.51) under boundary conditions (2.52)-(2.55) is given by

$$y(x,t) = \sum_{i=1}^{\infty} \phi_i(x) q_i(t) \tag{2.56}$$

where

$$\phi_i(x) = A_i\bar{\phi}_i \tag{2.57}$$
$$q_i(t) = B_i \cos\omega_i t + D_i \sin\omega_i t \tag{2.58}$$

with

$$\bar{\phi}_i = \cosh\frac{\beta_i x}{L} - \cos\frac{\beta_i x}{L} - \gamma_i\left(\sinh\frac{\beta_i x}{L} - \sin\frac{\beta_i x}{L}\right)$$

The coefficients and variables are defined: $A_i = \left[\frac{1}{\int_0^L \bar{\phi}_i^2(x)dx + \frac{M_t}{\rho}\bar{\phi}_i^2(L)}\right]^{\frac{1}{2}}$, β_i are positive solutions of $1+\cosh\beta\cos\beta + \frac{M_t\beta}{\rho L}(\sinh\beta\cos\beta - \cosh\beta\sin\beta) = 0$, γ_i is given by $\gamma = \frac{\cosh\beta+\cos\beta}{\sinh\beta+\sin\beta}$ with corresponding β_i, and $\omega_i = \frac{\beta_i^2}{L^2}\sqrt{\frac{EI}{\rho}}$.

Let $y_0(x,t_0)$ and $\dot{y}_0(x,t_0)$ denote the "initial" deflection and velocity profiles of the flexible beam (note: t_0 here is not zero. Since we are considering the motion of the system in the largest invariant set in the set of $\dot{V}=0$, the "initial" moment t_0 should denote the moment when the system enters the invariant set, instead of the initial operating moment), we have

$$y_0(x,t_0) = \sum_{i=1}^{\infty} \phi_i(x)(B_i \text{cos}\omega_i t_0 + D_i \text{sin}\omega_i t_0) \tag{2.59}$$
$$\dot{y}_0(x,t_0) = \sum_{i=1}^{\infty} \phi_i(x)(-B_i \text{sin}\omega_i t_0 + D_i \text{cos}\omega_i t_0)\omega_i \tag{2.60}$$

Define

$$I_{j,1} := \int_0^L y_0(x,t_0)\phi_j(x)dx + \frac{M_t}{\rho}y_0(L,t_0)\phi_j(L)$$

$$I_{j,2} := \frac{1}{\omega_j}\left[\int_0^L \dot{y}_0(x,t_0)\phi_j(x)dx + \frac{M_t}{\rho}\dot{y}_0(L,t_0)\phi_j(L)\right]$$

From the Expansion Theorem in (Meirovitch, 1975), we know that the infinite summation in (2.56) is *absolutely and uniformly* convergent. This implies that if we substitute (2.59) and (2.60) into $I_{j,1}$ and $I_{j,2}$, the corresponding integrals can be calculated termwise, i.e.,

$$\begin{aligned} I_{j,1} &= \int_0^L y_0(x,t_0)\phi_j(x)dx + \frac{M_t}{\rho}y_0(L,t_0)\phi_j(L) \\ &= \sum_{i=1}^{\infty}(B_i\text{cos}\omega_i t_0 + D_i\text{sin}\omega_i t_0)\left[\int_0^L \phi_i(x)\phi_j(x)dx + \frac{M_t}{\rho}\phi_i(L)\phi_j(L)\right] \\ I_{j,2} &= \frac{1}{\omega_j}\left[\int_0^L \dot{y}_0(x,t_0)\phi_j(x)dx + \frac{M_t}{\rho}\dot{y}_0(L,t_0)\phi_j(L)\right] \\ &= \sum_{i=1}^{\infty}\frac{\omega_i}{\omega_j}(-B_i\text{sin}\omega_i t_0 + D_i\text{cos}\omega_i t_0)\left[\int_0^L \phi_i(x)\phi_j(x)dx + \frac{M_t}{\rho}\phi_i(L)\phi_j(L)\right] \end{aligned}$$

By invoking the orthogonal condition

$$\rho\int_0^L \phi_i\phi_j dx + M_t\phi_i(L)\phi_j(L) = \begin{cases} 0 & i \neq j \\ \rho & i = j \end{cases} \tag{2.61}$$

$$EI\int_0^L \phi_i^{''}\phi_j^{''}dx = \begin{cases} 0 & i \neq j \\ \omega_i^2\rho & i = j \end{cases} \tag{2.62}$$

we have

$$\begin{aligned} I_{j,1} &= B_j\text{cos}\omega_j t_0 + D_j\text{sin}\omega_j t_0 \\ I_{j,2} &= -B_j\text{sin}\omega_j t_0 + D_j\text{cos}\omega_j t_0 \end{aligned}$$

Thus, the coefficients are given by

$$\begin{aligned} B_i &= I_{i,1}\text{cos}\omega_i t_0 - I_{i,2}\text{sin}\omega_i t_0 \\ D_i &= I_{i,1}\text{sin}\omega_i t_0 + I_{i,2}\text{cos}\omega_i t_0 \end{aligned}$$

in which the subscripts j have been replaced by i to keep symbolic compatibility with (2.58).

Up to now, we have not invoked the truncation in the number of flexible modes.

From modal analysis, we know that those modes with comparatively low frequencies are dominant, and thus we arrive at the following reasonable approximation (*Model truncation is introduced here*!)

$$y(x,t) = \sum_{i=1}^{N} \phi_i(x) q_i(t) \tag{2.63}$$

where N is the number of flexible modes we use to approximate the deflection. Substituting (2.63) into (2.50) gives

$$\sum_{i=1}^{N} EI\phi_i''(0) q_i(t) - k_p[\theta - \theta_f] = 0$$

which can be further written as

$$\sum_{i=0}^{N} \alpha_i q_i(t) = 0 \tag{2.64}$$

where

$$\begin{aligned}
\alpha_0 &= 1, \quad q_0 = k_p[\theta_f - \theta] \\
\alpha_i &= EI\phi_i''(0) \\
&= \frac{A_i\beta_i^2}{L^2}\left[\cosh\frac{\beta_i x}{L} + \cos\frac{\beta_i x}{L} - \gamma_i\left(\sinh\frac{\beta_i x}{L} + \sin\frac{\beta_i x}{L}\right)\right]_{x=0} \\
&= 2\frac{A_i\beta_i^2}{L^2} \neq 0 \\
& \quad i = 1, 2, \cdots, N
\end{aligned}$$

For the summation in (2.64), it is easy to check that the inner product defined in (Shifman, 1993) (Proposition A.3, pp987) is applicable

$$< q_i,\, q_j >:= \lim_{T\to\infty} \frac{1}{T}\int_0^T q_i(t) q_j(t) dt = \begin{cases} = 0, & i \neq j \\ \neq 0, & i = j \end{cases}$$

Now we can conclude from (2.64) that each q_i ($i = 0, 1, \cdots, N$) must be zero, subsequently

$$\theta = \theta_f, \quad y(x,t) = 0$$

This implies that the truncated system is at the final position, i.e., the final position is the largest invariant set in the set of $\dot{V} = 0$. Now, let us check the controller (2.18) again. Since the LaSalle's Theorem is only valid for autonomous system, we must keep in mind that to make the $f(t)$ feedback meaningful in improving vibration control, it should not be an explicit time function but rather a function of some vibration variables, which can be represented by the flexible modes. As for the integral part (which is an explicit function of time), according to the proof above, we can define an extra state as $x_a = \int_0^t |\dot{\theta}(\sigma)| f(\sigma) d\sigma$, which obviously approaches a constant as time intends to infinity. The constant may not be zero but it does not need to be. The boundedness of the constant is guaranteed from the boundedness of the Lyapunov function. Now, it is clear that if we regard the flexible modes and x_a as system states, the closed-loop system is an autonomous system. Consequently, from LaSalle's Theorem, the joint motion and the N flexible modes are asymptotically stable with respect to the final position. □

Bibliography

Bayo, E. (1987). A finite element approach to control the end-point motion of a single-link flexible robot. *J. of Robotics Systems*, **4**(1), 63–75.

Book, W.J. (1990). Modelling, design, and ontrol of flexible manipulator arms: a tutorial review. *Proc. 29th Conf. Decision & Control*, Honolulu, Hawaii. pp. 500–506.

Ge, S. S, , T. H. Lee and C. J. Harris (1998*a*). *Adaptive neural network control of robot manipulators*, World Scientific. London.

Ge, S. S., T. H. Lee and G. Zhu (1996*a*). Energy-based robust controller design for multi-link flexible robots. *Mechatronics*, **6**(7), 779–798.

Ge, S. S., T. H. Lee and G. Zhu (1997). Non-model-based position control of a planar multi-link flexible robot. *Mechanical Systems and Signal Processing*, **11**(5), 707–724.

Ge, S. S., T. H. Lee and G. Zhu (1998*b*). Asymptotically stable end-point regulation of a flexible scara/cartesian robot. *IEEE/ASME Trans. Mechatronics*, **3**(2), 138–144.

Ge, S. S., T. H. Lee and G. Zhu (1998*c*). Improved regulation of a single-link flexible manipulator with strain feedback. *IEEE Trans. on Robotics and Automation*, **14**(1), 179–185.

Ge, S.S., T.H. Lee and G. Zhu (1996*b*). Genetic algorithm tuning of lyapunov-based controllers: An application to a single-link flexible robot system. *IEEE Transactions on Industrial Electronics*, **43**(5), 567–674.

Hale, J.K. (1969). Dynamical systems and stability. *J. Mathematical Analysis and Applications*, **26**, 39–59.

Hastings, G.G. and W.J. Book (1987). A linear dynamic model for flexible robotic manipulators. *IEEE Control System Magazine*, **7**, 61–64.

Kanoh, H., S. Tzafestas, H.G. Lee and J. Kalat (1986). Modelling and control of flexible robot arms. *Proc. 25th Conf. Decision & Control*, Athens, Greece. pp. 1866–1870.

Khalil, H.K. (1992). *Nonlinear Systems*, Macmillan Publishing.

Khorrami, F. and S. Jain (1993). Nonlinear control with end-point acceleration feedback for a two-link flexible manipulator: Experimental results. *J. Robotic Systems*, **10**(4), 505–530.

Korolov, V.V. and Y.H. Chen (1988). Robust control of a flexible manipulator arm. *Proc. 1988 IEEE Int. Conf. Robotics and Automation*, Philadelphia, USA. pp. 159–164.

Kotnik, P., S. Yurkovich and Ü. Özgüner (1988). Acceleration feedback control for a flexible manipulator arm. *J. Robotic Systems*, **5**(3), 181–196.

Krishnan, H. (1988). *Bounded Input Discrete-Time Control of a Single-Link Flexible Beam.* Master Thesis, Department of Electrical Engineering. University of Waterloo.

Lee, T. H., S. S. Ge and Z. P. Wang (2001). Adaptive robust controller design for multilink flexible robots. *Mechatronics*, **11**(8), 951–967.

Luo, Z. H. (1993). Direct strain feedback control of flexible robot arms: New theoretical and experimental results. *IEEE Transaction on Automatic Control*, **38**(11), 1610–1622.

Luo, Z.H., N. Kitamura and B.Z. Guo (1995). Shear force feedback control of flexible robot arms. *IEEE Trans. Robotics and Automation*, **11**(5), 760–765.

Meirovitch, L. (1975). *Elements of Vibration Analysis*, McGraw-Hill. Inc.

Rattan, K. S. and V. Feliu (1992). Feedforward control of flexible manipulators. *Proc. 1992 IEEE Int. Conf. Robotics and Automation*, Nice, France. pp. 788–793.

R.H. Cannon, Jr. and E. Schmitz (1984). Initial experiments on the end-point control of a flexible one-link robot. *Int. J. Robotics Research*, **3**(3), 62–75.

Sakawa, Y., F. Matsuno and S. Fukushima (1985). Modeling and feedback control of a flexible arm. *J. of Robotic Systems*, **2**(4), 453–472.

Schultz, W.C. and V.C. Rideout (1961). Control system performance measures: Past, present and future. *IEEE Trans. Automatic Control*, **22**, 22–35.

Shifman, J.J. (1993). Lyapunov functions and the control of the euler-bernoulli beam. *Int. J. Control*, **57**(4), 971–990.

Siciliano, B. and W.J. Book (1988). A singular perturbation approach to control of lightweight flexible manipulators. *Int. J. Robotics Research*, **7**(4), 79–90.

Spector, V.A. and H. Flashner (1990). Modelling and design implications of noncollocated control in flexible systems. *ASME J. Dynamic Systems, Measurement, and Control*, **112**, 186–193.

Tzes, A. and S. Yurkovich (1993). An adaptive input shaping control scheme for vibration suppression in slewing flexible structures. *IEEE Trans. Control Systems Technology*, **1**(2), 114–121.

Wang, D. and M. Vidyasagar (1989). Transfer functions for a single flexible link. *Proc. IEEE Int. Conf. Rob. and Auto.*, pp. 1042–1047.

Yigit, A.S. (1994). On the stability of pd control for a two-link rigid-flexible manipulator. *ASME J. Dynamic Systems, Measurement, and Control*, **116**, 208–215.

Zuo, K. and D. Wang (1992). Closed loop shaped-input control of a class of manipulators with a single flexible link. *Proc. 1992 IEEE Int. Conf. Robotics and Automation*, Nice, France. pp. 782–787.

Chapter 3

Trajectory Planning and Compliant Control for Two Manipulators to Deform Flexible Materials

Coordinating two robot manipulators to handle flexible materials has a wide range of applications in the manufacturing industry. However, this problem has not been seriously addressed until recently. The two robot manipulators have to follow complicated trajectories to maintain a minimum interaction force with the flexible beam. These trajectories are very complicated and not suitable for real time systems. In this chapter, three approximation methods of the optimal trajectories and a compliant control scheme are introduced. The first one uses a piece-wise linear approximation of the optimal trajectories, while the second one uses adaptive piece-wise linear approximation. The third method applies a continuous approximation of the optimal trajectories using an ellipsoid. Finally, a compliant motion scheme is proposed to reduce the interaction forces and moments in the first method. The stability of the proposed system is investigated. Experimental results encourage the proposed schemes.

3.1 Introduction

Coordination of multiple robot manipulators has been addressed by several researchers [1,2,3,4,5] As a matter of fact, many tasks either cannot be done using a single robot manipulator or the performance can be improved by using multiple robot manipulators. One of these tasks is handling deformable objects. Researchers have addressed several aspects of multi-robots system; however, handling deformable objects has not received attention until recently.

Deformable objects can be found in many of the manufacturing industries which deal with flexible frames or beams, such as vehicles and aerospace industries. Handling of non-rigid bodies has been performed indirectly by using special tools such as vacuum pad [6]. This places an extra loading effort on the end-effector. If, on the other hand, the material is to be handled by two robot manipulators, special tools are not needed.

Zheng and Chen [7] proposed the use of minimization of forces and moments exerted on the end-effectors as a criterion of the trajectory planning. They studied the trajectory planning of two manipulators to deform a flexible beam. Optimal motion trajectories were derived to generate zero interaction moments. While the beam is being deformed,-forces and moments should be applied to the beam. For example, in the manufacturing of Printed Wiring Boards (PWB) (Fig. 3.1), a book is formed from a large number of elastic sheets and laminated to form a PWB. Some of the sheets are covered with electronic circuits and should be carefully aligned. For the alignment purpose, each sheet is equipped with four holes that should be aligned with four alignment pins on the assembly bed. It is difficult to align the four holes simultaneously because of the small tolerance between the pins and the holes. Thus, the manipulators should bend the sheet to align the middle holes with the corresponding pins. Then, they unfold the sheet to align the rest of the holes.

Despite the fact that deformation of beams is a traditional problem in elasticity [18], it has other perspectives in robotics. In order to deform a beam, two manipulators are required. Trajectory planning for the end-effectors while the manipulators are bending the beam is not a simple task. In this scope, trajectory planning refers to the end-effectors position and orientation as functions of time. After grasping the beam, the two end-effectors have to bend the beam for one reason or another.

The optimal trajectories are complicated and not suitable for implementation in real time systems. In fact, the optimal trajectories require the evaluation of two elliptic integrations which is very time

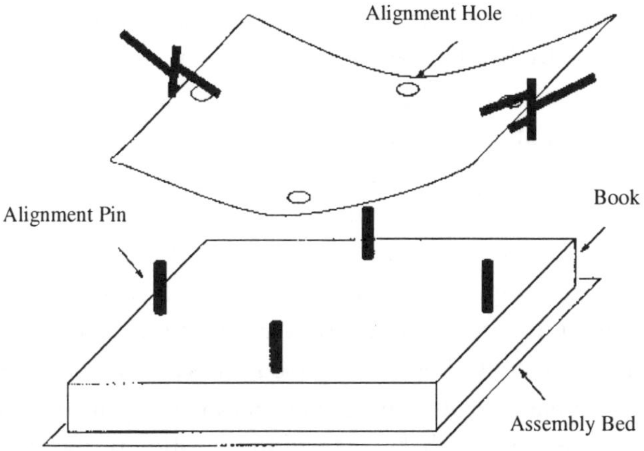

Figure 3.1: Two end-effectors aligning a flexible sheet to form a PWB

consuming [7]. Thus, alternative approximated trajectories are required to simplify the computations involved in the optimal ones. However, the approximation for these trajectories should be done with some caution. This is because these trajectories determine the force and moment exerted on each end-effector. Actually, the force relation between the case of optimal trajectories and that of the worst case approximation is four to one [7].

The objective of this chapter is to reduce the computation time for defining the optimal trajectory of the end-effectors. We propose three approximation methods for the optimal trajectories. The first scheme uses a piece-wise linear approximation, while the second method utilizes the adaptive piece-wise linear approximation scheme. In the third scheme, we attempt to approximate the optimal trajectories by a continuous function of time. Finally, a compliant control scheme is used to minimize the forces and moments exerted on the end-effectors in the case of piece-wise linear approximation.

In a compliant motion, instead of rejecting or resisting the external forces, the manipulator should accommodate them[8]. The compliant control scheme is attractive in this scope for more than one reason. First, the nature of the bending process motivates this algorithm; when two persons try to bend a beam cooperatively, they will sense the forces and moments exerted on their hands, and then modify the position and orientation by backing up if the force is large, or moving on if it is fair. The second reason is the fact that the robots are equipped with an internal position control system that cannot be accessed or modified to accept a force control method. Thus an external control loop should be designed to make use of the information about the interaction forces and provides a force control.

The compliant control was addressed by several researchers. The core of the compliance motion is the concept of mechanical impedance [9] in which both the dynamic behavior and the position of the manipulator are controlled. Kazerooni et al. [8] addressed the compliant motion in the frequency domain as a stabilizing dynamic compensator. In their approach, the relation between the interaction forces and endeffector position is constant over a certain frequency range. Force control can be effectively achieved using adaptive compliant control schemes [10]. Compliant motion can be used in the coordination of two robots to perform complicated tasks. Tao et al. [11] proposed the use of a compliant control to coordinate two robots operated in a master/slave mode. There are several applications of the compliant motion in the teleoperation systems [12, 13, 14]

Stability analysis of several compliant control schemes has been addressed by several researchers. The stability of the system depends on the desired impedance ranges. The impedance of the environment greatly affects the stability analysis of the system [15,16] Kazerooni [17] provided a nonlinear approach to analyze the stability of the compliant motion. His analysis is based on the small gain theorem and Nyquist criterion.

This chapter is organized as follows. The next section reviews the method of optimal trajectories and presents the three approximation methods. The compliant control scheme is presented in the third section. The stability of the system is addressed in the fourth section. Experimental investigation is presented in the fifth section. Finally, the chapter is concluded in the sixth section.

3.2 Efficient Trajectory Planning Using Optimal Bending Model

Trajectories which are based on the optimal bending model are open loop trajectories without any feedback from the environment [7]. It is found that both position and orientation of the end effectors need to follow the optimal trajectory in which the two end points of the beam coincide with its zero moment (ZM) points and the deflection angle of the beam and the orientation of the two end-effectors are the same [7]. Under these conditions, the end-effectors are exposed to minimum forces and zero moments since the end-effectors are holding the ZM points. The force is given by:

$$F = \frac{(2R(\alpha))^2 EI_z}{L^2} \tag{3.1}$$

where E is the stiffness of the beam, Iz is the moment of inertia of the beam crosssection, L is the length of the beam, α is the bending angle of the beam, and $R(\alpha)$ *is* calculated from the following elliptic integral (Fig. 3.2) [7]:

$$R(\alpha, \phi_1) = \int_0^{\phi_1} \frac{d\phi}{\sqrt{1 - k\sin^2\phi}} \tag{3.2}$$

when $k = \sin^2(\frac{\alpha}{2})$ and $\phi_1 = \frac{\pi}{2}$. Note that $R(\alpha, \phi_1)$ is a function of α and ϕ_1.

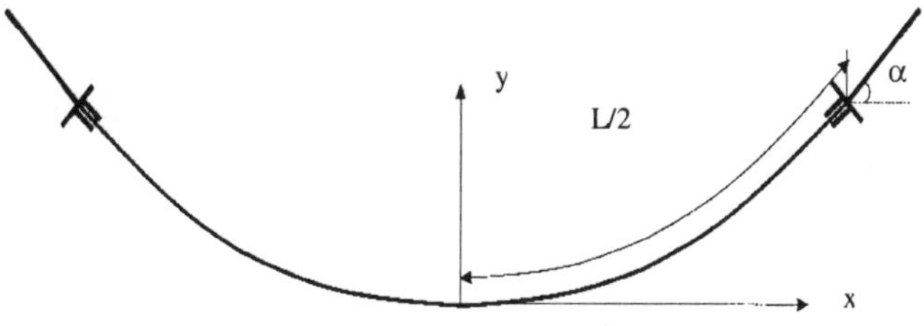

Figure 3.2: Two manipulators handle a flexible beam

The optimal trajectories are functions of the orientation of the end-effectors a, and should satisfy:

$$x = L(\frac{P(\alpha)}{R(\alpha)} - \frac{1}{2}) \quad (3.3)$$

and

$$y = \frac{1}{R(\alpha)} \sin(\frac{\alpha}{2}) \quad (3.4)$$

where $P(\alpha,\phi_1)$ is calculated from another elliptic integral given by the following equation [7]:

$$P(\alpha,\phi_1) = \int_0^{\phi_1} \sqrt{1 - k\sin^2\phi}\,d\phi \quad (3.5)$$

An optimal trajectory, based on (3.3) and (3.4), *is* shown in Fig. 3.3.

It is clear that the manipulator should calculate two elliptic integrals at each step of motion, which makes the process complicated and not suitable for real time applications. To simplify the computation, we propose three approximation methods.

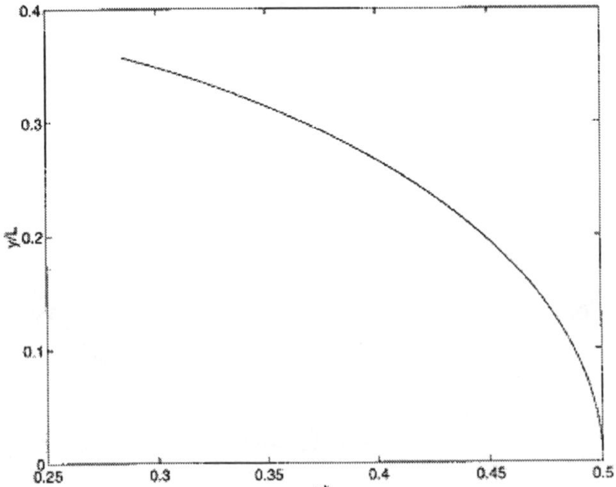

Figure 3.3: Optimal trajectories

3.2.1 *Piece-wise linear approximation of the optimal trajectories*

By using piece-wise linear approximation, the target bending angle α is divided into a number of smaller angles. The optimal trajectories are approximated by linear functions during the motion between any set of two contiguous points. In this way, the total computation time can be greatly reduced. To describe such trajectories, we only need to select the bending rate and the resolution of the approximation. The trajectory of the bending angle is linear and can be expressed as:

$$\alpha(t) = kt \tag{3.6}$$

where k is bending rate in rad/second. If a is divided into N points, then the trajectories for the position of the end effector between each pair of end points are approximated by linear segments. The end points are generated using (3.3) and (3.4).

Fig. 3.4 shows the simulation result of bending a beam by 30 degrees. This bending angle is divided into segments each of 5.7 degrees. The normalized moment is given by

$$\text{Normalized moment} = \frac{ML}{EI_z} \tag{3.7}$$

The peaks at the very beginning are due to the sensitivity of the angle to errors in the approximation. Furthermore, the peaks in force and moment are expected since for a small bending angle, approximation errors will drive the system to the case where the two end-effectors are pushing toward each other without any significant change in the orientation. From Fig. 3.4, we can see that piece-wise linear approximation guarantees minimum forces and moments only at the end points of each piece. In the middle of the linear pieces, both forces and moments can be very high.

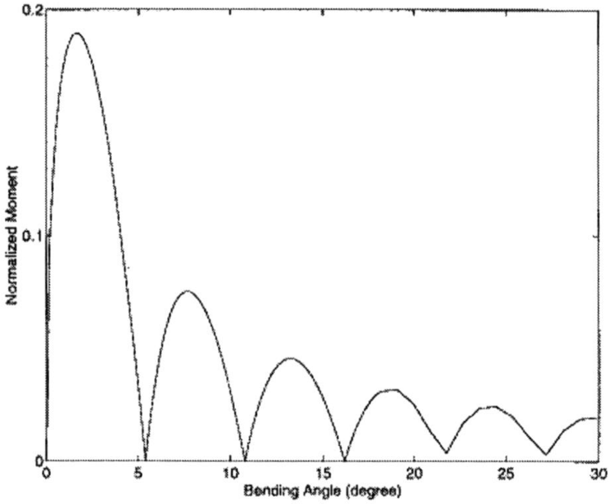

Figure 3.4: Normalized moment exerted on the end-effector under piece-wise linear approximation

3.2.2 *Adaptive piece-wise linear approximation of the optimal trajectories*

In the piece-wise linear approximation, the moment and force have different peaks as seen in Fig. 3.4. In order to reduce the peaks, an adaptive method is introduced here. If we use variable intervals of approximation, the performance of the approximation scheme will be improved.

The optimal trajectories as shown in Fig. 3.3 are nonlinear; thus, any uniform piece-wise linear approximation will have variable peaks in the approximation error. To reduce large magnitude of the peaks, we propose the concept of adaptive piece-wise linear approximation. Basically, this method can be described by the following algorithm:

1. Starting with an interval $[t_k, t_{k+1}]$, approximate the optimal trajectories, (3.3) and (3.4), using the piece-wise linear approximation method.

2. Calculate the maximum value of the force in this interval.

3. If the force exceeds the prescribed value, reduce t_{k+1} and go to step one, else stop.

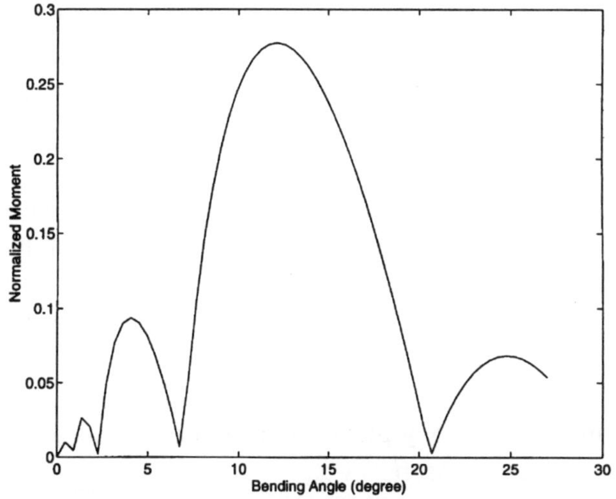

Figure 3.5: Normalized moment exerted on the end-effector under adaptive piece-wise linear approximation

The algorithm is executed off-line to calculate the points at which the prescribed maximum force is reached. After that, the set of points (x_k, y_k) are used to generate the end-effector trajectories using the same scheme used in the piece-wise approximation method.

This method is employed to approximate the optimal trajectories. The prescribed maximum normalized force is specified to be 15 normalized units. Fig. 3.5 shows the moment for the adaptive approximation scheme. We can see that the distribution of the intervals is not uniform. Moreover, the approximation interval gets larger as we increase the deflection angle. This process is due to the behavior of the optimal trajectories as shown in Fig. 3.3. The disadvantages of this method is that the computation time may still become excessive because a trial and error method is involved.

3.2.3 *Approximation of the optimal trajectories by a continuous function of time*

The two piece-wise linear approximation methods have different approximation error peaks and do not reflect the behavior of the optimal trajectories given by equations (3.3) and (3.4). Thus, we need an approximation method that can behave continuously as the optimal trajectories. These situations motivate us to think of a continuous approximation of the optimal trajectories.

Consider numerical solutions of (3.3) and (3.4). This solution represents the trajectory in the x-y plane, Fig. 3.3. The behavior of this trajectory is similar to the behavior of an ellipsoid, because the numerical solution of the optimal trajectories is very close to an ellipse in the x-y plane. So, as a final step, we try an ellipsoid type curve fitting.

An ellipsoid in the x-y plane can be expressed as

$$x = a\cos(\beta\alpha) + x_0$$
$$y = b\sin(\beta\alpha) + y_0 \tag{3.8}$$

where a is the radius along the x axis, b is the radius along the y axis, (x_0, y_0) is the center of the ellipse in the xy-plane, and β is a constant weight on the deflection angle α.

Some physical aspects of the deformation process can be used to find the parameters in (3.8). We know that when the deflection angle α is zero the following is true, $(\frac{x}{L}, \frac{y}{L})$ = (0.5, 0). Another point that can be used to simplify the equations is the relation between force and the deflection angle. We can use this relation to approximate the parameter β in (3.8). Substituting (3.4) in (3.1), we get

$$F = \frac{4\sin^2(\frac{\alpha}{2})EI_z}{y^2} \tag{3.9}$$

If we replace y by (8), in the above equation, and take the limit as $a \to 0$ we get:

$$\beta = \frac{1}{\pi b} \tag{3.10}$$

Substituting (3.10) in (3.8), we get:

$$x = (0.5 - x_0)\cos(\frac{\alpha}{\pi b}) + x_0 \tag{3.11}$$

and

$$y = b\sin(\frac{\alpha}{\pi b}) \tag{3.12}$$

We can see that only two points can be used to determine the unknowns in (3.11) and (3.12). We use these equations to find the performance of the system under continuous approximation. Fig. 3.6 shows the performance of the system for a target deflection angle of 30 degrees. We can see that the performance is better than both piece-wise linear approximation and adaptive approximation. However, for a large bending angle, the approximation error increases as shown in Fig. 3.6. Moreover, the performance depends on which points are used to solve for the unknowns in (3.11) and (3.12).

The results in Fig. 3.4, Fig. 3.5, and Fig. 3.6 have different shapes. This shows the different effects of the three approximation methods. We can see that the first two methods show periodic peaks in the moments. This is because the approximation is peace-wise, and only at the two end points of a piece the trajectory is optimal. The third approximation method, however, is continuous approximation of the entire trajectory. Therefore, the interaction moment increases without peaks.

3.3 Compliant Control

The previous section introduces some approximation methods of the optimal trajectory. These methods are suitable for real time but still have several disadvantages. In this section, we will use a compliant control scheme to improve the performance of the piece-wise linear approximation method. In compliant control, the force exerted on the end-effector is accommodated rather than resisted [8]. To accomplish this, the dynamic behavior of the manipulator is modulated either by modifying the robot joints' servo gains or by using an external control loop that produces the same results [12]. The latter has the advantage of high performance position tracking while keeping the force control in effect. Moreover, the transition from free motion to a constrained one is done naturally without any additional cost.

In this chapter, an external control loop is used such that the interaction forces drive a mechanical admittance to alter the commanded trajectories using a negative feedback scheme as shown in Fig. 3.7. If there is no contact between the manipulator and the environment, the system will precisely track the commanded position. If a contact is made, the external feedback loop will be excited, and the robot will comply with the interaction forces by a deviation of δx in the end-effector's position.

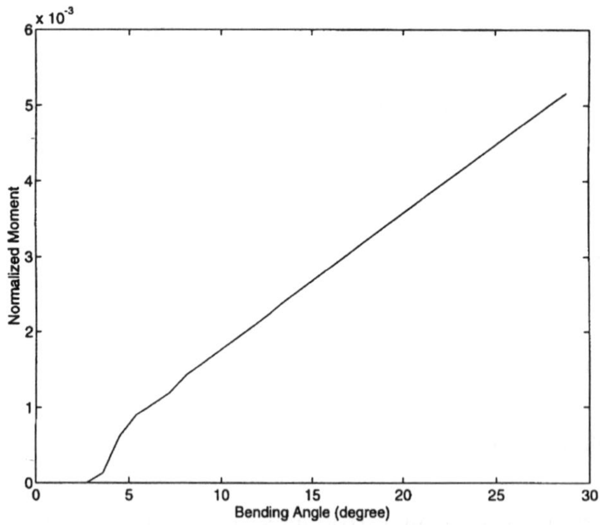

Figure 3.6: Normalized moment exerted on the end-effector under continuous approximation

The relation between the deviation δx from the commanded position and the interaction forces is given by:

$$B\delta\dot{x} + K\delta x = F \tag{3.13}$$

where B is the friction of the damper, and K is the stiffness of the spring [12]. By applying Laplace transform on (13), we get

$$(Bs + K)\delta x(s) = F(s) \tag{3.14}$$

The quantity *(Bs + K)* represents the mechanical impedance.

In the case of bending a flexible object, the forces and moments exerted on the endeffectors comprise the bending forces and moments and the gravitational effects. The manipulators should comply with the forces and moments due to the bending process and reject the effects of the gravity. If the manipulators comply with the gravity, the object will always move down which is unacceptable. Thus, in our system δx is a vector that represents the deviation in the motion along the x-axis and the rotation about the z-axis.

Consider the case where B and K are given by:

$$B = \begin{bmatrix} b_1 & 0 \\ 0 & b_2 \end{bmatrix} \tag{3.15}$$

and

$$K = \begin{bmatrix} k_1 & 0 \\ 0 & k_2 \end{bmatrix} \tag{3.16}$$

Then, the compliant controller can be written as two separate filters: one for the motion along x-axis and one for the rotation about z-axis. Under this assumption, we have

$$\delta x = \frac{F}{b_1 s + k_1} \tag{3.17}$$

and

$$\delta\alpha = \frac{M}{b_2 s + k_2} \tag{3.18}$$

Let b represent both b_1 and b_2 and k represent both k_1 and k_2. The impulse response of the filter for both force and moment is:

$$h(t) = \frac{1}{b}\exp(-\frac{k}{b}t) \tag{3.19}$$

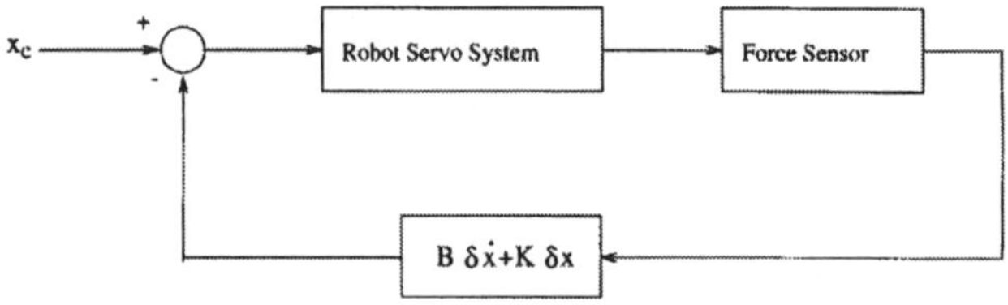

Figure 3.7: Compliant control Scheme

A discrete filter of the same impulse response has the following transfer function:

$$H(z) = \frac{\frac{1}{b}z}{z - \exp(-\frac{k}{b}T)} \tag{3.20}$$

where T is the sampling period. Thus, the discrete filter can be written as:

$$H(z) = \frac{\frac{1}{b}z}{z - \exp(-\frac{k}{b}T)} \tag{3.21}$$

One can see that the computation involved in (3.21) is very simple, because $\exp(-\frac{k}{b}T)$ is a constant. In fact only two multiplications and one addition are involved. Therefore, the time delay introduced by the computation is negligible.

Consider an approximation of the optimal trajectories. The compliant controller alters this approximation according to (3.13) to reduce the forces and moments exerted on the end-effectors due to the bending process. The deviation due to the bending moment is added to the bending angle a. Thus, if the moment is not zero, the deflection angle of the beam will increase and this reduces moments and forces exerted on the end-effectors. The motion of the end-effectors along the x-axes is deviated from the approximated trajectories by adding the filter output to the x-component of the trajectories at each point. The later will result in reducing the approximation error and thus reducing the bending forces and moments.

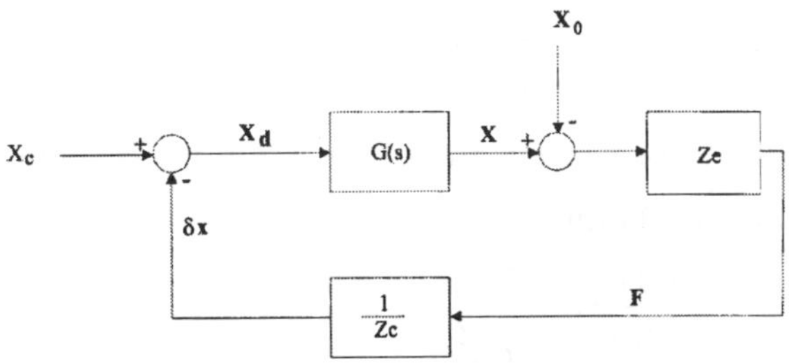

Figure 3.8: Closed loop system block diagram

3.4 Stability Analysis

The stability of the compliant motion scheme depends on many factors. The model of the robot positioning system is very important in the stability analysis. Most of the researchers used second order linear control systems to model the position of the end-effector as a function of the command [15, 16, 17].

We can model the end-effector positioning system along each axis as a second order system [16]. Let x be one of these components. Then the end-effector position x as a function of the desired position x_d (see Fig. 3.8) can be expressed as:

$$\frac{x(s)}{x_d(s)} = G(s) = \frac{\omega_n^2}{s^2 + 2\xi\omega_n + \omega_n^2} \tag{3.22}$$

where ω_n is the natural frequency and ξ is the damping ratio.

From Fig. 3.8, the relation between the desired position and the commanded one

is:

$$x_d = x - \delta x \tag{3.23}$$

Moreover, the interaction force is transformed into deflection using the compliant controller impedance Z_c. However, this force is related to the end-effector position by the environment impedance Z_e. Substituting δx in (3.23) we obtain:

$$x_d = x_c + \frac{Z_e(x(s) - x_0(s))}{Z_c} \tag{3.24}$$

where x_0 is the impedance center of the environment. Substituting (3.24) in (3.22) weobtain:

$$\frac{x(s)}{x_d(s)} = \frac{G(s)}{1 + G\dfrac{Z_e}{Z_c}} \tag{3.25}$$

To study the stability, we will assume that the environment impedance is dominated by the stiffness [10]. Thus, in our case we have

$$Z_e = k_e \tag{3.26}$$

and

$$Z_c = bs + k \tag{3.27}$$

Substituting the above in (3.25), we obtain:

$$\frac{x(s)}{x_d(s)} = \frac{\omega_n^2(bs + k)}{bs^3 + (2\xi\omega_n b + k)s^2 + (b\omega_n^2 + 2\xi\omega_n k)s + \omega_n^2(k_e + k)} \tag{3.28}$$

This equation can be used to study the stability of the compliant motion scheme. Using Routh's stability criterion and assuming that the impedance parameters are not negative, we obtain:

$$2b^2\xi\omega_n^3 + 4\xi^2\omega_n^2 bk + 2\xi\omega_n k^2 - \omega_n^2 bk_e > 0 \tag{3.29}$$

It is clear that the stability of the system for a given set of the impedance parameters (*b*,*k*) depends on the stiffness of the environment. For two robots handling a flexible material, this stiffness is determined by the stiffness of the beam. If the stiffness of the beam is very large, then a large value of the controller stiffness is required. Also, generally, from (3.29), small values of the controller parameters, damping and stiffness, can result in instability. In our analysis of the stability, we neglected any time delay in the control loop. Analysis of the effect of time delay in the control loop can be found in [16].

In the experimental part of this chapter, two PUMA 560 robots will be used. If we consider some values of the parameter in (3.22), we can get some limits on the choices of the parameters in the controller. The natural frequency of the PUMA 560 robot is not lower than 2Hz, and $2\xi\omega\omega_n > 25$ rad/sec [11]. Substituting these lower limits in (3.29) with the stiffness of the environment being estimated to be k_3 =11500 N/mm, we obtain:

$$k > 550 \tag{3.30}$$

and

$$b > 6.92 \tag{3.31}$$

As we will see in the next section, these limits are very close to that found experimentally.

3.5 Experimental Study

The system consists of two PUMA 560 robots and a supervisor computer. One of the two robots is equipped with a multi-axis force/torque sensor (Intelligent Multi-axis Force/Torque Sensor System). The sensor is connected to a local controller that calculates the force/torque components from the raw data and provides a temperature compensation and a filter. The supervisor computer communicates with the sensor controller via a serial port.

The experiments (Fig. 3.9) resembled the production of PWB example as mentioned in the introduction of this chapter. In both of the experiments the robots had to move to a pre-specified position and picked up the aluminum sheet. After that, the robots moved to another specified position where the bending process occurred. The sheet was bent to make the two middle holes aligned with pins first. Finally, the robots extended the sheet to align the rest of the holes.

Figure 3.9: Two PUMA robots align an aluminum sheet

Since the moment will be zero in the optimal case, we will consider it as a criterion for the system performance. We first ran the system without compliant control. The used trajectories were obtained in [7]. However, to control the robot motion, we used a piece-wise linear approximation of the optimal trajectories. Thus, only at the endpoints of every linear piece, the bending moment was zero. Between two end- points, the bending moment could be very high.

We only used the piece-wise linear approximation of the optimal trajectories in our experiment. This was considered based on the fact that it has the worst performance. Hence, if the compliant motion controller improves the performance of this approximation method, it shall do better for the other two methods.

The data from the force sensor was very noisy, this was due to the vibration of the end-effector during the robot motion. To get rid of the noise a low pass filter (LPF) with cutoff frequency of 10 Hz was added. The filter was first implemented using infinite impulse response (IIR) filter. However, this filter produced a phase lag which affected the stability of the compliant controller. Thus a finite impulse response(FIR) filter was used to smooth the noise.

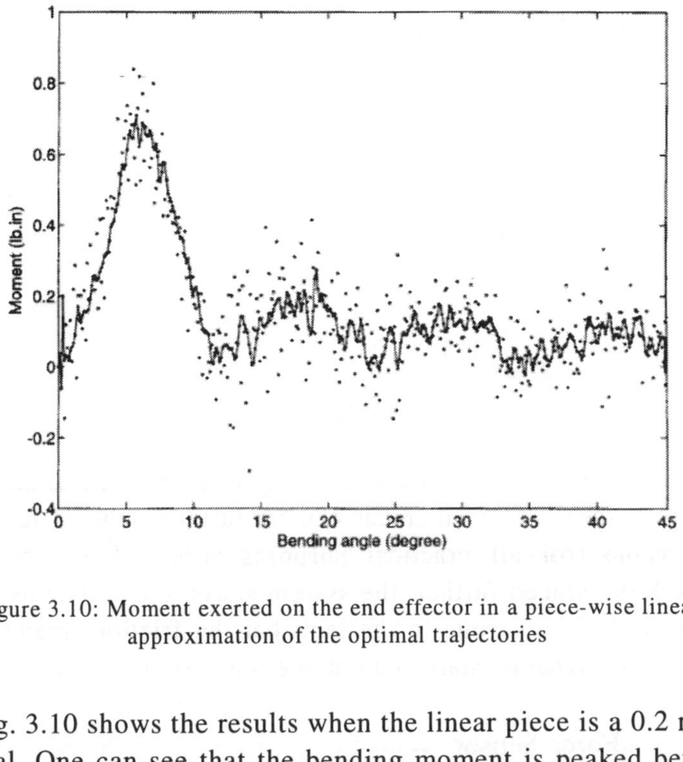

Figure 3.10: Moment exerted on the end effector in a piece-wise linear approximation of the optimal trajectories

Fig. 3.10 shows the results when the linear piece is a 0.2 radian interval. One can see that the bending moment is peaked between two end-points. It is clear that this result is similar to the simulation result in Fig. 3.4.

The PUMA robot is positionally controlled with each joint having its own controller. The joint processor receives a command every 28ms. Thus we can send a position command to joint servos every 28ms (ALTER mode in VAL II language). The gravitational effects, friction, and higher order dynamics start to have more effects as the frequency of the commanded position increases and explain the observed vibrations.

The compliant control in (3.14) was implemented using a digital- equivalent (impulse invariant method) scheme similar to that in (3.21). We added compliance in two directions. The first direction was along the bending force, while the other was along the bending moment. The force sensor was mounted at a distance L_x, from the center of the gripper (see Fig. 3.11). Thus, the forces exerted on the ed-effector would cause a moment on the force sensor. Since this moment was not exerted on the end-effector, we subtracted it from the force sensor torque measurements. With a sampling rate of 35.7Hz, the parameters were selected to be:

$$B = \begin{bmatrix} 7 & 0 \\ 0 & 0.5 \end{bmatrix} \tag{3.32}$$

and

$$K = \begin{bmatrix} 575 & 0 \\ 0 & 10000 \end{bmatrix} \tag{3.33}$$

The compliant controller clearly reduced moment as shown in Fig. 3.12. Ideally, this moment can be further reduced to a very small value (for all practical purposes zero). However, If the matrix B is reduced further, the system starts to oscillate as shown in Fig. 3.13. This is due to the fact that the friction, gravity, and higher order dynamic start to dominate the system characteristics.

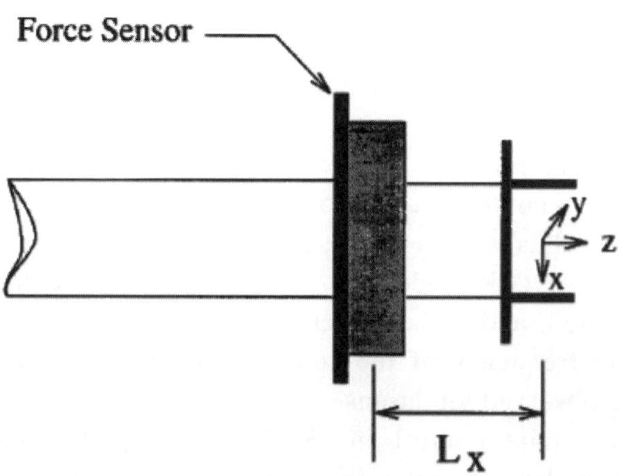

Figure 3.11: Force sensor

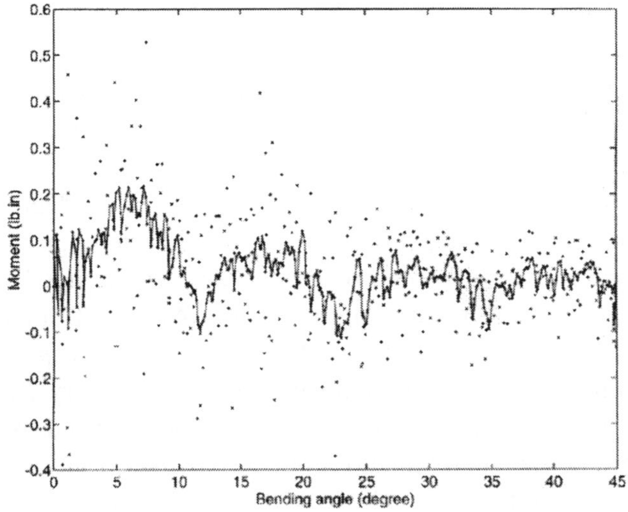

Figure 3.12: Moment exerted on the end effector in a piece-wise linear approximation of the optimal trajectories with compliant control

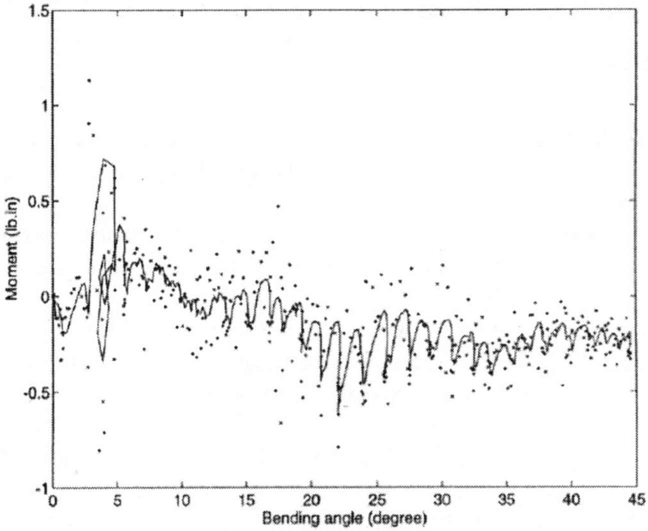

Figure 3.13: Moment exerted on the end effectors with compliant controller outside the stability boundaries

3.6 Conclusions

This chapter introduces efficient trajectories planning for two robots to handle flexible objects. We addressed the piece-wise linear approximation which, standing alone, could not be used as an efficient approximation for the optimal trajectories. Adaptive piece-wise approximation was used as another method to approximate the optimal trajectories. The performance of the system was better than that of the linear piece-wise approximation. After that, a continuous approximation was provided based on the physical aspects of the addressed problem. This approximation method was very close to the optimal trajectories. Finally, compliant control was used to alter the piece-wise linear approximation of the optimal trajectories. The compliant controller was able to reduce the moments and forces exerted on the end-effectors.

An experiment using two PUMA 560 robots was carried out and verified the proposed algorithms. Furthermore, stability of the compliant controller was investigated experimentally.

References

1. Thomas E. Alberts and Donald I. Soloway, "Force control of a multi-arm robot system," Proc. 1988 IEEE Int. Conf. on Robotics and Automation, Philadelphia, PA, April 14-19, pp. 1490-1496.
2. Fan-Tien Cheng and David E. Orin, "Efficient algorithm for optimal force distribution - the compact dual LP method," IEEE Trans. Robotics and Automation, Vol 6, No. 2, April 1990, pp. 178-187.
3. Ian D. Walker, Robert A. Freeman, and Steven I. Marcus, "Dynamic task distribution for multiple cooperating robot manipulators," Proc. 1988 IEEE Int. Conf. on Robotics and Automation, Philadelphia, PA, April 14-19, pp. 1288-1290.
4. Jian M. Tao and J. Y. S. Luh, "Robust position and force_ control for a system of multiple redundant- robots," Proc. 1992 IEEE Int. Conf. on Robotics and Automation, Nice, France, May 12-14, 1992, pp. 2211-2216.
5. T. J. Tarn, N. Xi, A. K. Ramadorai, and A. K. Bejczy, "A versatile experimental system for dual-arm planning and control," Proc. 1994 IEEE Int. Conf. on Robotics and Automation, San Diego, CA, May 8-13, 1994, pp. 1515-1520.

6. Douglas E. Ruth and Prasanna Mulgaonkar, "Robotic lay-up of prepreg composite plies," Proc. 1990 IEEE Int. Conf. on Robotics and Automation, Cincinnati, Ohio, May 13-18, 1990, pp. 1296-1300.
7. Yuan F. Zheng and Ming Z. Chen,"Trajectory planning for two manipulators to deform flexible beams," Robotics and Autonomous Systems, 12(1994), pp. 55-67.
8. H. Kazerooni, T. B. Sheridan, and P. K. Houpt, "Robust compliant motion for manipulators, Part I-II," IEEE Journal of Robotics and Automation, vol. RA-2, no. 2, 1986, pp. 83-105.
9. N. Hogan, "Impedance control: an approach to manipulation, Part I-III," ASME Journal of Dynamic Systems, Measurements and Control, vol. 107, 1985, pp. 1-24.
10. R. Colbaugh and A. Engelmann, "Adaptive compliant motion control of manipulators: theory and experiments," Proc. 1994 IEEE Int. Conf. on Robotics and Automation, San Diego, California, May 8 - 13, 1994, pp. 2719-2726.
11. Jain M. Tao, J. Y. S. Luh, and Yuan F. Zheng, "Compliant coordination control of two moving industrial robots," IEEE Trans. Robotics and Automation, vol. 6, no. 3, June 1990, pp. 322-330.
12. Paul S. Schenker, Antal K. Bejczy, Won S. Kim, and Sukhan Lee, "Advance man-machine interfaces and control architecture for dexterous teleoperations," IEEE Oceans'91, October 1-3, Honolulu, Hi," Underwater Robotics" Session.
13. H. Das, H. Zak, W. S. Kim, A. K. Bejczy, and P. S. Schenker "Performance experiments with alternative advanced teleoperator control modes for a simulated solar maximum satelite repair," 5th Annual Space operations, Appl. Research Symp., July 9-11, 1991, Houston, TX.
14. A. K. Bejczy and Z. F. Szakaly "A harmonic motion generator (HMG) for telerobotic applications," Proc. 1991 IEEE Int. Conf. on Robotics and Automation, Sacramento, CA, April 9-11, pp. 2032-2039.
15. Yangsheng Xu and Richard P. Paul "On position compensation and force control stability of a robot with compliant wrist," Proc. 1988 IEEE Int. Conf. on Robotics and Automation, Philadelphia, PA, April 14-19, 1988, pp. 1173-1178.
16. Dale A. Lawrence "Impedance control stability properties in common implementations," Proc. 1988 IEEE Int. Conf. on Robotics and Automation, Philadelphia, PA, April 14-19, 1988, pp. 1185-1190.
17. H. Kazerooni and T. I. Tsay "Stability criteria for robot compliant maneuvers," Proc. 1988 IEEE Int. Conf. on Robotics and Automation, Philadelphia, PA, April 14-19, 1988, pp. 1166-1172.
18. S.F. Borg, Fundamentals of Engineering Elasticity, D. Van Nostrand Company, Inc. 1962.

Chapter 4

Force Control of Flexible Manipulators

This chapter discusses modeling methodology and force control scheme of constrained flexible manipulators. Modeling and force control of a constrained one-link flexible arm are treated in the first section. In the second section modeling and hybrid position/force control of constrained n link flexible manipulators are considered. In the third section, a dynamic model and robust cooperative controllers of two one-link flexible arms are discussed. Important points for the modeling of constrained flexible manipulators are as follows: "1. The end-point of the flexible manipulator is constrained and a boundary condition is nonhomogeneous" and "2. The exact model can be described as a distributed parameter system of an infinite dimensional system. An approximated finite-dimensional model should be derived for designing controllers." In the controller design, an important point is as follows: "To suppress the spillover instability caused by the residual modes which are neglected at the controller design phase, a robust controller should be constructed."

4.1. Introduction

Robotic manipulators will play important roles in future space missions[1]. Because of the important demand for low-energy consumption and limitation of carrying capacity of space rockets, links of the space manipulators are required to be light as well as other space structures. As a result of the elasticity of the arms and the structures, undesirable low frequency vibrations may occur.

Modeling and vibration control of flexible systems have received a great deal of attention in recent years[2-12]. This problem has arisen, in particular, in the area of space and industrial robots with lightweight and flexible links.

In the case of constructing large space structures, for example a space station, by using space robot manipulators, it is necessary to control not only the position and vibration of the manipulators but also force exerted by hand on an object. Force control of rigid manipulators has been studied by many researchers[13-17] On the basis of an approximate linear model, the stability of force-controlled flexible manipulators is analyzed[18,19]. A stability analysis of a constrained flexible manipulator which is controlled as a rigid manipulator is discussed[20]. A quasi-static hybrid controller of the two-degree-of-freedom flexible manipulators by using the static relation between the elastic deformation and the contact force is developed[21,22]. Simulation results of the dynamic hybrid position/force control of a two-link manipulator with a flexible second link are reported[23].

Many benefits can be obtained by applying multiple cooperating manipulators to industrial manufacturing and space missions. A typical example is in a flexible assembly. For dexterous manipulation, multiple cooperating manipulators are necessary. Many studies related to the cooperative control of multiple rigid manipulators have been reported[24-27]. There are quite few papers which treat cooperative control of flexible rnanipulators[28,29].

In this chapter we discuss modeling methodology and force control scheme of constrained flexible manipulators. There are some difficulties of the modeling and the controller design of constrained flexible manipulators. As the end-point of the flexible manipulators is constrained on the given surface, a natural boundary condition is not homogeneous but nonhomogeneous. The exact model can be described as a distributed parameter system of an infinite dimensional system. As the dimension of the controller is finite, an approximated finite-dimensional model, which satisfies the nonhomogeneous boundary condition, should be derived for designing a controller. In the controller design phase, in order to suppress the spillover instability caused by the residual modes which are neglected at the controller design phase, a robust controller should be constructed. In the first section, the distributed-parameter model and the robust force controller of a constrained one-link flexible arm are considered. The second section is devoted to the derivation of the dynamic model of constrained n -link flexible manipulators and the hybrid position/force controller design. The third section addresses modeling and robust cooperative controllers of two one-link flexible arms holding a common rigid object.

We define that a prime ' denotes the derivative with respect to the spatial variable and a dot $^{\bullet}$ denotes the time derivative, in this chapter.

4.2. Force Control of a One-Link Flexible Arm

4.2.1. *Introduction*

In this section, a distributed parameter model of a constrained one-link flexible arm is derived and robust force controllers are constructed on the basis of a finite-dimensional modal model. Since the tip of the flexible arm contacts with a given constraint surface, a constraint condition should be satisfied. By using Hamilton's principle, we derive dynamic equations of the joint angle, the vibration of the flexible arm, and the constraint force. The derived boundary condition is nonhomogeneous. We introduce a new variable and derive a homogeneous boundary condition. By solving the related eigenvalue problem, the eigenvalues and the corresponding eigenfunctions are obtained. Using trancated eigenfunction expansion yields a finite-dimensional modal model. On the basis of the derived model the stability of the force feedback and the impedance control is analyzed. To compensate the spillover instability an optimal controller with low-pass property and a robust H_∞ controller are constructed. Simulation results confirm that the controllers perform remarkably well.

4.2.2. *Modeling of a Constrained Flexible Arm*

4.2.2.1. *Distributed parameter model*

We consider a one-link flexible arm which is rotated by a motor in a horizontal plane. The flexible arm of length L, having uniform mass density ρ per unit length, uniform flexural rigidity EI, is clamped on the vertical shaft of the motor at one end and has a concentrated mass m at the other end. Let J be the moment of inertia of the rotor of the motor. Let θ be the angle of rotation of the motor. In Fig. 4.1, a pair (i, j) of orthogonal unit vectors, which are fixed at the vertical shafts of the motor is shown. They are given by

$$i=[\cos\theta,\sin\theta]^T,\ j=[-\sin\theta,\cos\theta]^T. \tag{4.1}$$

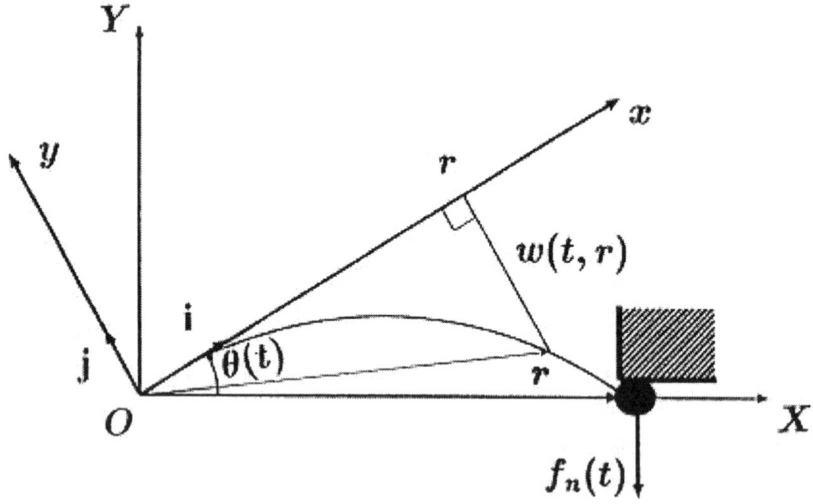

Fig. 4.1. A flexible manipulator

Let $\tau(t)$ be the torque developed by the motor. The flexible arm will deform as shown in Fig. 4.1. Let $w(t, r)$ denote the transverse displacement at time t and at a spatial point r $(0 < r \leq L)$, and let $w_e(t) = w(t, L)$ denote the transverse displacement at the end point. For the flexible arm we use the Euler-Bernoulli model, for which rotary-inertia and shear-deformation effects can be ignored. Because the end point of the manipulator is constrained, a reaction force f_n is generated along the normal direction n of the constraint surface. The orthogonal projection of the reaction force on the unit vector $-i$ acts as an axial force $Q(t)$ of the flexible manipulator. Let P be the position vector of the end point of the arm, and let r be the position vector of the flexible arm at a general position r.

As the root of the flexible arm is clamped on the vertical shaft of the motor, the geometric boundary condition is as follows:

$$w(t,0)=0, \qquad w'(t,0)=0. \tag{4.2}$$

The position vector P, r and their derivatives can be expressed as

$$P = Li - w_e j\,,\;\; \dot{P} = (L\dot{\theta} - \dot{w}_e)j + w_e\dot{\theta}\,i \quad ,$$
$$r = ri - w(t,r)j, \quad \dot{r} = (r\dot{\theta} - \dot{w}(t,r))j + w(t,r)\dot{\theta}\,i \tag{4.3}$$

Let *(X, Y)* designate an inertial Cartesian coordinates variable in a plane. In Fig. 4.1 the constraint surface is described as

$$\Phi(X,Y) = Y = 0. \tag{4.4}$$

By using Eq. (4.1) and Eq. (4.3) the position vector $P = [X, Y]^T$ can be expressed as

$$X = L\cos\theta + w_e \sin\theta, \quad Y = L\sin\theta - w_e\cos\theta. \tag{4.5}$$

The constraint condition (4.4) is rewritten as

$$\phi(\theta, w_e) = L\sin\theta - w_e\cos\theta = 0. \tag{4.6}$$

We assume $\theta \cong 0$. From the relation $Q = f_n \sin\theta$ and the assumption we can set $Q = 0$.

Now, the total kinetic energy T and the potential energy V of the arm are given by

$$T = \frac{1}{2}\int_0^L \rho\dot{r}^T\dot{r}dr + \frac{1}{2}m\dot{P}^T\dot{P} + + \frac{1}{2}J\dot{\theta}^2 \quad ,$$
$$V = \frac{1}{2}\int_0^L EI(w''(t,r))^2 dr \tag{4.7}$$

The virtual work δW is given by

$$\delta W = N_g \tau\delta\theta \tag{4.8}$$

where N_g is a gear ratio of the motor. Let λ be a Lagrange multiplier associated with the constraint. The constraint force for the end point of the arm, i.e., the contact force between the end-effector and the constraint surface, can be expressed in terms of the Lagrange

Multiplier λ. The law governing the motion of mechanical systems is the extended Hamilton's principle which can be written as follows:

$$\int_{t_1}^{t_2} (\delta T - \delta V + \delta W + \lambda \delta \phi) dt \equiv 0. \tag{4.9}$$

Substituting Eq. (4.6)- Eq. (4.8) into Eq. (4.9) and introducing small internal viscous damping of the arm material, we obtain the equation of the joint angle

$$J\ddot{\theta} + EIw''(t,0) = N_g \tau + \lambda \frac{\partial \phi}{\partial \theta} + \lambda \frac{\partial \phi}{\partial w_e} L \tag{4.10}$$

and vibration of the flexible link:

$$\ddot{w}(t,r) + \delta \frac{EI}{\rho} \dot{w}''''(t,r) + \frac{EI}{\rho} w''''(t,r) = r\ddot{\theta} \tag{4.11}$$

$$w(t,0) = 0, w'(t,0) = 0, w''(t,L) = 0 \tag{4.12}$$

$$EI\{\frac{m}{\rho} w''''(t,L) + w'''(t,L)\} = -\lambda \frac{\partial \phi}{\partial w_e} \tag{4.13}$$

where δ is a small damping constant. The natural boundary condition of the last equation of Eq. (4.12) represents the equilibrium of moments at the end point of the flexible arm. The other natural boundary condition of Eq. (4.13) represents the equilibrium of forces that act on the flexible arm along j direction. The right side of Eq. (4.13) implies the external force from the constraint surface. We can find that the natural boundary condition (4.13) is not homogeneous but nonhomogeneous.

4.2.2.2. *Dynamics of amplifiers of the motor*

If an amplifier of the motor is of speed-feedback type, then the torque developed by the motor is given by

$$N_g \tau(t) = k(V_{ref}(t) - k_f \dot{\theta}(t)) \tag{4.14}$$

where V_{ref} is the input-speed reference voltage to the amplifier, k is a large feedback gain constant and k_f is a constant[30]. Substituting Eq. (4.10) into Eq. (4.14) yields

$$V_{ref}(t) = k_f \dot{\theta}(t) + \frac{1}{k}\left\{ Ju(t) + EIw''(t,0) - \lambda \frac{\partial \phi}{\partial \theta} - \lambda \frac{\partial \phi}{\partial w_e} L \right\}, \quad (4.15)$$

$$u(t) = \ddot{\theta}(t).$$

As we set the initial rotational speed is zero ($\dot{\theta}(0) = 0$) and k is large enough, Eq. (4.15) can be approximated as

$$V_{ref}(t) = k_f \dot{\theta}(t) = k_f \int_0^t u(t)dt \,. \quad (4.16)$$

In the case of a torque control type amplifier, let the joint driving torque τ be given as

$$N_g \tau = -\lambda \frac{\partial \phi}{\partial \theta} - \lambda \frac{\partial \phi}{\partial w_e} L + EIw''(t,0) + Ju \quad (4.17)$$

where the value of $w''(t, 0)$ can be measured by using a strain gage sensor. Substituting Eq. (4.17) into the joint angle equation (4.10) gives the relation

$$\ddot{\theta}(t) = u(t). \quad (4.18)$$

Thus, the angular acceleration $\ddot{\theta}(t)$ can be regarded as the input for the system.

4.2.2.3. *Relation between constraint force and Lagrange multiplier*

The relation between the constraint force f_n and Lagrange multiplier λ is given by

$$J_c^T f_n n = \lambda \begin{bmatrix} \dfrac{\partial \phi}{\partial \theta} \\ \dfrac{\partial \phi}{\partial w_e} \end{bmatrix} \quad (4.19)$$

where J_c is a Jacobian matrix between the Cartesian coordinate and the generalized coordinate $[\theta \quad w_e]^T$. Calculating the both sides of Eq. (4.19) gives

$$J_c^T f_n n = \begin{bmatrix} -L\sin\theta + w_e\cos\theta & \sin\theta \\ L\cos\theta + w_e\sin\theta & -\cos\theta \end{bmatrix}^T f_n \begin{bmatrix} 0 \\ 1 \end{bmatrix},$$

$$\lambda \begin{bmatrix} \dfrac{\partial\phi}{\partial\theta} \\ \dfrac{\partial\phi}{\partial w_e} \end{bmatrix} = \lambda \begin{bmatrix} L\cos\theta + w_e\sin\theta \\ -\cos\theta \end{bmatrix}.$$

From this equation we obtain a simple relation between f_n and λ:

$$f_n(t) = \lambda(t). \tag{4.20}$$

4.2.2.4. *Derivation of a homogeneous boundary condition*

As the natural boundary condition (4.13) is nonhomogeneous, it is difficult to treat it directly. We must translate the nonhomogeneous boundary condition into a homogeneous boundary condition. Let us define

$$f(t) \equiv -\lambda \frac{\partial\phi}{\partial w_e}.$$

Using Eq. (4.20) and the assumption $\theta \cong 0$ gives

$$f(t) f_n \cos\theta \cong f_n(t) = \lambda(t). \tag{4.21}$$

We introduce a new variable

$$v(t,r) = w(t,r) - f(t)(-\frac{L}{2EI}r^2 + \frac{1}{6EI}r^3) \tag{4.22}$$

and derive a homogeneous boundary conditions. By using Eq. (4.22) the vibration equation (4.11)-(4.13) can be rewritten as

$$\ddot{v}(t,r) + \delta\frac{EI}{\rho}\dot{v}''''(t,r) + \frac{EI}{\rho}v''''(t,r) + \ddot{f}(t)\left(-\frac{L}{2EI}r^2 + \frac{1}{6EI}r^3\right) = r\ddot{\theta} \tag{4.23}$$

$$v(t,0)=0, v'(t,0)=0, v''(t,L)=0, \qquad \frac{m}{\rho}v''''(t,L)+v'''(t,L)=0. \quad (4.24)$$

We can find that the obtained boundary conditions (4.24) are homogeneous.

4.2.3. *Finite-dimensional Modal Model*

4.2.3.1. *Eigenvalue problem*

Let us consider the eigenvalue problem related to the partial differential equation (4.23) and the boundary conditions (4.24), which is given by

$$\begin{aligned} &\alpha\varphi''''(r)=\gamma\varphi(r), \quad (0\le r\le L) \\ &\varphi(0)=\varphi'(0)=\varphi''(L)=0, \quad \frac{m}{\rho}\varphi''''(L)+\varphi'''(L)=0 \end{aligned} \quad (4.25)$$

where $\alpha = EI/\rho$. Let us introduce a parameter β such that

$$\gamma=\alpha(\frac{\beta}{L})^4.$$

By using the boundary conditions (4.24), we see that the boundary value problem (4.25) has solutions, if β_i satisfies

$$1+\cosh\beta\cos\beta+\frac{m}{\rho L}\beta(\sinh\beta\cos\beta-\cosh\beta\sin\beta)=0. \quad (4.26)$$

Let β_i be the solutions of Eq. (4.26) such that $0<\beta_1<\beta_2<\cdots$. Then the eigenvalues are given by

$$\gamma_i=\alpha(\frac{\beta_i}{L})^4, \qquad i=1,2,\cdots.$$

The corresponding eigenfunctions can be calculated as

$$\varphi_i(r)=\frac{1}{C_i}\{\cosh\frac{\beta_i r}{L}-\cos\frac{\beta_i r}{L}-\frac{\cosh\beta_i+\cos\beta_i}{\sinh\beta_i+\sin\beta_i}(\sinh\frac{\beta_i r}{L}-\sin\frac{\beta_i r}{L})\} \quad (4.27)$$

where C_i are arbitrary constants. Let us define an inner product

$$<\zeta_1,\zeta_2>=\int_0^L \zeta_1(r)\zeta_2(r)dr+\frac{m}{\rho}\zeta_1(L)\zeta_2(L)$$

and set

$$C_i=\left\{L+\frac{\rho}{m}\left(\frac{L}{\beta_i}\right)^2\left(\frac{1+\cosh\beta_i\cos\beta_i}{\sinh\beta_i+\sin\beta_i}\right)^2\right\}^{\frac{1}{2}},$$

then the set $\{\varphi_i(\cdot)\}$ of the eigenfunctions forms an orthonormal system in a Hilbert space H defined by this inner product.

2.3.2. *Finite-dimensional modal model*

On the basis of the truncated eigenfunction expansion

$$v(t,r)=\sum_{i=1}^{N}\frac{1}{\gamma_i}v_i(r)\varphi_i(r) \tag{4.28}$$

A finite-dimensional model is derived. In Eq. (4.28), $v_i(t)$ is the solution of

$$\ddot{v}_i(t)=-\delta r_i\dot{v}_i(t)-r_iv_i(t)-d_i\ddot{f}(t)+b_iu(t), \tag{4.29}$$

$$u(t)=\ddot{\theta}(t) \tag{4.30}$$

where

$$d_i=\gamma_i<-\frac{L}{2EI}r^2+\frac{1}{6EI}r^3,\quad \varphi_i(r)>=-\frac{1}{\rho}\varphi_i(L),$$

$$b_i=\gamma_i<r,\quad \varphi_i(r)>=\frac{EI}{\rho}\varphi''_i(0).$$

In the constraint condition (4.6) using the assumption $\theta\cong 0$ and N mode approximation gives

$$L\theta - w_e = L\theta(t) - \{\frac{\varphi_1(L)}{\gamma_1}v_1(t) + \cdots + \frac{\varphi_N(L)}{\gamma_N}v_N(t)\} + \frac{L^3}{3EI}f(t) = 0 \quad (4.31)$$

and

$$L\ddot{\theta}(t) = \left\{\frac{\varphi_1(L)}{\gamma_1}\ddot{v}_1(L) + \cdots + \frac{\varphi_N(L)}{\gamma_N}\ddot{v}_N(L)\right\} - \frac{L^3}{3EI}\ddot{f}(t). \quad (4.32)$$

Using Eq. (4.29), Eq. (4.30), Eq. (4.32) yields

$$\ddot{v}_i(t) + d_i(t)\ddot{f}(t) = -\delta\gamma_i\dot{v}_i(t) - \gamma_i v_i(t) + b_i u(t), \quad (4.33)$$

$$\ddot{f}(t) = -\frac{1}{c}\sum_{j=1}^{N}\varphi_j(L)(v_j(t) + \delta\dot{v}_j(t)) + \frac{g}{c}u(t) \quad (4.34)$$

where

$$c = \sum_{i=1}^{N}\frac{d_i}{\gamma_i}\varphi_i(L) + \frac{L^3}{3EI}, \quad g = \sum_{i=1}^{N}\frac{b_i}{\gamma_i}\varphi_i(L) - 1.$$

Let θ_d and f_d be the desired joint angle and the desired value of f related to the desired contact force f_{nd}, respectively. From the constraint condition (4.31) the relation of them is given by

$$\theta_d = -\frac{L^2}{3EI}f_d, \quad f_d = -f_{nd}.$$

Combining Eq. (4.32) and Eq. (4.33) and using Eq. (4.30) gives the state equation

$$\dot{x}_f = A_f x_f + B_f u \quad (4.35)$$

where

$$x_f = \begin{bmatrix} f - f_d & \dot{f} & v_1 & \cdots & v_N & \dot{v}_N \end{bmatrix}^T \; A_f = \begin{bmatrix} A_{11} & A_{12} \\ 0 & A_{22} \end{bmatrix},$$

$$B_f = \begin{bmatrix} B_{f1} \\ B_2 \end{bmatrix},$$

$$A_{22} = \begin{bmatrix} 0 & 1 & \cdots & 0 & 0 \\ -\gamma_1 + a_{11} & \delta(-\gamma_1 + a_{11}) & \cdots & a_{1N} & \delta a_{1N} \\ \vdots & \vdots & \cdots & \vdots & \vdots \\ 0 & 0 & \cdots & 0 & 1 \\ a_{1N} & \delta a_{1N} & \cdots & -\gamma_N + a_{NN} & \delta(-\gamma_N + a_{NN}) \end{bmatrix},$$

$$B_{f1} = \begin{bmatrix} 0 \\ \frac{g}{c} \end{bmatrix} \quad B_2 = \begin{bmatrix} 0 \\ \mathrm{b}_1 - \frac{d_1 g}{c} \\ \vdots \\ 0 \\ \mathrm{b}_N - \frac{d_N g}{c} \end{bmatrix}, \qquad a_{ij} = \frac{d_i \varphi_j(L)}{c}.$$

By using the force sensor the value of $f(t)$ can be measured. From Eq. (4.21) the output equation can be written as

$$y_f = f(t) = C_f x_f \tag{4.36}$$

where

$$C_f = [1 \ 0 \ \cdots \ 0].$$

Substituting Eq. (4.34) into Eq. (4.33) and eliminating the term of f gives another state equation

$$\dot{x}_\theta = A_\theta x_\theta + B_\theta u \tag{4.37}$$

where

$$x_\theta = \begin{bmatrix} \theta - \theta_d & \dot{\theta} & v_1 & \dot{v}_1 & \cdots & v_N & \dot{v}_N \end{bmatrix}^T,$$

$$A_\theta = \begin{bmatrix} A_{11} & 0 \\ 0 & A_{22} \end{bmatrix}, B_\theta = \begin{bmatrix} B_{\theta 1} \\ B_2 \end{bmatrix}, B_{\theta 1} = \begin{bmatrix} 0 \\ 1 \end{bmatrix}.$$

The joint angle and angular velocity can be measured. By using the strain gage sensor and the force sensor the moment at the root of the arm $w''(t, 0)$ and the contact force f_n can be measured, respectively. On the basis of the relation

$$v''(t,0) = w''(t,0) + \frac{L}{EI} f(t) \tag{4.38}$$

and the output from the strain gage sensor and the force sensor, we can calculate the value of $v''(t, 0)$. The output equation can be written as

$$y = \begin{bmatrix} \theta & \dot{\theta} & v''(t,0) \end{bmatrix}^T = C_\theta x_\theta \tag{4.39}$$

where

$$C_\theta = \begin{bmatrix} 1 & 0 & 0 & 0 & \cdots & 0 & 0 \\ 0 & 1 & 0 & 0 & \cdots & 0 & 0 \\ 0 & 0 & \dfrac{\varphi''_1(0)}{\gamma_1} & 0 & \cdots & \dfrac{\varphi''_N(0)}{\gamma_N} & 0 \end{bmatrix}.$$

Our problem is to construct a controller that accomplishes the condition

$$f \to f_d, \quad \dot{f} \to 0, \quad (\theta \to \theta_d, \dot{\theta} \to 0), \quad v_i \to 0, \quad \dot{v}_i \to 0 \quad (i = 1, \cdots, N, \cdots) \tag{4.40}$$

4.2.3. *Stability Analysis*

On the basis of the finite-dimensional modal model, the stability of the force feedback is analyzed. We consider the direct force feedback (P control) for Eq. (4.35)

$$u = -\frac{c}{g} k_f (f - f_d) \qquad (k_f > 0) \tag{4.41}$$

and the force state feedback (PD control)

$$u = -\frac{c}{g} \left\{ k_{f1}(f - f_d) + k_{f2}\dot{f} \right\} \quad (k_{f1} > 0 \quad k_{f2} > 0) \tag{4.42}$$

In Eq. (4.42) we assume that the derivative $\dot{f}$ of the force can be measured directly. From Eq. (4.35) we find that the both feedback of Eq.

(4.41) and Eq. (4.42) can not ensure the global stability of the closed-loop system in general. In the next section the stability of the closed-loop system is analyzed numerically by using the root locus technique. Front Eq. (4.37) we find that the joint variable feedback (PD Control)

$$u = -k_{\theta 1}(\theta - \theta_d) - k_{\theta 2}\dot{\theta} \quad (k_{\theta 1} > 0 \quad k_{\theta 2} > 0) \qquad (4.43)$$

can ensure the closed-loop stability. Though the rigid mode can be stabilized, the vibration caused by the uncertainty of the link flexibility can not be diminished. The feedback (4.43) is an impedance control. We find that the direct force feedback can not ensure the closed-loop stability and the impedance control can not diminish the vibration but keep the closed-loop stability. We must construct the controller which accomplishes the convergence of the force to the desired value and the vibration absorption. The observer based state feedback law may be able to accomplish the objective. As the original system is infinite dimensional and the controller is finite dimensional, the spillover instability may occur because of the residual modes which are neglected at the controller design phase.

4.2.4. Construction of Robust Controllers

4.2.4.1. *Controller with low-pass property*

Let us consider the state equation for the control mode

$$\dot{x} = Ax + Bu \quad y = Cx \qquad (4.44)$$

where $x \in R^N, u \in R^m, y \in R^r$ We introduce two kind of filters as shown in Fig. 4.2.

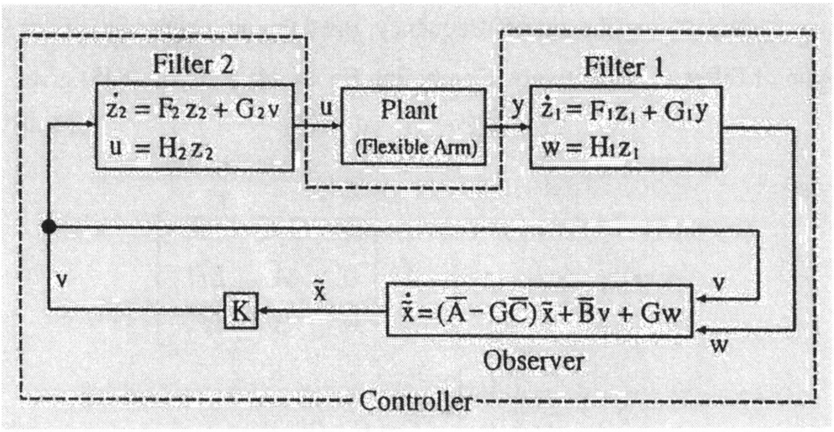

Fig.4.2. Optimal controller with low-pass property

Filter 1 and Filter 2 compensate the observation spillover and the control spillover, respectively. Filter 1 is designed so as to exclude the influence of the high frequency modes which is included in the output data. Filter 2 is designed so as to cut off the high frequency gain of the controller and not to excite the residual modes. The state space equation for Filter 1 and Filter 2 can be represented as

$$\begin{aligned} \dot{z}_1 &= F_1 z_1 + G_1 y \quad w = H_1 z_1 \\ \dot{z}_2 &= F_2 z_2 + G_2 v \quad u = H_2 z_2 \end{aligned}, \tag{4.45}$$

$$F_i = \begin{bmatrix} 0 & I \\ -\Omega_i^2 & -2\Omega_i \Xi_i \end{bmatrix},\ G_1 = \begin{bmatrix} 0 \\ I \end{bmatrix},\ H_1 = [\Sigma_1 \ \ 0],\ G_2 = \begin{bmatrix} 0 \\ \Sigma_2 \end{bmatrix},\ H_2 = [I \ \ 0],$$

$$\Omega_1 = diag\ \{w_{11}, \dots, w_{1r}\}, \quad \Omega_2 = diag\ \{w_{21}, \dots, w_{2m}\},$$

$$\Xi_1 = diag\{\xi_{11}, \cdots, \xi_{1r}\}, \quad \Xi_2 = diag\{\xi_{21}, \cdots, \xi_{2m}\},$$

$$\Sigma_1 = diag\{\sigma_{11}, \cdots, \sigma_{1r}\}, \quad \Sigma_2 = diag\{\sigma_{21}, \cdots, \sigma_{2m}\}.$$

w_{ij}, ξ_{ij} and σ_{ij} are the cut-off frequency, the damping coefficient and the gain of Filter i, respectively. Combining Eq. (4.44) and Eq. (4.45) gives

$$\dot{\bar{x}} = \bar{A}\bar{x} + \bar{B}v \quad w = \bar{C}\bar{x} \tag{4.46}$$

$$\bar{x} = \begin{bmatrix} z_1^T & x^T & z_2^T \end{bmatrix}^T \quad \bar{A} = \begin{bmatrix} F_1 & G_1 C & 0 \\ 0 & A & BH_2 \\ 0 & 0 & F_2 \end{bmatrix}$$

$$\bar{B} = \begin{bmatrix} 0 \\ 0 \\ G_2 \end{bmatrix}, \quad \bar{C} = [H_1 \;\; 0 \;\; 0]$$

The frequency-shaped regulator problem[31] is equivalent to the minimization problem of the corresponding performance index

$$J = \int_0^\infty (\bar{x}^T Q \bar{x} + v^T R v) dt \qquad Q \geq 0, \quad R > 0$$

where $Q \geq 0, \quad R > 0$. By solving the related Riccati equation, we obtain the optimal feedback control law

$$v = -R^{-1}\bar{B}^T P \bar{x} = K\bar{x}. \tag{4.47}$$

As the state vector x_2 cannot be measured directly, we construct an observer.

$$\dot{\tilde{x}} = (\bar{A} - G\bar{C})\tilde{x} + \bar{B}v + Gw,$$
$$v = K\tilde{x} \tag{4.48}$$

The estimated value $\tilde{x}$ are used instead of the immeasurable state value $\bar{x}$.

4.2.4.2. *Robust controller on the basis of H_∞, control theory*

The transfer function $P_N(s)$ corresponding to the nominal model Eq. (4.44) can be represented as

$$P_N(s) = C(sI - A)^{-1}B. \tag{4.49}$$

Though the original system of a distributed parameter system is infinite dimensional, the controller is designed on the basis of the finite-dimensional truncated system. Considering the modeling error caused by using the finite-dimensional truncated expansion, a robust controller on the basis of H_∞, control theory is constructed. Let m be the number of the residual modes that we consider. The truncated eigenfunction expansion can be rewritten as follows:

$$v(t,r) = \sum_{i=1}^{N+m} \frac{1}{\gamma_i} v_i(t)\varphi_i(r). \tag{4.50}$$

Using Eq. (4.50) gives the $(2N+m)+2$ dimensional state equation and the output equation

$$\begin{aligned} \dot{x}(t) &= A_{N+m}x(t) + B_{N+m}u(t) \\ y(t) &= C_{N+m}x(t) \end{aligned} \tag{4.51}$$

and the corresponding transfer function

$$P_{N+m}(s) = C_{N+m}(sI - A_{N+m})^{-1}B_{N+m}. \tag{4.52}$$

The modeling error can be represented as

$$\Delta P(s) = P_{N+m}(s) - P_N(s). \tag{4.53}$$

We assume that $\Delta P(s)$ is stable and satisfies following inequality

$$\|\Delta P(jw)\| < |W(jw)| \quad \text{for } {}^\forall w \in [0,\infty) \tag{4.54}$$

where $\|P(jw)\| = \sqrt{\lambda_{\max}[P^T(-jw)P(jw)]}$, $\lambda_{\max}$ denotes the largest eigenvalue and *W(s)* is a weighting function. We can easily obtain a controller *K(s)* with the following properties

1) The controller *K(s)* stabilizes $P_N(s)$.
2) $|W(jw)|\left\|K(jw)[I - P_N(jw)K(jw)]^{-1}\right\| < 1$ for ${}^\forall w \in [0,\infty)$

by using an algorithm of H_∞ control problem[32]. The control input can be represented as

$$u = K(s)y \tag{4.55}$$

4.2.5. *Simulation*

In this section, we present some results of numerical computation on the basis of the obtained model and the proposed control law. The physical parameters were taken as follows:
$E = 2.06\times10^{11}[N/m^2], I = 9.91\times10^{-12}[m^{-4}], \rho = 4.05\times10^{-1}[kg/m]$, $L = 2[m], m = 0.32[kg], \delta = 5.87\times10^{-4}[s]$. We set the desired force $f_{nd} = 0.5[N]$. The corresponding desired angle is $\theta_d = 0.327[rad]$. We took the first five modes as a simulation model. Figures 4.3 and 4.4 show the root loci of the closed-loop systems for Eq. (4.35) with the direct force feedback P feedback in Eq. (4.41) and PD feedback in Eq. (4.42), respectively.

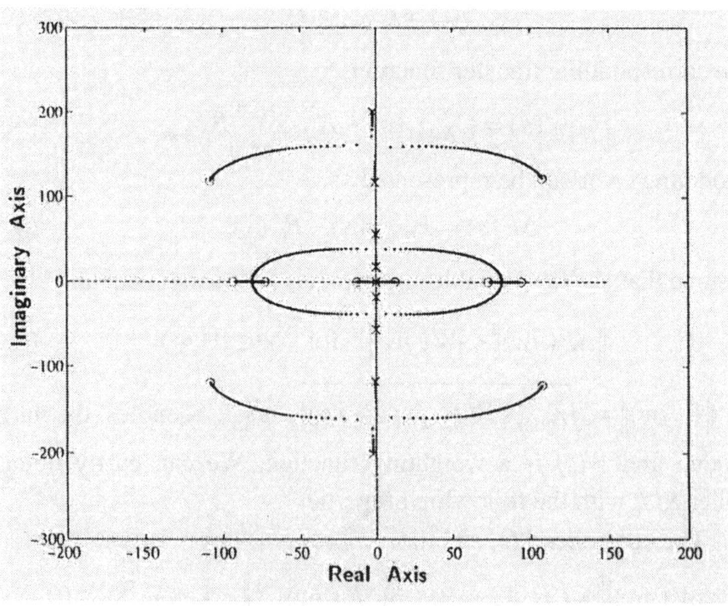

Fig. 4.3. Root locus for direct force feedback (P-controller)

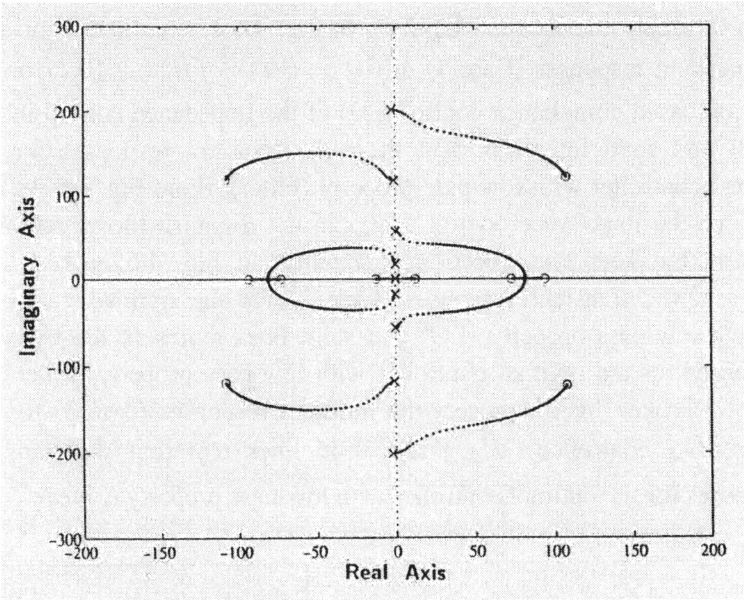

Fig. 4.4. Root locus for direct force feedback (P-controller)

In Fig. 4.3 and 4, O and × denote the poles and the zeros of the closed-loop system, respectively. As the force sensor is located at the end of the arm and the actuator of the motor is located at the root of the arm, the system is non-collocated and non-minimum phase. Some poles of the closed-loop system move to the unstable zeros and the direct force feedback can not ensure the global stability of the closed-loop system. For designing the robust controllers we used the state equation (4.37) and (4.39) corresponding to $N = 3$ and $m = 2$. A robust H_∞, control as a high authority control was applied to the closed-loop system of Eq. 4.37 for the low authority control of the joint PD feedback. We set the filter parameters as $w_0 = 160, \xi_0 = 0.7$ and the weighting function in the design parameter of the H_∞ controller as

$$W(s) = 0.0001 \times (\frac{s+0.5}{s^2+0.8s+0.3} + 0.0015)$$

so as to satisfy the inequality (4.54). In Fig. 4.5 broken lines represent the transient responses (Case 1) of $\theta(t)$, $\dot{\theta}(t)$, $f(t)$ $v''(0,t)$ for the position based compliance control (PD) of the impedance control in Eq. (4.43) and solid lines represent the same transient responses for the optimal controller with low-pass property (Filter). From Fig. 4.5 we can find that the impedance control (PD) can not diminish the vibration of the arm but keep the closed-loop stability. In Fig. 4.6 broken lines represent the transient responses (Case 2) for the optimal controller without low-pass property (LQ) and solid lines represent the transient responses for the optimal controller with low-pass property (Filter). In Fig. 4.7 broken lines represent the transient responses (case 3) for the robust H_∞ controller (H_∞) and solid lines represent the transient responses for the optimal controller with low-pass property (Filter).

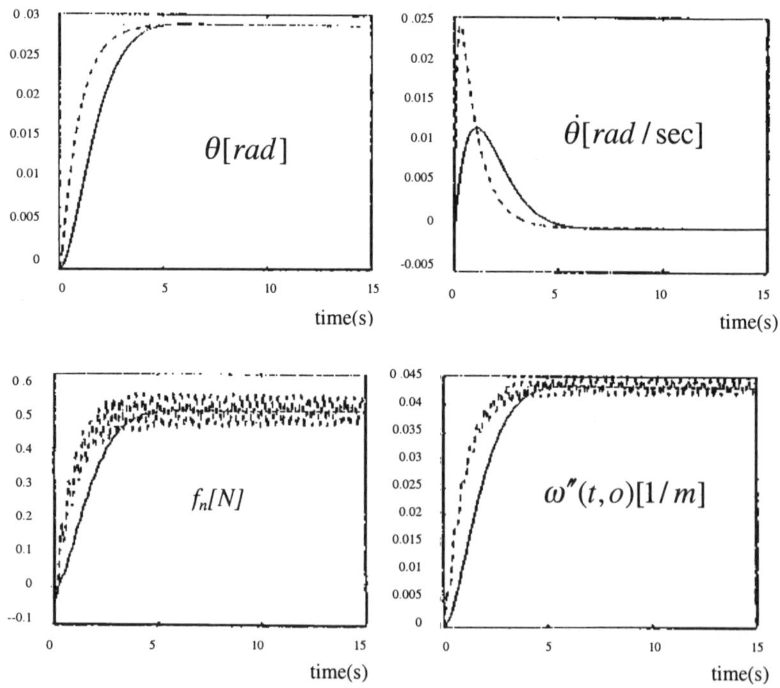

Fig. 4.5. Transient responses for Case 1 (---PD -----Filter)

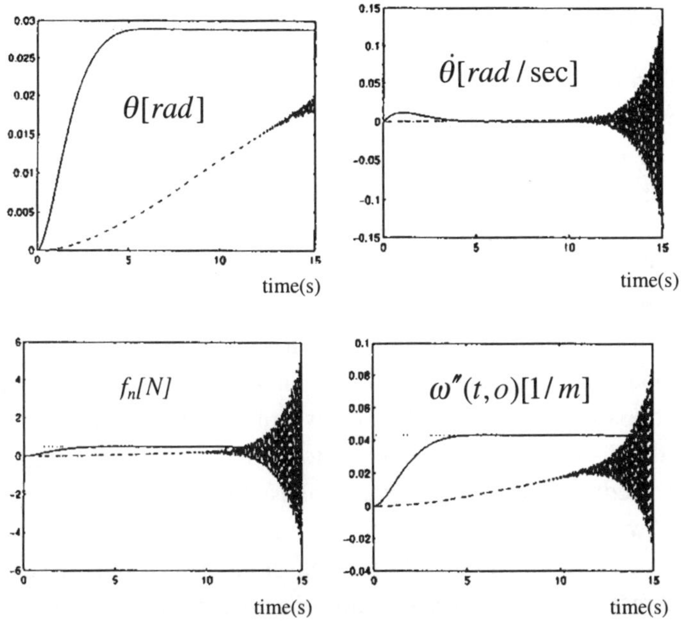

Fig. 4.6. Transient responses for Case 2 (---LQ -----Filter)

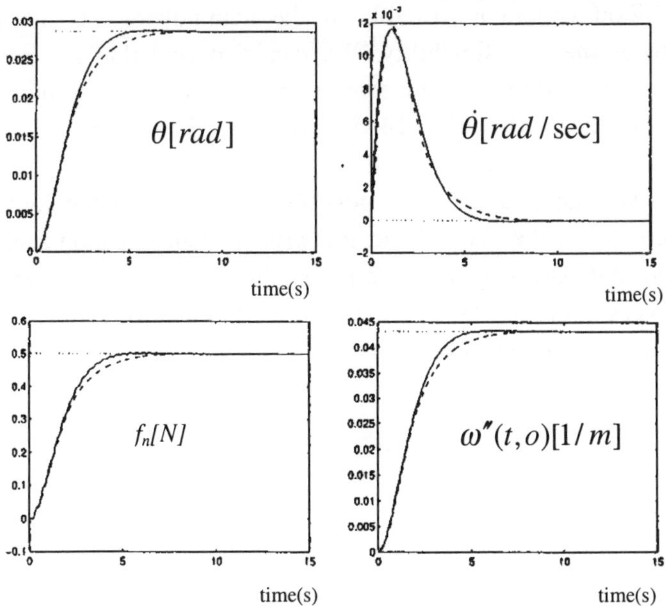

Fig. 4.7. Transient responses for Case 1 (--- H_∞ -----Filter)

4.2.6. *Concluding Remarks*

We have considered the stability of the force feedback and the robust force control of one-link arms considering the distributed link flexibility. We derived dynamic equations of the joint angle, the vibration of the arm, and the contact force by using Hamilton's principle. As the end-point of the flexible arm is constrained on the given constraint surface, the natural boundary condition, which implies the equilibrium of the force at the end-point, is nonhomogeneous. We introduced a new variable and derived a homogeneous boundary condition. By solving a standard boundary value problem the eigenvalues and the corresponding eigenfunctions were obtained. On the basis of a finite-dimensional modal model of the distributed-paramenter systems, we analyzed the stability of the force feedback and constructed robust controllers of an optimal controller with low-pass property and an H_∞ controller. A set of simulations has been carried out. From the simulation results we can find that the direct force feedback can not ensure the closed-loop stability. This result implies that the direct force sensor feedback is not useful and the closed-loop system is unstable, if the manipulator has the slight uncertainty of the link flexibility. We can also find that the position based compliance controller (impedance controller) can not diminish the vibration but keep the closed-loop stability. The proposed robust controllers perform remarkably well.

Though we can find that it is possible to accomplish the force and vibration control of the one-link flexible arm without vibration sensors[33], for example strain gages and accelerometers, there are no space to explain it in detail. Work is in progress in our laboratory to demonstrate the validity of the proposed controller by experiments.

4.3. Dynamic Hybrid Position/ Force Control of Constrained n-link Flexible Manipulators

4.3.1. *Introduction*

In this section, a modeling methodology and a hybrid position/force control law of constrained n-link flexible manipulators are proposed. Since the tip of the flexible manipulator contacts with a given constraint surface, a constraint condition should be satisfied. We approximate the elastic deformations of the flexible links by means of B-spline functions. By using the Lagrange multiplier method, we derive dynamic equations of joint angles, vibrations of the flexible links, and constraint forces. In the previous usual models[5-8] of multi-link flexible manipulators, the region of the applicability of them is restricted to the manipulators which move in the free space and the exact natural boundary conditions are not considered. The proposed model is useful for the constrained flexible manipulators and the constraint forces are concerned with the natural boundary conditions of the vibration equations. On the basis of the singular perturbation method[6,20,23,34], the obtained model is reduced to the slow and fast subsystems. A composite controller for the dynamic hybrid position/force control of the flexible manipulators is designed. Simulation results confirm that the controller performs remarkably well.

4.3.2. *Modeling of a Constrained n-link Flexible Manipulator*

We consider an n-link flexible manipulator as shown in Fig. 4.8 during contact with a known rigid object fixed in the robot work environment.

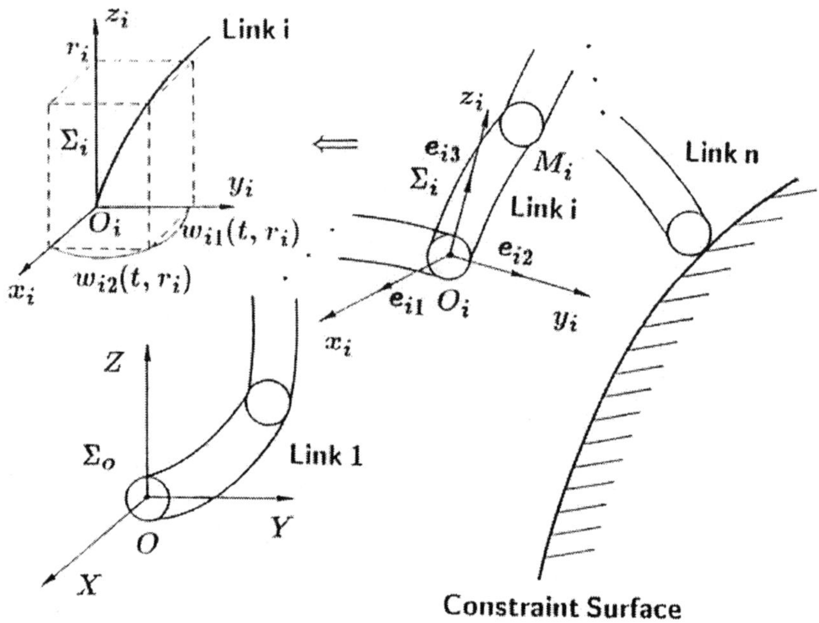

Fig. 4.8. A constrained n-link flexible manipulator

Let us define the coordinate frame Σ_i which is fixed at the i-th link, the principal axes x_i, y_i, z_i and the unit vectors e_{i1}, e_{i2}, e_{i3} along the principal axes as shown in Fig. 4.8. Let θ_i be the angle of rotation of motor i. The flexible link i of length L_i, having uniform mass density ρ_i, uniform flexural rigidity $E_i I_{i1}(x_i - axis), E_i I_{i2}, (y_i - axis)$ is clamped on the vertical shaft of the rotor of the motor i at one end and has a concentrated mass M_i, at the other end. Let r_i be the distance of an arbitrary point on the z_i axis from O_i. Because of the elasticity the links will deform as shown in Fig. 4.8. Let $w_{i1}(t, r_i)$ and $w_{i2}(t, r_i)$ be the elastic deformations along the e_{i1} and e_{i2} directions, respectively. Let us define $h_i = L_i / N_i$ and $h_{ij} = j \cdot h_i$.

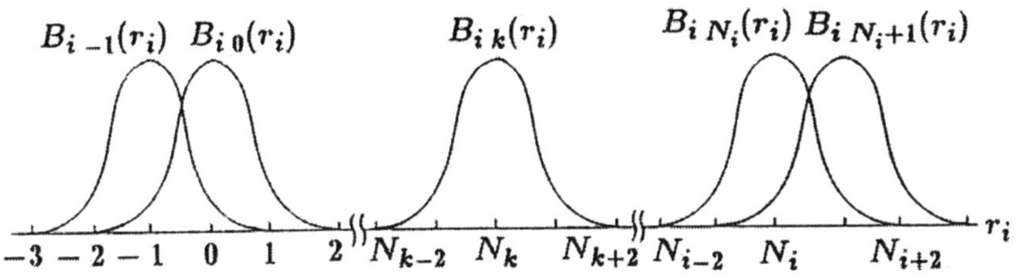

Fig. 4.9. B-spline functions

We approximate the elastic deformations as

$$w_{ij}(t,r_i)=\sum_{k=-1}^{N_i+1} a_{ijk}(t)B_{ik}(r_i) \qquad \left(i=1,\cdots,n,\ j=1,2\right) \quad (4.56)$$

where the subscript i means the number of the link, j means the direction of the elastic deformation along x_i axis (j = 1) or y_i axis (j = 2), k indicates the position on the z_i axis and $B_{ik}(r_i)$ is the B-spline function which is defined as follows:

$$B_{ik}(r_i)=\begin{cases} \dfrac{1}{6h_i^3}(r_i-h_{i\,k-2})^3, & if\ \ r_i\in[h_{i\,k-2}\ \ h_{i\,k-1}]; \\ \dfrac{1}{6h_i^3}\{h_i^3+3h_i^2(r_i-h_{i\,k-1})+3h_i\,(r_i-h_{i\,k-1})^2-3(r_i-h_{i\,k-1})^3\}, & if\ \ r_i\in[h_{i\,k-1}\ \ h_{i\,k}]; \\ \dfrac{1}{6h_i^3}\{h_i^3+3h_i^2(h_{i\,k+1}-r_i)+3h_i\,(h_{i\,k+1}-r_i)^2-3(h_{i\,k+1}-r_i)^3\}, & if\ \ r_i\in[h_{i\,k}\ \ h_{i\,k+1}]; \\ \dfrac{1}{6h_i^3}(h_{i\,k+2}-r_i)^3, & if\ \ r_i\in[h_{i\,k+1}\ \ h_{i\,k+2}]; \\ 0 & otherwise. \end{cases}$$

(4.57)

Figure4.9 shows the shapes of the B-spline functions. Let $[X,Y,Z,\alpha,\beta,\gamma]^T$ be the position and the orientation of the end-effector. Since the tip of the flexible manipulator contacts with a given constraint surface, a constraint condition should be satisfied. The constraint condition can be represented as

$$\Phi(X,Y,Z,\alpha,\beta,\gamma)=0 \tag{4.58}$$

where Φ is an m_c-dimensional vector. As the position and orientation of the end-effector can be represented as a function of $\theta=[\theta_1,\cdots,\theta_n]^T, w_e=[w_{11e},w_{12e},\cdots,w_{n1e},w_{n2e}]^T$, $w'_e=[w'_{11e},w'_{12e},\cdots,w'_{n1e},w'_{n2e}]^T$, the constraint condition (4.58) can be rewritten as

$$\Phi(\theta,w_e,w'_e)=0 \tag{4.59}$$

where $w_{ije}(t)=w_{ij}(t,L_i)$ and $w'_{ije}(t)=w'_{ij}(t,L_i)$. The equilibrium of the forces which act on the tip of the flexible link i along $e_{ij}(j=1,2)$ directions can be represented as

$$\begin{aligned} &M_i\ddot{P}_i^T e_{ij} - E_i I_{ij} w'''_{ije} + E_{i+1} I_{i+1\ j} w'''_{i+1\ j\ o} e^T_{i+1\ j} e_{ij} \\ &= Q_i w'_{ije} + \frac{\partial \phi^T}{\partial w_{ije}}\lambda \end{aligned} \tag{4.60}$$

and the equilibrium of the moments around e_{ij} directions can be represented as

$$E_{i+1} I_{i+1\ j} w''_{i+1\ j\ o} e^T_{i+1\ j} e_{ij} + J_{i+1\ j} B\dot{w}^T_{ije} e_{ij} + E_i I_{ij} w''_{ije} = \frac{\partial \phi^T}{\partial w'_{ije}}\lambda \tag{4.61}$$

where P_i is the position vector of the end point of the flexible link i with respect to the inertia coordinate frame $\Sigma_0, J_{i1}(J_{i2})$ is the moment of inertia of the rotor of the motor i with respect to x_i axis (y_i axis), ${}^B w_i$ is the angular velocity vector with respect to $\Sigma_0, \lambda=[\lambda_1,\cdots,\lambda_{m_c}]^T$ is the Lagrange multiplier associated with the constraint and $w'_{ije}(t)=w'_{ij}(t,L_i), w''_{ije}(t)=w'_{ij}(t,L_i)$, $w'''_{ije}(t)=w'''_{ij}(t,L_i), w'''_{ijo}(t)=w'''_{ij}(t,0), w''_{ijo}(t)=w''_{ij}(t,0)$. In the case of the n-th link the equilibrium of the moments around e_{nj} directions can be expressed as

$$E_n I_{nj} w_{nje}'' = 0 \qquad (j = 1, 2). \tag{4.62}$$

The equations (4.60)-(4.62) are the natural boundary conditions. As the flexible links are clamped on the vertical shaft of the rotor of the motor i, the geometric boundary conditions are given by

$$w_{ij}(t, 0) = 0, \qquad \frac{\partial w_{ij}}{\partial r_i}(t, 0) = 0. \tag{4.63}$$

Substituting Eq. (4.56) into Eq. (4.62) and Eq. (4.63), and using the properties of the B-spline functions, we obtain

$$\begin{aligned} &a_{i\,j\,-1}(t) = c_{-1} a_{ij1}(t), \qquad a_{i\,j\,0}(t) = c_0 a_{ij1}(t), \\ &a_{\mathrm{n}\,j\,N_n+1}(t) = c_{N_n-1} a_{\mathrm{n}\,j\,N_n-1}(t) + c_{N_n} a_{\mathrm{n}\,j\,N_n}(t) \end{aligned} \tag{4.64}$$

where

$$c_{-1} = -\frac{B_{i1}'(0)}{B_{i-1}'(0)}, \quad c_0 = \frac{B_{i-1}(0)B_{i1}'(0) - B_{i-1}'(0)B_{i1}(0)}{B_{i-1}'(0)B_{i0}(0)},$$

$$c_{N_n-1} = -\frac{B_{n\,N_n-1}''(L_n)}{B_{n\,N_n+1}''(L_n)}, \quad c_{N_n} = -\frac{B_{n\,N_n}''(L_n)}{B_{n\,N_n+1}''(L_n)}.$$

The equations (4.60) and (4.61) or (4.62) can be regarded as equations of $a_{i\,j\,N_j}(t)$ and $a_{i\,j\,N_j+1}(t)$ $(i = 1, \cdots, n,\ j = 1, 2)$.

Now, the total kinetic energy T of the manipulator is given by

$$T = \frac{1}{2}\sum_{i=1}^{n}\left\{\int_0^{L_i} \rho_i \dot{r}_i^T \dot{r}_i dr_i + M_i \dot{P}_i^T \dot{P}_i + w_i^T J_i w_i\right\} \tag{4.65}$$

where r_i is the position vector of the flexible link i at a general position with respect to Σ_0, w_i is the angular velocity vector with respect to the coordinate frame Σ_i, J_i is the inertia matrix of the rotor of the motor i with respect to Σ_i. The potential energy V is given by

$$V = V_g + V_s \tag{4.66}$$

$$V_g = \sum_{i=1}^{n} \{ \int_0^{L_i} \rho_i g^T r_i dr_i + M_i g^T P_i \},$$

$$V_s = \frac{1}{2} \sum_{i=1}^{n} \sum_{j=1}^{2} \int_0^{L_i} \{ E_i I_{ij} (\frac{\partial^2 w_{ij}}{\partial r_i^2})^2 + Q_i (\frac{\partial w_{ij}}{\partial r_i})^2 \} dr_i$$

where $g = [0, 0, g]^T$ and g is the acceleration of gravity. By using Eq. (4.64) the approximation (4.56) of the deformation can be rewritten as

$$w_{ij}(t, r_i) = (c_{-1} B_{i-1}(r_i) + c_0 B_{i\,0}(r_i)) a_{i\,j\,1}(t) + \sum_{k=1}^{N_i+1} a_{i\,j\,k}(t) B_{i\,k}(r_i) \quad (i = 1, \cdots, n-1,\ j = 1, 2),$$

$$w_{nj}(t, r_n) = (c_{-1} B_{i-1}(r_i) + c_0 B_{i\,0}(r_i)) a_{i\,j\,1}(t) + \sum_{k=1}^{N_n} a_{n\,j\,k}(t) B_{n\,k}(r_n)$$

$$+ c_{N_n-1} a_{n\,j\,N_n-1}(t) B_{n\,N_n-1}(r_n) + c_{N_n} a_{n\,j\,N_n}(t) B_{n\,N_n}(r_n) \qquad (j = 1, 2). \tag{4.67}$$

Let us define the vector α related to the link deformations as follows:

$$\begin{aligned} a &= [a_1, a_2, \cdots, a_N]^T \doteq [a_{11}^T, a_{12}^T, \cdots, a_{n1}^T, a_{n2}^T]^T \\ a_{ij} &= [a_{ij1}, \cdots, a_{ijN_i+1}]^T \ (i = 1, \cdots, n-1, j = 1, 2), \\ a_{nj} &= [a_{nj1}, \cdots, a_{njN_n}]^T \ (j = 1, 2) \end{aligned} \tag{4.68}$$

where $N = \bar{N}_n - 2, \bar{N}_i = \bar{N}_{i-1} + 2(N_i + 1), \bar{N}_0 = 0.$ The constraint condition (4.59) can be rewritten as

$$\phi(\theta, \alpha) = 0. \tag{4.69}$$

Let us define $q = [\theta^T, \alpha^T]^T$ and Jacobian matrices

$$\dot{r}_i = J_i(r_i)\dot{q}, \qquad \dot{P}_i = J_{ei}\dot{q}, \qquad w_i = J_{wi}\dot{q}. \tag{4.70}$$

By using Eq. (4.70) we obtain lagrangian L as follows:

$$
\begin{aligned}
L &= T - V \\
&= \frac{1}{2} v \sum_{i=1}^{n} \dot{q}^T \left\{ \int_0^{L_i} \rho_i J_i^T (r_i) J_i (r_i) dr_i + M_i J_{ei}^T J_{ei} + J_{wi}^T J_i J_{wi} \right\} \dot{q} \\
&- \sum_{i=1}^{n} g^T \left(\int_0^{L_i} \rho_i r_i dr_i + M_i P_i \right) \\
&- \frac{1}{2} \sum_{i=1}^{n} \sum_{j=1}^{2} a_{ij}^T \left\{ \int_0^{L_i} E_i I_{ij} b_i''(r_i) b_i''(r_i)^T dr_i + Q_i \int_0^{L_i} b_i'(r_i) b_i'(r_i)^T dr_i \right\} a_{ij} \quad (4.71) \\
&\doteq \frac{1}{2} \sum_{k=1}^{n} \sum_{p=1}^{n} H_{11}^{(kp)} \dot{\theta}_k \dot{\theta}_p + \sum_{k=1}^{n} \sum_{p=1}^{N} H_{12}^{(kp)} \dot{\theta}_k \dot{a}_p + \frac{1}{2} \sum_{k=1}^{N} \sum_{p=1}^{N} H_{11}^{(kp)} \dot{a}_k \dot{a}_p \\
&- V_g - \frac{1}{2} \sum_{k=1}^{N} \sum_{p=1}^{N} H_s^{(kp)} a_k a_p .
\end{aligned}
$$

where $b_i(r_i) = [B_{i1}(r_i), \cdots, B_{i\,N_i+1}(r_i)]^T$ and $H^{(ij)}$ is the (i, j) component of the matrix H. We consider the dissipation energy

$$
\begin{aligned}
D &= \frac{1}{2} \sum_{i=1}^{n} \sum_{j=1}^{2} \int_0^{L_i} d_{ij} (\dot{w}_{ij}(t, r_i))^2 dr_i \\
&= \frac{1}{2} \sum_{i=1}^{n} \sum_{j=1}^{2} \dot{a}_{ij}^T \int_0^{L_i} d_{ij} b_i (r_i) b_i^T (r_i) dr_i \dot{a}_{ij} \qquad (4.72) \\
&\doteq \frac{1}{2} \sum_{k=1}^{N} \sum_{p=1}^{N} D^{(kp)} \dot{a}_k \dot{a}_p .
\end{aligned}
$$

The constraint condition (4.69) can be approximated as

$$
\begin{aligned}
\phi(\theta, w_e, w_e') &\cong \phi(\theta, 0) + \sum_{i=1}^{n} \sum_{j=1}^{2} \left(\frac{\partial \phi}{\partial w_{ije}} w_{ije} + \frac{\partial \phi}{\partial w_{ije}'} w_{ije}' \right) \\
&= \phi(\theta, 0) + \sum_{i=1}^{n} \sum_{j=1}^{2} \left\{ \frac{\partial \phi}{\partial w_{ije}} b_i (L_i)^T + \frac{\partial \phi}{\partial w_{ije}'} b_i' (L_i)^T \right\} a_{ij} \qquad (4.73) \\
&\doteq \phi(\theta, 0) + \Theta(\theta) a
\end{aligned}
$$

where $v^{(k)}$ is the k-th component of a vector v. Let $u=[u_1,\cdots,u_n]^T$ be a driving torque vector. The equation of joint angles $\theta_i (i=1,\cdots,n)$ can be derived as

$$\begin{aligned}
&\sum_{k=1}^{n} H_{11}^{(ki)}\ddot{\theta}_k + \sum_{k=1}^{N} H_{12}^{(ki)}\ddot{\alpha}_k + \sum_{p=1}^{n}\sum_{k=1}^{n}\left(\frac{\partial H_{11}^{(ip)}}{\partial \theta_k} - \frac{1}{2}\frac{\partial H_{11}^{(kp)}}{\partial \theta_i}\right)\dot{\theta}_k\dot{\theta}_p \\
&+\sum_{p=1}^{n}\sum_{k=1}^{N}\left(\frac{\partial H_{11}^{(ip)}}{\partial \alpha_k} + \frac{\partial H_{12}^{(ik)}}{\partial \theta_p} - \frac{\partial H_{12}^{(kp)}}{\partial \theta_i}\right)\dot{\alpha}_k\dot{\theta}_p \\
&+\sum_{p=1}^{N}\sum_{k=1}^{N}\left(\frac{\partial H_{12}^{(ip)}}{\partial \alpha_k} - \frac{1}{2}\frac{\partial H_{22}^{(kp)}}{\partial \theta_i}\right)\dot{\alpha}_k\dot{\alpha}_p - \frac{\partial V_g}{\partial \theta_i} \\
&= u_i + \left(\frac{\partial \phi}{\partial \theta_i} + \frac{\partial}{\partial \theta_i}\Theta a\right)^T \lambda
\end{aligned} \tag{4.74}$$

and the equation of vibration $a_i (i \neq \bar{N}_k - 1, \bar{N}_k (k=1,\cdots,n-1), \bar{N}_n)$

$$\begin{aligned}
&\sum_{k=1}^{n} H_{12}^{(ki)}\ddot{\theta}_k + \sum_{k=1}^{N} H_{22}^{(ki)}\ddot{\alpha}_k + \sum_{p=1}^{n}\sum_{k=1}^{n}\left(\frac{\partial H_{12}^{(ip)}}{\partial \theta_p} - \frac{1}{2}\frac{\partial H_{11}^{(kp)}}{\partial \alpha_i}\right)\dot{\theta}_k\dot{\theta}_p \\
&+\sum_{p=1}^{N}\sum_{k=1}^{n}\left(\frac{\partial H_{12}^{(ki)}}{\partial \alpha_p} + \frac{\partial H_{22}^{(pi)}}{\partial \theta_k} - \frac{\partial H_{12}^{(kp)}}{\partial \alpha_i}\right)\dot{\theta}_k\dot{\alpha}_p \\
&+\sum_{p=1}^{N}\sum_{k=1}^{N}\left(\frac{\partial H_{22}^{(ki)}}{\partial \alpha_p} - \frac{1}{2}\frac{\partial H_{22}^{(kp)}}{\partial \alpha_i}\right)\dot{\alpha}_k\dot{\alpha}_p - \frac{\partial V_g}{\partial \alpha_i} \\
&-\sum_{k=1}^{N} H_s^{(ki)}\alpha_k - \sum_{k=1}^{N} D^{(ki)}\dot{\alpha}_k - \sum_{k=1}^{m_c} \Theta^{(ik)}\lambda^{(k)} = 0
\end{aligned} \tag{4.75}$$

Combining Eq. (4.60) – Eq. (4.62), Eq. (4.74) and Eq. (4.75) gives the equation of constrained n-link flexible manipulators as follows:

$$\begin{aligned}
&\begin{bmatrix} M_{11}(\theta,a) & M_{12}(\theta,a) \\ M_{21}(\theta,a) & M_{22}(\theta,a) \end{bmatrix}\begin{bmatrix} \ddot{\theta} \\ \ddot{a} \end{bmatrix} = \begin{bmatrix} f_1(\theta,\dot{\theta},a) \\ f_2(\theta,\dot{\theta},a) \end{bmatrix} + \begin{bmatrix} g_1(\theta,\dot{\theta},a,\dot{a}) \\ g_2(\theta,\dot{\theta},a,\dot{a}) \end{bmatrix} \\
&+\begin{bmatrix} \frac{\partial \phi^T}{\partial \theta}(\theta) + h_1(\theta,a) \\ h_2(\theta) \end{bmatrix}\lambda + EI\begin{bmatrix} O \\ Ka \end{bmatrix} + \begin{bmatrix} u \\ O \end{bmatrix}
\end{aligned} \tag{4.76}$$

where $EI = \max_{ij} E_i I_{ij}$. In the case that the k-th link is rigid, it is possible to obtain the dynamic model related to Eq. (4.76) by setting

$$w_{kj}(t, r_i) = 0 \quad (j = 1, 2).$$

The model proposed by Low[7] of flexible manipulators as a distributed parameter system doesn't clear the exact natural boundary conditions. In many references[6,18-20] the assumed mode method was used. In the previous usual models of multi-link flexible manipulators, the region of the applicability of them is restricted to the manipulators which move in the free space and the exact natural boundary conditions are not considered. The exact natural boundary conditions of the constrained flexible manipulator are concerned with the constraint force. As the tip of the flexible manipulator contacts with the constraint surface, we can't treat the flexible links as clamped-free beams. The proposed model (4.76) is useful for the constrained flexible manipulators.

4.3.3. *Construction of Hybrid Controller*

4.3.3.1. *Singular perturbation method*

By using the singular perturbation method[34], we derive the slow subsystem and the fast subsystem corresponding to the full system (4.76). Let define

$$\begin{bmatrix} M_{11} & M_{12} \\ M_{21} & M_{22} \end{bmatrix}^{-1} = \begin{bmatrix} H_{11} & H_{12} \\ H_{21} & H_{22} \end{bmatrix}, \alpha = \mu K^{-1}\xi. \tag{4.77}$$

The full system is rewritten as

$$\ddot{\theta} = H_{11}(\theta) f_1(\theta, \dot{\theta}) + H_{12}(\theta) f_2(\theta, \dot{\theta})$$

$$+ \left\{ H_{11}(\theta) \left(\frac{\partial \phi^T}{\partial \theta}(\theta, \mu K^{-1}\xi) + h_1(\theta, \mu K^{-1}\xi) \right) + H_{12}(\theta) h_2(\theta, \mu K^{-1}\xi) \right\} \lambda$$

$$+ H_{11}(\theta) g_1(\theta, \dot{\theta}, \mu K^{-1}\xi, \mu K^{-1}\dot{\xi})$$

$$+ H_{12}(\theta) g_2(\theta, \dot{\theta}, \mu K^{-1}\xi, \mu K^{-1}\dot{\xi}) + H_{12}(\theta)\xi + H_{11}(\theta)\tau, \tag{4.78}$$

$$\mu K^{-1}\ddot{\xi} = H_{21}(\theta)f_1(\theta,\dot{\theta}) + H_{22}(\theta)f_2(\theta,\dot{\theta})$$
$$+\left\{H_{21}(\theta)\left(\frac{\partial\phi^T}{\partial\theta}(\theta,\mu K^{-1}\xi) + h_1(\theta,\mu K^{-1}\xi)\right) + H_{22}(\theta)h_2(\theta,\mu K^{-1}\xi)\right\}\lambda$$
$$+H_{21}(\theta)g_1(\theta,\dot{\theta},\mu K^{-1}\xi,\mu K^{-1}\dot{\xi})$$
$$+H_{22}(\theta)g_2(\theta,\dot{\theta},\mu K^{-1}\xi,\mu K^{-1}\dot{\xi}) + H_{22}(\theta)\xi + H_{21}(\theta)\tau. \quad (4.79)$$

Formally, setting μ = 0 and solving for ξ in Eq. (4.79), we obtain

$$\bar{\xi} = -H_{22}^{-1}(\bar{\theta},0)\Big[H_{21}(\bar{\theta},0)f_1(\bar{\theta},\dot{\bar{\theta}},0) + H_{22}(\bar{\theta},0)f_2(\bar{\theta},\dot{\bar{\theta}},0)$$
$$+\left\{H_{21}(\bar{\theta},0)\left(\frac{\partial\phi^T}{\partial\theta}(\bar{\theta},0) + h_1(\bar{\theta},0)\right) + H_{22}(\bar{\theta},0)h_2(\bar{\theta},0)\right\}\lambda \quad (4.80)$$
$$+H_{21}(\bar{\theta},0)g_1(\bar{\theta},\dot{\bar{\theta}},0,0) + H_{22}(\bar{\theta},0)g_2(\bar{\theta},\dot{\bar{\theta}},0,0) + H_{22}(\bar{\theta},0)u_S\Big]$$

where the overbars are used to indicate that the system with $u = 0$ is considered $u_S = \bar{\tau}$. The state variable $\bar{\xi}$ corresponds to a static elastic deformation for the slow time scale. Substituting Eq. (4.80) into Eq. (4.78) with $\mu = 0$ and using the relations $M_{11}^{-1} = H_{11} - H_{12}H_{22}^{-1}H_{21}$, $g_1(\bar{\theta},\dot{\bar{\theta}},0,0) = 0$ yields the slow subsystem which is equivalent to the dynamic equation of the rigid manipulator[6,23]

$$M_{11}(\bar{\theta},0)\ddot{\bar{\theta}} = f_1(\bar{\theta},\dot{\bar{\theta}},0) + \frac{\partial\phi^T}{\partial\theta}(\bar{\theta},0)\lambda + u_S. \quad (4.81)$$

Let us define

$$\mu = \frac{1}{EI},\ \varepsilon = \sqrt{\mu},\ X_1 = \theta,\ X_1 = \dot{\theta},\ z_1 = K^{-1}\xi,\ z_2 = \varepsilon K^{-1}\dot{\xi}. \quad (4.82)$$

The full system (4.78) and (4.79) can be rewritten as

$$\dot{X}_1 = X_2,$$
$$\dot{X}_2 = H_{11}f_1 + H_{12}f_2 + \left\{H_{11}\left(\frac{\partial\phi^T}{\partial\theta} + h_1\right) + H_{12}h_2\right\}\lambda$$
$$+H_{11}g_1 + H_{12}g_2 + H_{12}Kz_1 + H_{11}\tau, \quad (4.83)$$

$$\varepsilon \dot{z}_1 = z_2$$

$$\varepsilon \dot{z}_2 = H_{21} f_1 + H_{22} f_2 + \left\{ H_{21} \left(\frac{\partial \phi^T}{\partial \theta} + h_1 \right) + H_{22} h_2 \right\} \lambda$$

$$+ H_{21} g_1 + H_{22} g_2 + H_{22} K z_1 + H_{21} \tau . \qquad (4.84)$$

To derive the fast subsystem, we introduce the fast time scale $s = \frac{t}{\varepsilon}$ and new fast variables

$$\eta_1 = z_1 - K^{-1} \bar{\xi} \, , \; \eta_2 = z_2 , \; u_f = \tau - \bar{\tau} . \qquad (4.85)$$

Using Eq. (4.85) the equation (4.83) of joint angles can be represented as

$$\frac{dX_1}{ds} = \varepsilon X_2$$

$$\frac{dX_2}{ds} = \varepsilon \left[H_{11} f_1 + H_{12} f_2 + \left\{ H_{11} \left(\frac{\partial \phi^T}{\partial \theta} + h_1 \right) + H_{12} h_2 \right\} \lambda \right.$$

$$\left. + H_{11} g_1 + H_{12} g_2 + H_{12} (K \eta_1 + \bar{\xi}) + H_{11} \tau \right] . \qquad (4.86)$$

From Eq. (4.86) we can find that $\varepsilon \to 0 \Rightarrow \frac{dX_1}{ds} = 0$, $\frac{dX_2}{ds} = 0 \Rightarrow X_1 =$ const., $X_2 =$ const. and the slow variables X_1 and X_2 are constant in the fast subsystem. Using Eq. (4.80), Eq. (4.84) and Eq. (4.85) gives the fast subsystem

$$\frac{d\eta_1}{ds} = \eta_2$$

$$\frac{d\eta_2}{ds} = \frac{d\eta_2}{dt} \frac{dt}{ds} = \varepsilon \dot{z}_2 = H_{22}(\bar{X}_1) K \eta_1 + H_{21}(\bar{X}_1) u_f . \qquad (4.87)$$

4.3.3.2. *Slow controller*

The constraint condition corresponding to the slow subsystem is

$$\phi(\overline{\theta}_1,\cdots,\overline{\theta}_n,0,\cdots.0)=0 \tag{4.88}$$

We assume that Eq. (4.88) can be represented as

$$\theta_2=\Omega(\theta_1) \tag{4.89}$$

$$\theta_1=\phi\left[\overline{\theta}_1,\cdots,\overline{\theta}_{n-m_c}\right]^T,\ \theta_2=\phi\left[\overline{\theta}_{n-m_c+1},\cdots,\overline{\theta}_n\right]^T.$$

Let us define

$$v=\begin{bmatrix} v_1 \\ v_2 \end{bmatrix}=\begin{bmatrix} \theta_1 \\ \theta_2-\Omega(\theta_1) \end{bmatrix},$$

$$T(v)=\begin{bmatrix} I_{n-m_C} & O_{(n-m_C)\times m_C} \\ \dfrac{\partial\Omega}{\partial v_1}(v_1) & I_{m_{C_1}} \end{bmatrix} \tag{4.90}$$

$$M(\theta)=T^TM_{11}T,\ F(\theta)=T^T(M_{11}\dot{T}\dot{v}-f_1),\ I_n=\left[E_1^T \quad E_2^T\right],$$

$$E_1=\left[I_{n-m_C} \quad O_{(n-m_C)\times m_C}\right],\ E_2=\left[O_{m_C\times(n-m_C)} \quad I_{m_C}\right].$$

Let the input of the slow subsystem be given as

$$u_s=(T^T)^{-1}\left[ME_1^T\{\ddot{v}_{1d}-K_v(\dot{v}_1-\dot{v}_{1d})-K_P(v_1-v_{1d})\}\right.$$
$$\left.+F-T^T\frac{\partial\phi^T}{\partial\theta}\lambda_d+E_2^TK_fE_2T^T\frac{\partial\phi^T}{\partial\theta}(\lambda-\lambda_d)\right] \tag{4.91}$$

where $K_P>0$, $K_v>0$, $K_f\neq -I_{m_C}$, $v_{1d}(=\theta_{1d})$ and λ_d are desired values of $v_1(=\theta_1)$ and λ, respectively. From the input torque (4.91) we can obtain the relation of the closed-loop system

$$v_1\to v_{1d},\ \lambda=\lambda_d. \tag{4.92}$$

The constraint force f is represented as

$$f=J_v\frac{\partial\phi^T}{\partial\theta}\lambda \tag{4.93}$$

where J_v is a Jacobian matrix between the Cartesian coordinate and the generalized coordinate $\left[\theta^T \quad w_e^T \quad w_e'^T\right]^T$. From (4.92) and (4.93) we obtain

$$f \to f_d \tag{4.94}$$

where f_d is a desired contact force. Equations (4.92) and (4.94) indicate that the input (4.91) accomplishes the control of the joint angles and the contact force.

4.3.3.3. *Fast controller*

Fast Subsystem (4.87) can be rewritten as

$$\frac{d}{ds}x = A(\overline{X}_1)x + B(\overline{X}_1)u_f \tag{4.95}$$

$$x = \begin{bmatrix}\eta_1^T & \eta_2^T\end{bmatrix}^T, \quad A = \begin{bmatrix} O & I \\ H_{22}(\overline{X}_1)K & O \end{bmatrix}, \quad B = \begin{bmatrix} O \\ H_{21}(\overline{X}_1) \end{bmatrix}.$$

The feedback control of the type

$$u_f = F_f(\overline{X}_1)x \tag{4.96}$$

will stabilize the fast subsystem (4.95). In the feedback law (4.96), the state vector x can't be measured directly. By cementing strain gage foils on m_i points $r_i = l_{ik} (i = 1,\cdots n, k = 1,\cdots, m_i)$ of the flexible link i , it is possible to measure the moments $w''_{i1}(t, l_{ik}),\ w''_{i2}(t, l_{ik})$. Because of the observability of the system, the enough number m_i of the points of strain gage sensors for the link i is $\left[\frac{N_i}{3} + \frac{2}{3}\right]$. The strain gage output $\tilde{y}_{ijk} (i = 1,\cdots n, j = 1,2, k = 1,\cdots, m_i)$ can be expressed as

$$l_{ik} = (3k-1)h_i, \qquad (1 \le k < m_i, 1 \le i \le n)$$

$$\tilde{y}_{ijk}(t) = E_i I_{ij}\{B''_{i\ 3k-2}(l_{ik})a_{ij\ 3k-2}(t)$$

$$+ B''_{i\ 3k-1}(l_{ik})a_{ij\ 3k-1}(t) + B''_{i\ 3k}(l_{ik})a_{ij\ 3k}(t)\},$$

$$l_{im_i} = L_i, \qquad (1 \le i < n)$$

$$\tilde{y}_{ijm_i}(t) = E_i I_{ij}\{B''_{i\ N_1-1}(L_i)a_{ij\ N_1-1}(t)$$

$$+B''_{i\ N_1}(L_i)a_{ij\ N_1}(t)+B''_{i\ N_1+1}(L_i)a_{ij\ N_1+1}(t)\} \quad (4.97)$$

$$l_{nm_n} = L_n - h_n$$

$$\tilde{y}_{njm_n}(t) = E_n I_{nj}\{B''_{n\ N_n-2}(l_{nm_n})a_{nj\ N_n-2}(t)$$

$$+B''_{n\ N_n-1}(l_{nm_n})a_{nj\ N_n-1}(t)+B''_{n\ N_n}(l_{nm_n})a_{nj\ N_n}(t)\}.$$

Let us define

$$\tilde{y} = [\tilde{y}_1^T, \cdots, \tilde{y}_n^T]^T = EI\tilde{C}a,$$

$$\tilde{y}_i = [\tilde{y}_{i1\ 1}, \cdots, \tilde{y}_{i1\ m_i}, \tilde{y}_{i2\ 1}, \cdots, \tilde{y}_{i2\ m_i}]^T. \quad (4.98)$$

Using Eq. (4.77), Eq. (4.82) and Eq. (4.85) yields

$$\tilde{y} = EI\tilde{C}a = \tilde{C}z_1 = \tilde{C}(\eta_1 + K^{-1}\bar{\xi}). \quad (4.99)$$

The output equation can be expressed as

$$y \doteq \tilde{y} - \tilde{C}K^{-1}\bar{\xi} = [\tilde{C} \quad O]\begin{bmatrix}\eta_1 \\ \eta_2\end{bmatrix} \doteq Cx. \quad (4.100)$$

As $\tilde{y}$ can be measured and $\bar{\xi}$ can be calculated by using Eq. (4.80), the vector

$$y = Cx = \tilde{C}\eta_1 = \tilde{y} - \tilde{C}K^{-1}\bar{\xi} \quad (4.101)$$

can be regarded as the output of the system. On the basis of Eq. (4.95) and Eq. (4.101) an observer can be constructed. The control input of the fast subsystem can be represented as

$$u_f = F_f(\bar{X}_1)z \quad (4.102)$$

In the feedback law (4.96), the estimate value z is used instead of the state value x. this feedback control law accomplishes the vibration absorption of the flexible link.

4.3.3.4. *Composite controller*

Combining the slow control input (4.91) and the fast control input (4.102) gives the motor input torque

$$u = u_f + u_s = F_f z + u_s \quad (4.103)$$

which achieves the dynamic hybrid position/force control of the flexible manipulator. The block diagram of the composite controller is shown in Fig. 4.10.

Tikhonov's theorem[34], a fundamental result in the singular perturbation theory, ensures that the state vectors of the full system can be approximated as

$$X_1 = \overline{X}_1 + O(\varepsilon),\ X_2 = \overline{X}_2 + O(\varepsilon)$$

$$z_1 = K^{-1}\overline{\xi} + \eta_1 + O(\varepsilon),\ z_2 = \eta_2 + O(\varepsilon) \qquad (4.104)$$

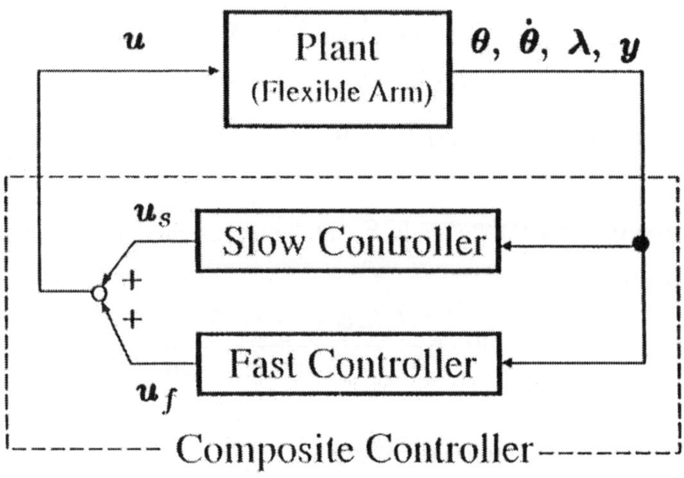

Fig. 4.10. Block diagram of composite controller

4.3.4. *Simulation*

4.3.4.1. *Two-degree-of-freedom flexible manipulator*

We consider a two-degree-of-freedom manipulator which is rotated by two motors in a horizontal plane as shown in Fig. 4.11. The first link is rigid while the second link is flexible. Following the same procedure in the sections 4.4.3.2. and 4.4.3.3., the slow and fast subsystems are derived and the composite controller is constructed. The natural boundary conditions corresponding to Eq. (4.60) and Eq. (4.62) are given by[21]

$$M_2 \ddot{P}^T j_2 - EIw_e''' = Qw_e' + \frac{\partial \phi}{\partial w_e} \lambda , \tag{4.105}$$

$$EIw_e'' = 0 . \tag{4.106}$$

The constraint rigid surface was taken as a straight line Y = aX + b. The constraint conditions corresponding to Eq. (4.58) and Eq. (4.59) are given by

$$\Phi(X,Y) = Y - aX - b = 0$$

$$\begin{aligned}\phi(\theta_1, \theta_2, \overline{w}_e) = L_1 \sin\theta_1 + L_2 \sin(\theta_1 + \theta_2) + \overline{w}_e \cos(\theta_1 + \theta_2) \\ - a\{L_1 \cos\theta_1 + L_2 \cos(\theta_1 + \theta_2) - \overline{w}_e \sin(\theta_1 + \theta_2)\} - b = 0 .\end{aligned} \tag{4.107}$$

Using Eq. (4.107) the angle θ_2 can be represented as a function of θ_1:

$$\theta_2 = \Omega(\theta_1) = \pi - \sin^{-1} \frac{L_1(a\cos\theta_1 - \sin\theta_1) + b}{\sqrt{(L_2 + a\overline{w}_e)^2 + (\overline{w}_e - aL_2)^2}} - \theta_1 - \tan^{-1} \frac{\overline{w}_e - aL_2}{L_2 + a\overline{w}_e}$$

the Lagrange multiplier is given as[23]

$$\lambda = -\frac{Q}{-a\cos(\theta_1 + \theta_2) + \sin(\theta_1 + \theta_2)} . \tag{4.108}$$

The relation between the contact force and the Lagrange multiplier is given as[21]

$$f_n = \lambda \sqrt{1 + a^2} . \tag{4.109}$$

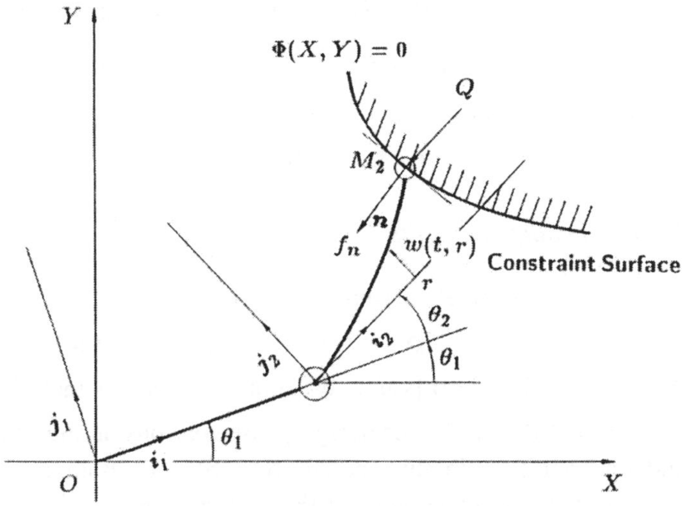

Fig. 4.11. A two-link flexible manipulator

Although the system matrices of the fast subsystem depend on the slow variable X_1, the fast controller is designed on the basis of the fixed model. We calculate the desired joint angle θ_{1d} with considering the elastic deformation w_e as following procedure. The static relation between the contact force and the elastic deformation is given as[21]

$$w_e(t) = \frac{1}{Q(t)}\left\{\lambda \frac{\partial \phi}{\partial w_e} L_2 + E_2 I_2 w''(t,0)\right\}.$$

On the basis of the output data from the force sensor and the strain gage sensor, the value w_e can be calculated. We can also use the following procedure to obtain the value of the elastic deformation w_e. The value $\overline{\xi}$ of Eq. (4.81) implies the static deformation of the flexible link. Using Eq. (4.56) and Eq. (4.77) yields the value of $\overline{w}_e$. It is possible to use the value of $\overline{w}_e$ instead of w_e. The relation between the end position of the manipulator $(X,Y)^T$ and the state variables θ_1, θ_2 w_e is given as

$$X = L_1 \cos\theta_1 + L_2 \cos(\theta_1 + \theta_2) - \overline{w}_e \sin(\theta_1 + \theta_2),$$
$$Y = L_1 \sin\theta_1 + L_2 \sin(\theta_1 + \theta_2) + \overline{w}_e \cos(\theta_1 + \theta_2). \quad (4.110)$$

On the basis of Eq. (4.110), it is easy to calculate the desired angles by using the standard inverse kinematics.

4.3.4.2. *Simulation results*

The physical parameters were taken as follows: L_1 = 1.2 [m], L_2 = 2.0 [m], EI =200 [kgm^2], ρ = 1.0 [kg/m], J_1 = 1.0 [kgm^2], J_2 = 0.1[kgm^2], M_2 = 0.5 [kg], δ = 0 [kg/s]. We took N=12 as a simulation model and set N = 6 at the controller design phase for considering modeling errors and uncertainties. In this model (N = 6) two strain gage sensors are needed for the observability of the system. The locations of them were taken as l_1 = 2h, l_2 = 5h and the outputs of them are represented as $\tilde{y}_1$, $\tilde{y}_2$, respectively. Though the system matrices of the fast subsystem depend on the slow variable *X1*, the fast controller is designed on the basis of the fixed model corresponding to θ_1 = 1.44 [rad], θ_2 = -1.99 [rad]. The flexible manipulator is controlled in such a way that the tip point of the manipulator moves along the given trajectory on the constraint surface with the desired speed 8.5 [cm/s] and that the contact force along the normal vector of the constraint surface is always kept to the desired value 2.0 [N].

Figure 4.12 indicates the desired manipulator configurations. Figure 4.13 indicates the transient responses θ_1, θ_2, $\dot{\theta}_1$, $\dot{\theta}_2$, f_n, $\tilde{y}_2$ of the feedback system when the controller is designed as a rigid manipulator (Slow Control). Figure 4.14 indicates the same transient responses for the proposed composite controller (Composite Control). Figure 4.15 indicates the transient responses θ_1, θ_2, $\dot{\theta}_1$, $\dot{\theta}_2$, f_n, $\tilde{y}_2$ for the slow control with an initial position error (0.1[m]). Figure 4.16 indicates the same transient responses for the composite control with the same initial position error. In Fig. 4.13-16, the solid lines represent the transient responses and the broken lines represent the desired values. It is clear from these figures that the proposed controller works well for the hybrid position/force control of the flexible manipulator.

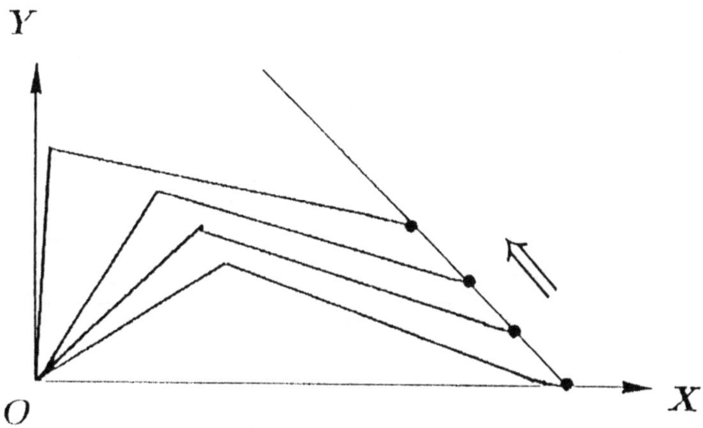

Fig. 4.12. Manipulator configurations

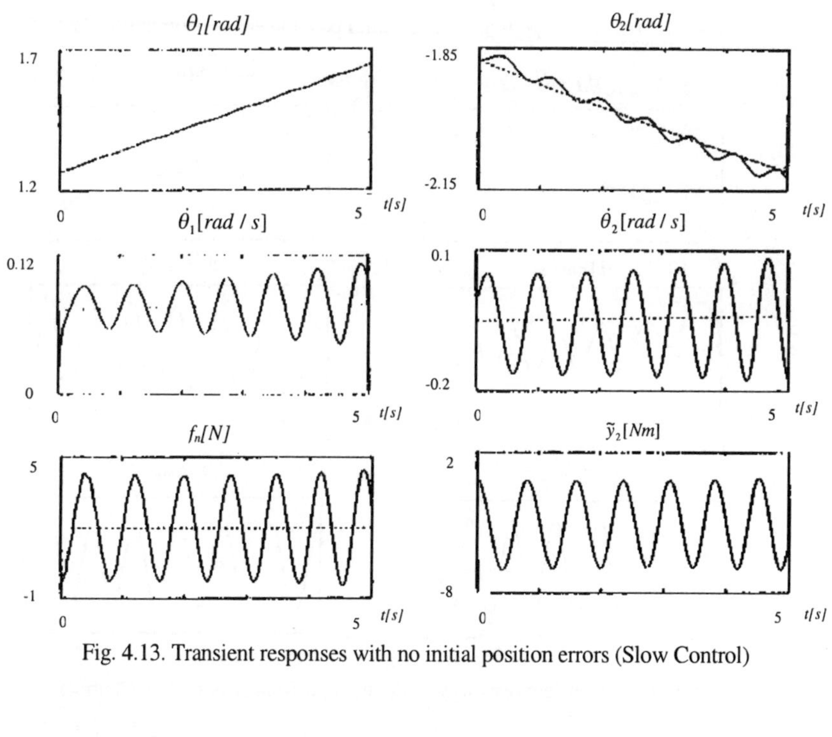

Fig. 4.13. Transient responses with no initial position errors (Slow Control)

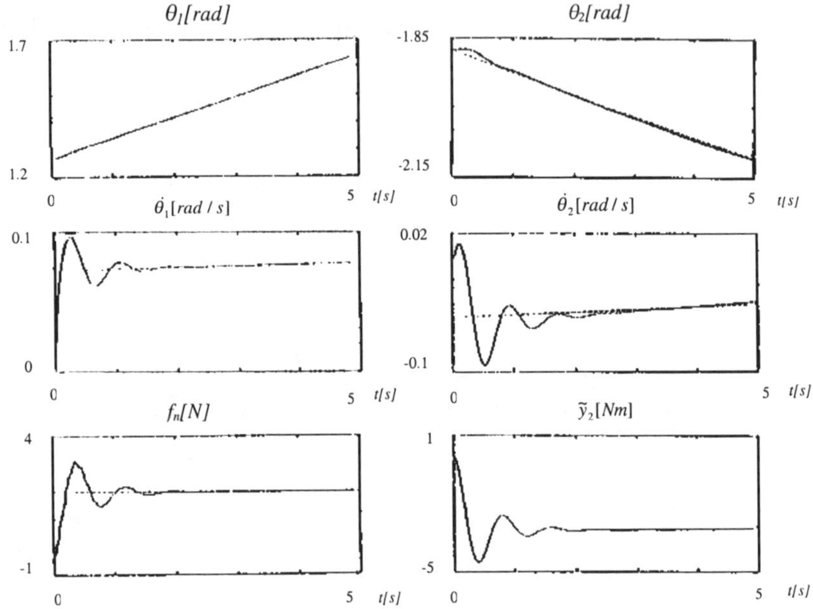

Fig. 4.14. Transient responses with no **initial** position errors (Composite Control)

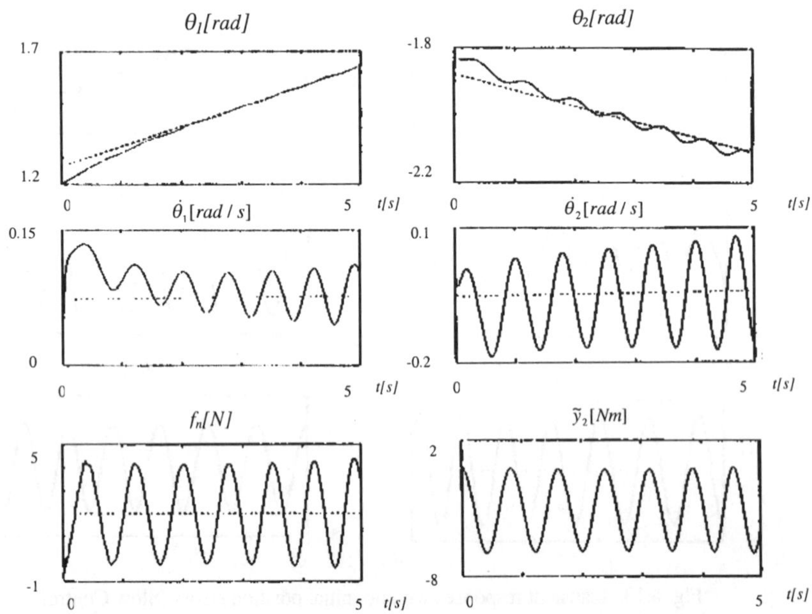

Fig. 4.15. Transient responses with initial position errors (Slow Control)

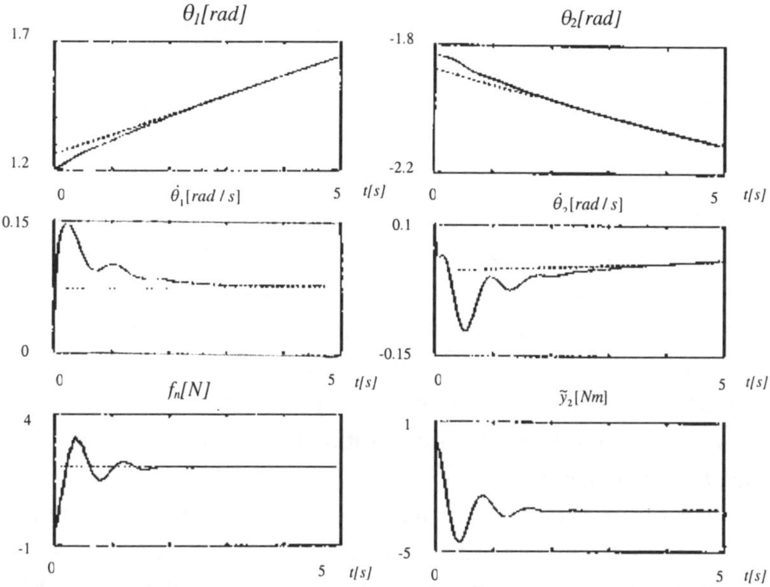

Fig. 4.16. Transient responses with initial position errors (Composite Control)

4.3.5. *Concluding Remarks*

We have considered the problem of dynamic hybrid position/force control of the n-link flexible manipulators. B-spline functions were used for an approximation of the elastic deformations of the flexible links. Since the constraint force and the axial force are considered in the boundary conditions of the flexible links, the proposed model is useful for the constrained flexible manipulator. On the basis of the singular perturbation method, the derived full model was reduced to the slow subsystem and the fast subsystem. A composite controller for the dynamic hybrid position/force control of the flexible manipulator was designed. Simulation results for the two-link flexible manipulator confirm that the controller performs remarkably well. Though the system matrices of the fast subsystem depend on the manipulator configuration, the controller for the fast subsystem is designed on the basis of the fixed matrices which are obtained for a specific manipulator configuration. It is easy to construct a robust controller for the uncertainties of the model by applying the same technique in the section 4.4.2.4.

Though the simulation results for the two-link flexible manipulator in the free workspace by using singular perturbation method have been reported[6], the effectiveness of the composite controller is not so remarkable. The frequency bands of joint angle dynamics and vibration dynamics for non-constrained flexible manipulators moving the free workspace are close. It may be difficult to reduce the full system of non-constrained flexible manipulators to the slow subsystem and the fast subsystem. The frequency of vibration of the constrained flexible manipulator is much higher than that of the non-constrained flexible manipulator. The dynamics of constrained flexible manipulators is a suitable system for applying the singular perturbation method. This is one of the reason that the composite controller is very useful for the hybrid position/force control of the constrained flexible manipulators.

A set of experiments for the planar two-link flexible manipulator has been carried out and the experimental results are reported[33]. The experimental results ensure the effectiveness of the derived model and the proposed dynamic hybrid controller.

4.4. Robust Cooperative Control of Two One-link Flexible Arms

4.4.1. *Introduction*

In this section, a dynamic model and robust control architecture are developed for the closed chain motion of two one-link flexible arms holding a rigid object in a horizontal workspace. Since the hands of the flexible arms hold a common rigid object, a constraint condition should be satisfied. By using the Lagrange multiplier method and Hamilton's principle, we derive the dynamic relations of joint angles, vibrations of the flexible links, and constraint force. The boundary condition of the vibration equation is nonhomogeneous. We introduce a new variable and derive a homogeneous boundary condition. By solving the related eigenvalue problem a finite dimensional modal model is derived. Because of the spillover instability caused by the residual modes which are neglected at the controller design phase, the closed-loop system for a controller without considering the residual modes may be unstable. To compensate the spillover instability an optimal controller with low-pass property31 and a robust H_∞ controller32 are constructed. A set of simulations has been carried out.

4.4.2. *Dynamics of Cooperative Two One-link Flexible Arms*

We consider two one-link flexible arms holding a common rigid object in a horizontal plane as shown in Fig. 4.17.

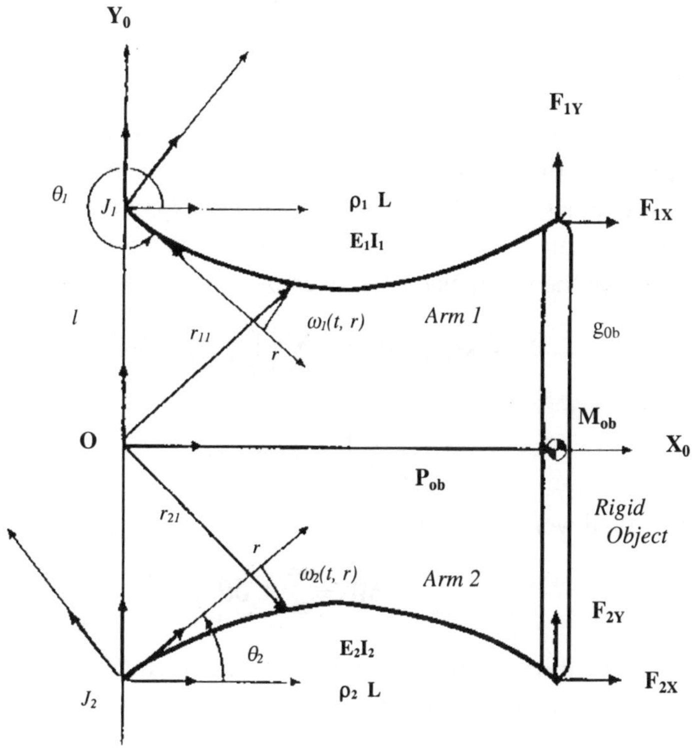

Fig. 4.17. Two flexible arms holding a rigid object

The flexible link of the arm i has, length L, uniform mass density ρ_i per unit length, and uniform flexural rigidity $E_i I_i$. The motor i rotates the flexible link i as shown in Fig. 4.17. The flexible link i is clamped on the rotor of the motor i at one end. The arms hold the rigid object, having length $2l$, mass M_{Ob} and moment of inertia I_{Ob}. Let J_i be the moment of inertia of the rotor of the motor i, θ_i be the angle of rotation of the link i, τ_i be the torque developed by the motor i, Let (X, Y) designate an inertial Cartesian coordinate variable in a plane. Since the hands of the flexible arms hold the rigid object, a constraint condition should be satisfied. The constraint condition for the position of the rigid object is expressed as

$$\phi(\theta_1,\theta_2,w_{1e},w_{2e}) = \begin{bmatrix} L\cos\theta_1 - L\ \cos\theta_2 - w_{1e}\sin\theta_1 - w_{2e}\sin\theta_2 \\ L\sin\theta_1 - L\ \sin\theta_2 + w_{1e}\cos\theta_1 + w_{2e}\cos\theta_2 \end{bmatrix} = 0 \quad (4.111)$$

where $w_{ie} = w_i(t,L)$. Let $\lambda = [\lambda_1 \quad \lambda_2]^T$ be the Lagrange multiplier vector associated with the constraint (4.111). The constraint force for the end point of the manipulator, i.e., the contact force between the hands of the manipulators and the object, can be expressed in terms of the Lagrange multiplier λ. Now, the total kinetic energy T and the potential energy $V$of the manipulator are given by

$$T = \frac{1}{2}M_{ob}\dot{P}_{ob}^T\dot{P}_{ob} + \frac{1}{2}\sum_{i=1}^{2}\left\{\int_0^L \rho_i \dot{r}_{i1}^T \dot{r}_{i1} dr + J_i\dot{\theta}_i^2\right\}$$

$$V = \frac{1}{2}\sum_{i=1}^{2}\int_0^L E_i I_i (w_i''(y,r))^2 dr\,.$$

The virtual work δW is given by

$$\delta W = \sum_{i=1}^{2}\tau_i \delta\theta_i\,.$$

Using Hamilton's principle and considering internal viscous damping of the Voigt type, we obtain the equation of rotation of motor 1

$$J_1\ddot{\theta}_1 - E_1 I_1 w_1''(t,0) = \tau_1 + \lambda^T \frac{\partial\phi}{\partial\theta_1} - \lambda^T \frac{\partial\phi}{\partial w_1 E} L \qquad (4.112)$$

the equation of vibration of link 1:

$$\ddot{w}_1(t,r) + \delta_1 \frac{E_1 I_1}{\rho_1}\dot{w}_1''''(t,r) + \frac{E_1 I_1}{\rho_1} w_1''''(t,r) = -r\ddot{\theta}_1 \qquad (4.113)$$

$$w_1(t,0) = 0,\ w_1'(t,0) = 0,\ w_1''(t,L) = 0, \qquad (4.114)$$

$$E_1I_1\left\{\frac{M_{ob}}{\rho_1}w_1^{''''}(t,L)+w_1^{'''}(t,L)\right\}=-\lambda^T\frac{\partial\phi}{\partial w_{1e}} \tag{4.115}$$

the equation of rotation of motor 2:

$$J_2\ddot{\theta}_2+E_2I_2w_2^{''}(t,0)=\tau_2+\lambda^T\frac{\partial\phi}{\partial\theta_2}+\lambda^T\frac{\partial\phi}{\partial w_{2e}}L \tag{4.116}$$

the equation of vibration of link 2

$$\ddot{w}_2(t,r)+\delta_2\frac{E_2I_2}{\rho_2}\dot{w}_2^{''''}(t,r)+\frac{E_2I_2}{\rho_2}w_2^{''''}(t,r)=r\ddot{\theta}_2 \tag{4.117}$$

$$w_2(t,0)=0\,,\ w_2^{'}(t,0)=0\,,\ w_2^{''}(t,L)=0\,, \tag{4.118}$$

$$E_2I_2w_2^{'''}(t,L)=-\lambda^T\frac{\partial\phi}{\partial w_{2e}} \tag{4.119}$$

where δ_1 , $\delta_2 > 0$ are damping constants. The relation between the grasping force $F_i=\left[F_{iX} \quad F_{iY}\right]^T$ and the Lagrange Multiplier λ is expressed as

$$\frac{\partial\phi^T}{\partial q_i}\lambda=J_{vi}^TF_i\,,\ q_i=\left[\theta_i \quad w_{ie}\right]^T \tag{4.120}$$

where J_{vi} is a Jacobian matrix

$$J_{vi}=\begin{bmatrix}\frac{\partial X_i}{\partial\theta_i} & \frac{\partial X_i}{\partial w_{ie}}\\ \frac{\partial Y_i}{\partial\theta_i} & \frac{\partial Y_i}{\partial w_{ie}}\end{bmatrix} \tag{4.121}$$

$$X_1=L\cos\theta_1-w_{1e}\sin\theta_1\,,Y_1=L\sin\theta_1+w_{1e}\cos\theta_1+l\,,$$
$$X_2=L\cos\theta_2+w_{2e}\sin\theta_2\,,Y_2=L\sin\theta_2-w_{2e}\cos\theta_2-l\,. \tag{4.122}$$

From Eq. (4.120) – Eq. (4.122) we obtain

$$F_1=\lambda\,,\ F_2=-\lambda\,. \tag{4.123}$$

Let us define

$$f_i(t) = -\lambda^T \frac{\partial \phi_i}{\partial w_{ie}}. \tag{4.124}$$

As the boundary conditions (4.115) and (4.119) are nonhomogeneous, we introduce a new variable

$$v_i(t,r) = w_i(t,r) - f_i(t)\left(-\frac{L}{2E_iI_i} r^2 + \frac{L}{6E_iI_i} r^3 \right) \tag{4.125}$$

and derive homogeneous boundary conditions. By using Eq. (4.125) the vibration equations can be rewritten as

$$\ddot{v}_i(t,r) + \delta_i \frac{E_iI_i}{\rho_i} \dot{v}_i''''(t,r) + \frac{E_iI_i}{\rho_i} \dot{v}_i''''(t,r)$$

$$+ \ddot{f}_i(t)\left(-\frac{L}{2E_iI_i} r^2 + \frac{L}{6E_iI_i} r^3 \right) = (-1)^i r\ddot{\theta}_i \tag{4.126}$$

$$v_1(t,r) = 0, \quad v_1'(t,0) = 0, \quad v_1''(t,L) = 0$$

$$\frac{M_{ob}}{\rho_i} \dot{v}_1''''(t,L) + v_1'''(t,L) = 0 \tag{4.127}$$

$$v_2(t,0) = 0, \quad v_2'(t,0) = 0, \quad v_2''(t,L) = 0, v_2'''(t,L) = 0 \tag{4.128}$$

4.4.3. *Finite Dimensional Modal Model*

By solving the eigenvalue problem related to the partial differential equations (4.126) and the boundary conditions (4.127) and (4.128), we can obtain the eigenvalues γ_{ij} and the corresponding eigenfunctions $\varphi_{ij}(r)$ (i = 1, 2, j = 1, 2, ...). We can see that the set { $\varphi_{1j}(\cdot)$ } and { $\varphi_{2j}(\cdot)$ } form the orthonormal systems in Hilbert spaces H_1 and H_2 defined by the inner products

$$<\varsigma_1, \varsigma_2>_1 = \int_0^L \varsigma_1(r)\varsigma_2(r)dr + \frac{M_{ob}}{\rho} \varsigma_1(L)\varsigma_2(L),$$

$$<\varsigma_1,\varsigma_2>_2=\int_0^L \varsigma_1(r)\varsigma_2(r)dr\,,$$

respectively. The solution of Eq. (4.126) can be expressed as

$$v_i(t,r)=\sum_{j=1}^{\infty}\frac{1}{\gamma_{ij}}v_{ij}(t)\varphi_{ij}(r)\,, \tag{4.129}$$

In Eq. (4.129) $v_{ij}(t)\,(i=1,2,\; j=1,2,\bullet\bullet\bullet)$ is the solution of

$$\ddot{v}_{ij}(t)=-\delta_i\gamma_{ij}\dot{v}_{ij}(t)-\gamma_{ij}v_{ij}(t)-d_{ij}\ddot{f}_i(t)+b_{ij}u_i(t) \tag{4.130}$$

$$u_i(t)\equiv\ddot{\theta}_i(t) \tag{4.131}$$

where

$$d_{ij}(t)=\gamma_{ij}<-\frac{L}{2E_iI_i}r^2+\frac{L}{6E_iI_i}r^3,\varphi_{ij}(r)>_i=-\frac{1}{\rho_i}\varphi_{ij}(L)\,,$$

$$b_{ij}(t)=(-1)^i\gamma_{ij}<r,\varphi_{ij}(r)>_i=(-1)^i-\frac{E_iI_i}{\rho_i}\varphi_{ij}''(0)\,.$$

As we can assume $\theta_1\cong 0$, $\theta_2\cong 0$ in the constraint condition (4.111)

$$\phi(\theta_1,\theta_2,w_{1e},w_{2e})\cong\begin{bmatrix} w_{1e}\theta_1+w_{2e}\theta_2 \\ L\theta_1-L_2\theta_2+w_{1e}+w_{2e}\end{bmatrix}=0$$

we obtain the approximated constraint condition

$$\begin{bmatrix}\theta_1\\ \theta_2\end{bmatrix}=\frac{1}{L}\begin{bmatrix}-w_{2E}\\ w_{1E}\end{bmatrix}. \tag{4.132}$$

Using the constraint condition (4.132) and the N mode approximation in Eq. (4.129) gives

$$L\theta_1+w_{2e}=L\theta_1+\left\{\frac{\varphi_{21}(L)}{\gamma_{21}}v_{21}+\cdots+\frac{\varphi_{2N}(L)}{\gamma_{2N}}v_{2N}\right\}-\frac{L^3}{3E_2I_2}f_2=0 \tag{4.133}$$

$$L\theta_2-w_{1e}=L\theta_2-\left\{\frac{\varphi_{11}(L)}{\gamma_{11}}v_{11}+\cdots+\frac{\varphi_{1N}(L)}{\gamma_{1N}}v_{1N}\right\}+\frac{L^3}{3E_1I_1}f_1=0 \tag{4.134}$$

and eliminating the terms of f_1, f_2 gives

$$\ddot{v}_{ij}(t) = -\delta_i \gamma_{ij} \dot{v}_{ij}(t) + \frac{\delta_i d_{ij}}{e_i} \sum_{k=1}^{N} \varphi_{ik}(L) \dot{v}_{ik}(t) - \gamma_{ij} v_{ij}(t)$$

$$+ \frac{d_{ij}}{e_i} \sum_{k=1}^{N} \varphi_{ik}(L) v_{ik}(t) + h_{i1j} u_1(t) + h_{i2j} u_2(t) \tag{4.135}$$

where

$$e_i = \sum_{j=1}^{N} \frac{d_{ij}}{\gamma_{ij}} \varphi_{ij}(L) + \frac{L^3}{3E_i I_i}, \quad g_i = \sum_{j=1}^{N} \frac{d_{ij}}{\gamma_{ij}} \varphi_{ij}(L),$$

$$h_{11j} = b_{1j} - \frac{d_{1j} g_1}{e_1}, h_{12j} = \frac{Ld_{1j}}{e_1}, h_{21j} = -\frac{Ld_{2j}}{e_2}, h_{22j} = b_{2j} - \frac{d_{2j} g_2}{e_2}.$$

Combining Eq. (4.135) and Eq. (4.131) yields the state equation

$$\dot{x} = Ax + Bu \tag{4.136}$$

where

$$x = \begin{bmatrix} x_1^T & x_2^T \end{bmatrix}^T, u = \begin{bmatrix} u_1 & u_2 \end{bmatrix}^T, x_1 = \begin{bmatrix} \theta_1 - \theta_{1d} & \dot{\theta}_1 & \theta_2 - \theta_{2d} & \dot{\theta}_2 \end{bmatrix}^T.$$

$$x_2 = \begin{bmatrix} v_{11} & \dot{v}_{11} & \cdots & v_{1N} & \dot{v}_{1N} & v_{21} & \dot{v}_{21} & \cdots & v_{2N} & \dot{v}_{2N} \end{bmatrix}^T$$

and θ_{id} (i = 1, 2) are the desired angle of rotation corresponding to the desired grasping force. The joint angle and angular velocity x_1 can be measured. By using the strain gage sensor and the force sensor the moment at the root of the arms $w_1''(t,0)$, $w_2''(t,0)$ and the constraint forces $F_1(t)$, $F_2(t)$ can be measured, respectively. On the basis of the relation between the constraint forces and the sate variables f_i

$$f_1(t) = -\lambda^T \frac{\partial \phi}{\partial w_{1e}} = \lambda_1 \sin\theta_1 - \lambda_2 \cos\theta_1$$

$$f_2(t) = -\lambda^T \frac{\partial \phi}{\partial w_{2e}} = \lambda_1 \sin\theta_2 - \lambda_2 \cos\theta 2$$

and Eq. (4.123), the values of f_i can be calculated. Let us define the output equation as

$$y = \begin{bmatrix} y_1^T & y_2^T \end{bmatrix}^T, y_1 = x_1, \; y_2 = \begin{bmatrix} v_1''(t,0) & v_2''(t,0) \end{bmatrix}^T = C_2 x_2 \qquad (4.137)$$

From Eq. (4.125) we can obtain

$$v_i''(t,0) = w_i''(t,0) + \frac{L}{E_i I_i} f_i(t). \qquad (4.138)$$

On the basis of Eq. (4.138) the values of $v_i''(t,0)$ ($i = 1, 2$) can be calculated by using output data from strain gage sensor and force sensor. Let the joint driving torques τ_i be given as

$$\tau_1 = -\lambda^T \frac{\partial \phi}{\partial \theta_1} + \lambda^T \frac{\partial \phi}{\partial w_{1e}} L - E_1 I_1 w_1''(t,r) + J_1 u_1, \qquad (4.139)$$

$$\tau_2 = \lambda^T \frac{\partial \phi}{\partial \theta_2} + \lambda^T \frac{\partial \phi}{\partial w_{2e}} L + E_2 I_2 w_2''(t,0) + J_2 u_2. \qquad (4.140)$$

Substituting Eq. (4.139) and Eq. (4.140) into the joint angle equations (4.112) and (4.116) gives

$$\ddot{\theta}_1(t) = u_1(t), \ddot{\theta}_2(t) = u_2(t) \qquad (4.141)$$

and the angular accelerations can be regarded as the system inputs. Our problem is to construct a controller that accomplishes the condition

$$\theta_i \to \theta_{id}, \dot{\theta}_i \to 0, v_{ij} \to 0, \dot{v}_{ij} \to 0, f_i \to f_{id}, f_i \to 0.$$

From the constraint conditions (4.133) and (4.134) the desired joint angle value θ_{id} and the desired value f_{id} related to the desired grasping force should satisfy the relation

$$\theta_{1d} = \frac{L^2}{3E_2 I_2} f_{2d}, \theta_{2d} = \frac{L^2}{3E_1 I_1} f_{1d}. \qquad (4.142)$$

4.4.4. *Construction of Controller*

On the basis of the finite-dimensional modal model (4.136) and (4.137), an observer based controller can be constructed. To suppress the spillover instability caused by the residual modes which are neglected at the controller design phase, a controller with low-pass property is designed based on the formulation as a frequency-shaped regulator problem as mentioned in the section 4.4.2.4..

The original system of a distributed parameter system is an infinite dimensional system, the controller is designed on the basis of the finite-dimensional system. Considering the modeling error caused by using the finite-dimensional truncated expansion, a robust controller on the basis of the H_∞ control theory is constructed as mentioned in the section 4.4.2.5.

4.4.5. *Simulation*

In this section, we present some results of numerical computation on the basis of the obtained model and the proposed control law. The physical parameters were taken as follows: E = 2.06 x 10^{11}[N/m^2], I = 1.13 x 10^{-10}[m^4], ρ = 1.14[kg/m], L = 1.5[m], M_{ob} = 1.20[kg], δ = 1.17X10^{-4}[s]. We set the desired grasping force f_{nd} = 5.0[N]. The corresponding desired angles are θ_{1d} = -1.62 × 10^{-1}[rad], θ_{2d} = 1.62 ×10^{-1}[rad]. The design parameters of the optimal controller with low-pass property were taken as w_0 = 800, ς_0 = 0.7 so as not to excite the residual modes. We took the first five modes as a simulation model. The optimal controller was designed for the system corresponding to the first three modes. The output values for y_1 and y_2 are calculated on the basis of the system for the first five modes. In H_∞ controller we set the weighting function as

$$W(s) = \frac{0.001s^2 + 0.001s + 0.0001}{0.03s^2 + 15s + 20}$$

In Fig. 4.18 (Case 1) broken lines represent the transient responses of $\theta(t)$, $\dot{\theta}(t)$, $f(t)$, $v''(t,0)$ for the system when the controller has only the angle and the angular velocity feedback (PD) and solid lines represent the same transient responses for the optimal controller with low-pass property (Filter). In Fig. 4.19 (Case 2) broken lines represent the transient responses for PD and solid lines represent the transient responses for the robust H_∞ controller (H_∞). In Fig. 4.20 (Case 3) broken lines represent the transient responses for the optimal controller without low-pass property (LQ) and solid lines represent the transient responses for Filter. Figure 4.21 shows the frequency responses of $|W(j\omega)|$ and $\|\Delta P(j\omega)\|$.

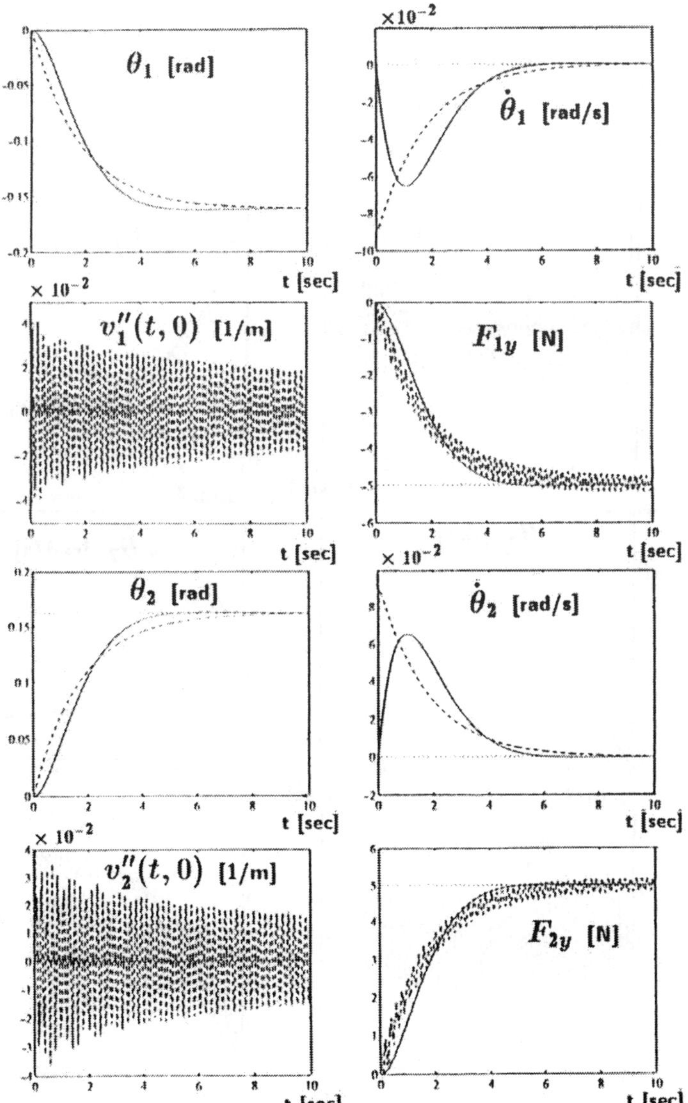

Fig. 4.18. Transient responses for case 1 (- - - PD -----Filter)

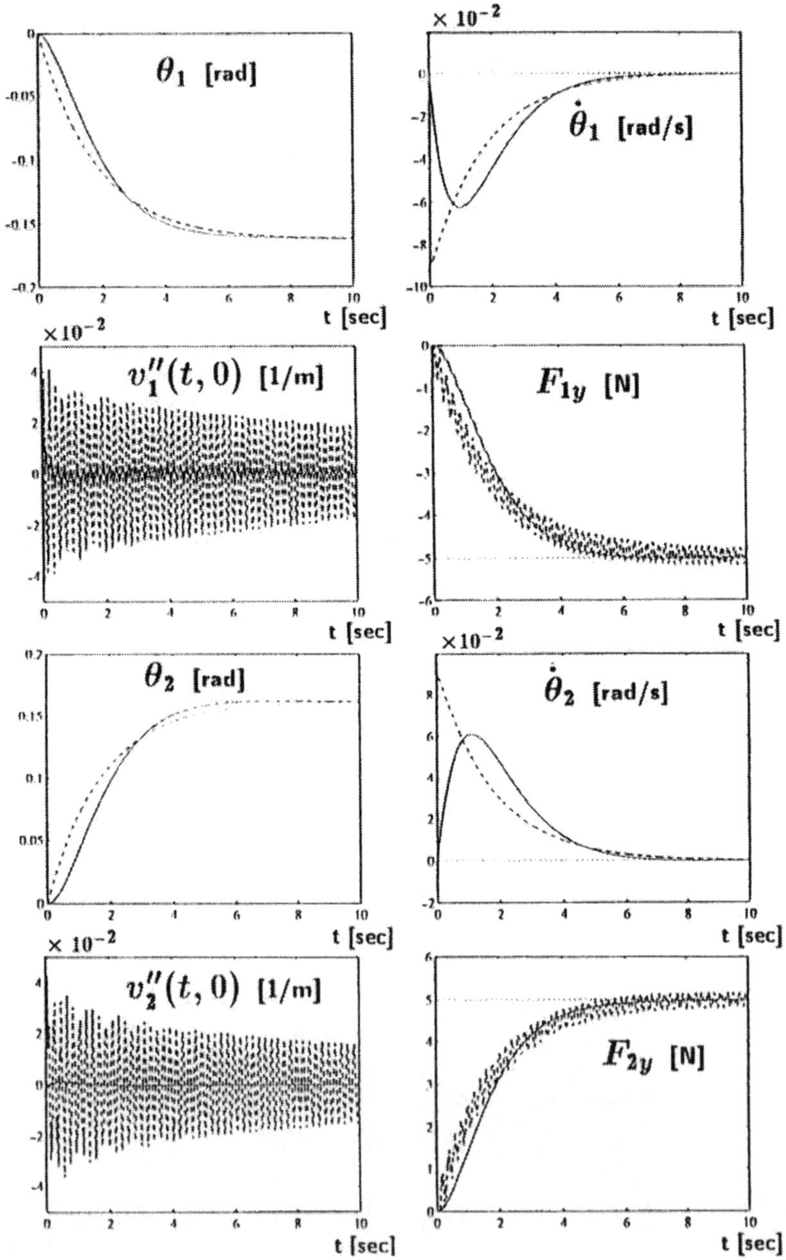

Fig. 4.19. Transient responses for case 2 (- - - PD ----- H_∞)

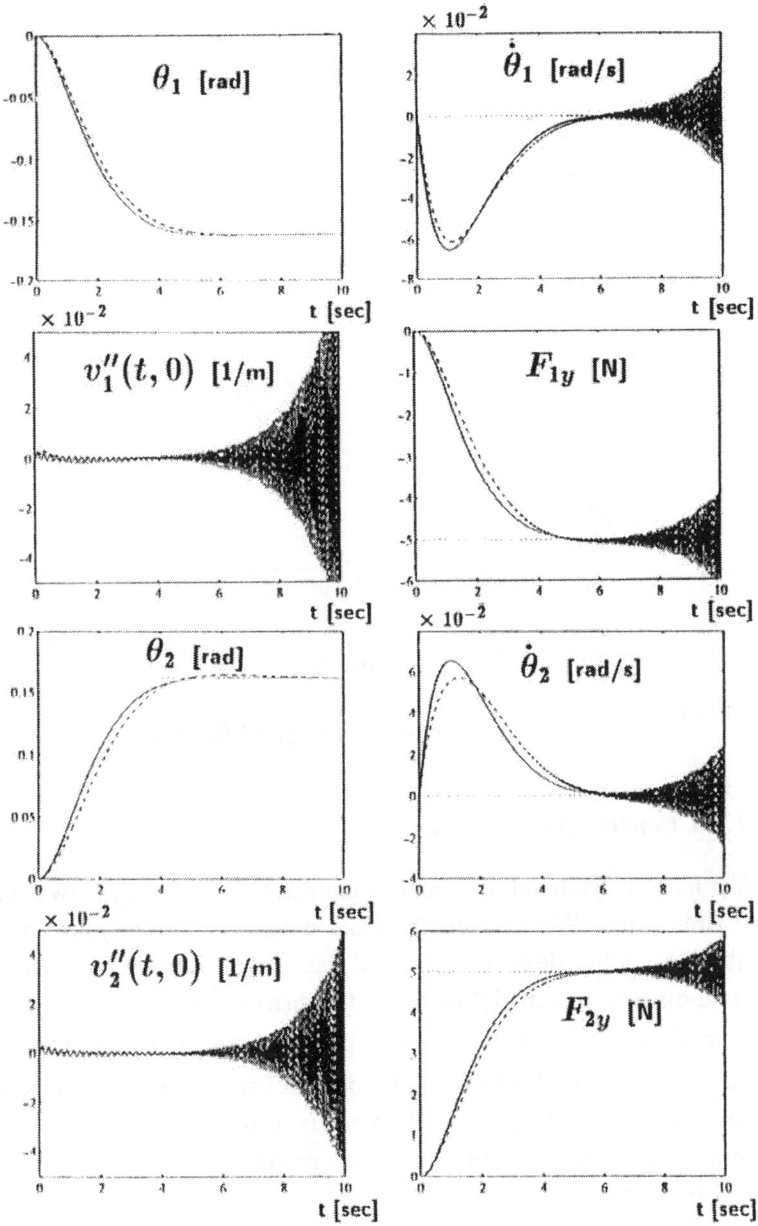

Fig. 4.20. Transient responses for case 3 (- - - LQ -----Filter)

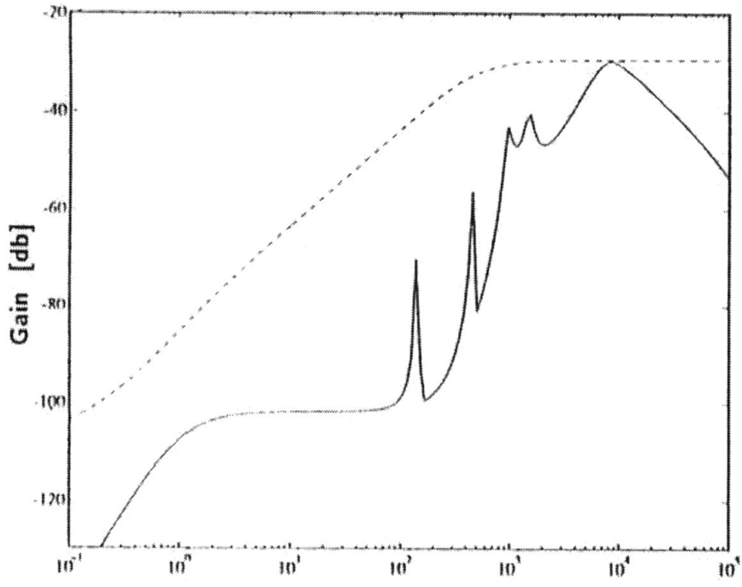

Fig. 4.21. $|W(j\omega)|$ and $\|\Delta P(j\omega)\|$.

4.4.6. *Concluding Remarks*

We have considered the robust cooperative control of two one-link flexible arms. We derived dynamic equations of the joint angle, the vibration of the flexible arms, and the contact force. On the basis of a finite-dimensional modal model of the distributed- parameter systems, an optimal controller with low-pass property and a robust H_∞ controller were constructed. A set of simulations has been carried out and responses for controllers were compared. Work is in progress in our laboratory to demonstrate the validity of the proposed robust controllers by experiments.

4.5. Conclusion

For application of manipulators to complex tasks and dexterous manipulations, it is often necessary to control not only position of the manipulator but also the force exerted by the hand on an object. For this purpose, impedance control strategies and hybrid control strategies of the rigid manipulators have been proposed. We can find the vibrations of the responses of the contact force, when the contact forces are controlled. The reason of the vibration may be the flexibility of links and/or joints. There are some papers treated the force control of flexible joint rnanipulators[36]. In this chapter, we have considered link flexibility and discussed the force control of flexible manipulators. From the stability analysis of the constrained one-link flexible arm, we can find that the direct force feedback can not ensure the global stability of the closed-loop system. We proposed modeling methodologies and robust controllers for constrained flexible manipulators.

Because of the page limitation, we can not refer to the quasi-static approaches[22,29]. A set of experiments of the quasi-static hybrid position/force control of constrained planar two-link flexible manipulator is carried out and the experimental results indicate the effectiveness of the proposed quasi-static controller[22]. An experimental study on the quasi-static control of cooperative multiple two-link flexible manipulators is reported[29]. Experimental results demonstrate the effectiveness of the quasi-static cooperative controller. We will propose the observer based force sensor feedback controller with low-pass property, which ensure the stability of the closed-loop system.

It is clear that the set of the flexible manipulators includes the set of the rigid manipulators. We will be able to find some hints for solving open and difficult problems of the rigid manipulators in the studies of the flexible manipulators. The research field of the force control of the flexible manipulators is new and attractive. The research topics for the handling of flexible objects is also very new and challenging[37-39]

The author would like to thank his graduate students Mr. Kazuo Yamamoto and Miss. Miwako Tanaka of Kobe University for their cooperation in the computer simulation and Prof. Masao Ikeda of Kobe

University and Prof. Yoshiyuki Sakawa of Kinki University for their encouragement.

The author dedicates this article to the memory of last his graduate student Mr. Motohiro Kisoi of Kobe University who died in KOBE EARTHQUAKE on January 17, 1995.

References

1. Y. Xu and T. Kanade, Space Robotics: Dynamics and Control, *Kluwer Academic Publishers, Boston*, 1993.
2. W. J. Book, O. Maizza-Neto and D. E. Whitney, Feedback Control of Two Beam, Two Joint Systems with Distributed Flexibility, *ASME J. Dynamic Systems, Measurement and Control*, 97(4), 424-431 (1975).
3. R. H. Cannon, Jr. and E. Schmitz, Initial Experiments on The End-point Control of A Flexible One-link Robot, *Int. J. Robotics Research*, 3(3), 62-75 (1984).
4. Y. Sakawa, F. Matsuno and S. Fukushima Modeling and Feedback Control of a Flexible Arm, *J. Robotic Systems*, 2(4), 453-472, (1985).
5. R. P. Judd and D. R. Falkenburg, Dynamics of Nonrigid Articulated Robot Linkages, *IEEE Trans. Automat. Contr.,* AC-30(5), 499-502 (1985).
6. B. Siciliano and W. J. Book, A Singular Perturbation Approach to Control of Lightweight Flexible Manipulators, Int. J. Robot. Res., 7(4), 79-90 (1988).
7. K. H. Low, A Systematic Formulation of Dynamic Equations for Robot Manipulators with Elastic Links, *J. Robotic Syst.*, 4(3),435-456, (1987).
8. E. Bayo, P. Papadopoulos, J. Stubbe and M. A. Serna, Inverse Dynamics and Kinematics of Multilink Elastic Robots: An Iterative Frequency Domain Approach, *Int. J. Robot. Res.*, 8(6), 49-62, (1989).
9. S. Cetinkunt and W. J. Book, Performance Limitations of Joint Variable-Feedback Controllers Due to Manipulator Structural Flexibility, *IEEE Trans. Robotics Automation,* RA-6(2), 219-231, (1990).
10. F. Matsuno and Y. Sakawa, A Simple Model of Flexible Manipulators with Six Axes and Vibration Control by Using Accelerometers, *J. Robotic Syst.*, 7(4), 575-597 (1990).
11. Z. H. Luo, Direct Strain Feedback Control of Flexible Robot Arms: New Theoretical and Experimental Results, *IEEE Trans. Automat. Contr.,* AC-38(11), 1610-1622 (1993).
12. F. Matsuno, T. Murachi and Y. Sakawa, Feedback Control of Decoupled Bending and Torsional Vibrations of Flexible Beams, *J. Robotic Syst.*, 11(5), 341-353 (1994).
13. M. H. Raibert and J. J. Craig, Hybrid Position/Force Control of Manipulators, *ASME J. Dyn. Sys. Meas. Control*, 103(2), 126-133, (1981).

14. N. Hogan, Impedance Control Part 1-Part 3, *ASME J. Dyn. Sys. Meas. Control,* 107, 1-24 (1985).
15. O. Khatib, A Unified Approach for Motion and Force Control of Robot Manipulators The Operational Space Formulation, *IEEE Trans. Robotics Automation,* RA-3(1), 43-53, (1987).
16. T. Yoshikawa,Dynamic Hybrid Position/Force Control of Robot Manipulators - Description of Hand Constrains and Calculation of Joint Driving Force, *IEEE Trans. Robotics Automation,* RA-3(5), 386-392 (1987).
17. N. H. McClamroch and D. Wang, Feedback Stabilization and Tracking of Constrained Robots, *IEEE Trans. Automat. Contr.,* AC-33(5), 419-426 (1988).
18. B. C. Chiou and M. Shahinpoor, Dynamic Stability Analysis of a One-Link Force Controlled Flexible Manipulator, *J. Robotics Syst.,* 5(5), 443-451, (1988).
19. B. C. Chiou and M. Shahinpoor, Dynamic Stability Analysis of a Two-Link Force Controlled Flexible Manipulator, *ASME J. Dyn. Sys. Meas. Control,* 112(6), 661-666 (1990).
20. J.K. Mills, Stability and Control Aspects of Flexible Link Robot Manipulators During Constrained Motion Tasks, *J. Robotic Syst.*, 9(7), 935-953 (1992).
21. F. Matsuno, Y. Sakawa and T. Asano, Quasi-Static Hybrid Position/Force Control of a Flexible Manipulator, *Proc. 1991 IEEE Int. Conf. on Robotics and Automation,* 2838-2843, (1991).
22. F. Matsuno, T. Asano and Y. Sakawa, Quasi-Static Hybrid Position/Force Control of Constrained Planar Two-Link Flexible Manipulators, *IEEE Trans. Robotics Automation,* RA-10(3), 287-297, (1994).
23. F. Matsuno and K. Yamamoto, Dynamic Hybrid Position/Force Control of a Two Degree-of-Freedom Flexible Manipulator, *J. Robotic Syst.*, 11(5), 355-366, (1994).
24. J. Y. Luh and Y. F. Zheng, Constrained Relations Between Two Coordinated Industrial Robots for Motion Control, *Int. J. Robot. Res.,* 6(3), 60-70 (1987).
25. C. R. Carignan and D. L. Akin, Cooperative Control of Two Arms in the Transport of an Inertial Load in Zero Gravity, *IEEE Trans. Robotics Automation,* RA-4(4), 414-419 (1988).
26. A. J. Koivo and M. A. Unseren, Reduced Order Model and Decoupled Control Architecture for Two Manipulators Holding a Rigid Object, *Trans. ASME J. Dyn. Syst. Meas. Control,* 113, 646-654 (1991).
27. I. D. Walker, R. A. Freeman, and S. I. Marcus, Analysis of Motion and Internal Loading of Objects Grasped by Multiple Cooperating Manipulators, *Int. J. Robot. Res.,* 10(4), 396-409 (1991).
28. F. Matsuno, K. Yamamoto and S. Kasai, Robust Cooperative Control of Two One-link Flexible Arms, *Proc. 1995 IEEE Int. Conf. on Robotics and Automation*, (1995)
29. F. Matsuno and M. Hatayama, Quasi-Static Cooperative Control of Two Two-link Flexible Manipulators, *Proc. 1995 IEEE Int. Conf. on Robotics and Automation,* (1995)

30. Y. Sakawa and F. Matsuno, Modeling and Control of a Flexible Manipulator with Parallel Drive Mechanism, *Int. J. Control*, 44(2), 299-313, (1986).
31. N. K. Gupta, Frequency-Shaped Cost Functionals : Extension of Linear-Quadratic Gaussian Design Methods, *J. Guidance and Control*, 3(6), 529-535, (1980).
32. M. Sampei, T. Mita and M. Nakamichi, An Algebraic Approach to H_∞ Output Feedback Control Problems, *Systems and Control Letters*, 14(1), 13-24, (1991).
33. F. Matsuno, M. Tanaka and M. Ikeda, Force and Vibration Control of a Flexible Arm Without Vibration Sensor, *Proc. 33rd IEEE CDC, 763-768,* (1994).
34. P. V. Kokotovic, Applications of Singular Perturbation Techniques to Control Problems, *SIAM Review*, 26(4), 501-550, (1984).
35. F. Matsuno, T. Asano and Y. Sakawa, Dynamic Hybrid Position/Force Control of n-Link Flexible Manipulators, *Proc. 33rd IEEE CDC,* 1803-1810, (1994).
36. K.P. Jankowski and H. S. ElMaraghy, Dynamic decoupling for hybrid control of rigid-/flexible-joint robots interacting with the environment, *IEEE Trans. Robotics Automation*, RA-8(5), 519-534, (1992).
37. C. Chen and Y. F. Zheng, Deformation Identification and Estimation of One-Dimensional Objects by Vision Sensors, *J. Robotic Syst.*, 9(5), 595-612, (1992).
38. J. K. Mills, Multi-Manipulator Control for Fixtureless Assembly of Elastically Deformable Parts, *Proc. 1992 Japan-U.S.A. Sympo. Flexible Automation*, 1565-1572, (1992).
39. F. Matsuno, N. Sakabe and M. Ikeda, Task Understanding and Optimal Strategies for Handling Flexible Beams by Using N-Link Manipulators, *Proc. IEEE/RSJ Int. Conf. Intelligent Robot and Systems,* 941-948, (1994).

Chapter 5

Experimental Study on the Control of Flexible Link Robots

In this chapter, the control of flexible link robots is examined. Modeling methods which are pertinent to control are briefly described and significant results from the control literature are discussed. Comparisons of some of these control techniques are conducted experimentally. The single flexible link and multi-link/multi-axis flexible link manipulators are both examined.

Introduction

One of the disadvantages of current industrial robots is that they have very high weight to payload ratios. Because of this, they tend to be slow and expensive. The reason for this is that these robots rely on the joint angles/displacements of the actuators in order to give endpoint positioning information. Thus, any deflections of the links due to elastic effects have a very detrimental effect on endpoint positioning. All the links must be made very rigid. This, in turn, implies that they must be massive.

An alternate approach is to allow elasticity in the links by letting the mass of the links be small. Such a manipulator is referred to as a *flexible link* robot. Then, by measuring the elastic deformations of the link and using a more sophisticated control algorithm, one should be able to control the endpoint of the robot with a relatively high degree of precision with minimal vibrations. This problem is the focus of this chapter. A similar area of research which is not discussed in this chapter is the issue of flexible joint robots [21] which arises when harmonic gears are used in a robot. Finally, it is important to note that even a rigid manipulator can exhibit structural vibrations if its performance is pushed past its specified limits. Thus, this work is relevant to present day manipulators.

Obviously, a flexible link robot brings an improvement in the weight to payload ratio of an industrial manipulator. This can translate to a cost savings as well. Another advantage is that the compliance in the links may allow force control in assembly tasks to be more easily done without the additional use of force/torque sensors. The deformation sensors (e.g., strain gages or cameras) would provide that information. This would be a real advance in the state of the art for industrial robots. However, before this can become a reality, the position control problem must be fully studied. The control objective is thus the endpoint tracking of smooth trajectories with minimal vibration.

It has been argued in the literature that such a flexible link robot would also be faster due to the quicker accelerations allowed by the lighter links. However, most control algorithms will not produce such a speed advantage due to the induced vibrations which take a significant amount of time to attenuate. In the end, there is little, if any, savings in time.

In this chapter, the position control of a single flexible link (SFL) manipulator is first discussed. This is a simplification of the general multi-link/multi-axis vibration problem. The SFL manipulator has been studied in detail in the literature. Pertinent models will be discussed as well as several of the more important control strategies proposed in the literature. A comparison will be made experimentally of several of these methods. The more general problem of multi-link/multi-axis flexible link (MMFL) manipulators will then be discussed. Again, modeling will be discussed briefly as well as several control strategies. Then, several of these strategies will be compared to each other experimentally on a 5-bar-linkage manipulator with a single flexible link. This is the simplest extension of the SFL manipulator to the general problem. Finally, a brief discussion of force control will conclude this chapter.

5.1 SFL Manipulator

5.1.1 *Modeling methods*

When modeling a SFL experiment, the goal is not to obtain a model which attains an exact match to experimental results. The goal is to produce a model which is simple enough for control purposes and which produces simulation results which are close to those observed experimentally. Any inaccuracies between the model and the actual experiment are taken care of by designing a controller which is robust to these uncertainties.

The single flexible link experiment is shown in Figure 5.1. It consists of a motor which applies a torque T to one end of a flexible link. The other end of the link is free. The motion of the link is horizontal, allowing gravitational effects to be ignored. The length of the link is l and the cross-section of the link is such that the bending stiffness is much higher vertically than it is horizontally.

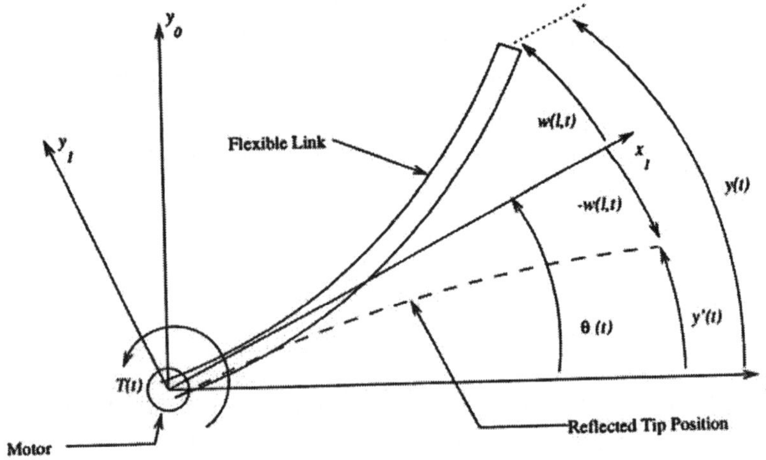

Figure 5.1

As is well known, the model of this link can be either represented as an Euler-Bernoulli or a Timoshenko beam equation with appropriate boundary equations [3]. Using Hamilton's principle and an Euler-Bernoulli beam model results in the following equations [9]

$$(I_h \rho l^3 / 3)\ddot{\theta}(t) + \rho \int_0^l x_1 \ddot{w}(x_1, t) dx_1 = T(t) \tag{5.1}$$

$$EIw''''(x_1, t) + \rho \ddot{w}(x_1, t) + \rho x_1 \ddot{\theta}(t) = 0$$

Where the dot indicates temporal derivatives and the primes are spatial derivatives. The I_h is the hub inertia, θ is the motor joint angle, ρ is the linear mass density, $w(x_1, t)$ is the elastic deflection along the length x_1 of the beam, and EI is the beam stiffness. These partial differential equations are subject to boundary conditions which depend on the location of the rotating frame $x_1 - y_1$. If the frame rotates so that it is always tangent to the beam at $x_1 = 0$, then the following *pseudo-clamped* boundary conditions are used

$$w(0,t) = w'(0,t) = 0 \tag{5.2}$$

$$w''(l,t) = w'''(l,t) = 0$$

It can be shown that, if the system input is considered to be the torque T, these partial differential equations cannot be stabilized theoretically using a finite-order lumped parameter controller and more sophisticated models must be used [22]. With the system input being the joint velocity $\dot{\theta}$, some simple control methods can be demonstrated to be successful [52].

Experimental verification has almost exclusively relied on finite dimensional models. The assumed modes [2] and finite element method [4] have both been used successfully. Although both techniques give equivalent results, the former will be examined here in more detail. In the assumed modes method, the elastic deformation $w(x_1,t)$ of the link is modeled as

$$w(x_1,t) = \sum_{i=1}^{n} q_i(t)\phi_i(x_1) \tag{5.3}$$

Where $q_i(t)$ are time-varying coefficients and the $\phi_i(x_1)$ are appropriate modal shape functions. The choice of these shape functions depends upon the location of the frame $x_1 - y_1$. If this frame is chosen to be tangent to the link at all times as shown in Figure 5.1, then the shape functions are chosen to be the clamped-free eigenfunctions of the Euler-Bernouilli (or Timoshenko) beam equations [10]. If the frame passes through the center of gravity of the flexible link at all times, the pinned-free eigenfunctions are chosen [1]. These choices were shown to be equivalent to within a linear state transformation in [9]. Other appropriate shape functions can be used as well.

In many cases, an output variable is chosen for the system. In most cases, the output $y(t)$ is considered to be the *net tip position* which is the sum of the rigid rotation, $l\theta(t)$, and the elastic deformation at the tip, $w(l,t)$ (see Figure 5.1). This is the output to be controlled unless otherwise stated. To measure this output variable, it is necessary to measure $w(l,t)$. The most common sensors employed are either cameras (linear CC D arrays) pointed at an LED mounted at the end of the link or strain gages mounted along the length of the beam. Both give identical information [8, 23].

5.1.2 *Control methods*

The control of the SFL robot has created a great deal of interest in the control theory field. It can be argued that it has become a benchmark problem for comparing the performance of newly developed control strategies. The reason for this is the inherent difficulty in controlling such a system. This is caused by several factors. First, this is mathematically an infinite-dimensional problem. Controllers generally need to be finite-order in order to be implementable (with the exception of time delays). As well, the internal damping in the beam is extremely difficult to model accurately resulting in a plant with parametric uncertainties. Finally, if the net tip deflection is chosen as the output, then the transfer function of the plant is non-minimum phase (ie, it contains unstable zeros). As will be seen, this will make it very difficult to implement some control strategies which are commonly used for conventional rigid link robots.

It is possible to formulate the control problem in the infinite dimensional setting of (5.1). When a model is employed which uses exclusively the partial differential equations, the control problem is usually handled using semigroup operator theory. In this technique, an attempt is made to provide a formulation which, to a large extent, emulates the finite dimensional case. First, a linear semigroup operator T_t is chosen as a mapping from the infinite dimensional state space X onto itself. This T_t is chosen to satisfy the conditions

$$T_t : X \rightarrow X, \quad T_0 = I \text{ (identity operator)}, \quad T_{t+s} = T_t T_s \qquad (5.4)$$

Note the similarity to the properties of the state transition matrix $\Phi(t_o,t)$ in the finite dimensional case. That is,

$$\Phi(t,t_o): X \to X, \quad \Phi(t_o,t_o) = I, \quad \Phi(t_2,t_o) = \Phi(t_2,t_1)\Phi(t_1,t_o) \quad (5.5)$$

Associated with T_t (in addition to additional conditions) is an infinitesimal generator A defined by $Az = \lim_{t\to 0^+} \frac{1}{t}(T_t - I)z$. It can then be shown that a solution of

$$\dot{x} = Ax \quad , \quad x(0) = x_o \quad , \quad x_o \in Domain(A) \qquad (5.6)$$

is $x(t) = T_t x_o$. Again, notice the similarity to the finite dimensional case. Be aware, however, that all the above takes place in an infinite dimensional setting. The above equation can be extended to the infinite dimensional problem

$$\dot{x} = Ax(t) + Bu(t), \quad t \ge 0, \quad x(0) = x_o, \quad y(t) = Cx(t) \; (5.7)$$

This problem is dealt with in detail in [18].

Cannon and Schmitz [1] are generally attributed as having first studied the problem of controlling a SFL robot experimentally. The control methodology used is the LQR approach described below. Since then, a multitude of methods have been studied. Some of the more important research is now described. All the models used in the forthcoming techniques are finite-order models.

The first category of control strategies to be examined in detail are the optimal control strategies. In Cannon and Schmitz's work, an LQR regulator is used with a deterministic estimator (see Figure 5.2). The objective is to find the state feedback K such that the cost function $J = \int x^t Qx + T^t RTdt$ is minimized for x being the states of the flexible model, T being the torque input, Q being positive semidefinite and R being positive definite. The Q and R matrices of this controller were chosen to give a desired rise time and settling time. This results in a controller with the same order as the plant. Since the states of a flexible manipulator are not readily available, an estimator must be used

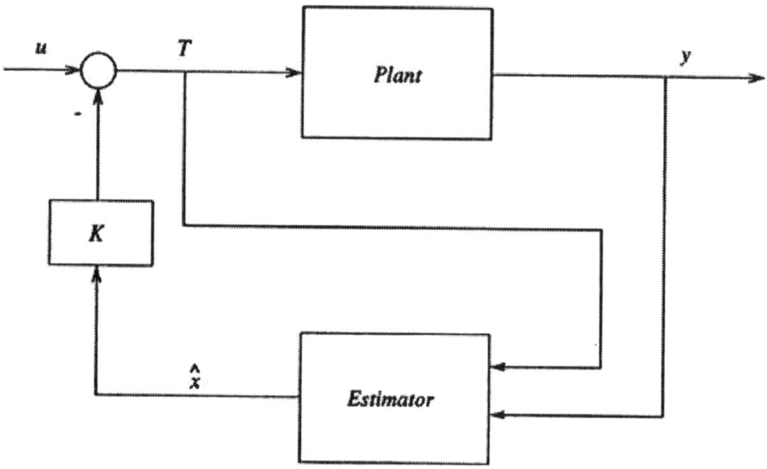

Figure 5.2

In [24, 6], $H_{\text{inf}\,inity}$-optimal control theory is employed to control the flexible link. For more detail on H_∞-control, see [17]. This is a frequency domain technique and allows a designer to achieve robustness with respect to unstructured uncertainties or perturbations. The H_∞ norm of a linear operator P is given by

$$\|P\|_\infty := \sup_{\|u\|_2 \leq 1} \|P(u)\|_2 = \sup_{\omega \in R} |P(j\omega)| \tag{5.8}$$

This is the induced norm from L_2 to L_2. The L_2 norm can be interpreted as being representative of the energy in the signal. There were both one and two parameter compensators designed. The one parameter compensator considered only the first mode of vibration of the flexible link (see Figure 5.3). The mapping from the exogenous input r to the regulated outputs z_i is defined to be P. The W_i are weighting matrices. The objective is to find the K which minimizes the H_∞ norm of P. In doing so, the weighting matrices have the following effects on the system response. W_1 affects the tracking and disturbance rejection, W_2 affects the actuator signal and is chosen to reduce its magnitude, and W_3 represents the modeling uncertainty due to unmodeled high order modes. The higher order modes were modeled as a multiplicative uncertainty. It turns out that one could specify a very small nominal damping and the controller would be robust to all higher values of damping ratios. This is important as it is very difficult to ascertain the exact value of damping in the beam. This design strategy reduces the possibility of the unmodeled modes becoming unstable under feedback. This phenomenon is known as spillover [12]. In the two parameter compensator (see Figure 5.4), the same design goal was retained for robustness to the unmodeled modes, tracking accuracy and actuator bounds. As well, robustness to torque disturbances was included. The explanation of this torque disturbance will be elaborated on in the next section. In this case, the goal is to find the K_1 and K_2 to minimize the H_∞ norm of the transfer function from the exogenous inputs r, d and n to the regulated outputs z_1, z_2 and z_3. The effect of W_1, W_2 and W_3 are the same as for the one parameter compensator. W_d weights the effect of the torque disturbances and W_n weights the measurement noise.

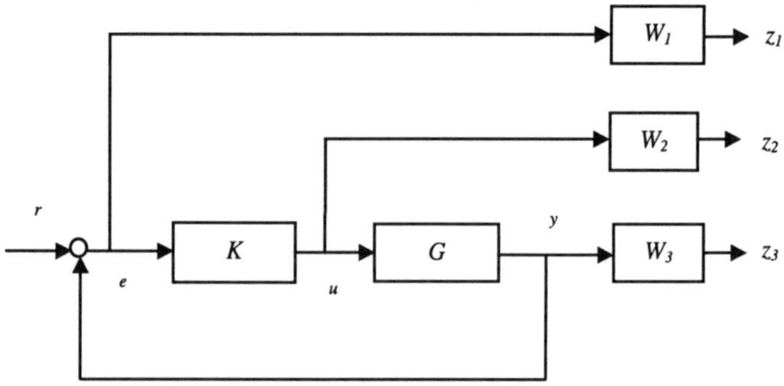

Figure 5.3

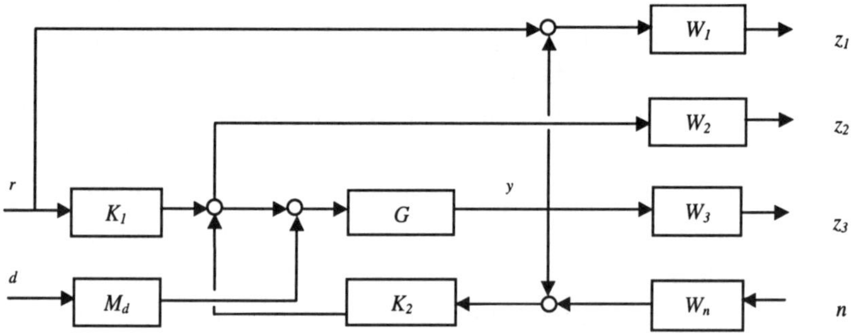

Figure 5.4

In all the previous optimal control strategies, the control laws were complicated. The order of the controller was roughly the same as the order of the flexible link model. This can create problems in the implementation of the control strategies on a digital controller.

Nonoptimal control strategies from the literature will now be discussed. It should be noted that a conventional PD controller is often a viable technique for controlling a single flexible link. A PD controller can be designed assuming the flexible link to be rigid and by considering the motor joint angle $\theta(t)$ to be the output. This is the type of control strategy used predominantly in industrial robots.

It is a collocated control problem and results in a transfer function which is minimum phase [1, 13]. It was shown in [13] that the use of a PD controller guarantees stability. An intuitive explanation of this stability result can be formulated as follows. The equations of a PD controller are essentially identical to those of a mechanical spring and damper. Thus, the PD controller at each joint can be replaced by a spring and damper. It is not hard to visualize that any robot manipulator with springs and dampers attached to every joint will always remain stable, no matter how flexible the links. However, transient performance and steady state error may not be acceptable. This controller will be referred to as the *rigid body* PD controller.

The idea of the spring and damper equivalent of a PD controller can be exploited in the development of passive control strategies. These control strategies all rely on the concept of passivity. A *passive* system $G(\cdot)$ is one which satisfies $\int G(u)u dt \geq 0$, or equivalently, does not create energy. Then, by the *passivity theorem* [14], any strictly passive controller with finite gain can be employed to stabilize the system. It is possible utilizing this result to employ very simple control strategies on the flexible link. However, when the transfer function from the motor torque to the net tip position is used, the transfer function is not passive (in fact, it is nonminimum phase). The challenge is to find an alternate transfer function which is passive. In [25], passivity was used to show that a PD controller which controls only the joint angle will stabilize the system. In [8], it was found that, although the transfer function from the torque input to the derivative of the net tip position was nonminimum phase, an alternate transfer function which employed the derivative of the *reflected tip* position (the reflected tip position $y'(t)$ is considered to be the rigid body deflection *minus* the elastic deformation and is shown in Figure 5.1) was minimum phase and *passive* provided the link was sufficiently stiff.[1] In [8], an extremely simple first order

[1] This is violated as the number of modes increases to a sufficiently large number. However, in practice, only the first few modes need to be included for experimental success

passive controller employing this alternate output was shown to stabilize the flexible link. This controller will be referred to as the *passive flexible* controller. In [25], a rigorous proof of the passivity of this transfer function was undertaken as well as alternate outputs which would result in passive transfer functions. Another paper which examined alternate measurement positions on the flexible link in order to achieve passivity was examined in [19].

Another controller which has generated a great deal of interest is the *shaped input* controller. Essentially, this employs a filter designed in the time domain to reduce vibration effects. Intuitively, assume one has a linear system which, under an impulse input, undergoes a purely sinusoidal oscillation. If the system is hit with another impulse 180 degrees out of phase with the first one, the vibrations will cancel by superposition (see Figure 5.5). For an arbitrary input, one simply convolutes this input with the impulse sequence to decrease vibrations. A simple version of this was presented in [26]. An open loop version of this technique with larger number of impulses to increase the controller robustness with respect to uncertain frequencies or damping was demonstrated in [15]. In theory, any number of modes can be handled. The contribution in [7] was to extend the shaped input control methodology to a closed loop scenario. The only assumption is that the fundamental frequency of the elastic vibrations were sufficiently higher than the bandwidth of the closed loop system employing a rigid body controller. This controller has increased robustness properties.

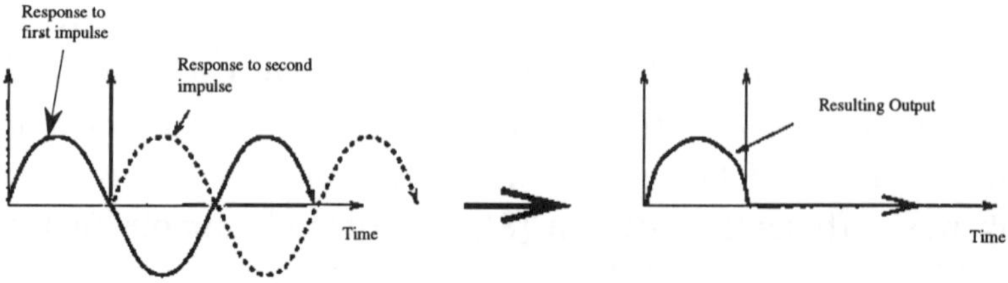

Figure 5.5

In rigid link robots, a popular method to control the robot motion is the inverse dynamics approach [27]. For a linear system $G(s)$, the method would work as follows. If $y(s) = G(s)u(s)$, then one could use the controller $u(s) = G^{-1}(s)u'(s)$ which would give the transfer function $y = u'$. The u' is therefore the desired trajectory. In this method, the open loop torques can be precomputed. Obviously, if $G(s)$ is strictly proper, then $G^{-1}(s)$ would be improper. This is still implementable if u' is sufficiently smooth. That is, if $G(s)$ is of relative degree n, u' must be n-times differentiable. If $G(s)$ is nonminimum phase, then $G^{-1}(s)$ will also have nonminimum phase poles. In other words, $G^{-1}(s)$ is unstable. This is exactly the case for flexible manipulators. In [28], the inverse dynamics approach is extended to flexible manipulators. In this methoa, a gaussian trajectory is used to compute the desired trajectory to ensure the smoothness condition. As well, by using a noncausal input torque, it is possible to get around the nonminimum phase pole problem. A causal torque close to this noncausal torque is then used. In [30], a much simpler method using a state space approach is used to achieve the same torque profile. It should be noted that this is an open loop control strategy.

In [29], a fuzzy logic controller (FLC) is implemented. It was noted that the optimal control techniques described above tended to create large initial deflections. For example, suppose a step input is to be tracked. Then, as soon as the input is applied, a large error exists which causes a large initial torque. However, the nonminimum phase characteristic results in the tip of the beam initially bending in the opposite direction to the torque, thus causing the error to increase rather than decrease [8]. The controller tends to then use more control effort to reduce this error. The net result of this is a large initial torque and ensuing vibrations. Based on this, it was easy to visualize heuristic laws which should be applied to reduce this effect. For example, initially, one does not want a large torque if the error is large. This is very difficult to implement using linear control strategies. However, it is very straightforward to implement this using a fuzzy logic controller. In [29], a two stage fuzzy Logic controller was shown to exhibit excellent performance. One stage controls the rigid motion while the second stage monitors the elastic deformations and modifies the output of the first stage to reduce the induced vibrations (see Figure 5.6). This approach was taken to greatly reduce the number of rules needed in the fuzzy associative memory.

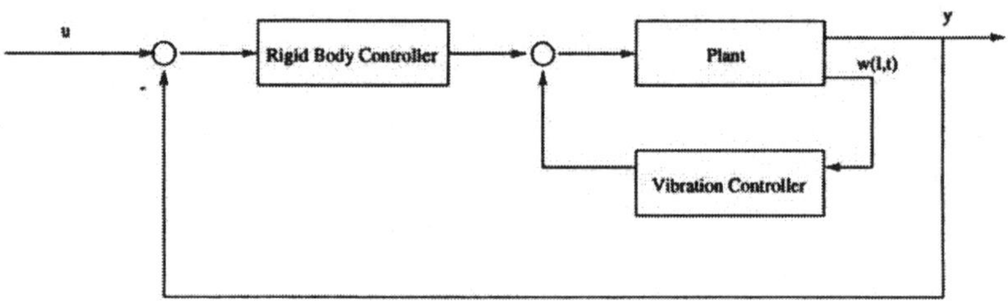

Figure 5.6

Another approach is to use singular perturbation theory [31]. This method uses the fact that the beam stiffness of the flexible link, *EI*, approaches infinity as the beam becomes more rigid. If one defines a variable $\varepsilon = 1/EI$, then the dynamic equations can be written in the form

$$\ddot{\theta} = f\left(\theta, \dot{\theta}, \varepsilon, z, \dot{z}, u\right) \tag{5.9}$$

$$\varepsilon^2 \ddot{z} = g(\theta, \dot{\theta}, \varepsilon, z, \dot{z}, u)$$

This is referred to, in [31], as a singularly perturbed model. It should be noted that the form of this equation is not quite what is standardly considered to be a singularly perturbed model [32]. Now, when ε approaches zero, the above equations reduce to those of the rigid link robot. To design a controller, one considers the equations where $\varepsilon = 0$ to be the slow subsystem. One then defines a fast time scale to be $\tau = t/\varepsilon$. As well, fast variables z_f are defined in (5.9) as the state $z_f = z - \bar{z}$ where $\bar{z}$ are the values of z in the slow subsystem. In the fast subsystem, the slow variables are considered to be constant. A composite control is then used where $u = u_{slow}(\theta,\dot{\theta}) - u_{fast}(\theta,\dot{\theta},z_f,\dot{z})$. The u_{slow} controls the slow subsystem and u_{fast} is designed so that the trajectories are always forced quickly to the slow subsystem (ie the fast subsystem variables approach the equilibrium point much quicker than the response time of the slow variables). When the slow subsystem (or more correctly, the slow manifold) is reached, then $u_{fast} = 0$. One then applies Tikhonov's theorem [33] to show that the desired trajectory is tracked to an error of order ε.

5.1.3 *Experimental comparisons*

Several different control strategies are compared experimentally in this section. The controllers will be divided into two classes: optimal and nonoptimal design strategies. It should be mentioned that it is desirable to keep the sampling period the same for each experiment for comparison purposes. The number of flexible modes considered in each design were therefore changed in order to keep the complexity of the resulting control laws approximately the same.

The actual experimental apparatus (see Figure 5.7) used is now described. It consists of a 1 meter long aluminum beam with a height of .025m and a width of .0034m. It employs a DC motor attached directly to the link without any gearing. Angular rotation of the motor is measured via an optical encoder. The elastic deflection is measured with a CCD array camera mounted at the hub of the motor and an LED attached to the end of the flexible link. The sampling times for the optimal controllers were 15 milliseconds while the sampling times for the nonoptimal controllers were 5 milliseconds.

Figure 5.7

During the course of these investigations, it was found that the most important perturbations in the experimental apparatus were unmodeled high order modes and input torque disturbances. It is suspected that these modeling problems will occur in any experimental apparatus. The higher order modes correspond to the modes in (2) which are neglected during the design of the control strategy. As mentioned, this can lead to instability due to spillover. The second disturbance is due to the wires coming from the camera to the computer. Again, most sensors will require similar wiring off of the end of the link. It turns out that these cables create a torque on the motor hub which can be very significant. In virtually every experimental run, the effect of the cable drag can be seen in the step response as a steady state error while the effect can be seen in the torque response as a DC offset.

For the optimal control strategies, a rigid link LQR is designed by assuming the flexible link is rigid. Thus, the model is the same as that of a standard DC motor with the beam inertia added to the motor hub inertia. Then, a flexible link LQR is designed using the finite order model described in Section 1.1. Finally, both one parameter H_∞ and two parameter H_∞ controllers were implemented.

The nonoptimal controllers implemented experimentally included a joint PD controller, a passive flexible controller, a loosed-loop shaped-input controller and a fuzzy logic controller.

First, the results of the optimal control strategies are shown. The input to be tracked is a step input. In Figure 5.8, 5.9 and 5.10, the net tip position, the tip deflection and the motor torques are shown for the rigid body LQR (A) and the flexible body LQR (B). In Figure 5.11, 5.12 and 5.13, the same graphs are shown for the one parameter H_∞-controller (A) and the two parameter H_∞-controller (B). The nonoptimal controllers are shown in Figures 5.14, 5.15 and 5.16 for the rigid body PD controller (A) and the passive flexible controller (B), and are shown in Figures 5.17,5.18 and 5.19 for the closed-loop shaped input controller (A) and the fuzzy logic controller (B).

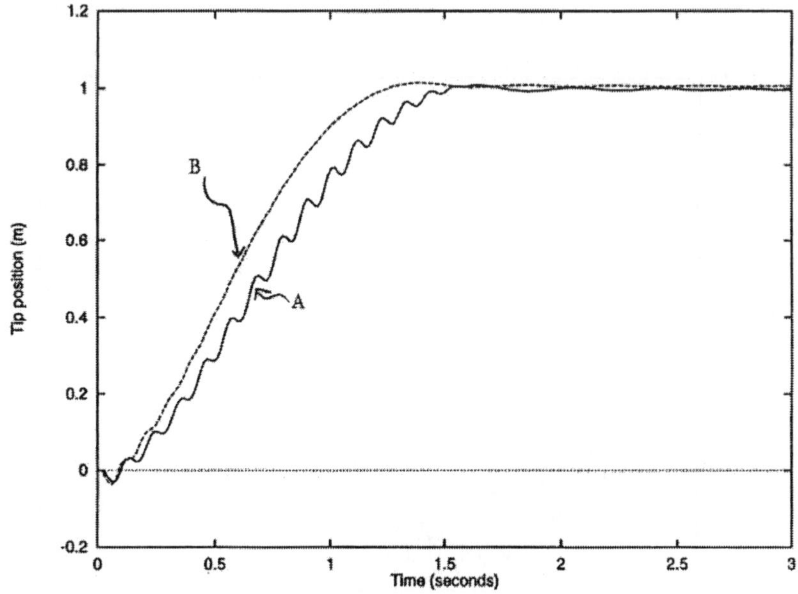

Figure 5.8

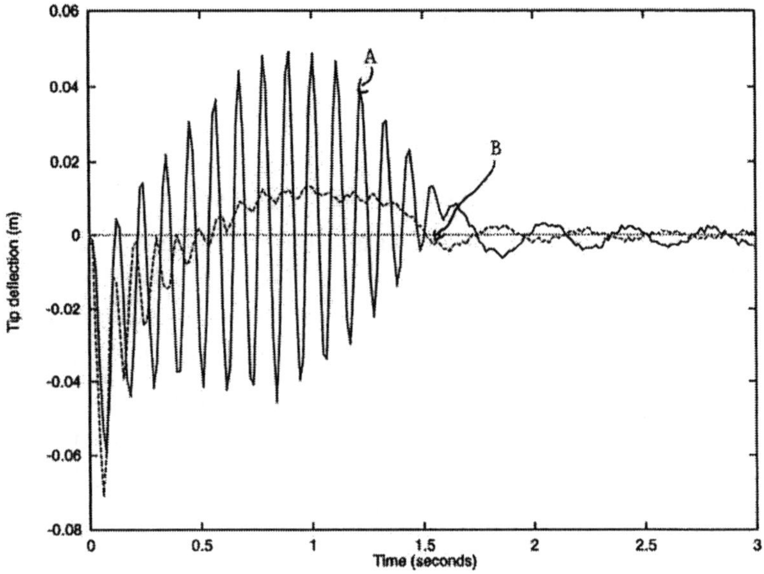

Figure 5.9

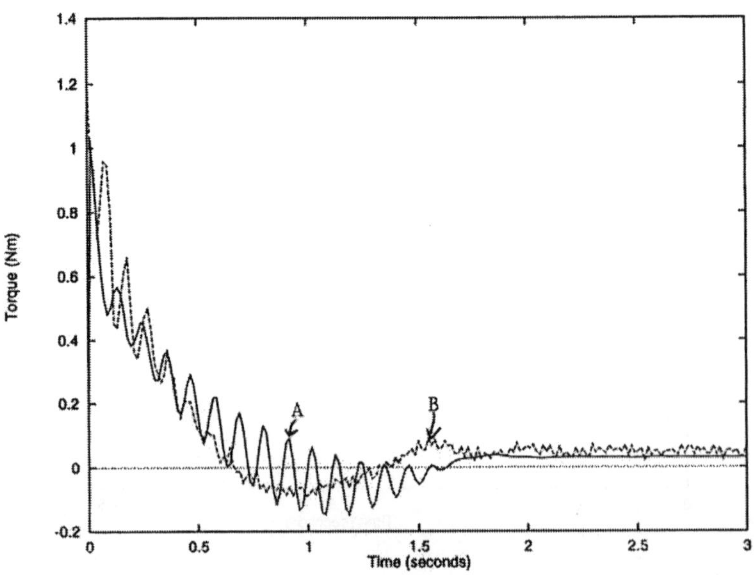

Figure 5.10

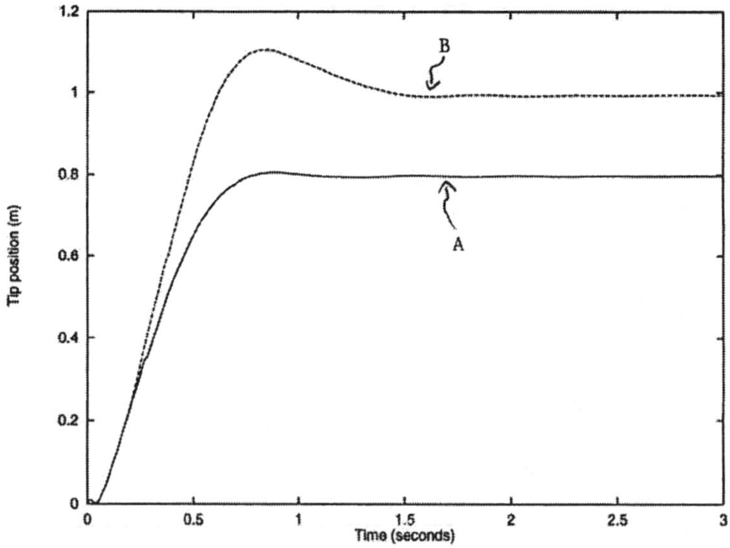

Figure 5.11

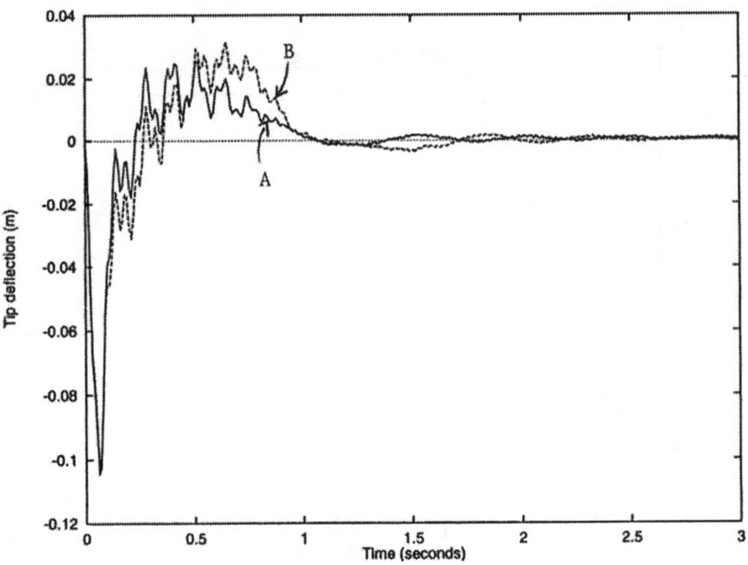

Figure 5.12

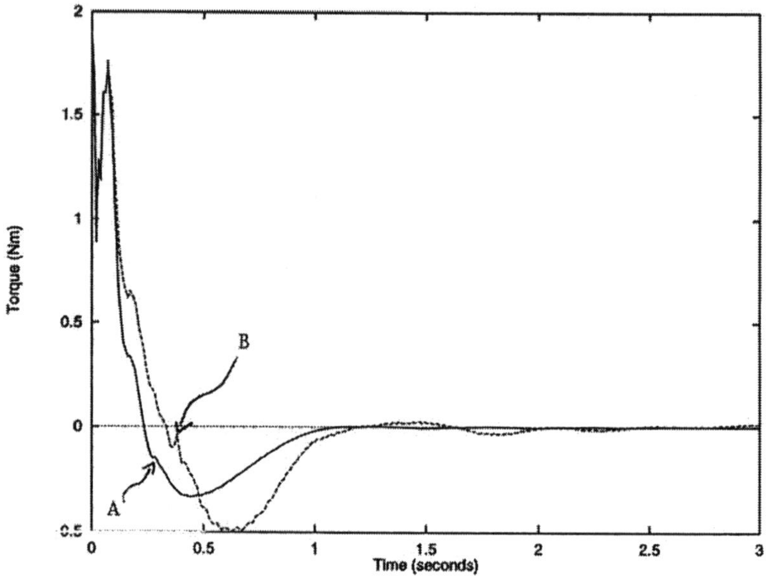

Figure 5.13

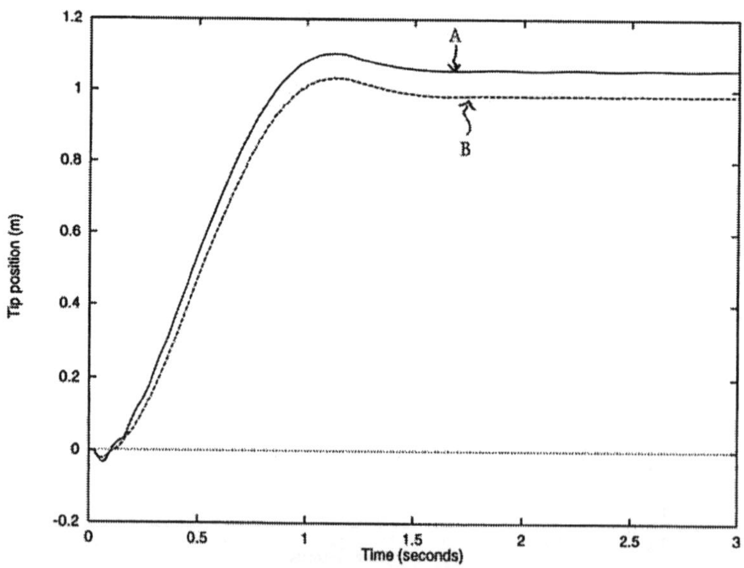

Figure 5.14

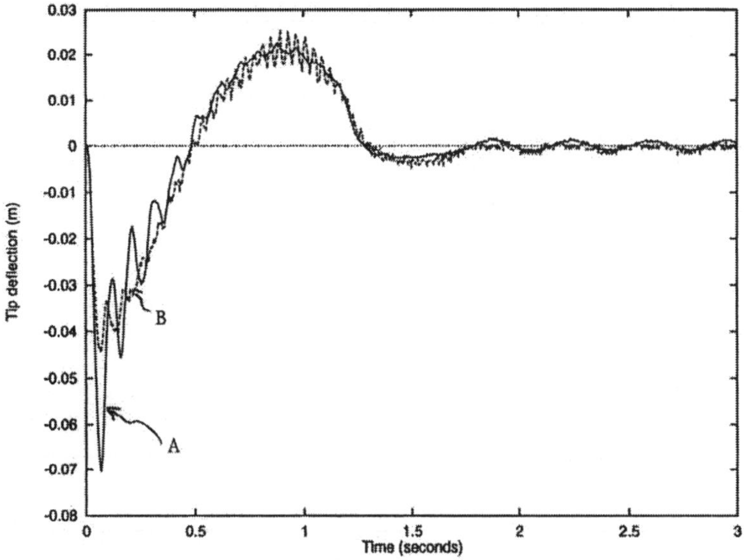

Figure 5.15

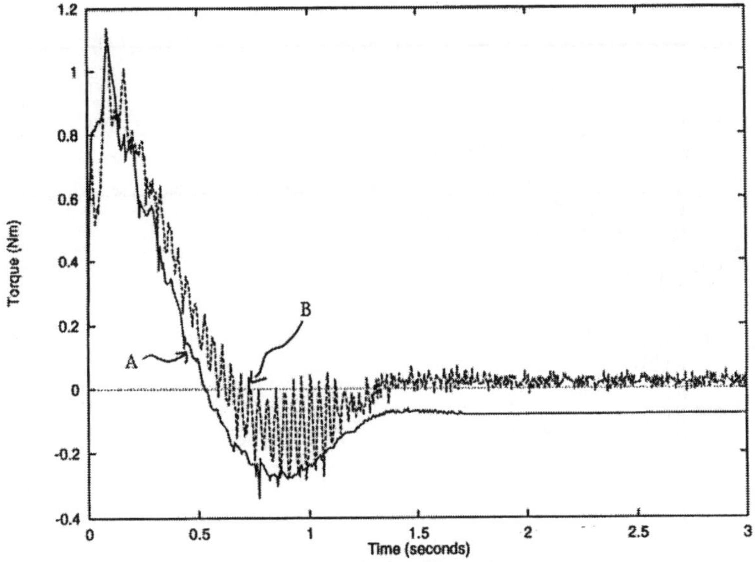

Figure 5.16

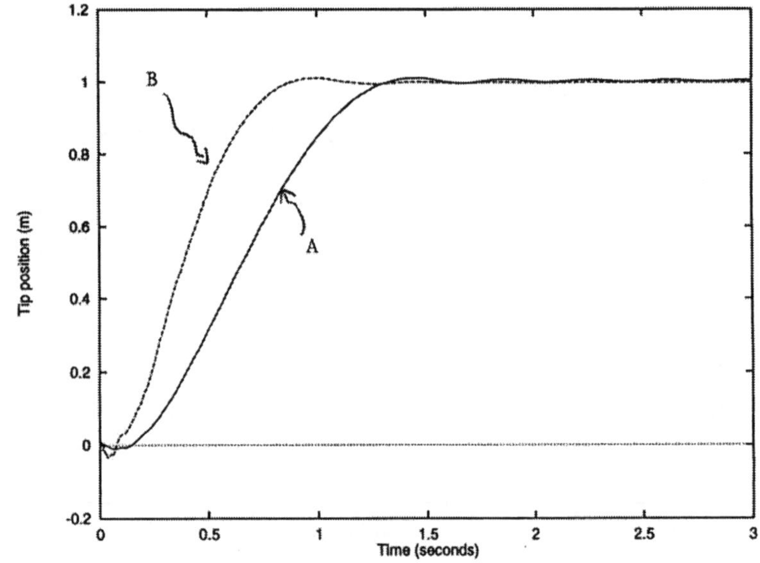

Figure 5.17

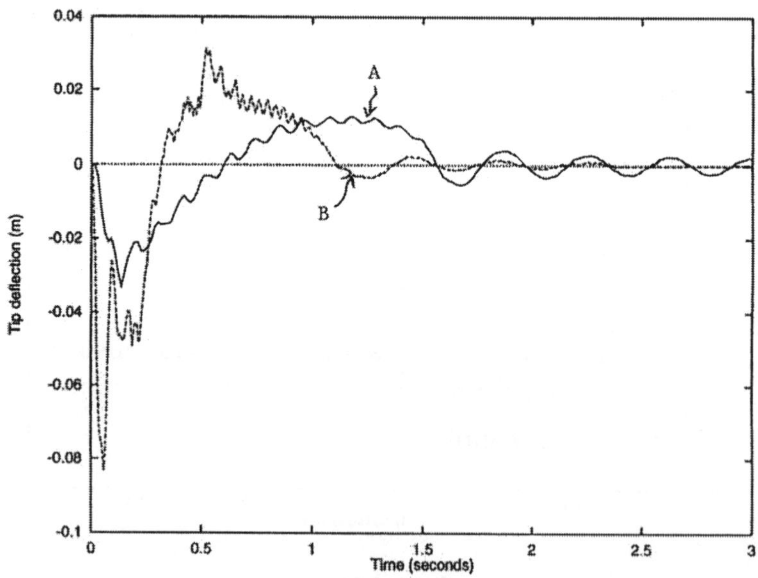

Figure 5.18

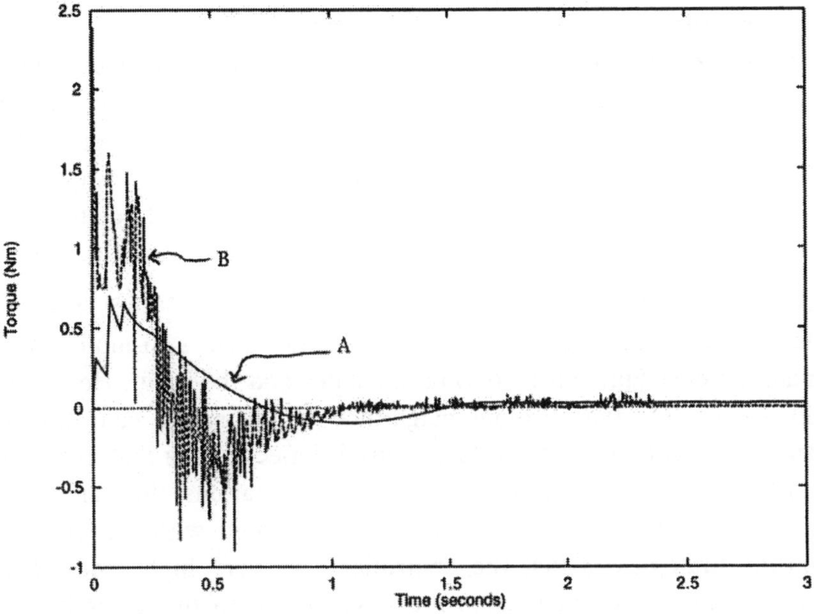

Figure 5.19

5.1.4 *Comparison of methodologies*

There are many interesting observations which can be made upon examination of the experimental results. However, it must be kept in mind that, for all of these control strategies, there are many design parameters which can be changed to give very different responses. Thus, these comparisons will only point out general trends indicated by the experimental results.

First, it is obvious that the rigid body LQR design is not suitable for this application and will not be discussed. This is because of its poor step response.

A comparison can be made between the optimal class of controllers and the nonoptimal class of controllers. The optimal control strategies gave higher torques, especially in the first half second. This results in large initial deflections being excited. These exceeded 20 cm. The controller does work quickly to damp out these vibrations however. The cause of this large torque and initial deflection is that the optimal control strategies try to reduce the norm of some quadratic cost function related to the error as quickly as possible. To do this, the link must be accelerated to its final position quickly, hence the large torques. In the case of the H_∞-optimal control strategies, the goal is to minimize the L_2 norm of the output signal. For the LQR regulator, the goal is to minimize a quadratic cost function involving the states and the input. However, this means that the absolute magnitude of the signal can still be large. The problem of these large initial deflections is that mechanical fatigue can occur. In most industrial applications, this can result in frequent failures. Another problem associated with the optimal control strategies is that the controllers are very high order. The order of all the optimal controllers was close to the order of the model considered. As well, as the links get stiffer, the frequency of the vibrations increases. Both of these factors imply that implementation problems will occur due to the higher sampling frequencies needed as well as the large amount of computations involved. These optimal control strategies are also linear strategies which, of course, make these methodologies difficult to extend to the nonlinear multi-link problem.

The advantages of these strategies are that they gave quicker responses than the nonoptimal approaches. Among the optimal control strategies, comparisons can also be made. The flexible body LQR regulator gave very good results with a simple design strategy. Two modes were sufficient to prevent spillover effects. The H_∞ control strategies were dependent on the choice of weighting functions corresponding to the relevant disturbances. As can be seen from the graphs, it was not easy to choose these weighting matrices to achieve reasonable responses. The two-parameter H_∞ controller did reduce the steady state error to zero most consistently among all the controllers during repeated trials.

The nonoptimal control strategies offer several advantages. First, the rigid body PD controller performed surprisingly well. The performance of the other two nonoptimal strategies was better than the rigid PD controller and was comparable to each other. These nonoptimal controllers were all much simpler in comparison to the optimal controllers. In fact, these controllers could all be implemented as analogue controllers. This implied that these control strategies could be used for flexible links with high stiffnesses. As well, the extension of these control strategies to multi-link manipulators appeared more promising due to both the simplicity of the control laws and the fact that stability was guaranteed using passivity or Lyapunov approaches. These two approaches are as valid for nonlinear systems as for linear systems. The torque requirements were low and much lower overall deflections were excited. This implies that mechanical fatigue problems would be greatly reduced. However, these controllers were slower in general than the optimal controllers.

There were also restrictions which applied to each nonoptimal control strategy. For the passive flexible controller, this technique will not work for a sufficiently flexible link. For the closed loop shaped input controller, there is the assumption that the vibration frequencies are higher than the closed loop rigid body bandwidth. As well, since this is predominantly a time domain filtering technique, there will be certain frequencies from the input which cannot be reproduced at the output.

The comparison of these experimental results seems to indicate that optimal control strategies, although greatly studied in the literature, may not be the most suitable techniques for controlling flexible structures. They are limited to linear systems and generally create unsatisfactorily high elastic deformations which can contribute to failure due to mechanical fatigue. The goal perhaps should be shifted to control strategies which are simple in structure, robust and are stable while maintaining low overall elastic deflections during the movement of the arm. The modeling of the disturbances is very critical as well. It was found in the course of this work that the effect of higher order modes must be considered. However, the effect of the cables from the sensor systems also had a great effect. They could create a torque disturbance which can easily be seen in the experimental results. In future work, these factors must be considered when designing robust controllers.

5.2 Multi-Link/Multi-Axis Flexible Link (MMFL) Manipulators

The more general problem of controlling flexible link manipulators can be complicated in the following manner. First, a manipulator may have several flexible links throughout its structure, making it a highly nonlinear distributed parameter control problem. In addition, despite the fact that the single flexible link only has lateral vibration in one plane, a general flexible link would be able to undergo lateral and longitudinal vibrations in any direction as well as having torsional vibrations.

5.2.1 *Modeling*

Not many papers approach the modeling of a general flexible link manipulator from a distributed parameter formulation. The best paper dealing with this is [35]. It uses Hamilton's principle to derive the dynamic equations for a large class of multi-link flexible manipulators. The resulting equations for these manipulators are partial differential-integral equations with a set of dynamic boundary conditions.

Many nonlinear models of MMFL manipulators used for control purposes employ the assumed modes or finite element approach. The main reason for this is that it allows the use of the Euler-Lagrange approach, and results in solutions that are finite dimensional. If sufficient modes are taken, then the results tend to be quite close to that of the partial differential equations. In fact, in [35], it was shown that the previously described partial-differential-integral equations reduce to the standard models when the appropriate assumptions are made. One of the first finite-dimensional models used was that of [36]. The general dynamic equations of serial multi-link manipulators consisting of rotary joints and flexible links were derived using the homogeneous transformation matrices of Devanit and Hartenberg.

Although there are many possible techniques to arrive at a finite-dimensional model, they all have end up having the form

$$\begin{bmatrix} M_{11}(q_r,q_f) & M_{21}(q_r,q_f) \\ M_{21}^T(q_r,q_f) & M_{22}(q_r,q_f) \end{bmatrix} \begin{bmatrix} \ddot{q}_r \\ \ddot{q}_f \end{bmatrix} + \begin{bmatrix} h_r(q_r,\dot{q}_r,q_f,\dot{q}_f) \\ h_f(q_r,\dot{q}_r,q_f,\dot{q}_f) \end{bmatrix} = \begin{bmatrix} T \\ 0 \end{bmatrix} \tag{5.10}$$

where q_r, are the rigid body generalized coordinates and q_f are the flexible vibration generalized coordinates. The M_{ij} terms are the inertia matrix terms whereas the h_i terms contain the coriolis, centripetal, damping, gravitational and system stiffness terms. The T contains all the motor torque inputs.

5.2.2 *Control*

Much of the work for MMFL manipulators has been motivated by the work on single flexible link manipulators. The research to be discussed in this section will be those which show great promise in a practical sense.

One possibility is to use a nonlinear state feedback to stabilize the MMFL robot. This also requires the use of a nonlinear observer to estimate the states of the dynamic equations. It is shown in [45] that such an approach is feasible and a rigorous stability proof is given. In [46], a similar technique using a LTI state feedback and an accelerometer-based observer is implemented experimentally on a 3 degree-of-freedom MMFL manipulator.

One of the most successful control techniques is the SI control of [15]. Because of the robustness of the filter to changes in frequency and damping, it is possible to attenuate a large range of frequencies and damping. As long as the nonlinearities in the system are not great, the nonlinear changes of frequency and damping can be accounted for. Although successful, this is an open loop technique which is susceptible to disturbances. The closed-loop SI technique, which was proposed in [7], was shown experimentally to effectively remove structural vibrations [16] in a MMFL manipulator.

Another effective method of controlling beam oscillations in a MMFL manipulator is a method proposed by [37]. In this method, a rigid body controller, with a feedforward control, is first found. Of course, implementing such a controller will excite link vibrations. These link vibrations are then precomputed and the torque profile is modified off-line in order to partially compensate for these vibrations. Now, when this new torque profile is used, vibrations will still be induced. The dynamic equations are now linearized about this desired trajectory and the resulting linear time-varying state space equations can be controlled using linear control theory. Of course, the difficulty in this method is the design of a controller for a time-varying linear system. However, if the linearization is only done with respect to the final, steady-state position, then the resulting state space equations are linear and are easy to control. Significant experimental results are also shown in [37].

The use of the Passivity theorem has also proved to be successful. Since one only needs to be concerned about the input-output mappings of both the controller and plant, the extension into the nonlinear setting is promising. As was discussed in Section 1.2, this can be used to show the stability of a flexible link manipulator controlled using joint-based PD control. In [25], an experimental verification was made of the technique on a 5-bar-linkage manipulator with a single flexible link. As was the case in the SFL problem, the control algorithms could be kept simple.

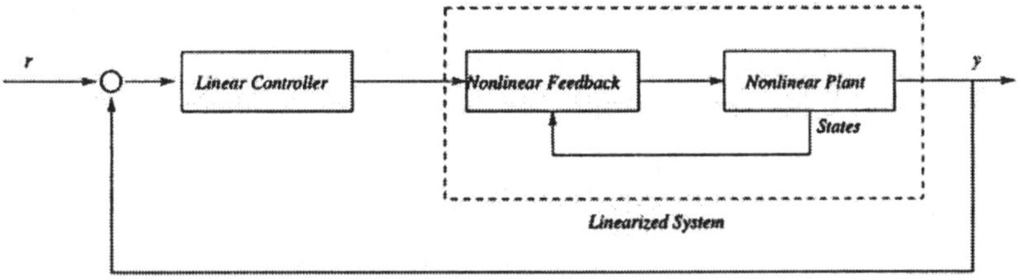

Figure 5.20

In rigid body control, one popular nonlinear control technique is that of the inverse dynamics approach. In this approach, one can consider that there are two feedback loops; an inner feedback loop which linearizes the dynamics, and an outer feedback loop which controls the linearized equations (see Figure 5.20). The inner loop is often referred to as feedback linearization [39]. In general, there are two approaches to the feedback linearization. The first approach is that of input-output feedback linearization. One takes successive time derivatives of the output until one can find an input which linearizes the input-output equations. For a single-input single-output plant, the input would be of the form $u = \alpha(x) + \beta(x)u'$ and the resulting equations would be

$$\frac{\partial^{v} y}{\partial t^{v}} = u' \tag{5.11}$$

where u' is a new input. This is a linear equation which can be controlled easily. If, however, there are n states to the dynamic equations, then there have been $n - v$ states that have been rendered unobservable. In other words, even if the above equations are stabilized, there are $n - v$ states for which there is no stability information. These unobservable dynamics are referred to as the zero dynamics. The stability of the zero dynamics determines the stability of the overall control strategy. This adds to the complexity of the input-output feedback linearization technique. The second approach to linearizing the dynamic equations is state linearization. In this technique, a nonlinear state feedback $u = \alpha(x) + \beta(x)u'$ and a nonlinear state transformation $z = \phi(x)$ is used to convert the nonlinear dynamic equations $\dot{x} = f(x,u)$ into a linear state space equation of the form

$$\dot{z} = Az + Bu' \tag{5.12}$$

where the number of states is the same in both representations. If one finds a $u' = r(z)$ to stabilize the linear system of (5.12), then the nonlinear feedback to stabilize the original equations is of the form $u(x) = \alpha(x) + \beta(x)r(\phi(x))$. In [40], it was shown this technique can be used for single flexible links with an appropriate choice of measurement positions on the beam, and, in [34], it was shown that these results can also be extended to a large class of planar manipulators. In [41], a special class of manipulators with a single flexible link was examined. Basically, this class of manipulators consists of vertical planar manipulators driven by one or two side motors. A base motor rotates the entire manipulator about a vertical axis to give three degrees of freedom. The last link is flexible and the cross section is chosen so that vibration is always in the horizontal direction. This is done by letting the height of the cross section be much greater than its width. There is no payload at the end of the last link. This is the simplest extension of the single flexible link to a multi-degree of freedom system. Examples of manipulators belonging to this class are the elbow, spherical, cylindrical, 5-bar-linkage and single flexible manipulators (See Figure 5.21). However, when the single flexible link is extended to this simple MMFL manipulator class, then both of these feedback linearization techniques fail. The input-output feedback linearization technique fails because the zero-dynamics are not asymptotically stable while the input-output feedback linearization fails because it is not possible to cancel out the nonlinear coriolis coupling between the vibration of the beam and motion orthogonal to the vibration of the beam caused by the rigid motion of the manipulator.

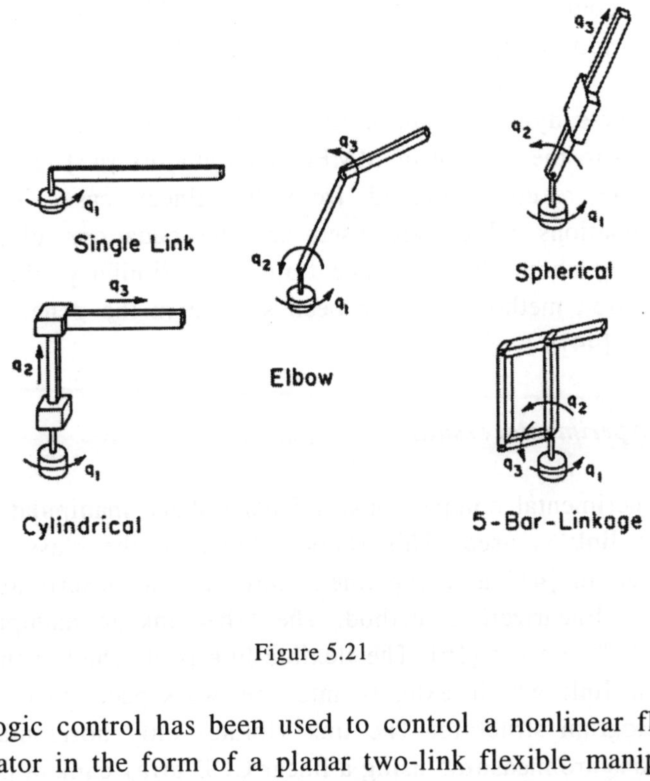

Figure 5.21

Fuzzy logic control has been used to control a nonlinear flexible manipulator in the form of a planar two-link flexible manipulator [42]. The rule base is first realized by considering each link separately. The rule base is then modified by considering how the links dynamically interact with each other. This results in improved performance. This fuzzy logic controller is referred to as the low level fuzzy logic controller. Further improvements are obtained by using a higher level supervisory fuzzy logic controller which would adjust the low level fuzzy logic controller between two different lower level fuzzy logic controllers, one designed for small slewing motions whereas the second was designed for large slewing motions.

Another approach is to use linear control techniques which can robustly stabilize a flexible link manipulator at all possible configurations. Such a study was done in [50], where an H_∞-optimal controller is used on a single flexible link attached to a shoulder joint to give the link both horizontal and vertical motion.

The inverse dynamics approach has also been used as a possible approach to the control of MMFL manipulators [43]. This is done using the results obtained from the linear case. The linear approximations which are used are the equations obtained by linearizing about the reference trajectory. Similarly, the singular perturbation method has also been studied in the general flexible link case [44].

5.2.3 *Experimental results*

For experimental comparisons, a 5-bar-linkage manipulator with a flexible link is used. This robot belongs to the class of robots described in [41] and explained above in the description of the feedback linearization method. The 5-bar-linkage manipulator of Figure 5.22 is used [16]. The flexible link is attached to the part of the last link which extends into the workspace past the rigid parallelogram structure. The link vibrates only horizontally. The vibrations are measured using a linear CCD array camera pointed at an LED mounted at the tip of the link.

Figure 5.22

A challenging cartesian trajectory of a 4 leaf-clover was chosen as the test trajectory. This trajectory is traced out in a vertical plane. A plot of the reference trajectory is shown in Figure 5.23. Plots of the tracking results for a PD Controller (See Figure 5.23), H_∞-optimal controller (see Figure 5.23), a passive flexible controller (see Figure

5.24) and a closed-loop shaped-input controller (see Figure 5.25) are shown and compared. Because no feedforward term was employed, no attempt was made to explicitly track the trajectory. The comparison will be made based primarily on the vibration suppression of the various techniques. First, the PID controller created a great deal of vibrations throughout the trajectory and resulted in unacceptable performance. The passive flexible controller gave excellent results with respect to vibration suppression and had good tracking characteristics. The H_∞ optimal controller also gave excellent vibration suppression but had sluggish tracking performance. Finally, the closed-loop shaped-input technique gave good vibration suppression and reasonable tracking ability. Overall, however, these results are promising and show that it is possible to control elastic effects in MMFL manipulators.

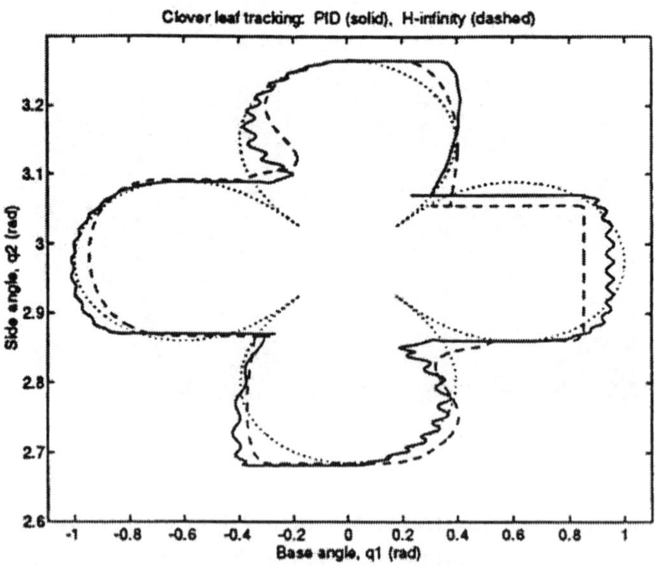

Figure 5.23

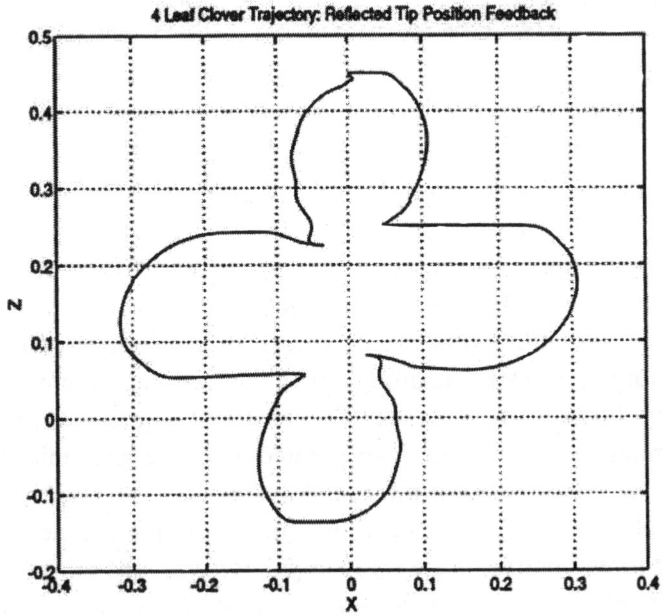

Figure 5.24

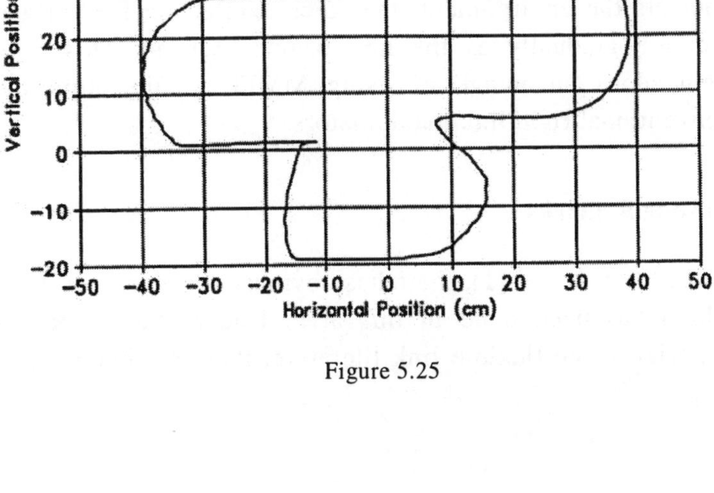

Figure 5.25

5.3 Force Control

Obviously, to be useful in most tasks, the flexible link manipulator must contact a payload or the environment. Thus, it is crucial be able to control forces exerted on the payload or environment, especially for assembly or pick-and-place operations.

Much work has already been studied in terms of force control of rigid link manipulators [49]. The research in the area of force control of flexible link manipulators is still not very developed. Modeling of a MMFL manipulator in contact with an environment is usually handled using a set of constraint equations [47]. These constraint equations are incorporated into the dynamic equations using a Lagrange multiplier and using the extended Hamilton's principle. The first paper dealing with this problem appeared in [48] where force control of a SFL manipulator was studied. In [47], with appropriate assumptions, a simpler set of equations referred to as quasi-static equations are used instead of the dynamic equations. Then, a hybrid position/force control strategy is verified experimentally for a MMFL robot. The paper deals with constrained motion only and does not study the transition problem between unconstrained position control and the constrained hybrid control cases. However, in a flexible manipulator, the "bounce" problem should not occur as readily as in the rigid manipulator case [51] because of the compliance in the flexible link. As the flexible link contacts the environment, the force exerted on the surface should increase gradually as the link begins to bend. This illustrates a tremendous advantage of using MMFL manipulators instead of conventional rigid link manipulators.

Closing Remarks

Since the paper of [1], there has obviously been a great deal of work which has been done in this area. Due to the sheer volume of material in the flexible link literature, there are bound to be works which other researchers consider important that I have missed. For this, I must apologize. The papers chosen here reflect my bias as to what has been the key papers in this area. This is by no means a complete survey of the area. However, it is hoped that this will give a flavor of some of the papers in this area of research. It is also seen that there is still a great deal of interesting research which can still be done to contribute to the eventual possibility of a commercially viable flexible link manipulator.

Acknowledgements

I would like to thank all of the graduate students who have worked with me over the years on the topic of flexible link manipulators. All of the experimental results shown in this paper are due to their contributions. I would also like to thank Kevin Krauel for his technical assistance in making our robots actually work!

References

1. R.H. Cannon and R. Schmitz, *International Journal of Robotic Research*, Vol. 3, No. 3, pp. 62-75.
2. L. Meirovitch, *Analytical Methods in Vibrations*, (The Macmillan Co., 1967).
3. S. Timoshenko, *Vibration Problems in Engineering*. (D. Van Nostrand Company, Inc., Princeton, New Jersey, 1955).
4. E. Bayo, *Journal of Robotic Systems*, Vol. 4, No. 1, (1987) pp. 63-75.
5. H. Krishnan, Proceedings of the IEEE Conference on Decision and Control, (1988), pp. 1619-1626.
6. T. Ravichandran, G. Pang and D. Wang, *Control-Theory and Applications*, Vol. 9, No. 4, (1993), pp. 887-908.
7. K. Zuo and D. Wang, 1992 IEEE Conference on Robotics and Automation, (Nice, France, 1992).
8. D. Wang and M. Vidyasagar, *International Journal of Robotics Research*, Vol. 11, No. 6, (1992), pp. 572-578.
9. F. Bellezza, L. Lanari and G. Ulivi, Proceedings of the IEEE Conference on Robotics and Automation, (1990), pp 734-739.
10. D. Wang and M. Vidyasagar, *International Journal of Robotics Research*, Vol. 10, No. 5, (1991), pp. 540-549.
11. C.T. Chen, *Linear System Theory and Design,* (Holt, Rinehart, and Winston, 1984).

12. A. Truckenbrodt, Proceedings of the IFAC Control Science and Technology Conference, (1981), pp. 1909-1914.
13. W.B. Gevarter, *AIAA Journal*, Vol. 8, No. 2, (1970), pp. 666-672.
14. M. Vidyasagar, Input-Output Analysis of Large Scale Interconnected Systems, (Springer-Verlag, Berlin, 1985).
15. N.C. Singer and W.P. Seering, *Transactions of the ASME*, Vol. 112, (1990), pp. 76-82.
16. V. Drapeau and D. Wang, 1993 IEEE *Conference on Robotics and Automation*, Vol. 3, (Atlanta, Georgia, 1993), pp. 216-221.
17. B. Francis, A Course in H_∞ Control Theory, (Springer-Verlag, Berlin, 1987).
18. R.F. Curtain, Infinite Dimensional Linear Systems Theory, Lecture Notes in Control and Information Sciences, (Springer-Verlag, 1974).
19. A. De Luca and L. Lanari, Proceedings of the Int. Symp. on Intelligent Robotics, (1991), pp. 224-235.
20. E. Bayo, R. Moraghar and M. Medics, Int. *J. off Robotics and Automation*, Vol. 3, No. 3, (1988).
21. M. W. Spong, Trans ASME J of Dyn. Sys. , Meas., and Contr., Vol. 109, (1987), pp. 310-319.
22. K. Morris, *Finite-dimensional Control of Infinite Dimensional Systems,* Ph.D. Thesis, University of Waterloo, (Canada, 1989).
23. G. Hastings, W. Book, *Robotics: Theory and Applications,* (ASME, 1986).
24. R. Mukherji, B. Francis, R. Kwong and J. MacLean, *Proceedings of the Mathematical Theory of Network Systems,* (Amsterdam, 1989).
25. M. Rossi, K. Zuo and D. Wang, *Proceedings of the IEEE Conference on Robotics and Automation,* (San Diego, CA, 1994), pp. 321-326.
26. G.H. Tallman and O.J. Smith, IRE Transactions on Automatic Control, (1958).
27. M.W. Spong and M. Vidyasagar, *Robot Dynamics and Control,* (New York, Wiley, 1989).
28. E. Bayo, R. Movaghar and M. Medus, *Int. J. of Robotics and Automation,* Vol. 3, No. 3, (1988).
29. E. Kubica and D. Wang, *1993 IEEE Conference on Robotics and Automation,* Vol. 2, (Atlanta, Georgia, 1993), pp. 236-241.
30. D-S. Kwon and W.J. Book, *Proc. 1990 American Control Conf.,* (San Diego, 1990).
31. B. Siciliano and W.J. Book, *Int. J. of Robotics Research,* Vol. 7, No. 4, (1988), pp. 79-90.
32. M. Vidyasagar, Nonlinear Systems Analysis, (Prentice Hall, 1993). 33. P.V. Kokotivic, SIAM Review, 26(4), (1989), pp. 501-550.
33. X. Ding, T.J. Tarn, and A.K. Bejczy, *Proceedings of the 27th IEEE Conference on Decision and Control,* (Austin, Texas, 1988), pp 52-57.
34. X. Ding, T.J. Tarn and A. K. Bejczy, *Proc. IEEE Int. Conf. on Robotics and Automation.* (Scottsdate, AZ., 1989), pp. 1678-1683.

35. W.J. Book, Int. J. of Robotics Research, Vol. 3, No. 3, (1984), pp. 87-101. 37. F. Pfeiffer, *Journal of Robotic Systems, 6(4),* (1989), pp. 407-416.
36. A. Isidori, Lecture Notes in Control and Information Sciences. 72 (Springer-Verlag, Berlin, 1985).
37. A. DeLuca, P. Lucibello and G. Ulivi, *Journal of Robotic Systems, Vol. 6,* (1989) pp. 325-344.
38. D. Wang and M. Vidyasagar, ASME Journal of Dynamic Systems, Measurement and Control Vol. 113, No. 4, (1991) pp. 655-661
39. V.G. Moudgal, K.M. Passino and S. Yurkovich, IEEE Transactions on Control Systems Technology, Vol. 2, No. 4, (1994).
40. E. Bayo, P. Papadopoulos, J. Stubbe and M. Serna, *Int. Journal of Robotics Research,* Vol. 8, No. 6, (1989), pp 49-62.
41. B. Siciliano, J.V.R. Prasad and A.J. Calise, *ASME J. of Dynamic Sys, Meas, and Cont.,* Vol. 114, (1992), pp. 70-77.
42. D. Wang and M. Vidyasagar, ASME Journal of Dynamic Systems, Measurement and Control Vol. 113, No. 4, (1991), pp. 662-668.
43. F. Matsuno and Y. Sakawa, *Journal of Robotic Systems, Vol.* 7, No. 4, (1990), pp. 575-597.
44. F. Matsuno, T. Asano and Y. Sakawa, IEEE Transactions on Robotics and Automation, Vol. 10, No. 3, (1994).
45. S.W. Tilley, R.H. Cannon and R. Kraft, *Robotics: Theory and Applications,* (ASME, 1986).
46. M. H. Raibert and J.J. Craig, *Trans. ASME J. Dyn. Sys. Meas. and Control,* Vol. 103, No. 2, (1981), pp. 126-133.
47. C. Trautman and D. Wang, 1995 Proceedings of the IEEE Conference on Robotic and Automation Accepted for Publication
48. J.K. Mills, Proc. IEEE Int. Conf. on Robotics and Automation, (Cincinatti, May 13-18, 1990).
49. Z.H. Luo, IEEE Transactios on Automatic Control, Vol. 38, (1993), pp. 1610-1622.

Chapter 6

Sensor Output Feedback Control of Flexible Robot Arms

In this chapter, we present sensor output feedback control laws for one-link flexible robot arms with rotational joints or prismatic joints driven by velocity-controlled motors. Specifically, we discuss mission function, strain feedback control, gain adaptive strain feedback control, and shear force feedback control. Emphasis has been laid on the stability analysis of various sensor feedback closed-loop systems. It is shown that, in addition to being easily implemented, strain feedback control and shear force feedback control introduce damping for motion/vibration control which results in good control performance. It is also shown that simple gain adaptive strain feedback control can cope with tip load variations of the flexible arm in maintaining good vibration control performance.

6.1. Introduction

In control of flexible robot arms, a typical method is to, first derive an infinite dimensional dynamic model governed by partial differential equations, and then reduce the infinite dimensional model to a finite dimensional one based on which some finite dimensional dynamic controllers such as optimal controllers and robust controllers are designed. Although dynamic controllers are designed with the expectation of increasing control performance for a well modeled system, it is widely recognized that a high dimensional controller may not be so easy to implement, that it is difficult to get an accurate model, and that the ignored high frequency dynamics may cause instability for the closed-loop control system.

In this chapter, we discuss various sensor output feedback control laws for flexible robot arms. Sensor output feedback control is attractive because of its easy implementation in the sense that

1) the dynamic model is only used for stability analysis, not for on-line controller design,
2) it is not necessary to know the exact parameters involved in the model,
3) there are not spillover problems since we deal with infinite dimensional system models.

In section 6.2, we present dynamic models, described by partial differential equations, of one-link flexible robot arms with rotational and prismatic joints, respectively. Although an industrial robot consists generally of multiple links, the study of one-link flexible robots is a first step toward a better understanding of essential properties of the problem and toward further development of a more general theory. For a flexible arm with and without a tip body, the governing partial differential equation is the same, but the associated boundary conditions are different. We will briefly mention these differences but we will consider control design only for the most simple case – a flexible arm without a tip body. The reader should find that all the ideas also apply to flexible robots with tip bodies.

In section 6.3, which is the main part of this chapter, we present sensor output control methods such as Mission Function Control, Strain Feedback Control, Gain Adaptive Strain Feedback Control, and Shear Force Feedback Control. Emphasis will be laid on the closed-loop stability analysis of various feedback laws.

Section 6.4 gives control experimental results and in section 6.5 conclusions are drawn.

6.2. Dynamic Models of One-link Flexible Robot Arms

6.2.1 Flexible arms with rotational joints

The flexible arm to be considered is a uniform beam which is fixed on a hub with rotational inertia J in the horizontal plane. The hub is rotated by a control motor by an angle $\theta(t)$ at time t. We assume that only bending vibration occurs in the horizontal plane and denote the bending displacement by y(t, x). The length, bending rigidity and mass per unit length of the flexible arm are denoted by l, EI and ρ, respectively. If the tip end of the flexible arm is free, then the partial differential equation governing the bending vibration can be written as

$$\begin{cases} \ddot{y}(t,x)+\dfrac{EI}{\rho}y''''(t,x)=-x\ddot{\theta}(t) \\ y(t,0)=y'(t,0)=0 \\ y''(t,l)=y'''(t,l)=0 \\ y(0,x)=y_0(x), \dot{y}(0,x)=y_1(x) \end{cases} \tag{6.2.1}$$

where $y_0(x)$ and $y_1(x)$ are initial displacement and initial velocity of flexible arm, respectively. Details can be found in the literature[1, 2]. Throughout this chapter, a dot $\dot{y}$ denotes time derivative $\partial y/\partial t$, and a prime y' denotes spatial derivative $\partial y/\partial x$. It can be seen from this model that the motion acceleration $\ddot{\theta}(t)$ of the motor affects the dynamic behavior of flexible arms.

Eq.(6.2.1) represents an Euler-Bernoulli model where the rotatory inertia and shear deformation of the arm are neglected. There are other more accurate models such as Timoshenko model in which both rotatory inertia and shear deformation are taken into consideration[3]. We shall be working with the Euler-Bernoulli model (6.2.1) since it is the simplest one and it retains the essential features of vibration of flexible arms.

On the other hand, the motion equation of the control motor is given by

$$J\ddot{\theta}(t) = \tau(t) + EIy''(t,0) \tag{6.2.2}$$

where $\tau(t)$ denotes control torque of the motor. The last term on the right hand side of the above equation reflects reaction torque of the flexible arm exerted on the motor.

So far we have discussed a dynamic model of a flexible arm with a free tip end. When the arm carries a rigid tip body, not only the bending vibration but also torsional vibration will occur. The dynamic model in this case becomes a little more complicated. Let $\Phi(t, x)$ denote the twisting angle of the arm at time t and at a distance x from the base. The dynamic model is derived as follows:

$$\begin{cases} \ddot{y}(t,x) + \dfrac{EI}{\rho} y''''(t,x) = -x\ddot{\theta}(t) \\ \ddot{\Phi}(t,x) - \dfrac{GJ}{\rho\kappa^2}\Phi''(t,x) = 0 \\ y(t,0) = y'(t,0) = 0, \Phi(t,0) = 0 \\ \begin{bmatrix} \ddot{y}(t,l) \\ \ddot{y}'(t,l) \\ \ddot{\Phi}(t,l) \end{bmatrix} + M^{-1} \begin{bmatrix} -EIy'''(t,l) \\ EIy''(t,l) \\ GJ\Phi'(t,l) \end{bmatrix} = -\begin{bmatrix} l \\ 1 \\ 0 \end{bmatrix} \ddot{\theta}(t) \end{cases} \tag{6.2.3}$$

where the positive definite matrix M is given by

$$M = \begin{bmatrix} m & 0 & me \\ 0 & J_z & J_{xz} \\ me & J_{xz} & J_x + me^2 \end{bmatrix};$$

GJ and $\rho\kappa^2$ are the torsional rigidity and the mass polar moment of inertia per unit length of the flexible arm, respectively; J_x, J_z, J_{xz} are moments of

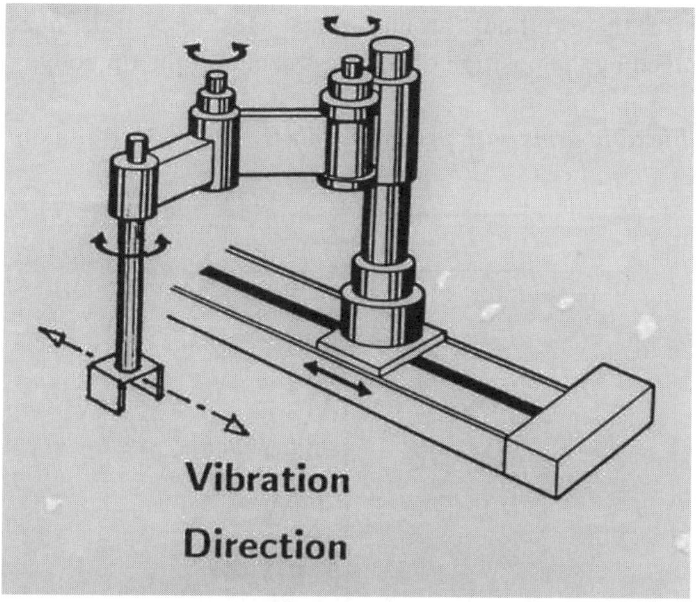

Fig.6.1 A SCARA robot with a long trip arm.

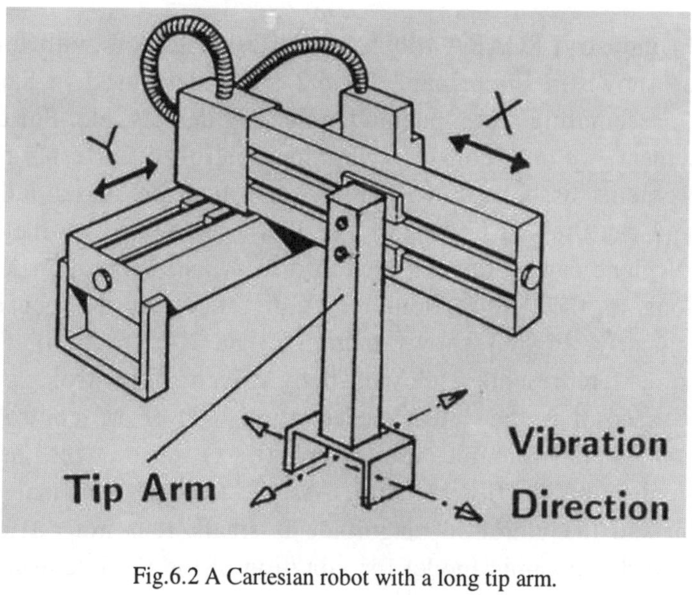

Fig.6.2 A Cartesian robot with a long tip arm.

inertia of the tip body around some axes; e > 0 is a constant determined by the position of the mass center of the tip body.

6.2.2 Flexible arms with prismatic joints

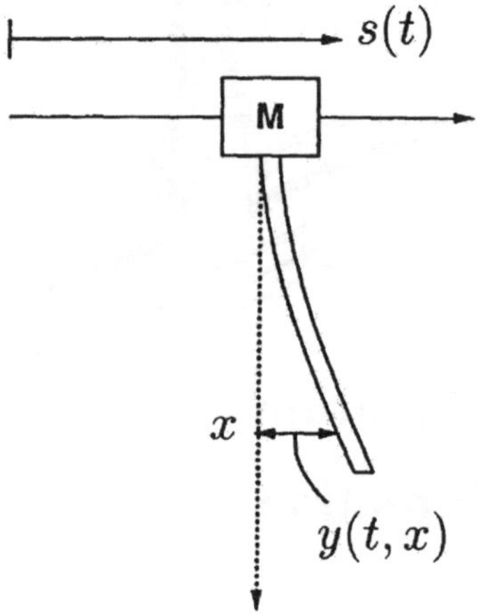

Fig.6.3 Vibration of the tip flexible arm of a Cartesian robot along X-axis.

In industry, SCARA robots or Cartesian Robots with long tip arms as shown in Fig.6.1 and Fig.6.2 are widely used in handling objects, assembling parts, packing, detecting defects, etc. For clarity of statement, we only consider vibration control of Cartesian robots. The problems associated with SCARA robots are essentially the same and can thus be treated in a similar way. Since any motion in the X-Y plane can be decomposed into its X and Y components, the vibrations in the X-direction and Y-direction can be considered independently. Fig.6.3 shows motion of an X-Y robot in the X-direction. M represents a moving body driven by control motor. It can be seen that the linear acceleration $\ddot{s}(t)$ of M controls the vibration in the flexible arm. Let y(t, x) denote the bending displacement of the flexible arm. Assume that the arm material is uniform and the vibration magnitude is small. It is not difficult to show that the dynamic model for vibration of the flexible arm in the X-direction can be written as[12]

$$\begin{cases} \ddot{y}(t,x)+\dfrac{EI}{\rho}y''''(t,x)=-\ddot{s}(t) \\ y(t,0)=y'(t,0)=y''(t,l)=y'''(t,l)=0 \\ y(0,x)=y_0(x),\dot{y}(0,x)=y_1(x) \end{cases} \qquad (6.2.4)$$

The only difference between model (6.2.1) and model (6.2.4) is that in Eq.(6.2.1) the rotational acceleration $\ddot{\theta}(t)$ affects the dynamic behaviour, while in Eq.(6.2.4) the linear acceleration $\ddot{s}(t)$ affects the dynamic behaviour. This difference, however, requires us to use different sensor feedback control laws as will be made clear later.

We have thus obtained 3 models (6.2.1), (6.2.3), and (6.2.4). From now on, we will consider model (6.2.1) and model (6.2.4), since it is already shown that model (6.2.3) can be treated in a similar fashion by appropriately choosing a Hilbert state space[4].

6.3. Sensor Output Feedback Control

6.3.1 Mission function control[5]

The mission function control law is derived by using the Lyapunov method. We will be working on model (6.2.1) together with (6.2.2). Let θ_r (a constant) be the desired position of the motor angle. We wish to determine the control torque $\tau(t)$ such that vibration in flexible arm will eventually die down and the motor position error $\theta(t)-\theta_r$ tend to zero. That is to say, we want to simultaneously control both motion and vibration. For this purpose, define a Lyapunov candidate V(t)

$$V(t)=\frac{1}{2}[a_1 J\dot{\theta}(t)^2+a_2\left[\theta(t)-\theta_r\right]^2$$
$$+a_3\left\{\int_0^l \rho\left[\dot{y}(t,x)+x\dot{\theta}(t)\right]^2dx+\int_0^l EI\left[y''(t,x)\right]^2dx\right\}] \quad \ldots \quad (6.3.1)$$

where a_i (i = 1,..., 3) are arbitrary positive constants. The third term on right hand side of Eq.(6.3.1) represents the kinetic energy of the flexible arm. Taking the time derivative of V(t) along solutions of Eqs.(6.2.1) and (6.2.2) and integrating by parts yield

$$\dot{V}(t) = \dot{\theta}(t)[a_1\tau(t) + (a_1 - a_3)EIy''(t,0) + a_2(\theta(t) - \theta_r)] \qquad (6.3.2)$$

If we choose control torque $\tau(t)$ according to

$$\tau(t) = \left(\frac{a_3}{a_1} - 1\right)EIy''(t,0) - \frac{a_2}{a_1}(\theta(t) - \theta_r) - \frac{a_4}{a_1}\dot{\theta}(t) \qquad (6.3.3)$$

then

$$\dot{V}(t) = -a_4\dot{\theta}(t)^2 \leq 0 \qquad (6.3.4)$$

In Eqs.(6.3.3) and (6.3.4), a_4 is also an arbitrary positive constant.

The control law (6.3.3) implies that a combined PD feedback of motion variables $\theta(t)$ and $\dot{\theta}(t)$ and a static feedback of bending strain EIy"(t, 0) guarantees closed-loop stability for arbitrary positive constants a_i (i =1, ... , 4). However, the control performance depends on a careful choice of these parameters. For example, if vibration suppression is of primary concern, we can choose large value for $\frac{a_3}{a_1}$. On the other hand, if good transient response of motion is desired, we can choose small value for $\frac{a_3}{a_1}$. The control system designer can determine the control law according to the control purpose (mission). This is why it is named "mission function control". One of drawbacks of mission function control is that it is difficult to achieve simultaneously good performance of motion control and vibration control as will be seen from simulation results.

It should be noted that if we choose $a_3 = a_1$, $a_2 = a_1 k_p$ and $a_4 = a_1 k_d$ then the control torque Eq.(6.3.3) becomes

$$\tau(t) = -k_p(\theta(t) - \theta_r) - k_d\dot{\theta}(t)$$

which is no more than a PD feedback control of motor motion.

Let us now look at the closed-loop equation which y(t, x) should satisfy. To this end, substituting Eq.(6.3.3) into Eq.(6.2.1) and Eq.(6.2.2) yields

$$\begin{cases} \ddot{y}(t,x) + \dfrac{EI}{\rho} y''''(t,x) + \dfrac{a_3}{a_1}\dfrac{EI}{J} xy''(t,0) = \dfrac{x}{J}\left[\dfrac{a_2}{a_1}(\theta(t) - \theta_r) + \dfrac{a_4}{a_1}\dot{\theta}(t)\right] \\ J\ddot{\theta}(t) + \dfrac{a_4}{a_1}\dot{\theta}(t) + \dfrac{a_2}{a_1}(\theta(t) - \theta_r) = \dfrac{a_3}{a_1} EIy''(t,0) \\ y(t,0) = y'(t,0) = y''(t,l) = y'''(t,l) = 0 \\ y(0,x) = y_0(x), \dot{y}(0,x) = y_1(x), \theta(0) = \theta_0, \dot{\theta}(0) = \theta_1 \end{cases} \tag{6.3.5}$$

The third term $\dfrac{a_3}{a_1}\dfrac{EI}{J} xy''(t,0)$ on the left-hand side of the first equation of Eq.(6.3.5) is introduced by the strain feedback in control law (6.3.3). Obviously, this term does not play a role of damping, but it can be shown that the existence of this term leads to an increase of vibration frequency of flexible arm. The damping effect is obtained through the coupled motion of the motor. In the next section, we suggest a method to directly introduce a damping term into the differential equation of y(t, x). For this purpose it is necessary to feedback the time rate of change of bending strain if a torque controlled motor is used, unlike the control law (6.3.3).

6.3.2 Strain feedback control

From the above observation, to introduce a damping term for vibration control, one possibility is to feedback the time rate of change of bending strain if we use torque controlled motors.

However, because the bending strain is easier to measure than the time rate of change of bending strain in practice, also because most motion control systems for industrial robots are controlled in joint space by a set of independent PID feedback controllers in velocity mode, we prefer to use a high gain velocity controlled motor in which the command voltage to the motor driver is proportional to the rotational velocity of the motor, i.e.,

$$v_{ref}(t) = k_f \dot{\theta}(t) \tag{6.3.6}$$

In Eq.(6.3.6), k_f is the back emf constant of DC control motor.

The bending strain EIy"(t,0) which is to be used in feedback can be measured by strain gauge foils properly cemented on both sides at the base of the beam. The strain feedback control law is given by

$$v_{ref}(t) = -k_d(\theta(t) - \theta_r) - k_i \int_0^t (\theta(\tau) - \theta_r) d\tau + kEIy''(t,0) \tag{6.3.7}$$

where $k_d > 0$, $k_i > 0$ are PI feedback gains and $k > 0$ is the strain feedback gain. Combining Eqs.(6.3.6) and (6.3.7) we obtain

$$\dot{\theta}(t) = -\frac{k_d}{k_f}(\theta(t) - \theta_r) - \frac{k_i}{k_f} \int_0^t (\theta(\tau) - \theta_r) d\tau + \frac{kEI}{k_f} y''(t,0). \tag{6.3.8}$$

Taking the time derivative of Eq.(6.3.8) and substituting $\ddot{\theta}(t)$ into Eq.(6.2.1) yields

$$\begin{cases} \ddot{y}(t,x) + \dfrac{kEI}{k_f} x\dot{y}''(t,0) + \dfrac{EI}{\rho} y''''(t,x) = \dfrac{x}{k_f}\left[k_d \dot{e}(t) + k_i e(t)\right] \\ y(t,0) = y'(t,0) = y''(t,l) = y'''(t,l) = 0 \\ k_f \ddot{e}(t) + k_d \dot{e}(t) + k_i e(t) = kEI\dot{y}''(t,0) \\ y(0,x) = y_0(x), \dot{y}(0,x) = y_1(x), e(0) = e_0, \dot{e}(0) = e_1 \end{cases} \tag{6.3.9}$$

where $e(t) = \theta(t) - \theta_r$. This is a set of coupled partial and ordinary differential equations describing coupled motion and vibration control. By functional analysis, we can show that, for some suitably chosen feedback gains k_d and k_i, the hybrid system (6.3.9) is exponentially stable. The reader is referred to section 3.4 for a similar derivation of the stability conditions. Here, in order to understand the basic features of strain feedback control, we consider a special case where $k_d = k_i = 0$, i.e., we control only vibration, while leaving motion of the motor uncontrolled. In this case, Eq.(6.3.9) becomes

$$\begin{cases} \ddot{y}(t,x) + \dfrac{kEI}{k_f} x\dot{y}''(t,0) + \dfrac{EI}{\rho} y''''(t,x) = 0 \\ y(t,0) = y'(t,0) = y''(t,l) = y'''(t,l) = 0 \\ \quad y(0,x) = y_0(x), \dot{y}(0,x) = y_1(x). \end{cases} \tag{6.3.10}$$

Clearly, the strain feedback introduces a term $\dfrac{kEI}{k_f} x\dot{y}''(t,0)$. The problems which remain are

1) Does there exist a solution to Eq.(6.3.10)?
2) Is the solution exponentially stable?

To answer the first question, let us define an operator A as

$$\begin{cases} D(A) = \{ y \mid y \in H, y''''' \in H, y(0) = y'(0) = y''(l) = y'''(l) = 0 \} \\ \quad Ay = \dfrac{EI}{\rho} y'''', \forall y \in D(A) \end{cases} \tag{6.3.11}$$

where $H = L^2(0,l)$ is the usual square integral function space, with inner product $<\cdot\ ;\ >$ and the induced norm $\|\cdot\|$. D(A) denotes the domain of operator A. It is well known that the operator A is positive definite with compact inverse[4]. Also let us define an operator Π by

$$\Pi_y = \frac{kEI}{k_f} xy''(x), \quad x \in (0,l), \forall y \in D(\Pi)$$

It is obvious that operator Π is neither self-adjoin nor positive definite. However, it can be shown that Π is A-symmetric and A-positive semi-definite[6]. Thus, Π can be decomposed as $\Pi = QA$

when restricted on D(A), with Q being a bounded, positive semi-definite operator. By making use of these properties of the above mentioned operators, equation (6.3.10) can be written as an abstract equation on H as follows

$$\begin{cases} \ddot{y}(t)+QA\dot{y}(t)+Ay(t)=0 \\ y(0)=y_0, \dot{y}(0)=y_1 \end{cases} \tag{6.3.12}$$

Let $H_1 = D(A^{1/2})\times H$ and $H_2 = D(A)\times D(A^{1/2})$. Then H_1 and H_2 are Hilbert spaces when equipped with the following inner products and the corresponding induced norms

$$\left\langle \begin{bmatrix} h_1 \\ h_2 \end{bmatrix}, \begin{bmatrix} \tilde{h}_1 \\ \tilde{h}_2 \end{bmatrix} \right\rangle_{H_1} = \left\langle A^{1/2}h_1, A^{1/2}\tilde{h}_1 \right\rangle + \left\langle h_2, \tilde{h}_2 \right\rangle$$

$$\left\langle \begin{bmatrix} h_1 \\ h_2 \end{bmatrix}, \begin{bmatrix} \tilde{h}_1 \\ \tilde{h}_2 \end{bmatrix} \right\rangle_{H_2} = \left\langle Ah_1, A\tilde{h}_1 \right\rangle + \left\langle A^{1/2}h_2, A^{1/2}\tilde{h}_2 \right\rangle$$

Now let $z(t)=\begin{bmatrix} A^{1/2}y(t) \\ A^{1/2}\dot{y}(t) \end{bmatrix}$. Then Eq.(6.3.12) can be rewritten as

$$\begin{cases} \dot{z}(t)=\tilde{A}z(t) \\ z(0)=\begin{bmatrix} A^{1/2}y_0 \\ A^{1/2}y_1 \end{bmatrix} \end{cases}$$

on H_1, where operator $\tilde{A}$ is defined by

$$D(\tilde{A})=\left\{ h=\begin{bmatrix} h_1 \\ h_2 \end{bmatrix} \mid h_2 \in D(A^{1/2}), A^{1/2}h_1 + QA^{1/2}h_2 \in D(A^{1/2}) \right\}$$

$$\tilde{A}h=\begin{bmatrix} I & 0 \\ 0 & -A^{1/2} \end{bmatrix}\begin{bmatrix} 0 & I \\ A^{1/2} & QA^{1/2} \end{bmatrix}\begin{bmatrix} h_1 \\ h_2 \end{bmatrix}, \forall h \in D(\tilde{A})$$

Such defined $D(\tilde{A})$ is dense in H_1, and $\tilde{A}$ is a closed operator. Thus, the adjoint operator $\tilde{A}^*$ of $\tilde{A}$ can be uniquely defined. Moreover, it can be shown that, for any $h \in D(\tilde{A})$ and $h^* \in D(\tilde{A}^*)$, there hold $\langle \tilde{A}h, h \rangle_{H_1} \le 0$ and $\langle \tilde{A}^* h^*, h^* \rangle_{H_1} \le 0$. Consequently, there exists a strongly continuous semigroup of contractions S(t) generated by $\tilde{A}$, and for initial conditions y_0 and y_1 satisfying $y_0, y_1 \in D(A)$ and $Ay_0 + QAy_1 \in D(A^{1/2})$, there exists a unique solution to Eq.(6.3.12) which can be expressed as[7]

$$\begin{bmatrix} y(t) \\ \dot{y}(t) \end{bmatrix} = T(t) \begin{bmatrix} y_0 \\ y_1 \end{bmatrix} \tag{6.3.13}$$

where

$$T(t) = \begin{bmatrix} A^{-1/2} & 0 \\ 0 & A^{-1/2} \end{bmatrix} s(t) \begin{bmatrix} A^{1/2} & 0 \\ 0 & A^{1/2} \end{bmatrix} \tag{6.3.14}$$

is a strongly continuous semigroup on H_2 (this is easily verified).

To show the exponential stability of Eq.(6.3.10), we introduce a new variable $w(t,x) = y''(t, l-x)$. Then Eq.(6.3.10) can be written as

$$\begin{cases} \ddot{w}(t,x) + \dfrac{EI}{\rho} w''''(t,x) = 0 \\ w(t,0) = w'(t,0) = w''(t,l) = 0 \\ w'''(t,l) = \dfrac{\rho k}{k_f} \dot{w}(t,l) \\ w(0,x) = y_0''(l-x), \dot{w}(0,x) = y_1''(l-x) \end{cases} \tag{6.3.15}$$

which is a dissipative boundary velocity feedback system. The exponential stability of Eq.(6.3.15) is proved by Chen et al. using the energy multiplier method[8]. Once w(t, r) is exponentially stable, the solution of Eq.(6.3.10) is also exponentially stable since $y(t,x)=\int_0^x\int_0^s w(t,r)drds$ and the integration operator is bounded. The above analyses imply that the strain feedback term $\frac{kEI}{k_f}x\dot{y}''(t,0)$ in Eq.(6.3.10) does play a role in damping.

6.3.3 Gain adaptive strain feedback control[9]

In the last section, we discussed strain feedback control for a flexible arm with free tip end. It can be verified that the stability results obtained there also hold for a flexible arm with tip mass or rigid tip body, no matter what their physical parameters are. However, the control performance (transient responses etc.) may degrade significantly due to large tip load variations which are often encountered in practical applications. In order to maintain high performance, we propose to adaptively tune the strain feedback gain, instead of keeping the feedback gain as a constant. Unlike the well-known model-based adaptive control where it is required to estimate the physical parameters of control plant so as to tune adaptively parameters of the controller, a simple adaptive gain tuning method is proposed in this section. Our policy is to keep the controller as simple as possible.

Again for simplicity, we consider only vibration control. Let the control voltage $v_{ref}(t)$ to the motor driver be

$$v_{ref}(t)=k(t)y''(t,0) \tag{6.3.16}$$

where y"(t, 0) is the measured bending strain and k(t) is a time varying strain feedback gain to be chosen. Substituting Eq.(6.3.16) into Eq.(6.3.6) and differentiating both sides of the resultant equation, we obtain

$$\ddot{\theta}(t)=\dot{k}(t)y''(t,0)+k(t)\dot{y}''(t,0) \tag{6.3.17}$$

where we use *k(t)* to denote $k(t)/k_f$.

We first consider the following integral type gain adaptive law

$$\begin{cases} \dot{k}(t) = \alpha[y''(t,0)]^2 \\ k(0) = k_0 \geq 0 \end{cases} \tag{6.3.18}$$

where $\alpha > 0$ is an arbitrary constant. Since $k_0 \geq 0$, it is easily seen that *k(t)* is a positive function. Intuitively, this adaptive law implies that if there exists vibration, then the value of *k(t)* should be increased. In practice, however, since *k(t)* is the integration of measured bending strain, and since there always exist some sensor noise and. steady state offset, it may happen that *k(t)* becomes very large as time tends to infinity, so that the closed-loop becomes unstable. To avoid this, we introduce a positive constant β, and consider the following modified gain adaptive law.

$$\begin{cases} \dot{k}(t) = \alpha[y''(t,0)]^2 - \beta k(t) \\ k(0) = k_0. \end{cases} \tag{6.3.19}$$

It is seen from this equation that if bending strain eventually becomes small, *k(t)* will decrease. In this way we can deal with the noise and steady state offset problems. We remark that β can be chosen as a small constant as long as *k(t)* does not diverge at steady state.

Another gain adaptive law that can be considered is the following proportional/integral type which we call the PI-type adaptive law

$$\begin{cases} k(t) = k_p(t) + k_i(t) \\ k_p(t) = \gamma[y''(t,0)]^2 \\ \dot{k}_i(t) = \alpha[y''(t,0)]^2 - \beta k(t) \end{cases} \tag{6.3.20}$$

where $\gamma > 0$ is a constant.

We analyze the closed-loop stability of strain feedback with the gain adaptive law given by Eq.(6.3.18). To this end, substituting

Eqs.(6.3.17)-(6.3.18) into Eq.(6.2.1), we get the following initial-boundary value equation

$$\begin{cases} \ddot{y}(t,x)+k(t)x\dot{y}''(t,0)+\dfrac{EI}{\rho}y''''(t,x)+\alpha x[y''(t,0)]^3=0 \\ y(t,0)=y'(t,0)=y''(t,l)=y'''(t,l)=0 \\ \dot{k}(t)=\alpha[y''(t,0)]^2 \\ y(0,x)=y_0(x),\ \dot{y}(0,x)=y_1(x), k(0)=k_0 \end{cases} \tag{6.3.21}$$

Eq.(6.3.21) represents a nonlinear system. It can be shown, using the nonlinear semigroup approach[10], that there exists a unique solution to Eq.(6.3.21). To show its stability, we consider a Lyapunov candidate

$$V(t)=\frac{1}{2}\int_0^l[\dot{y}''(t,x)]^2dx+\frac{EI}{2\rho}\int_0^l[y''''(t,x)]^2dx+\frac{\alpha}{4}[y''(t,0)]^4 \tag{6.3.22}$$

The time derivative of V(t) along solutions of Eq.(6.3.21) gives

$$\dot{V}(t)=\int_0^l\dot{y}''(t,x)\ddot{y}''(t,x)dx+\frac{EI}{\rho}\int_0^l y''''(t,x)\dot{y}''''(t,x)dx+\alpha\dot{y}''(t,0)[y''(t,0)]^3$$
$$=-k(t)[\dot{y}''(t,0)]^2\le 0. \tag{6.3.23}$$

where we made use of the following boundary conditions

$$\dot{y}''(t,l)=0$$
$$\dot{y}'''(t,l)=0$$
$$y''''(t,0)=-\frac{\rho}{EI}\ddot{y}(t,0)=0$$
$$y'''''(t,0)=-\frac{\rho}{EI}\left(k(t)\dot{y}''(t,0)+\alpha[y''(t,0)]^3\right).$$

To conclude the asymptotical stability, we need to invoke the LaSalle's theorem. We will not go into the details. It should be pointed out that even exponential stability of Eq.(6.3.21) has been proved[11].

For the case of PI-type gain adaptive law given by Eq.(6.3.20) (with $\beta = 0$), the closed-loop equation is found to be

$$\begin{cases} \ddot{y}(t,x) + x(k(t) + 2\gamma[y''(t,0)]^2)\dot{y}''(t,0) + \dfrac{EI}{\rho} y''''(t,x) + \alpha x[y''(t,0)]^3 = 0 \\ y(t,0) = y'(t,0) = y''(t,l) = y'''(t,l) = 0 \\ k(t) = k_0 + \gamma[y''(t,0)]^2 + \alpha \int_0^t [y''(\tau,0)]^2 d\tau \\ y(0,x) = y_0(x), \dot{y}(0,x) = y_1(x) \end{cases} \tag{6.3.24}$$

The function Eq.(6.3.22) also serves as a Lyapunov function for Eq.(6.3.24), since if we take time derivative of V(t) along solutions of Eq.(6.3.24), we get

$$\dot{V}(t) = -\left(k(t) + 2\gamma[y''(t,0)]^2\right)[\dot{y}''(t,0)]^2 \le 0 \tag{6.3.25}$$

From the above discussions, we see that the closed-loop solution of gain adaptive direct strain feedback control decays as time passes by.

6.3.4 Shear force feedback control[12]

In this section, we consider sensor output feedback control problems for flexible robot arms with prismatic joints. We shall be working on the model given by Eq.(6.2.4). For flexible arms with rotational joints, it has been shown that direct strain feedback introduces damping for vibration control. For flexible arms with prismatic joints, we propose to feedback shear force, instead of bending strain. To be more precise, suppose we can measure shear force, which is proportional to $y'''(t, 0)$, at the root end of the flexible arm. Suppose we can also control the motion of the motor such that

$$\dot{s}(t) = -ky'''(t,0) \tag{6.3.26}$$

where k is an arbitrary constant feedback gain. Therefore, we have

$$\ddot{s}(t) = -k\dot{y}'''(t,0). \tag{6.3.27}$$

Substituting Eq.(6.3.27) into Eq.(6.2.4) yields the shear force feedback controlled closed-loop system equation:

$$\begin{cases} \ddot{y}(t,x) + \alpha y''''(t,x) - k\dot{y}'''(t,0) = 0, \\ y(t,0) = y'(t,0) = y''(t,l) = y'''(t,l) = 0 \\ y(0,x) = y_0(x), \dot{y}(0,x) = y_1(x) \end{cases} \tag{6.3.28}$$

where $\alpha = EI/\rho$. At this point, we have at least two questions which need to be clarified.

1) How to measure the shear force $EIy'''(t,0)$?

2) Is the initial-boundary value equation (6.3.28) exponentially stable?

As for the first question, there is no commercial shear force sensor known to us. However, the bending strain $EIy''(t,0)$ can be easily measured by cementing strain gauge foils at the arm's root end. Similarly, by cementing strain gauge foils at location $x = \varepsilon$ (ε is a small constant), close to the root end x = 0, we can measure $y''(t, \varepsilon)$,. Therefore, one method for measuring shear force, $y'''(t, 0)$, is to approximate it by $(y''(t,\varepsilon) - y''(t,0))/\varepsilon$. This method is adopted in our experiments. Although it is commonly recognized that derivation may introduce noise, it has been found, by experiments, that such treatment is quite acceptable and the noise is not severe enough to cause problems.

We now discuss the second question. First of all, notice that when k = 0, i.e., there is no shear force feedback, Eq.(6.3.28) is a conservative system whose energy theoretically never decays. Our objective is to show that the third term $-k\dot{y}'''(t,0)$ on the left-hand side of the first equation of Eq.(6.3.28) is a damping term and shear force feedback control can exponentially decay the solution of Eq.(6.3.28). The existence of a unique solution to Eq.(6.3.28) can not be easily inferred from the existing literature. Fortunately, it can be shown, using the concept of A-dependent operators6, that there does exist a unique solution to Eq.(6.3.28) for any $k > 0$ and smooth initial conditions y_0, y_1. We claim that this solution is exponentially stable. To this end, introduce a new variable

$$w(t,x) = y''(t,l-x) \tag{6.3.29}$$

as before. Assume that the initial conditions are smooth enough that the solution of Eq.(6.3.28) admits continuous special derivatives of up to the 6th order. Taking twice the special derivative of Eq.(6.3.29) and using the corresponding boundary conditions, we find that Eq.(6.3.28) can be transferred into the following initial-boundary value equation:

$$\begin{cases} \ddot{w}(t,x) + \alpha w''''(t,x) = 0 \\ w(t,0) = w'(t,0) = 0 \\ w''(t,l) = -\dfrac{k}{\alpha}\dot{w}'(t,l) \\ w'''(t,l) = 0 \\ w(0,x) = y_0''(l-x), \\ \dot{w}(0,x) = y_1''(l-x). \end{cases} \tag{6.3.30}$$

Energy stored in Eq.(6.3.30) is dissipative provided $k > 0$. To see this, define an energy function for Eq.(6.3.30) as

$$E(t) = \frac{1}{2}\int_0^l [\dot{w}(t,x)]^2 dx + \frac{\alpha}{2}\int_0^l [w''(t,x)]^2 dx.$$

Then the time derivative of E(t) along the solution of Eq.(6.3.30) is given by

$$\dot{E}(t) = \int_0^l \dot{w}(t,x)\ddot{w}(t,x)dx + \alpha\int_0^l w''(t,x)\dot{w}''(t,x)dx$$
$$= -k[\dot{w}'(t,l)]^2 \leq 0$$

Actually the energy dissipates exponentially, which had been long unproved by using the energy multiplier method, but is finally proved in the literature[13]

With w(t,x) shown to be exponentially stable, it is claimed that the solution y(t, x) of Eq.(6.3.28) is also exponentially stable since y(t, x) can be expressed as

$$y(t,x) = \int_0^x \int_0^s w(t,r)drds$$

and the integration operator is bounded.

We have thus far discussed control of vibration using strain feedback for flexible arms with rotational joints and shear force feedback for flexible arms with prismatic joints. Stability analyses indicate that these feedback laws introduce damping for vibration control. In practical applications, it is necessary to control not only vibrations but also the position of the control motor. In this case, the control law is given by Eq.(6.3.7) for flexible arms with rotational joints. Similarly, for flexible arms with prismatic joints, we control the motor such that the velocity of the moving body satisfies

$$\dot{s}(t) = -k_d(s(t) - s_r) - k_i\int_0^t (s(\tau) - s_r)d\tau - ky'''(t,0) \qquad (6.3.31)$$

where $k_d \geq 0$ and $k_i \geq 0$ are PI feedback gains and $k \geq 0$ is the shear force feedback gain. s_r is the desired position of the moving body. Taking the time derivative of Eq.(6.3.31) and substituting the resultant equation into Eq.(6.2.4), we obtain

$$\begin{cases} \ddot{y}(t,x) - k\dot{y}'''(t,0) + \alpha y''''(t,x) = k_d\dot{e}(t) + k_i e(t) \\ y(t,0) = y'(t,0) = y''(t,l) = y'''(t,l) = 0 \\ \ddot{e}(t) + k_d\dot{e}(t) + k_i e(t) = -k\dot{y}'''(t,0) \\ y(0) = y_0, \dot{y}(0) = y_1, e(0) = e_0, \dot{e}(0) = e_1 \end{cases} \tag{6.3.32}$$

where $e(t) = s(t) - s_r$. This equation is very similar to Eq.(6.3.9) in form, so we will only discuss Eq.(6.3.32). We are interested in deriving conditions on the feedback gains k, k_d and k_i which guarantee exponential stability of Eq.(6.3.32). Possible ways to do this are to use the Lyapunov method or the positive real theory. Unfortunately, we tried these methods without much success. It was not until recently that we succeeded in showing that if the feedback gains satisfy

$$k_i < \frac{\lambda_1}{2} \; and \; \frac{4l}{\alpha} k_i k < k_d < \frac{\alpha}{kl} \tag{6.3.33}$$

then the closed-loop hybrid system (6.3.32) is exponentially stable. In Eq.(6.3.33) Al represents the first (smallest) eigenvalue of the positive definite operator $A = \alpha\frac{\partial^4}{\partial x^4}$. We give a sketch of the proof of Eq.(6.3.33). For this purpose, let us first rewrite Eq.(6.3.32) as

$$\begin{cases} \ddot{y}(t,x) - k\dot{y}'''(t,0) + \alpha y''''(t,x) + k_d k y'''(t,0) = k_d[\dot{e}(t) + k y'''(t,0)] + k_i e(t) \\ y(t,0) = y'(t,0) = y''(t,l) = y'''(t,l) = 0 \\ \ddot{e}(t) + k_d\dot{e}(t) + k_i e(t) = -k\dot{y}'''(t,0) \\ y(0) = y_0, \dot{y}(0) = y_1, e(0) = e_0, \dot{e}(0) = e_1 \end{cases}$$

(6.3.34)

Then, we consider its dominant equation:

$$\begin{cases} \ddot{\overline{y}}(t,x) - k\dot{\overline{y}}'''(t,0) + \alpha \overline{y}''''(t,x) + k_d k\overline{y}'''(t,0) = 0 \\ \overline{y}(t,0) = \overline{y}'(t,0) = \overline{y}''(t,l) = \overline{y}'''(t,l) = 0 \\ \ddot{\overline{e}}(t) + k_d \dot{\overline{e}}(t) + k_i \overline{e}(t) = -k\dot{\overline{y}}'''(t,0) \\ \overline{y}(0) = \overline{y}_0, \dot{\overline{y}}(0) = \overline{y}_1, \overline{e}(0) = \overline{e}_0, \dot{\overline{e}}(0) = \overline{e}_1 \end{cases} \tag{6.3.35}$$

It can be shown that if

$$\alpha > k_d kl, \tag{6.3.36}$$

then the energy associated with system (6.3.35) decays exponentially, i.e.,

$$E(t) \le M_1 e^{-w_1 t} E(0), \quad \textit{for some } M_1 \ge 1,\ w_1 > 0 \tag{6.3.37}$$

where

$$E(t) = \int_0^l [(\overline{y}'''(t,x))^2 + (\dot{\overline{y}}''(t,x))^2]dx + \overline{e}^2(t) + \dot{\overline{e}}^2(t).$$

Now, let

$$\begin{cases} \overline{y}_1(t,x) = \frac{1}{2}\left\{\sqrt{\alpha}\,\overline{y}''(t,x) + [\dot{\overline{y}}(t,x) - k\overline{y}'''(t,0)]\right\} \\ \overline{y}_2(t,x) = \frac{1}{2}\left\{\sqrt{\alpha}\,\overline{y}''(t,x) - [\dot{\overline{y}}(t,x) - k\overline{y}'''(t,0)]\right\} \\ \overline{e}_1(t) = \overline{e}(t), \overline{e}_2(t) = \dot{\overline{e}}(t) + k\overline{y}'''(t,0), \end{cases} \tag{6.3.38}$$

and $\overline{Y} = (\overline{y}_1(t,x), \overline{y}_2(t,x), \overline{e}_1(t), \overline{e}_2(t))^T \in H = L^2(0,l) \times L^2(0,l) \times R^2$. Then, Eq. (6.3.35) can be written as

$$\dot{\overline{Y}}(t) = A\overline{Y}(t) \tag{6.3.39}$$

where operator $A : D(A) \subset H \to H$ is defined by

$$A\begin{bmatrix}\phi_1\\ \phi_2\\ e_1\\ e_2\end{bmatrix}=\begin{bmatrix}-\sqrt{\alpha}\phi_2''+k_d[\phi_1(0)-\phi_2(0)]/2\\ \sqrt{\alpha}\phi_1''-k_d[\phi_1(0)-\phi_2(0)]/2\\ e_2+\phi_1(0)-\phi_2(0)\\ -k_ie_1-k_de_2-k_d[\phi_1(0)-\phi_2(0)]\end{bmatrix}, \quad (6.3.40)$$

$$D(A)=\{(\phi_1,\phi_2,e_1,e_2)^T\in H|\phi_1(l)+\phi_2(l)=$$
$$\phi_1'(l)+\phi_2'(l)=\phi_1'(0)-\phi_2'(0)=0,$$
$$\phi_1(0)-\phi_2(0)=-\frac{k}{\sqrt{\alpha}}[\phi_1'(0)+\phi_2'(0)]\ \} \quad (6.3.41)$$

It is not difficult to show that operator A is an infinitesimal generator of a C_0-semigroup e^{At} on H with the exponential decay

$$\left\|e^{At}\right\|\le M_d e^{-w_dt}, \quad \textit{for some } M_d\ge 1, w_d>0. \quad (6.3.42)$$

Our idea is to use this operator A to rewrite Eq.(6.3.32) as an abstract equation on H as follows,

$$\dot{Y}(t)=(A+B)Y(t) \quad (6.3.43)$$

where

$$B(\phi_1(x),\phi_2(x),e_1,e_2)^T=(\frac{k_ie_1+k_de_2}{2},\frac{-k_ie_1-k_de_2}{2},0,0)^T$$

$$Y(t)=(y_1(t,x),y_2(t,x),e_1(t),e_2(t))^T\in H$$

$$y_1(t,x)=\frac{1}{2}\{\sqrt{\alpha}\,y''(t,x)+[\dot{y}(t,x)-ky'''(t,0)]\}$$

$$y_2(t,x)=\frac{1}{2}\{\sqrt{\alpha}\,y''(t,x)-[\dot{y}(t,x)-ky'''(t,0)]\}$$

$$e_1(t)=e(t), e_2(t)=\dot{e}(t)+ky'''(t,0).$$

It is seen that B is a compact operator. Therefore $A+B$ generates a C_0-semigroup. Furthermore, this semigroup is exponentially stable if

$$s(A+B)<0 \tag{6.3.44}$$

where $s(A+B)=\sup\{\operatorname{Re}\lambda | \lambda \in \sigma(A+B)\}$ is the spectral bound of $(A+B)$, since e^{At} is already exponentially stable and B is compact. By exploiting the spectrum of $A+B$, we see that if

$$k_d \geq 4lk_i k/\alpha, \quad k_i < \lambda_1/2 \tag{6.3.45}$$

then $\operatorname{Re}\lambda < 0$, for any $\lambda \in \sigma(A+B)$. Combining Eq. (6.3.45) with Eq. (6.3.36) yields Eq.(6.3.33).

Fig.6.4 A one-link flexible robot with a rotational joint.

6.4. Experimental Results

In order to verify the various control methods presented in section 6.3, we designed a one-link flexible robot with a rotational joint as shown in Fig.4. A stainless steel flexible beam ($l = 0.825$m) is driven by a high gain velocity controlled direct drive motor in the horizontal plane. The physical parameters which appear in the dynamic model (6.2.1)-(6.2.2) are given below.

EI : $(= 6.87N \cdot m^2)$bending rigidity;

ρ : $(= 0.8kg/m)$ mass per unit length of the beam;

J : $(= 0.94kg \cdot m^2)$moment of inertia of the motor and the beam;

k_f : (= 0.667) back emf constant.

• *Mission Function Control Experimental Results*

In Eq.(6.3.3), we put $a_1 = 1$, and choose an appropriate *kp* such that

$$a_2 = k_p, \quad a_3 = 1, \quad a_4 = k_d = 2\sqrt{k_p}.$$

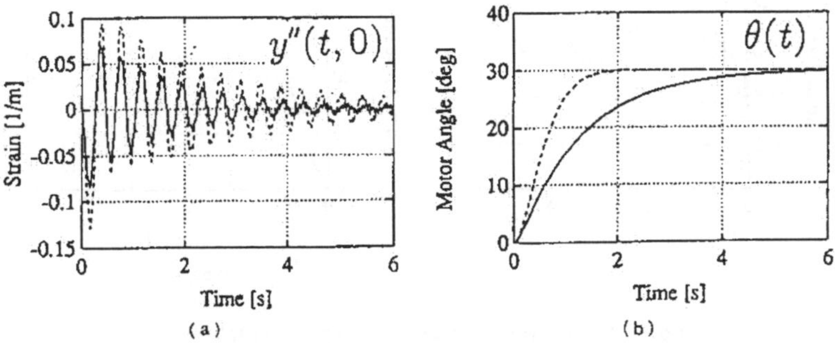

Fig.6.5 PD feedback control experimental results

Then the mission function control is actually a PD control given by

$$\tau(t) = -k_p(\theta(t) - \theta_r) - k_d\dot{\theta}(t)$$

where θ_r = 30 deg is the desired position of the motor. The dashed lines in Fig.5(a) and (b) represent control responses of the bending strain y"(t,0) and motor angle $\theta(t)$, respectively. To achieve better vibration control performance rapid damping), k_p is kept unchanged, and k_d is increased to $4\sqrt{k_p}$. The control results for these parameters are shown in Fig.5 by the solid lines. It can be seen that good vibration control performance is achieved at the price of sacrificing the motion control performance. Actually, this is apparent — the slower the motor rotates, the less vibration occurs.

Let us now see what happens if we include bending strain in feedback law as shown in Eq.(6.3.3). The dashed lines in Fig.6.6(a) and (b) are the same as in Fig.5 ($a_1 = a_3 = 1$, without strain feedback). We keep a_1= 1 and other feedback gains unchanged, but increase a_3.

We see that damping is enhanced for vibration, but again the settling time becomes longer and overshoot is observed in the response of motor motion, i.e., motion control performance becomes worse. So it is hard to get simultaneously good performances for both motion and vibration control in mission function control.

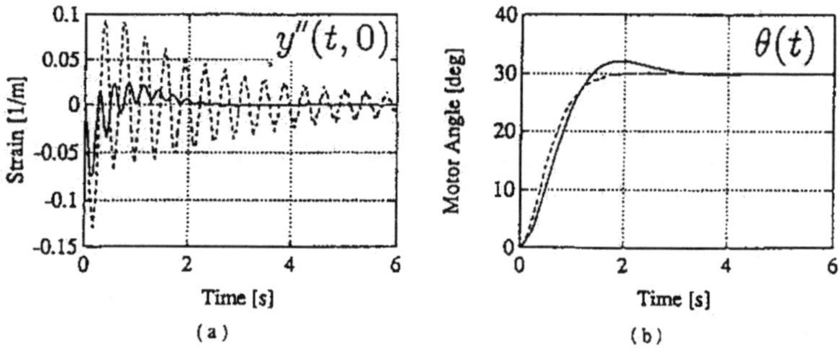

Fig.6.6 Mission function control experimental results.

- *Strain Feedback Control Experimental Results*

The control law is given in Eq.(6.3.7). Let θ_r = 30 deg, $k_d = 2\sqrt{k_p}$, and k_p be the same as in mission function control. The experiment without strain feedback, i.e., k= 0 in Eq.(6.3.7), is first conducted. The dashed lines in Fig.7 show responses of the bending strain and motor angle in this case. These are the same as in Figs.5 and 6. Now we choose the strain feedback gain to be k = 0.6 and keep other feedback gains k_d and k_p unchanged. Experimental results are plotted in Fig.6.7(a)(b) with the solid lines. Clearly, the vibration is well suppressed, and the response of the motor angle is almost unchanged (the solid line and the dashed line in Fig.6.7(b) are almost indistinguishable). Comparing Fig.6.6 with Fig.6.7, we see that strain feedback control can result in simultaneous good performance for motion/vibration control. This is because in strain feedback control we introduced damping for vibration and motion separately.

• *Gain Adaptive Strain Feedback Control Experimental Results*

To demonstrate the effectiveness of gain adaptive strain feedback control, we compared control performance with tip load variations, for both fixed constant

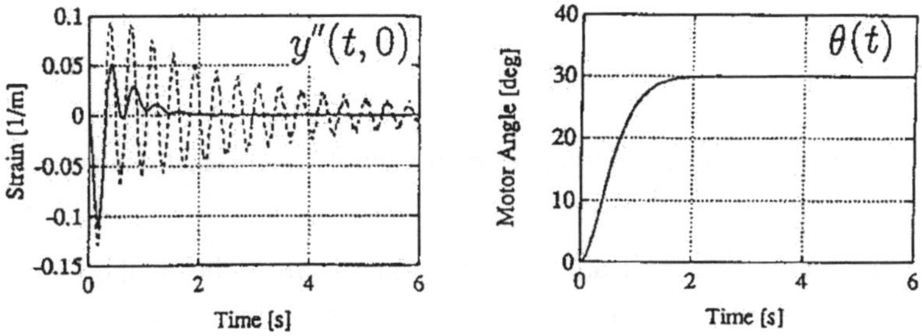

Fig.6.7 Strain feedback control experimental results.

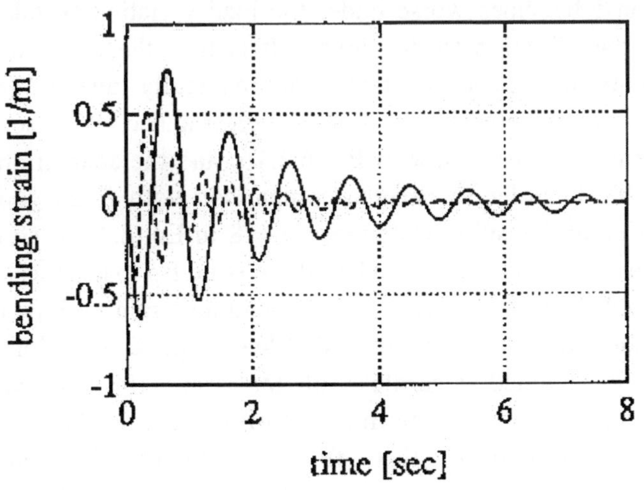

Fig.6.8 Change of responses of bending strain under the tip load variation.

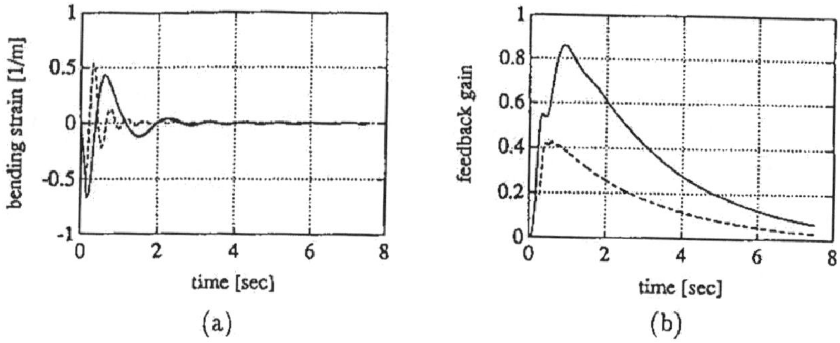

Fig.6.9 Gain adaptive strain feedback control experimental results.

feedback gain and adaptive feedback gain. The dashed line in Fig.6.8 shows the response of bending strain with constant strain feedback gain *k(t)* = 0.4 and without arm's tip load. If the strain feedback gain is kept unchanged, but a tip load weighted 0.6kg is attached, we get the response shown in Fig.8, with the solid line. Obviously, in both cases the closed-loop system remains stable, but the response becomes worse under the load variation (it takes more time for the vibration to die down). It is thus the purpose of gain adaptive strain feedback control to automatically tune the feedback gain so as to maintain good control performance. We use, in this experiment, the control law in Eq.(6.3.16) and the adaptation law in Eq.(6.3.19). In order to make a comparison with fixed gain strain feedback control results, we choose $\alpha = 8$ in Eq.(6.3.19) so that the peak value of *k(t)* is near 0.4 when there is no tip load. β is set to be 0.4. Fig.6.9(a) shows the response of bending strain with (solid line) and without (dashed line) tip load. Clearly, good vibration control results are obtained regardless of the arm's tip load variation. Fig.6.9(b) corresponds to the time histories of the feedback gain *k(t)*. It is interesting to observe that, after the tip load has been added, the feedback gain becomes larger, but the value of *k(t)* remains acceptably low.

• *Shear Force Feedback Control Experimental Results*

To demonstrate shear feedback control, an X-Y Cartesian robot with a long flexible arm as shown in Fig.6.10 has been established. In experiments, we control the motor in such a way that the velocity of the moving body is given by Eq.(6.3.31) where $s_r = 85mm$. Notice that we use the same control law Eq.(6.3.31) for control of the X and Y axes independently. The PI gains k_d and k_i are set to be

k_d=79.20V/m, k_i=0.50V/m/s

in order to complete the motion control in less than 2 seconds without overshoot in the time response of position of the moving body. After k_d and k_i are determined, the shear force feedback gain *k* can be chosen by trial and error. We begin by choosing a small *k*. If the vibration suppression is not so satisfactory, we increase *k*. For *k* = 0.85, the output of the X-axis shear force sensor is displayed with the solid line in Fig.6.11(a). Note that in practice there exists a small constant, depending on the physical parameters of the robot mechanical device such as the radius of the driving wheel, before the feedback gain k in Eq.(6.3.26). In Eq.(6.3.26), these physical constants have been taken to be 1. When all these factors are taken into consideration, there exists a wide range of feedback gains *k*, k_d, k_i such that Eq.(6.3.26) is satisfied (i.e., the closed-loop system is stable)

The dashed line in Fig.6.11(a) is the sensor output when only PI control is implemented (without shear force feedback). It can be seen that there is not much noise in these sensor outputs, and that the vibration decays very rapidly when shear force information is used for feedback. What is more attractive is that the time responses of the moving distance of the moving body do not change much as can be seen from Fig.6.11(b) where the solid line (corresponding to Pl+shear force control) and the dashed line (corresponding to PI control) almost coincides.

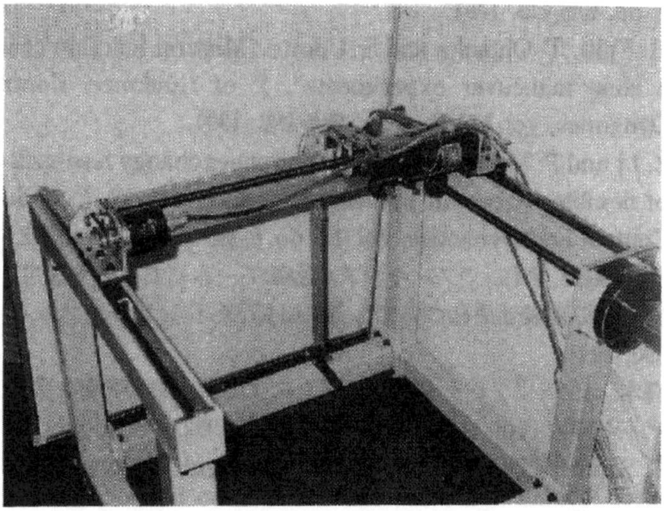

Fig.6.10 A Cartesian robot with a long flexible tip arm.

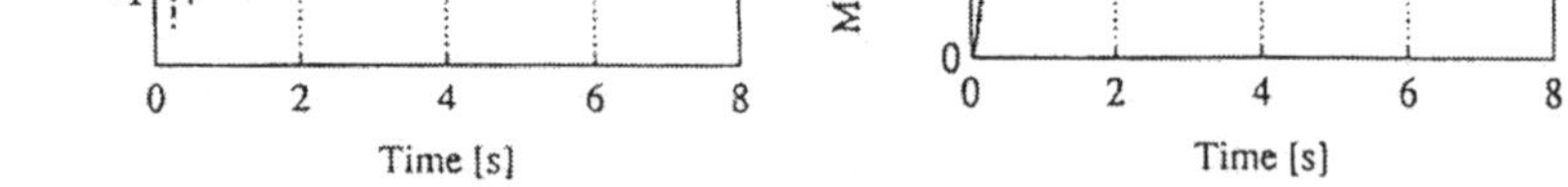

Fig.6.11 Shear force feedback control experimental results for the Cartesian robot

6.5. Conclusions

In this chapter, we presented 1) strain feedback control and gain adaptive strain feedback control laws for one-link flexible arms with rotational joints; 2) shear force feedback control for flexible arms with prismatic joints. Our main emphasis has been on the stability analysis of the closed-loop system with these sensor output feedback control laws, since the stability issue is not only important for practical purpose, but also interesting theoretically. It is shown that the closed-loop equations are always exponentially stable for either strain feedback or shear force feedback control (without motion control). Furthermore, a sufficient condition is given which guarantees closed-loop exponential stability of the simultaneous motion/vibration control of flexible arms. Control experiments verify these theoretical results. Finally, it is remarked that, to our surprise, we recently found that shear force feedback is also effective for vibration control of flexible robot arms with rotational joints[14].

Acknowledgements

The author thanks Dr. B. Z. Guo, Beijing Institute of Technology, for many useful discussions and contributions.

References

1. R. H. Cannon, Jr. and E. Schmitz, Int. J. Robotics Research, 3(1984) 62.
2. Y. Sakawa, F. Matsuno and S. Fukushima, J. RobotZc Systems, 2(1985) 453.
3. F.-Y. Wang and G. Guan, J. Sound and Vibration, 171(1994) 433.
4. Y. Sakawa and Z. H. Luo, IEEE Trans. Automatic Control, 34(1989) 970.
5. H. Fujii, T. Ohtsuka and S. Udou, J. of Guidance, Control and Dynamics, 14(1986)986.
6. Z. H. Luo, IEEE Trans. Automatic Control, 38 (1993) 1610.
7. A. Pazy, Semigroups of Linear Operators and Applications to Partial Differential Equations(Springer-Verlag, New York, 1983).
8. G. Chen, M. C. Delfour, A. M. Krall and G. Payre, SIAM J. Contr. Opt.,25(1987) 526.
9. Z.H.Luo and Y.Sakawa, Proc. IEEE TENCON'93/Beijing, 4(1993) 199.
10. V.Barbu, Analysis and Control of Nonlinear Infinite Dimensional Systems(Academic Press, New York, 1993).
11. B. Z. Guo and Z. H. Luo, Nonlinear Analysis, (1995), to appear.
12. Z. H. Luo, N. Kitamura and B. Z. Guo, IEEE Trans. Robotics and Automation(1995), to appear.

13. G. Chen, S. G. Krantz, D. W. Ma, C. E. Wayne and H. H. West, LN in Pure and Applied Math., 108(Marcel Dekker, 1987).
14. Luo and B. Z. Guo, IEEE Trans. Automatic Control, (1995), submitted.

Chapter 7

On GA Based Robust Control of Flexible Manipulators

This chapter addresses the issues related to the design of robust controllers using genetic algorithms (GA) for lightweight, one-link flexible manipulators working under dynamic environments and other uncertain influences. First, a design procedure based on improved GA is proposed to tune parameters in PID controllers to achieve mixed H_2/H_∞ optimal performances for flexible robotic manipulators. Graphic method, local search strategy, refusal strategy and renewal strategy are used to solve successfully constrains imposed in the design problem. Second, by selecting sensitivity weight functions properly using the GA method, a mixed sensitivity H_∞ controller is developed to ensure robustness of manipulator control systems for varying payloads and other modeling uncertainties. Numeric simulation has been conducted and the results have demonstrated the effectiveness of the proposed method.

7.1 Introduction

Due to its significant benefits, such as fast speed, lightweight, less energy consumption, and wide range of applications, flexible manipulators have been the one of focuses in robotic research for many years [15]. Many results in modeling and controlling the dynamics of flexible manipulators are published [1-14, 21,24]. However it is difficult to develop a precise model for flexible manipulators due to their structural flexibility. It is well understood that the performances of many known controllers for flexible manipulators shows a strong dependence to the precision of the mathematical model used for control, thus is sensitive to variations in model parameters, construction, materials, as well as operating environments. To a certain extend, this dependence has negatively impacted their usefulness for many applications.

This chapter is focused on developing a robust controller using genetic algorithms (GA) for a one-link flexible manipulator working under various uncertainties, such as payload variation, parameter changing, and modeling error, etc. The work reported here is our first step toward a more comprehensive study of systematic investigation of applying GA-based optimization technology for designing robust control systems for flexible robotic manipulators.

We start with establishing a mathematical model of flexible manipulators as a nominal plant for robust control design. This model is derived from the classical Euler-Bernouli beam theory with tip payload and then simplified using the unconstrained model analysis method. For control design, two mixed H_2/H_∞ robust strategies are used to ensure both system performance and robust stability. Controller and uncertainty modeling parameters are used as design variables to meet the robustness requirement performance for a flexible manipulator. The optimal solution for those design variables is obtained by various GA strategies and the numerical analysis from two design examples has indicated that GA based technique is likely to produce good design results even cases when the standard gradient based technique fails.

The chapter is organized as follows. Section 7.2 presents the model of one-link flexible manipulators and its approximation. Section 7.3 discusses the general H_2/H_∞ robust PID control design problem. Section 7.4 proposes an improved GA method for solving the mixed H_2/H_∞

robust PD control problem and a numerical design example is included here. Section 7.5 addresses the problem of design mixed sensitivity H_∞ controllers for flexible manipulators and the corresponding GA solution and number design examples are given in Section 7.6. Finally, Section 7.7 concludes the chapter with some remarks for the future work.

7.2 Dynamic Model of One-link Flexible Manipulators

The flexible manipulator considered here is a one-link arm mounted in the horizontal plane with a tip mass payload. The arm consists of a beam of length L fixed on a rigid hub as showed as Fig. 7.1, where E, I, ρ, A, M_p, a_c, and I_H are Young's modulus, the moment of inertia of the cross-sectional area, beam density, cross-sectional area of beam, the mass payload, the distance of the tip mass center to the end of the beam along x axis, and the hub inertia, respectively. The motion of the flexible manipulator system is described by rigid rotation θ of the hub and flexible displacement w of the beam. The total displacement y of the manipulator is defined as:

$$y = x\theta(t) + w(x,t)$$

For simplicity of the presentation, both rotatory inertia and shear deformation of the beam are neglected because they have very little influence on the fundamental vibration modes. Thus we will use the Euler-Bernoulli model to describe the dynamic behavior of the flexible manipulator [6]. Accordingly, the governing equations of beam deformation and hub rotation are as following:

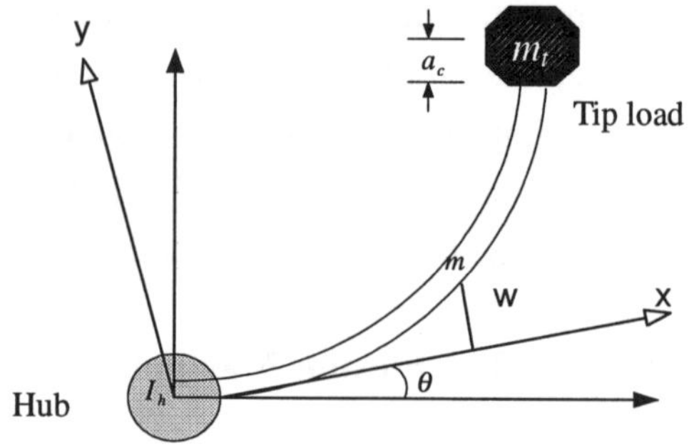

Fig. 7.1. Illustration of a flexible manipulator

$$EI\left(\frac{\partial^4 y}{\partial x^4}\right) + \rho S \frac{\partial^2 y}{\partial t^2} = 0$$

$$\tau - I_H \ddot{\theta} + EI\left(\frac{\partial^2 y}{\partial x^2}\right)_{x=0} = 0$$

With the four boundary conditions:

$$y_{x=0} = 0 \quad ; \qquad \left(\frac{\partial y}{\partial x}\right)_{x=0} = \theta(t) \quad ; \qquad EI\left(\frac{\partial^2 y}{\partial x^2}\right)_{x=L} + a_c M_p \ddot{y}' = 0 \quad ;$$

$$EI\left(\frac{\partial^3 y}{\partial x^3}\right)_{x=L} = M_p(\ddot{y} + a_c \ddot{U}')$$

Furthermore, the corresponding equations of harmonic vibration and the boundary conditions can be written as follows [2]:

$$EIY'''' + \rho S s^2 Y = 0$$
$$T - I_H s^2 \Theta + EIY''|_{X=0} = 0$$
$$Y(x,s) = 0, \quad Y'(x,s) = \Theta(s), \quad x = 0$$
$$Y''(x,s) + a_c M_p s^2 Y = 0, \quad Y'''(x,s) = M_p(s^2 Y + a_c s^2 Y'), \quad x = L$$

One can obtain the exact model for flexible manipulators in both time and frequency domains based on those two sets of equations [24]. An approximate but simple model can be derived using the unconstrained assumed modal method. In this case, the solution of the flexible motion can be represented as the linear combination of admissible functions $\Phi_i(x)$ multiplied by time dependent generalized coordinates q_i, where the admissible function $\Phi_i(x)$ should satisfy all the essential boundary conditions, that is

$$y = \sum_{i=1}^{n} \Phi_i q_i \,.$$

Based on this approach and using Lagrange's method, the state space model for flexible manipulators is [24]

$$\dot{X} = AX + BU_{in}$$

where U_{in} is a vector of input torque, and

$$X = \begin{bmatrix} q_0 \\ \dot{q}_0 \\ q_1 \\ \dot{q}_1 \\ \vdots \\ \\ q_n \\ \dot{q}_n \end{bmatrix} \quad A = \begin{bmatrix} 0\,1 & & & & & \\ 0\,0 & & & & & \\ & 0 & 1 & & & \\ & -\omega_1^2 & -2\zeta_1\omega_1 & & & \\ & & & \cdots & & \\ & & & \cdots & & \\ & & & & 0 & 1 \\ & & & & -\omega_n^2 & -2\zeta_n\omega_n \end{bmatrix} \quad B = \begin{bmatrix} 0 \\ 1 \\ 0 \\ \dfrac{\dot{\Phi}_1(0)}{l_{s1}} \\ \vdots \\ 0 \\ \dfrac{\dot{\Phi}_n(0)}{l_{sn}} \end{bmatrix}$$

Where n is the number of vibration modes used in the approximation, and ω_i, ζ_i are the corresponding fundamental vibration frequencies and damping coefficients at each mode.

In this chapter we use tip position of the beam as system output and then output equation can be written as follows:

$$Y = [y(L)] = \left[\sum_{i=1}^{n} \Phi_i q_i\right] = [C]X = [0 \ 0 \ \Phi_1(L) \ 0 \cdots \Phi_n(L) \ 0]X$$

The corresponding transfer functions in the frequency domain can be obtained easily.

7.3 Mixed H_2/H_∞ Robust PID Control

A H_∞ robust controller will maintain the system stability and accurate control under plant parameter perturbation and other uncertain disturbance. As we all know, the optimal control based on H_2 norm can produce good performance but sometime its robustness is not good. Therefore, a mixed H_2/H_∞ control method is considered here that combines H_2 with H_∞ to provide the both robust stability and acceptable performance not only for nominal systems but also the systems with bounded perturbations.

Assuming under perturbation $\Delta G(s)$ the flexible manipulator plant $G(s)$ is to be controlled by PD controller C(s), the closed loop system is shown in Fig. 7.2.

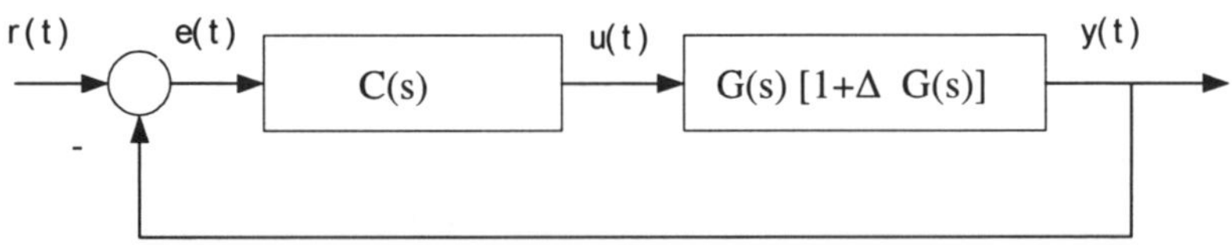

Fig 7.2. PD control system with flexible manpulator plant under perturbation

PD controller:

$$C(s) = K_P + K_D s; \tag{7.1}$$

Uncertain disturbance $\Delta G(s)$ is stable, and is bounded by $\xi(j\omega)$:

$$\|\Delta G(j\omega) \le |\xi(j\omega)\|, \quad \forall \omega \in [0, \infty). \tag{7.2}$$

The following inequality ensures the robust stability under the controller C(s) not only in the nominal closed loop system but also in the system of uncertain disturbance:

$$\left\| \frac{G(s)C(s)\xi(s)}{1+G(s)C(s)} \right\|_{\infty} \leq 1; \qquad (7.3)$$

The H_{∞} norm of the nominal model $G(s)$ is defined as:

$$\|G(s)\|_{\infty} = \sup_{\omega \in [0,\infty)} |G(j\omega)| \qquad (7.4)$$

Under the constraint (7.3), system in Fig.7.1 under perturbation is also asymptotically stable. However, in control system design robust stability alone usually is not enough, required performance must be considered too. One of the most widely used performance indices in industry is the error energy, i.e., the integral squared error criterion. Its expression can be formulated as:

$$\mathrm{J}=\int_0^{\infty} e^2(t)dt \qquad (7.5)$$

The optimal H_2 integral performance control design is to minimum the error energy [4].

To ensure both robust stability and optimal performance in this case, we propose the following mixed H_2/H_{∞} control design method:

$$\mathrm{J(ISE)}= J_m = \min \int_0^{\infty} e^2(t)dt \qquad (7.6)$$

Therefore, the flexible manipulator control design consists of finding a PD controller that minimizes the H_2 integral performance index in equation

(7.5) while subject to the inequality constrain (7.3) for the H_∞ norm of the closed loop system under the disturbance.

7.4 Improved GA Method for Mixed H_2/H_∞ Robust PD Control Design

The minimum of integral squared error J(ISE) can be computed by using Parseval's theorem with the aid of residual theorem as in [5-6], which involves significant computation and sometime without obtaining the analytic solution, due to the nature of flexible manipulators. Thus, to simplify computation of J_m, error $e(t)$ is calculated based on its definition:

$$e(t) = y_{or}(t) - y_o(t) \tag{7.7}$$

where $y_{or}(t)$ is the expected output, and $y_o(t)$ the actual output. Furthermore, the computation of $\int_0^\infty e^2(t)dt$ is reduced from infinite to finite time T.

The H_∞ constraint means that PD controller satisfies inequality constrain (7.3) for the H_∞ norm of the flexible manipulator plant with uncertain perturbation:

a) From Definition (7.4), constraint in (7.3) can be expressed by:

$$\left\| \frac{G(s)C(s)\xi(s)}{1+G(s)C(s)} \right\|_\infty = \sup_{\omega\in[0,\infty)} \sqrt{\frac{p(\omega)}{q(\omega)}} \Rightarrow \sup_{\omega\in[0,\infty)} \frac{p(\omega)}{q(\omega)} = \sup_{\omega\in[0,\infty)} \Theta(\omega) \le 1 \tag{7.8}$$

b) Since $\sup\limits_{\omega\in[0,\infty)}$ means the maximum during $[0,\infty)$, so the peak must satisfy:

$$\frac{d\Theta(\omega)}{d\omega} = 0 \Leftrightarrow p(\omega)d\frac{q(\omega)}{d\omega} - q(\omega)\frac{dp(\omega)}{d\omega} = 0 \Leftrightarrow \prod_{i=1}^{n}(\omega - \alpha_i) = 0 \quad (7.9)$$

c) Therefore, the final constraint can represented as:

$$\sqrt{\max_{\alpha_i} \frac{p(\alpha_i)}{q(\alpha_i)}} \leq 1. \quad (7.10)$$

The GA-based optimization is are widely used in solving optimal design problems because they are powerful, inherent parallel, and global search algorithms based on the mechanics of natural selection and natural genetics [7]. Since the criterion to be optimized here is a complicated integral, we use the GA to adjust the PD gain parameters in the mixed controller. However, during the solving process, two kinds of constraints make the GA application very difficult: the first are inequality constrains in the parameter space and the second are constrains imposed on by robust control design since the criterion that must be optimized is of integral type. To overcome this situation, graphic method, local search strategy, refusal strategy and renewal strategy are used to solve hard constrains. Simulation results indicate those methods and strategies make GA quickly and accurately converge to the global optimal solution.

The improved GA based algorithm is implemented in the following steps:

Step 1:

Design variables are K_P (proportional constant) and K_D (differential constant), constraints can be obtained via the Routh-Hurwitz Criterion specifying the parameter domain of Δ (K_P、 K_D) that guarantees the stability of the flexible manipulator system.

The parameters space consists of several inequality constraints. In solving the design problem using GA, most of classical methods are complicated and time-consuming. The method here eliminates infeasible solutions using a graphic method and other strategies whose details are in the design example below.

Step 2:

Building the optimization model using the H_2 optimal criterion:

$$J_m = J_m(K_P, K_D) = \min \int_0^\infty e^2(t)dt .$$

Step 3:

Coding method: real number coding is used here to simplify the problem.

Step 4:

The evaluation method for the individual fitness: the fitness function F(x) is defined the same as the cost function f(x):

$$f(x) = F(x) = J_m = \int_0^\infty e^2(t)dt \approx \sum_{t=0}^{T} (1 - y_{step})^2 .$$

Step 5:

Designing the GA operators: we used the improved GA so that it can search the allowable solutions quickly, accordingly α_i (i=1, …, n) can be obtained via (7.8) and (7.9), the detailed description of those special operators can be found in [8].

Step 6:

Optimizing the H_∞ constraint: proper generation created from Step 5 must be checked by H_∞ constraint (7.10), individuals will be refused if it doesn't meet (7.10) according to a refusal strategy, then new individuals are created to replace the refusals via a renewal strategy to guarantee the stability of each generation.

Step 7:

Compute J_m in (7.5) to determine if a satisfactory optimal and robust controller is obtained. If not, then repeat the procedure in Step5 to Step7 until a solution is reached.

The flowchart for the procedure outlined is illustrated in Fig.3.

Design Example I:

The following data are used in the design example:

L=0.45m; E=2 x10^{11}N/m^2I=8.23 x$10^{-13}$$m^4$; I_h=0.0039kgm^2/s^2; m=0.06kg;

M_p=0.1kg; R_a=2.6; K_m=7.67x10^3Nm/A; K_g=70.

Where R_a, K_m, and K_g are Armature resistance, Motor torque consistence, Gear ratio respectively [17].

After tedious computation, we can obtain the following simple reduced-order model for the flexible manipulator:

$$G(s) = \frac{-1.163s + 11.85}{s^2 + 5.17s + 11.85};$$

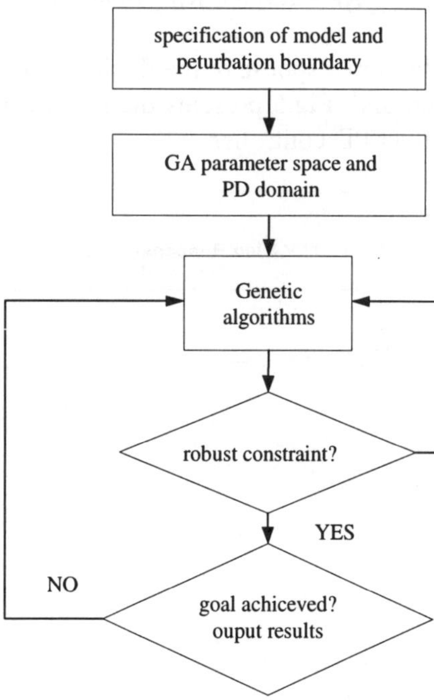

Fig. 7.3. Flowchart of procedure

Assuming the prescribed boundary of uncertainties and disturbances in the system can be expressed by the equation:

$$\xi(s) = \frac{1}{s + 0.01}.$$

Combining the nominal model of the flexible manipulator and the uncertain perturbation boundary, using the Routh-Hurwitz Criterion, the allowable space of variables K_P, K_D can be derived as the closed area by the following three lines:

$$K_D < 0.8596;\ K_P < (11.85\ K_D + 6.333)/1.163\ ;\ K_P > 0$$

Which is obviously a closed triangle. Computer simulation is carried out using the above algorithm. Based on the local search strategy, by separating the parameter space into several regular little areas. better optimal solutions are obtained on each of the regular areas. The parameters for the final optimal robust controller are as following:

$$K_P = 2.5768,\ K_D = 0.4437;$$

Fig. 7.4 shows the step response of the flexible manipulator under the optimal PD controller, and Fig.5 presents the step response of the unity feedback system without PD controller.

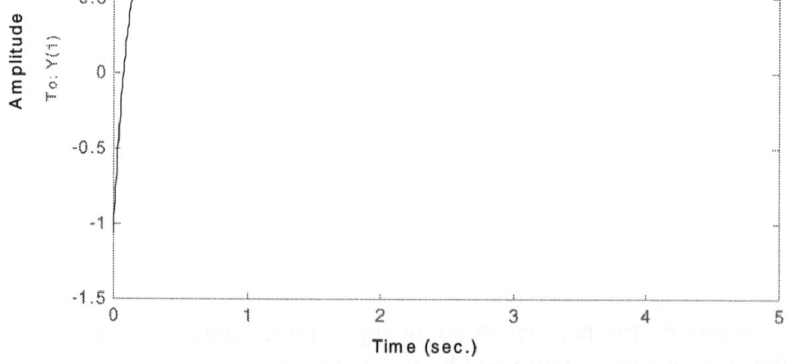

Fig.7.4. Step Response with the Optimal PD

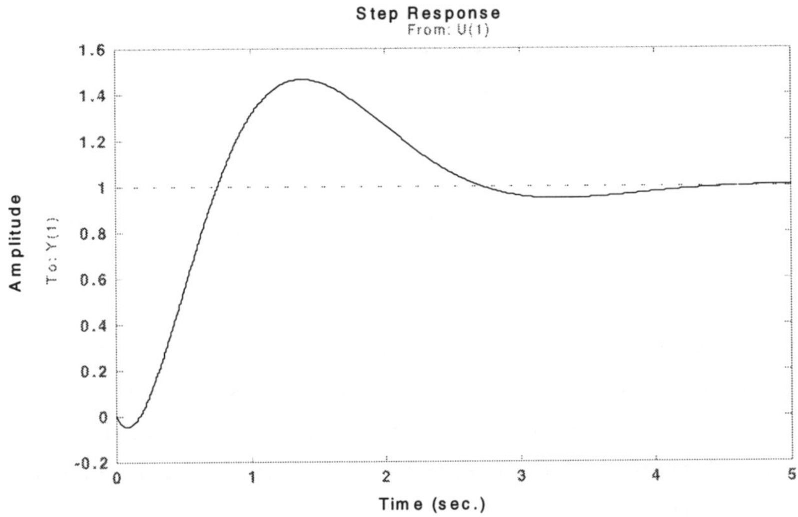

Fig.7.5. Step Response of the Unity Feedback System without PD Controller

These results show that the optimal robust controller can satisfy the system requirements well and guarantee the robustness of the system stability by applying the synthesized methodology to design the control under the uncertainties, by solving the complicated constraint problems using the improved GA strategies.

7.5 Mixed Sensitivity H_∞ Control Problem

In Section 7.2, a nominal plant representation of a one-link flexible manipulator established with nominal parameters. But there will be errors between the model responses and the real system responses under different conditions. In this section we just consider the cases of tip mass variations and variations in some system parameters. Those variations will affect the whole system in terms of effective frequency, damping ratio, frequency responses, etc. These effects can be treated as interference and perturbation to the nominal plant model.

The variations towards system parameters can be experimentally obtained or theoretically determined. We adopt latter to investigate the effects in the frequency domain, for example, to see what is happening if the tip mass changes ±200%, and so on.

To guarantee the robust stability and performance, here a mixed sensitivity H_∞ controller is considered. The controller is based on the feedback of tip position of the flexible beam. A 3-block mixed S/R/T scheme is used for system/controller design, as shown in Fig. 7.6, where u_1, u, e, y are reference input, control input, errors and system output, respectively, G is the nominal plant and K is the controller. S is the sensitive function, and T is the complementary sensitivity function. They can be defined,

$$S = (I + GK)^{-1},\ R = \frac{K}{(I + GK)},\ T = \frac{GK}{I + GK} = I - S$$

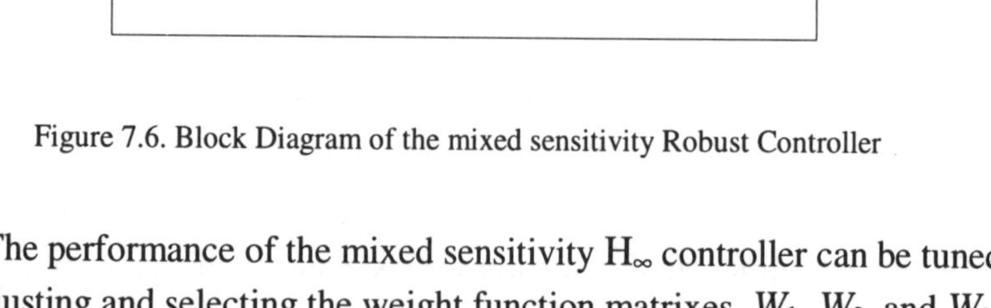

Figure 7.6. Block Diagram of the mixed sensitivity Robust Controller

The performance of the mixed sensitivity H_∞ controller can be tuned by adjusting and selecting the weight function matrixes, W_1, W_2, and W_3, as shown in Fig. 7.2. In the frequency domain, loop shaping is carried out using those frequency-dependent weighting function matrixes until shaped plants, W_1S, W_2R and W_3T, satisfy the desired shapes.

Now the standard form of the mixed sensitivity problem considering weight function matrixes can be defined as

$$\begin{bmatrix} W_1 e \\ W_2 u \\ W_3 y \\ \cdots\cdots \\ e \end{bmatrix} = \begin{bmatrix} W_1 & -W_1 G \\ 0 & W_2 \\ 0 & W_3 G \\ \cdots\cdots & \cdots\cdots \\ I & -G \end{bmatrix} \begin{bmatrix} u_1 \\ u \end{bmatrix}$$

$$u = K e$$

Then the close-loop transfer function matrix can be written as follows:

$$P = \begin{bmatrix} W_1 S \\ W_2 R \\ W_3 T \end{bmatrix}$$

The design problem is to find a controller K that can stabilize the system and satisfy

$$\min \|P\|_\infty \text{ or } \|P\|_\infty \leq \gamma$$

Note that the analysis of the nominal plant and variations of system parameters indicates that additional gain or other form of pre-compensators is needed to improve performances such closed loop bandwidth, robustness, etc.

Generally, W_1 is chosen as a low-pass filter to satisfy the performance of the system sensitivity in the low frequency stage. The response speed of the nominal plant is slow and a gain, R_1, to the input channel can improve this situation. And the disturbance caused by variations of tip mass and so on tend to be in a low frequency range. A perfectly selected W_1 will mitigate the effect imposed to the system. Therefore, W_1 is designed as follows:

$$W_1 = \frac{R_1}{R_4 s + 1}$$

where R_1, the gain, and R_4, are parameters to be optimized. In addition, the pole is used for high frequency roll-off ratio which guarantees robust stability of the system.

On the other hand, W_3 is selected as a high-pass form to restrict high-frequency characteristics of T, the complementary sensitivity function, and to limit system bandwidth. It is defined here as

$$W_3 = \frac{s^2 + 1}{R_3}$$

where R_3 is another parameter to be optimized.

In addition, we select W_2 as real value so it can be used as a regulation gain to ensure a relatively wide range of perturbations in a low or middle frequency range.

7.6 GA Method for Mixed Sensitivity H_∞ Control Design

In this section, a GA-based method is used to solve the optimal design problem for mixed sensitivity controller. The genetic operators applied here are Roulette Wheel selection, Arithmetical crossover, and non-uniform mutation.

The key to the successful search is the fitness function that characterizes how well each particular member solves the given problem. Here we adopt ISE standard, shown in Section 7.3.

Here GA is adopted to optimize the parameters in the three weight function matrixes and then simultaneously search corresponding optimal controller K. The procedure mainly consists of:

Step 1:
Initialize all the parameters in the iterative operation;

Step 2:
Set the generation number = 1. Randomly generate an initial population of R_1, R_2, R_3, R_4.

Step 3:
Determine the current population for GA operations;

Step 4:
Apply the three genetic operators to generate a new generation population; in each operation, a system stability check is carried out before generating a new controller. Any population that cannot pass the check will be discarded and, in the same time, a new, randomly generated population is created;

Step 5:
If the termination condition is satisfied, stop and return the best individuals. Otherwise, increase the generation number by 1 and go to Step 3.

The real-coded genetic algorithm procedure is fairly simple but powerful. It tends to search for a global optimal solution. Furthermore, the well-selected parameters, Ps, selection probability, Pc, crossover probability and Pm, mutation probability, can obviously help in finding optimal solutions.

Design Example II:

The following data are used in the design example:

$L = 1.0m,\ EI = 0.227,\quad \rho = 2.73\times10^{3}\ Kg/m^{3}\ ,\quad M_P = 0.05\,Kg\ ,$

$I_H = 0.0174\,Kgm^2$. Utilizing the assumed mode model, the beam vibration equation is truncated to the first 2 modes and used in the example. Other parameters include

$$Pc = 0.5,\ Pm = 0.2.$$

Fig. 7.7 presents the Bode diagram of the one-link flexible manipulator with the optimal H$_\infty$ controller and Fig. 7.8 shows its corresponding response to a unit step. The performance of the controller is acceptable with changes in parameters.

Fig. 7.9 illustrates the search process by the GA method.

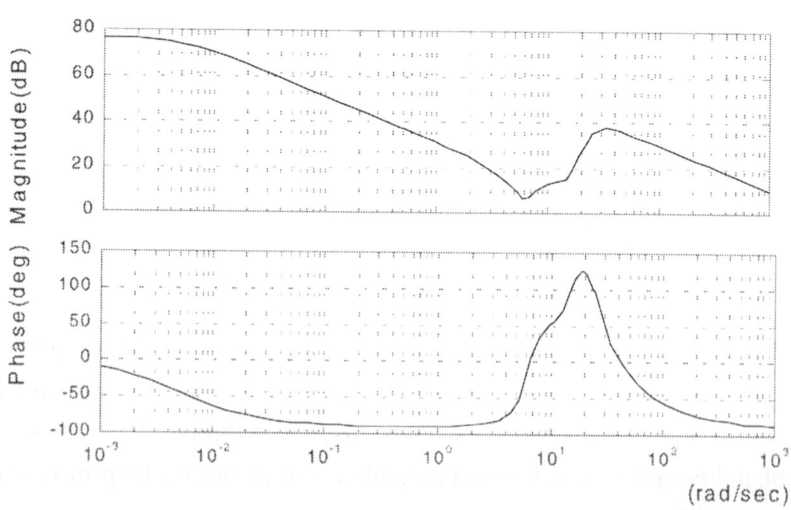

Fig 7.7-1. Bode diagram of the optimal H$_\infty$

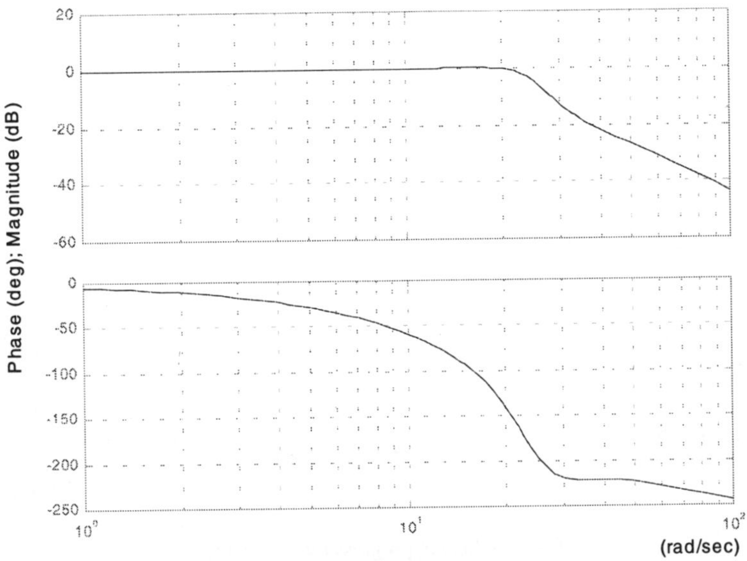

Fig 7.7-2. Bode diagram of the one-link flexible manipulator with the optimal H_∞ controller

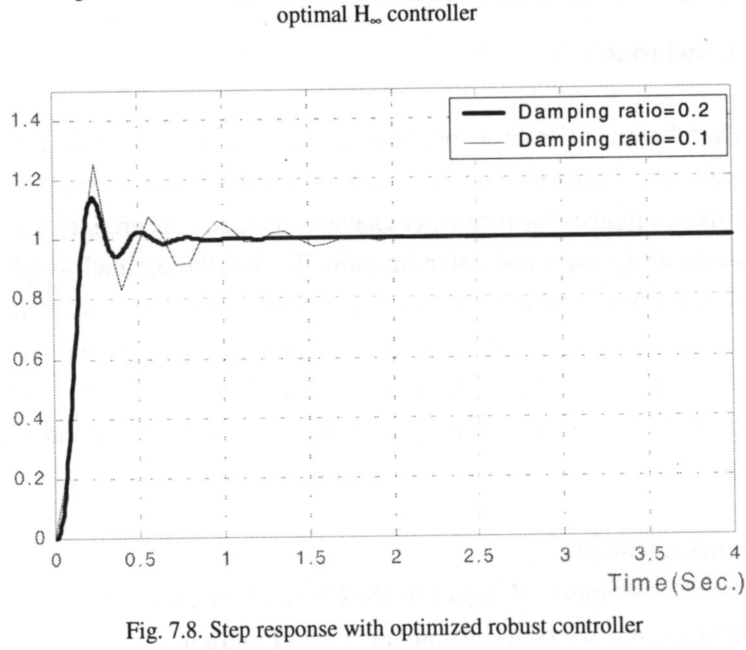

Fig. 7.8. Step response with optimized robust controller

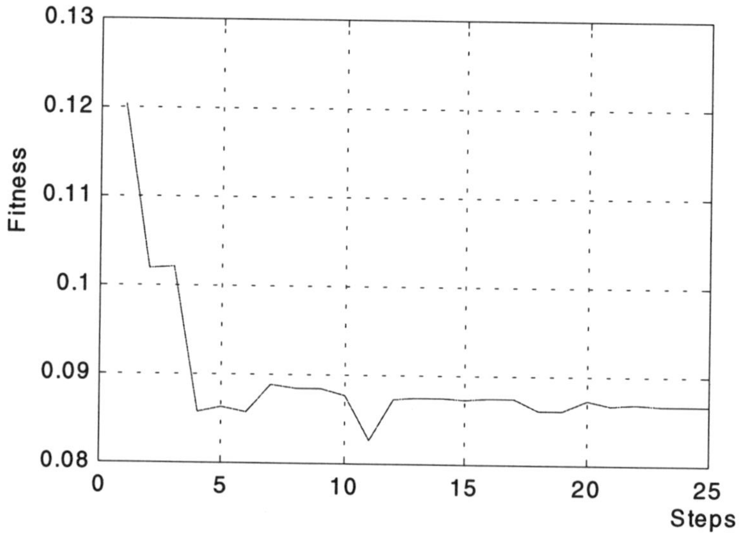

Fig. 7.9. Search progress by GA method

7.7 Conclusion

The design of robust position controllers for one-link flexible manipulators discussed in this chapter using various robust mixed performance indices for parameter changes, dynamic uncertainty, such as variations in tip mass and damping ratio. To find the optimal weighting function matrixes and optimal controller, real-coded GA is adopted and used for solving the corresponding optimization problem. Initial numerical design examples have demonstrated the effectiveness of the proposed design method. Currently further research is undergoing to improve the proposed method and make it more feasible.

Acknowledgements

This work was supported in part by the Chinese National Natural Science Foundations, the Oversea Outstanding Talent Program.

References

1. M. Boutaghou and A. G. Erdman, A Unified Approach for the Dynamics of Beams Undergoing Arbitrary Spatial Motion, Journal of Vibration and Acoustics-Transactions of the ASME, Vol. 113, No. 4, pp. 494-507, 1991.
2. R. H. Cannon, Jr. and E. Schmitz, Precise Control of Flexible Manipulators, Robotics Research: The First Int. Symposium, MIT Press, Cambridge, MA, pp. 841-861, 1984.
3. S. Cetinkunt and W. L. Yu, Closed Loop Behavior of a Feedback Controlled Flexible Arm: A Comparative Study, Int. J. of Robotics Res., Vol. 10 No. 3, pp. 263-275, 1991.
4. A. De Luca, P. Lucibello, and G. Ulivi, Inversion Techniques for Trajectory Control of Flexible Robot Arms, Journal of Robotic Systems, Vol. 6, No. 4, pp. 325-344, 1989.
5. A. De Luca, B. Siciliano, Trajectory Control of Non-linear One Link Flexible Arm, Int. Journal of Control, Vol. 50, pp. 1699-1715, 1989.
6. M. Kelemen and A. Bagchi, Modeling and feedback Control of a Flexible Arm of a Robot for Prescribed Frequency-domain Tolerances, IFAC Journal Automatica, Vol. 29, No. 4, pp. 899-909, 1993.
7. F. Khorrami and Umit Ozguner, Perturbation Method in Control of Flexible Link Msnipultors, IEEE Int. Conf. On Robotics and Automation, Vol. 1, pp. 310-315, 1988.
8. D. Wang and M. Vidyasagar, Passive Control of a Stiff Flexible Link, International Journal of Robotics Research, Vol. 11, No. 3, pp. 572-578, 1992.
9. F. Y. Wang, Optimum Design of Vibrating Cantilevers: A Classical Problem Revisited, J. of Optimization Theory and Applications, Vol. 84, No. 3, 1995.
10. F. Y. Wang, On the Extreme Fundamental Frequencies of One-ink Flexible Manipulators, Int. Journal of Robotics Research, Vol. 13, No. 2, pp. 162-170, 1994.
11. Fei-Yue Wang and Jeffery L. Russell, "Minimum-Weight Robot Arm for a Specified Fundamental Frequency," *Journal of Robotic Systems*, Jan 1997.
12. F. Y. Wang and Jeffery L. Russell, Optimum Shape Construction of flexible Manipulators with Total Weight Constraint, IEEE Transactions on Systems, Man and Cybernetics, Vol. 25, No. 3, 1995.
13. F. Y. Wang and G.G. Guan, Influences of Rotary Inertia, shear Deformation and Tip load on Vibrations of Flexible Manipulators, Journal of Sound and Vibrations. Vol. 171, No. 4, pp. 433-452, 1994.
14. Book, W. J., "Recursive Lagrangian Dynamics of Flexible Manipulator Arms", The International Journal of robotic Research, Vol. 3, No. 3, pp. 87-101, 1984.
15. Wang Shuxin, Yuan Jintian, Shi Jurong, Liu Youwu, Flexible manipulator model construct theory and control methodology research summary, Robotics, 2002, 24(1):86-92. (Chinese)
16. Fei-Yue Wang and Oliver Kwan (1993), Influence of Rotatory Inertia. Shear Deformation and Loading on Vibration Behaviors of Flexible Manipulators, scheduled in Journal of Sound and Vibrations, to appear in Vol. 169, No. 2.

17. P. Chaichanavong and D. Banjerdpongchai. A Case Study of Robust Control Experiment on One-link Flexible Robot Arm, Dept.of EE Stanford CA 94305 USA.
18. C.N. Calistru. Mixed PID Robust Control via Genetic Algorithms and MATHEMATICA Facilities, Dept. of Automatic Control & Industrial Informatics.
19. Zhou Ming, Sun Shudong, GA Theory and Application, National Defence Industry Publishing House, 1999. (Chinese)
20. Lingli Cui, Zhiquan Xiao, Research on the mechanism of real-coded genetic operation, to appear in Pattern Recognition and Artificial Intelligence.
21. R.P. Sutton, G.D. Halikias, A.R. Plummer and D.A. Wilson, "Robust Control of a Lightweight Flexible Manipulator under the Influence of Gravity", Proceedings of the 1997 IEEE International Conference on Control Applications, Hartford, CT, 1997, pp. 300-305.
22. Wu Xudong, Xie Xueshu, "Weighting Function Matrix Selection in H_∞ Control", Journal of Tsinghua University (Sci. & Tech.), Vol. 27. No. 1, 1997, pp. 27-30.
23. Xue Ding Yu, "Design and Analysis of Feedback Control System — MATLAB Application", 2000.
24. Guo-Ben Yang and Max Donath, "Dynamic Model of One-link Robot Manipulator with Both Structural and Joint Flexibility", 1988, IEEE, pp. 476-481.

Chapter 8

Analysis of Poles and Zeros for Tapered Link Designs

This chapter analyzes the pole and zero locations of a linearly-tapered Euler-Bernoulli beam pinned at one end and free at the other end. Of particular interest is the location of zeros of the transfer function from torque applied at the pin to displacement of the free end. When tapered beams are used as the links of light-weight robots, the existence of nonminimum phase (right half plane) zeros complicates the robot control problem. Tapering the beam gives the robot designer an additional design parameter when establishing the flexible dynamics. The pole and zero locations are determined from a transfer matrix model which is the exact solution for a uniform beam. The approximate results for a tapered model result from segmentation of the beam into segments of different but constant cross sections. The relative position of poles and zeros varies significantly as the rate of taper changes, which will have consequences on feedback stability and noncausal effects in inverse dynamics.

8.1 Introduction

8.1.1 *Problem Definition*

As research for new applications for industrial robots proceeds, one major area of research is in robot flexibility. Traditionally, industrial robots are designed with stiff links, so the dynamics of the links can be assumed negligible in positioning the robot. In theory then as the robot moves, the links remain straight and do not bend. The tip position of the robot can be found geometrically from joint position at any given moment. In flexible robotics the links are no longer assumed to be rigid. As the robot moves, the links flex which can cause unwanted vibrations in the robot. These vibrations can cause error in positioning the tip of the robot.

Some of the applications motivating research in this field are assembly of space structures, inspection of large structures, and nuclear waste retrieval. When transporting things to outer space, weight is always a concern. Light-weight robots designed for space applications will be flexible and must be controlled as such. Large structures like airplanes and submarines require careful inspection to insure detection of flaws. The inspections can be laborious and repetitive which is ideal work for a robot. The large workspace dictates the links be as light as possible resulting in flexible links. An emerging area of research is remote handling of nuclear waste. Existing nuclear waste storage facilities are no longer safe and the waste needs to be removed and restored in safer containers. The old containers are very large, while the access is usually quite small. Again, a light-weight slender robot with a large workspace is required. All of these applications are driving the research in the field of flexible robotics.

A common problem with flexible systems is how to control the system accurately to position the end-point. Rigid link robots are typically collocated systems; that is, the actuators and sensors are located at the same location (ie., a joint). With a flexible system this is not always the case. Most flexible systems are noncollocated. The system output (actuator torque) is generally located at the base of the system, while the output (tip position) is located at the end of the system. Noncollocated systems exhibit nonminimum phase behavior which results directly from the system zeros in the right-half of the s-plane (RHP zeros).

Controller design for collocated systems has been heavily researched and is well understood compared to controller design for noncollocated systems. In noncollocated systems, uncertainties from model inaccuracies and modal truncation present fundamental problems with system performance and stability[20]. The fundamental difference between collocated and noncollocated systems is the presence of RHP zeros. To advance controller design for noncollocated systems, research needs to be conducted into the factors that affect the location of these RHP zeros. This research targets the relationship between RHP zeros and structural design.

8.1.2 *Review of Related Research*

Although research on RHP zeros is limited, there has been some notable research done in the past. Some of the research deals directly with the problems presented by nonminimum phase systems, while other research examines different techniques to change the system characteristics from nonminimum phase to minimum phase.

In 1988, Nebot and Brubaker[15] experimented with a single-link flexible manipulator. The manipulator was constructed from thin plates connected by several bridges along their length. This provided flexibility in the horizontal plane, while maintaining stiffness in the vertical plane and torsional mode. They analytically determined the location of the first six zeros and determined three of them to be RHP zeros. They concluded these RHP zeros pose a formidable constraint in the controller design task.

In 1989, Spector and Flashner[21] investigated the sensitivity effects of structural models for noncollocated control systems. They considered a pinned-free beam with discrete endpoint mass and inertia. They used transfer matrices to analyze the system. From the results they concluded the following. First, accurate dynamic modeling is critical in noncollocated control design. Poor modeling can result in interchanging the pole/zero order which produces phase errors resulting in closed-loop instability. Second, accurate modeling of zero location near the system bandwidth is critical in modeling noncollocated systems. Third, zeros are more sensitive to perturbations in system parameters and boundary conditions than modal frequencies.They suggest more research attention be given to modeling system zeros in noncollocated systems.

In 1990, Spector and Flashner[20] again studied modeling and design implications pertinent to noncollocated control. A similar system was used, a pinned-free beam without endpoint mass, only the system was analyzed using wave number plane theory. They also studied the effects of varying sensor/actuator separation distance. Most conclusions are identical to those drawn from the previous paper. In addition, they concluded all noncollocated systems are nonminimum phase above some finite frequency (the location of the lowest RHP zero dictates this frequency), and this frequency increases as sensor/actuator distance increases. Again they recommend more research into the modeling of zeros in noncollocated systems.

The physical interpretation of the zeros of flexible mechanical systems is more difficult than the system poles, which are the natural frequencies. Miu[14] explained the zeros of simple flexible systems as the natural frequencies of constrained subsystems of the overall system. For a single beam the real zeros, both positive and negative, are the poles of the constrained subsystem consisting of the beam between the sensor and actuator. The complex zeros are the natural frequencies (complex poles) of a constrained subsystem consisting of the beam segments outside the sensor and actuator. The special case of sensor and actuator collocation gives all complex zeros alternating between the poles. The special case of a torque actuator at one end of a beam in bending and position sensor at the other end of that beam gives all real zeros. Miu further explains the real zeros as arising from the ability of a beam in bending to store energy locally in nonpropogating waves.

Also in 1990, Park and Asada[16,17] investigated a minimum phase flexible arm with a torque actuation mechanism. Basically they used a cable mechanism to transfer the torque actuation point from the base to the tip of the arm. Since the sensor and actuator are located at the same point, the system is minimum phase. They concluded the inverse dynamics solution does not diverge because the RHP zeros are relocated to the LHP by the torque transmission mechanism. Also end-point feedback control can be stabilized for this system with simple a P-D controller. Unfortunately, implementation of the transmission device on multi-link systems could be difficult.

In 1991, Park, Asada, and Rai[1] expanded their previous work on a minimum phase flexible arm with a torque transmission device. In this research they integrate structure and control design using Finite Element Analysis (FEA) to design the shape of the arm while constraining pole and zero location. Essentially, they use the FEA program to generate a design that will increase the fundamental natural frequency and use the torque transmission device to eliminate the RHP zeros. A prototype of the new system had not been tested at that point, and the main contribution was a method to evaluate nonuniform beams for design applications.

8.1.3 *Proposed Method of Approach*

The underlying issue in noncollocated control is how to deal with the RHP zeros in the control algorithm. A major step in solving the problem is understanding what design parameters can be used to change the location of these RHP zeros. This research targets the relationship between RHP zero location and structural design. Specifically, how do changes in the shape of the structure (link) affect the location of these zeros?

Traditionally links are designed with uniform properties along the length because analytic solutions to this problem exist. A link with variable cross-section cannot be solved analytically, but with aid of a computer a numerical approximation can be found. The key to an accurate numerical solution is a good model of the system.

8.2 Nonminimum Phase Systems

8.2.1 *System Characteristics*

As mentioned before, a system is considered nonminimum phase if there are system zeros or poles located in the right half of the complex plane. Figures 8.2.1a and 8.2.1b graphically express the difference between minimum phase and nonminimum phase systems.

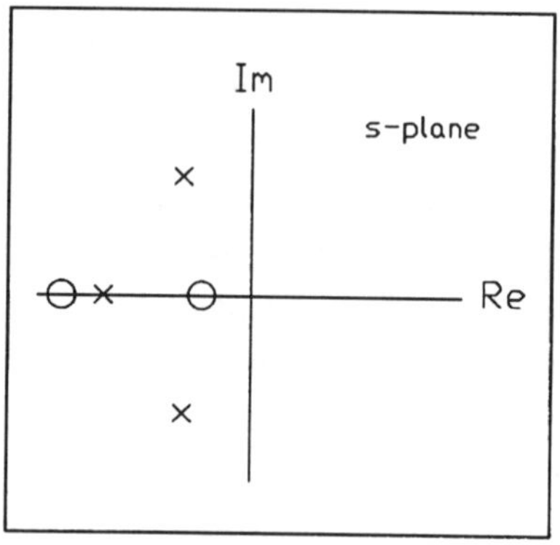

Figure 8.2.1a Minimum phase
Pole/zero pattern

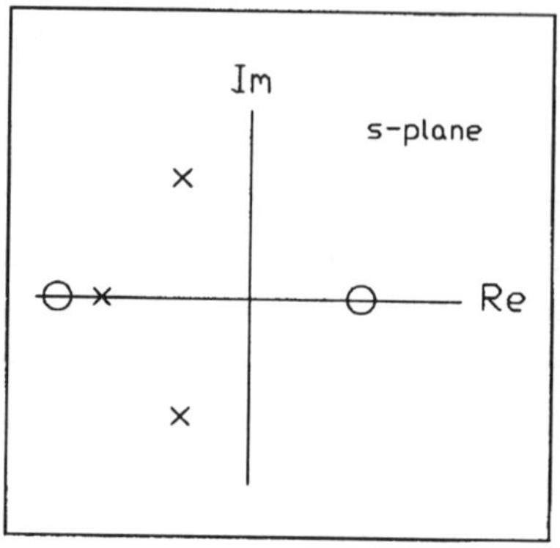

Figure 8.2.1b nonminimum phase
Pole/zero pattern

This is the case in the continuous-time domain. In the discrete-time domain (z-transform), the nonminimum phase zeros would lie outside the unit circle.

Often RHP zeros are called unstable zeros, but this is not good terminology. RHP zeros do not cause the plant to go unstable. Poles in the RHP will cause the system response to exponentially increase resulting in instability, but zeros do not cause this. It is the controller design that can cause the zeros to have an effect on system stability. For example, when using an inverse dynamics algorithm, the RHP zeros will become unstable poles in the inverse system. Now the controller has unstable poles which can cause the entire system to go unstable.

A noticeable characteristic of a nonminimum phase system is the time response to a step input. Figure 8.2.2 shows the difference between minimum phase (MP) and nonminimum phase (NMP) response of the tip position for a single-link flexible manipulator.

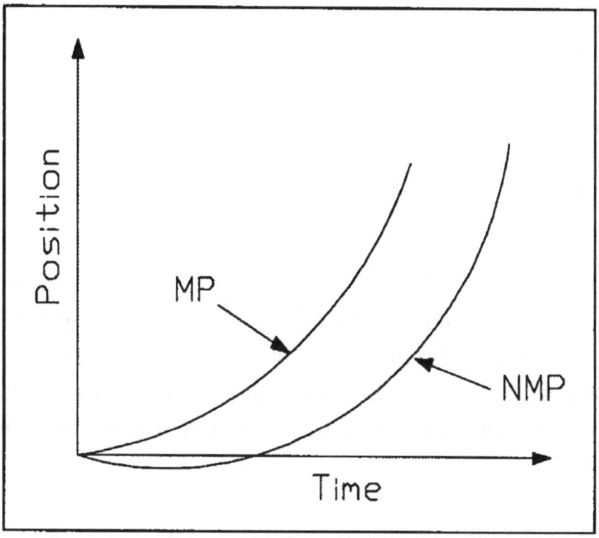

Figure 8.2.2: MP vs. NMP Time Response

Notice the tip of the NMP system initially starts to move in the direction opposite to the command. This type of response can be verified in Park and Asada's paper[1].

It has been stated that RHP zeros are indicative of a NMP system, but what physical phenomenon is responsible for NMP behavior? Miu[14] explains the zeros in terms of energy absorbed by a substructure of the flexible system. Real zeros result from absorption due to nonpropogating waves that can absorb energy through local mechanisms. Spector and Flashner[20] concluded that NMP behavior is an inescapable result of the finite wave propagation speed of elastic deformation in the structure. This wave propagation speed directly results in a time delay between system input and the corresponding system output. The time delay affects the system by reducing the phase margin. If the phase lag from the time delay exceeds the system phase margin (at the cutoff frequency), the system will be unstable.

These are some of the more prominent characteristics of nonminimum phase systems. Of interest in this research is the control of nonminimum phase systems and how to advance the research in this area. The following section describes some of the current techniques used to control nonminimum phase systems.

8.2.2 *Control of Nonminimum Phase Systems*

One method of controlling a nonminimum phase system studied by Misra13 in 1989 is augmenting a nonminimum phase plant to make the overall system minimum phase. He used a "feedthrough" compensator so the augmented system was minimum phase. A feedthrough compensator was added so the poles of the compensated system move to the minimum phase zeros.

In 1987, Bayo2 presented a structural finite element technique based on Bernoulli-Euler beam theory for open-loop control of flexible manipulators. The differential equations of motion are integrated in the frequency domain to determine the necessary torques for desired tip motion. The computed torque reproduced the desired trajectory without any overshoot, but closed-loop control was not investigated.

Another control algorithm investigated at Georgia Tech by Kwon and Boo[5,9,10] used an inverse dynamic method to deal with a NMP flexible arm. The method is similar to Bayo's, only integration was carried out in the time domain. The dynamic equations of motion for a flexible manipulator can be written as:

$$\begin{bmatrix} M_{rr} & M_{rf} \\ M_{rf}^T & M_{ff} \end{bmatrix} \begin{Bmatrix} \ddot{q}_r \\ \ddot{q}_f \end{Bmatrix} + \begin{bmatrix} D_{rr} & D_{rf} \\ D_{rf}^T & D_{ff} \end{bmatrix} \begin{Bmatrix} \dot{q}_r \\ \dot{q}_f \end{Bmatrix} + \begin{bmatrix} 0 & 0 \\ 0 & K_{ff} \end{bmatrix} \begin{Bmatrix} q_r \\ q_f \end{Bmatrix} = \begin{bmatrix} B_r \\ B_f \end{bmatrix} \tau \qquad (8.2.1)$$

where,

q_r - Rigid body motion coordinate

q_f - Flexible motion coordinate

After some manipulation the inverse dynamics equations can be obtained in the following form:

$$\begin{aligned} \dot{X}_i &= [A_i]X_i + [B_i]\dot{q}_{ir} \\ \tau &= [C_i]X_i + [F_i]\dot{q}_{ir} \end{aligned} \qquad (8.2.2)$$

where,

$$X_i = [q_f, q_f]^T \qquad q_{ir} = [q_r, q_r]^T$$

For the forward dynamic equations, the input is torque, and the outputs are all states. For the inverse dynamic system, the input is end-point desired trajectory, and the output is torque. The problem addressed is how to integrate these equations since the matrix $[A_i]$ has positive real poles. The RHP poles come from the RHP zeros in the original system. Their approach is to relax the solution range to include noncausal solutions allowing a unique stable solution of the inverse dynamic equations.

To better understand the inverse dynamics solution, some terminology needs to be defined. According to Kwon[9], a causal system is one in which the system output (impulse response) occurs after the system input (impulse). An anticausal system has the output (backward impulse response) before an input is applied. A noncausal system is a combination of both a causal system and an anticausal system.

To illustrate these concepts Figure 8.2.3 shows the motion of a flexible arm moving from point A to point B.

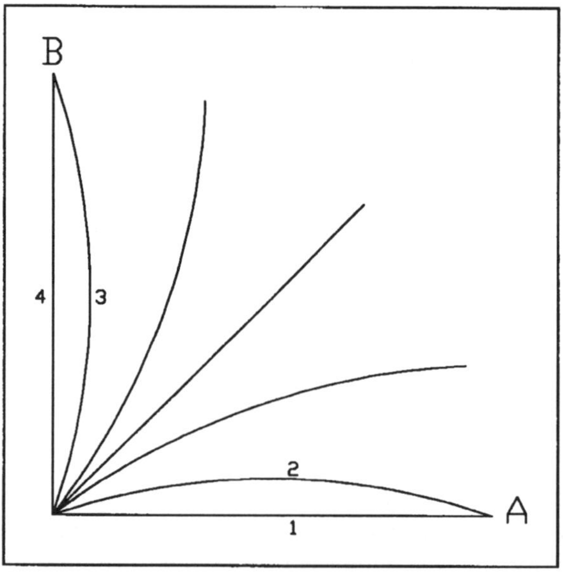

Figure 8.2.3: Flexible Link Motion

The two areas of interest on this curve are the start of motion and the end of motion. Motion starts as the arm moves from position 1 to 2, but the end-point does not move. The torque provided is applied to preshape the beam. This is the anticausal part of the inverse solution. The torque (output of the inverse system) occurs before the end-point (input to the inverse system) moves. When motion stops, the arm moves from position 3 to 4, again the end-point does not move. This represents the causal part of the inverse solution. The tip has stopped moving, but the torque continues to be applied. The torque applied between positions 4 and 5 is used to release the stored energy in the arm.

Since the motion can be divided into causal and anticausal parts, the solution to Eq. 8.2.2 can be divided into both causal and anticausal parts. Of interest to this research is the anticausal solution. The poles of the anticausal system are unstable and a direct result of the RHP zeros from the forward dynamic system. The ability to place these RHP zeros would be equivalent to placing the poles of the inverse anticausal problem. This would give the designer some freedom in choosing the location of the anticausal poles, and allow the system to be designed for specific needs. One benefit could be minimizing the time of preshaping and the amount of energy provided by the actuator to preshape the beam before tip motion begins.

8.3 Transfer Matrix Method

8.3.1 *Transfer Matrix Theory*

Transfer matrices describe the interaction between two serially connected elements. These elements can be beams, springs, rotary joints, or many others. In 1979 Book, Majette, and Ma [6] and Book[4] (1974) used transfer matrices to develop an analysis package for flexible manipulators. They used transfer matrices to serially connect different types of elements to model the desired manipulator. Of interest in this paper is how to connect similar types of transfer matrices (beam elements) to model a beam with different cross-sectional area. Pestel and Leckie[18] provide an in depth discussion of transfer matrix derivations and applications.

Transfer matrices can be mathematically expressed by Eq. 8.3.1. The state vector u_i is given by the state vector u_{i-1} multiplied by the transfer matrix B.

$$u_i = [B_i]u_{i-1} \tag{8.3.1}$$

When elements are connected serially, the states at the interface of two elements must be equal. By ordered multiplication of the transfer matrices, intermediate states can be eliminated to determine the transfer matrix for the overall system.

The concept of state vector in transfer matrix theory is not to be confused with the state space form of modern control theory. The state equation in modem control theory relates the states of the system as a function of time. In transfer matrix theory the state equation relates the states as a function of position. The independent variable in transfer matrix theory is frequency, not time. The elements of the matrix B depend on the system driving frequency; therefore, the states will change as the system frequency changes. The transfer matrix B essentially contains the transformed dynamic equations of motion that govern the element in analytic form. Therefore, analytical solution of the transfer matrix alone does not involve numerical approximations. This is desirable since numerical approximations introduce error into the solution.

8.3.2 *Modeling of a Nonuniform Beam*

A single-link manipulator as pictured in Figure 8.3.1 can be thought of as a beam with torque applied at one end and free at the other end. There are several steps to determine the RHP zeros and imaginary poles of this system. First, develop a model for the beam. Second, determine the appropriate boundary conditions. Third, determine the system input and output. Forth, solve for the system zeros. The following sections will discuss each of these steps in more detail.

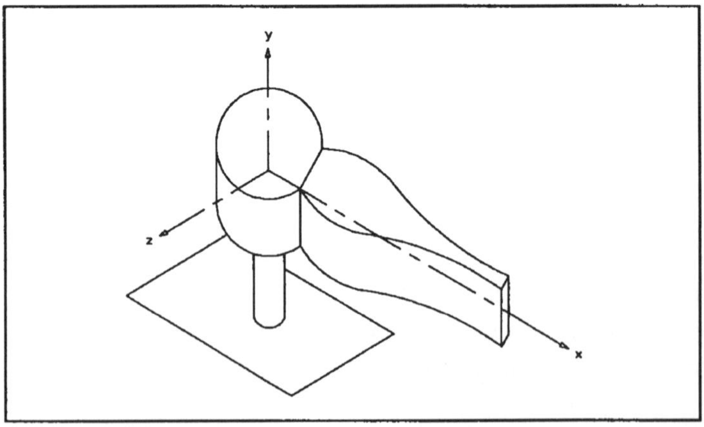

Figure 8.3.1: Single-Link, Flexible Manipulator

8.3.2.1 *Element Approach to Modeling*

A link with nonuniform cross-sections can be modeled as a series of discrete elements. While the shape of these elements is similar, the size can vary to allow for changes in cross-section. The appropriate element to model a flexible link is an Euler-Bernoulli beam element. The Euler-Bernoulli model neglects the effects of rotary inertia and shear deformation in the element[12]. This assumption is generally valid for modeling beams whose length is roughly ten times the height. Flexible manipulators have long, slender links which are appropriately modeled under the Euler-Bernoulli assumption.

Transfer matrices are derived from the equation of motion for a given element. For a uniform Euler-Bernoulli beam element, the equation of motion transformed to the frequency domain has the form:

$$\frac{d^4 w(x,\omega)}{d x^4} = \frac{\mu \omega^2}{EI} w(x,\omega) \tag{8.3.2}$$

where,

μ = mass density per unit length
ω = frequency in radians/second
E = Young's modulus
I = Cross sectional area moment of inertia

Notice the equation is fourth order thus requiring four states to describe the solution in transfer matrix form. The state vector for the Euler-Bernoulli element is:

$$u = \begin{bmatrix} -w \\ \psi \\ M \\ V \end{bmatrix} = \begin{bmatrix} displacement \\ slope \\ moment \\ shear\ force \end{bmatrix} \tag{8.3.3}$$

The first two elements of the state vector are displacements (w and ψ) while the last two elements are forces (V and M). This arrangement of states is characteristic of transfer matrix theory. Figure 8.3.2 shows how these are defined for transfer matrix theory.

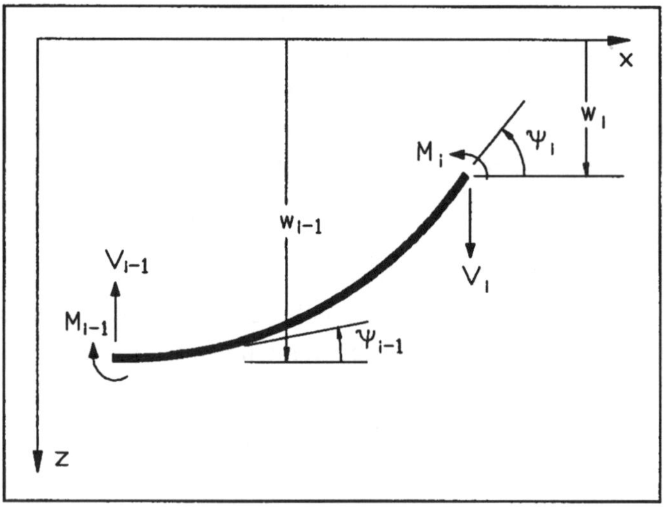

Figure 8.3.2: State Variable and Sign Conventions

An analytic solution to Eq. 8.3.2 can be found when the element has uniform properties (ie. constant cross-section, mass density, and stiffness). Eq. 8.3.4 gives the transfer matrix for a uniform Euler-Bernoulli element. Each element of Eq. 8.3.4 is a function of frequency and must be reevaluated as the frequency of interest changes.

$$TM = \begin{bmatrix} C_0 & 1C_1 & aC_2 & a1C_3 \\ \frac{\beta^4 C_3}{l} & C_0 & \frac{aC_1}{l} & aC_2 \\ \frac{\beta^4 C_2}{a} & \frac{\beta^4 lC_3}{a} & C_0 & 1C_1 \\ \frac{\beta^4 C_1}{a1} & \frac{\beta^4 C_2}{a} & \frac{\beta^4 C_3}{1} & C_0 \end{bmatrix} \tag{8.3.4}$$

where,

$$C_0 = \frac{1}{2}(\cosh\beta + \cos\beta) \tag{8.3.5}$$

$$C_1 = \frac{1}{2\beta}(\sinh\beta + \sin\beta) \tag{8.3.6}$$

$$C_2 = \frac{1}{2\beta^2}(\cosh\beta - \cos\beta) \tag{8.3.7}$$

$$C_3 = \frac{1}{3\beta^3}(\sinh\beta - \sin\beta) \tag{8.3.8}$$

and

$$\beta^4 = \frac{\omega^2 l^4 \mu}{EI} \tag{8.3.9}$$

$$a = \frac{l^2}{EI} \tag{8.3.10}$$

With the transfer matrix for the fundamental beam elements, one can combine these elements serially to generate a model for the link. Figure 8.3.3 illustrates how a simple model can be constructed for a tapered beam. Although only two elements are considered here, more elements can be added to better approximate the shape of the link.

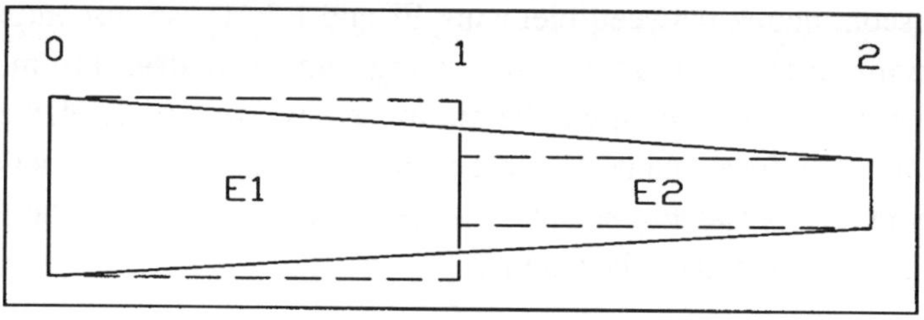

Figure 8.3.3: Simple Model of a Tapered Beam

Element El can be represented by the equation:

$$u_1 = [B_1]u_0 \tag{8.3.11}$$

Similarly for element E2,

$$u_2 = [B_2]u_1 \tag{8.3.12}$$

Since the states at interface u_1 are the same for both elements, u_1 can be eliminated to obtain an overall transfer matrix for the beam:

$$u_2 = [B_2][B_1]u_0 \tag{8.3.13}$$

Eliminating one state simply illustrates the point that this multiplication can be carried out to eliminate all intermediate states in a model with more elements.

As previously mentioned, transfer matrices themselves are not numerical approximations. The transfer matrix for an Euler-Bernoulli beam contains the analytic solution for a uniform beam element. It is not an assumed modes solution. The approximation made in using transfer matrix theory involves the modeling of the beam and solution of the equations. To generate the model of a link with variable cross-section, the size of the elements must vary. The interface of two different size elements will be discontinuous. In Figure 8.3.3, interface 1 is discontinuous between elements El and E2. These discontinuities are the major approximation when using transfer matrices to model a beam. This approximation can be minimized by using more elements to model a nonuniform beam. As more elements are added to the model, the discontinuities between elements will decrease thus reducing the effects of this approximation on the results.

Transfer matrix theory is similar to Finite Element Analysis (FEA). In FEA, first the system must be discretized. Then an appropriate interpolation function must be selected to describe each element (i.e. element stiffness). Next the system matrices must be assembled to produce a set of linear algebraic equations. Finally the linear equations are solved to get an approximate solution to the system under consideration.

Like FEA, when using transfer matrices the system must be first discretized into a finite number of elements. Unlike FEA though, there is no approximate interpolation function needed to describe each element. Each matrix contains the analytic equations describing the element. The two methods also differ in the method of solution of the numerical system of equations (for this application). As will be explained later, a root finder is used to determine the location of poles and zeros. An equation is extracted from the overall transfer matrix based on the desired input and output and the boundary conditions. The root finder then searches this equation to determine the location of poles and zeros. Although both are numerical methods to find an approximate solution to a continuous system, transfer matrix theory does reduce some approximations by using exact solutions to the partial differential equations to describe the individual elements.

8.3.2.2 *Boundary Conditions*

The second step in finding the RHP zeros and imaginary poles of a system is applying the appropriate boundary conditions. As Figure 8.3.4 shows, there are several boundary conditions that can be applied to model a flexible link. The clamped-free condition corresponds to a rigid coordinate attached at the hub. The pinned-pinned condition corresponds to a rigid coordinate which passes through the end-point of the manipulator. In this research, the pinned-free boundary condition was chosen to model the flexible link. This corresponds to a rigid coordinate passing through the center of mass of the beam. This boundary condition was chosen because it naturally describes a flexible link and because it allows easy comparison with previous research by Spector and Flashner[20,21] who also used these boundary conditions to model the flexible link.

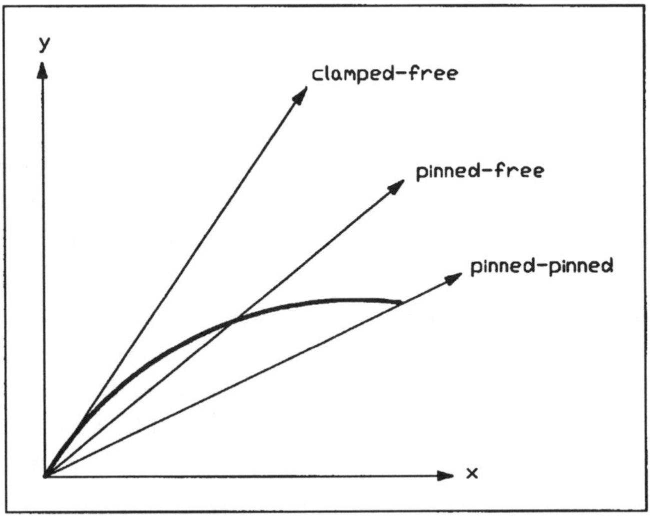

Figure 8.3.4: Boundary Conditions for a Flexible Link

A pinned-free boundary condition implies that:

At x=0 (base): w=0 $M=O$ (pinned)

At x=L (tip): V=0 $M=O$ (free)

These boundary conditions are applied to the overall transfer matrix for the system and the appropriate state variables are set to zero.

$$\begin{bmatrix} -w \\ \psi \\ 0 \\ 0 \end{bmatrix}_{x=L} = \begin{bmatrix} B_{11} & \cdots & B_{14} \\ \vdots & \ddots & \vdots \\ B_{41} & \cdots & B_{44} \end{bmatrix} \begin{bmatrix} 0 \\ \psi \\ 0 \\ V \end{bmatrix}_{x=0} \tag{8.3.14}$$

8.3.2.3 *System Input and Output*

When the system zeros are of interest, one must chose the system input and output. Unlike the natural frequency calculation which depends only on the boundary conditions, the location of system zeros will change as the input/output relationship changes. To illustrate this point, consider a single-link flexible manipulator modeled as a continuous system. Figure 8.3.5 shows the pole zero pattern of two different input/output relationships for the same system. Figure 8.3.5.a shows the transfer function between the joint angle, $\theta(s)$, and joint torque, $\tau(s)$, to be minimum phase. This is expected since these two are collocated. Figure 8.3.5.b shows the transfer function between tip position, X(s), and joint torque, $\tau(s)$, to be nonminimum phase. The RHP zeros are a result of the noncollocated output relationship. Since this research targets the location of RHP zeros the system output is tip position, and the system input is joint torque. Considering the system input and output, the overall system transfer matrix will have the form:

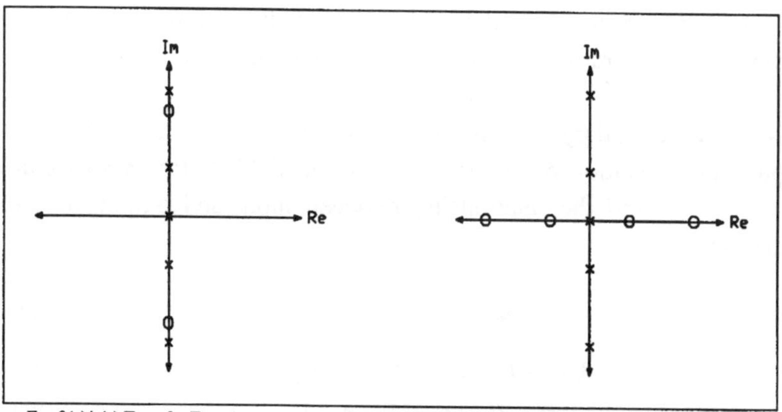

a: For θ(s)/τ(s) Transfer Function b: For X(s)/τ(s) Transfer Function

Figure 8.3.5 Pole/Zero Patterns For Different Input/Output Relationships

$$\begin{bmatrix} -w \\ \psi \\ 0 \\ 0 \end{bmatrix}_{x=L} = \begin{bmatrix} B_{11} & \cdots & B_{14} \\ \vdots & \ddots & \vdots \\ B_{41} & \cdots & B_{44} \end{bmatrix} \begin{bmatrix} 0 \\ \psi \\ \tau \\ V \end{bmatrix}_{x=0} \tag{8.3.15}$$

In the above equation, WL is the system output which corresponds to tip position, and τ is the system input corresponding to joint torque at the base of the manipulator.

8.3.2.4 *Zero Function*

The zeros of a system are defined as the frequencies that result in zero system output for an arbitrary system input. To determine the system zeros, one must know a) the system input, b) the system output, and c) the relationship between the input and the output. This can be expressed in an equation of the form:

$$INPUT = \begin{pmatrix} TRANSFER \\ FUNCTION \end{pmatrix} * OUTPUT \tag{8.3.16}$$

For an arbitrary input to the system, the only way to guarantee zero output is for the transfer function to be zero at the given frequency.

Given the boundary conditions chosen in Section 8.3.2.2 and the input/output relationship chosen in Section 8.3.2.3, Eq. 8.3.15 can be expanded to find the relationship between input and output. The four equations are:

$$-w_L = B_{12}\psi_0 + B_{13}\tau + B_{14}V_0 \tag{8.3.17}$$

$$\psi_L = B_{22}\psi_0 + B_{23}\tau + B_{24}V_0 \tag{8.3.18}$$

$$0 = B_{32}\psi_0 + B_{33}\tau + B_{34}V_0 \tag{8.3.19}$$

$$0 = B_{42}\psi_0 + B_{43}\tau + B_{44}V_0 \tag{8.3.20}$$

Since ψ_L is not of interest, Eqs. 8.3.17, 8.3.19, and 8.3.20 can be solved for the relationship between w_L and τ :

$$w_L = -\left[\frac{B_{12}B_{44}B_{33} - B_{12}B_{34}B_{43} + B_{13}B_{34}B_{42} - B_{13}B_{44}B_{32} + B_{14}B_{43}B_{32} - B_{14}B_{33}B_{42}}{B_{34}B_{42} - B_{44}B_{32}}\right]\tau \tag{8.3.21}$$

Where the B_{ij} are elements of the overall transfer matrix in Eq. 8.3.15. When the function inside the brackets is zero (for a given frequency), the output will always be zero regardless of the input; therefore, the zeros of the bracketed term are the system zeros.

$$f(\omega)=\left[\frac{B_{12}B_{44}B_{33}-B_{12}B_{34}B_{43}+B_{13}B_{34}B_{42}-B_{13}B_{44}B_{32}+B_{14}B_{43}B_{32}-B_{14}B_{33}B_{42}}{B_{34}B_{42}-B_{44}B_{32}}\right] \quad (8.3.22)$$

To search for RHP zeros, one must consider what type of frequency to input into Eq. 8.3.22. Using the relationship which defines the Laplace variable, s

$$s = j\omega \quad (8.3.23)$$

one can easily determine co should have the form:

$$\omega = 0 - jb \qquad where \quad 0 \leq b \leq \infty \quad (8.3.24)$$

Purely imaginary negative values of co will result in purely real positive values of s. Thus searching Eq. 8.3.22 with frequencies of the form of Eq. 8.3.24 one can find the location of the RHP zeros. Assuming the damping factor to be zero, it can be shown that the elements of the transfer matrix are real for a purely complex frequency. If the elements of the transfer matrix are real, the zeros function, Eq. 8.3.22, will also be real.

8.3.2.5 *Natural Frequency Function*

Although the location of RHP zeros is of primary concern in this research, knowledge of pole location will help in analysis of the results. Since the system damping is ignored, the poles will lie on the imaginary axis of the s-plane in complex conjugate pairs. The location of these poles can be determined by simply searching the positive imaginary axis of the s-plane. Considering the applied boundary conditions, one can extract two homogeneous equations from Eq. 8.3.14 to get the homogeneous system:

$$\begin{Bmatrix} 0 \\ 0 \end{Bmatrix} = \begin{bmatrix} B_{32} & B_{34} \\ B_{42} & B_{44} \end{bmatrix} \begin{Bmatrix} \psi \\ V \end{Bmatrix} \quad (8.3.25)$$

The poles (eigenvalues) of the system are those values of ω which make the determinant of the sub-transfer matrix in Eq. 8.3.25 equal to zero (see Book, et al.[6] for a detailed explanation). For a two by two matrix this determinant is simply:

$$g(\omega) = B_{32}B_{44} - B_{34}B_{42} \tag{8.3.26}$$

Referring to Eq. 8.3.23, one finds that Eq. 8.3.26 is the denominator of the input/output transfer function which is to be expected. To find the values of the purely complex poles, one must search Eq. 8.3.26 for its roots. According to the definition of s, ω must have the form:

$$\omega = b + j\,0 \tag{8.3.27}$$

Searching over a range of values for b will give the poles in that range. With the zero and natural frequency functions determined, the problem remains to implement a computer solution to find the RHP zeros and imaginary poles.

8.3.2.6 *Mode Shapes For Pinned-Pinned Boundary Condition*

To implement the results of this research in the inverse dynamic control algorithm developed by Kwon and Book[9], the mode shapes must be determined for pinned-pinned boundary conditions. The natural frequencies for a tapered link are easily determined with the tapered link algorithm. Pinned-pinned boundary conditions change the frequency determinant which changes the search function.

The tapered design was chosen so that A=0.6 in. and B=0.3 in. (R=2.0). As described earlier, the beam has L=40 in., H=1 in., and properties of aluminum. For a discussion of mode shape generation using transfer matrices see Majette[11].

The state matrix consists of the state vectors at each interface for the given natural frequency. The chosen design has twenty elements; therefore, the state matrix will have twenty-one columns and four rows. Recall from Section 3 that the state vector is described by Eq. 8.3.3. Figures 8.3.6 and 8.3.7 present the mode shapes for the first and second natural frequencies respectively.

8.3.3 *Computer Algorithm*

Like Finite Element Analysis, the solution to pole/zero location of a flexible link using transfer matrix theory is computationally intensive. As the number of elements in the model increases, so does the number of computations. With the availability of computers today, the problem is fairly easy to solve if the proper algorithm can be implemented. Previous research by Book and others [6,11] used transfer matrices to model systems and this provided some insight on how to realize a computer solution using transfer matrices, especially the DSAP[6] package. The program structures are purposely similar to aid in combining the programs for future research.

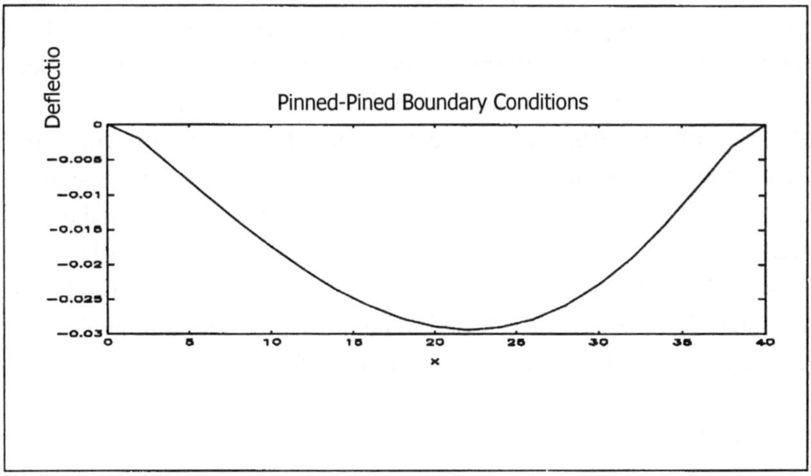

Figure 8.3.6: First Mode Shape For Tapered Link

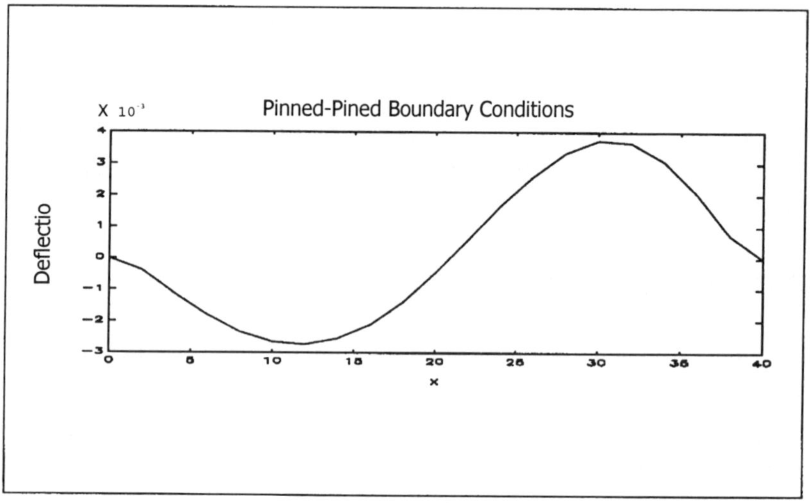

Figure 8.3.7: Second Mode Shape For Tapered Link

8.4 Results

The results of the zero and pole locations found from the Transfer Matrix approach are presented in this section as a collection of examples. Each example investigates a different aspect of the relationship between RHP zero location and structural link design. As previously noted, pole location is often of interest to the designer; therefore, this information is presented for each example. Unless otherwise specified, several dimensions remain the same from one example to the next (referred to as nominal dimensions). The overall length of the beams is 40 inches, and the height (which remains constant over length) is 1 inch. The material properties are selected to be those of aluminum: modulus of elasticity, E, is 10E6 psi, and the density is 9.55E-2 lb/in^3.

8.4.1 *Validity of Results*

$$\omega_n = (\beta_n l)^2 \left(\frac{EI}{\rho A l^4} \right)^{\frac{1}{2}} \tag{8.4.1}$$

Before examining the relationship between RHP zeros and link design, the validity of the computer algorithm to determine zero/pole location must first be checked. Since analytic solutions exist for the location of poles for a uniform beam, the results from the Transfer Matrix approach are compared to the analytic solution to determine the accuracy of the algorithm. The vibrations text by Rao[19] contains the analytic solution for pole location of a pinned-free beam under lateral vibration. The poles are determined from the following equation:

For a pinned-free beam, the values of $\beta_n l$ are:

$$\beta_1 l = 3.926602$$
$$\beta_2 l = 7.068583$$
$$\beta_3 l = 10.210176$$

For a uniform beam with width=0.5 inches and nominal properties as given above, the pole locations are presented in Table 8.4.1 along with the results from the Transfer Matrix approach.

Table 8.4.1: the Transfer Matrix vs. Analytical Solution

Pole	Transfer Matrix	Analytical Solution
1	14.23	14.23
2	46.12	46.12
3	96.23	96.23

The results generated from the Transfer Matrix approach show excellent correspondence to the analytic values. However analytic calculation of zeros is not as simple of a task since the boundary conditions are no longer homogeneous, and texts lack tabulated results

It must be noted that the results presented in this section will not include the two poles lying at the origin. These poles are a result of the rigid body mode of the system. Keep in mind the location of the poles will be presented as a real number, but they actually are located on the s-plane in complex conjugate pairs along the imaginary axis. The zeros are also presented as real numbers, and they lie on the real axis as reflected pairs about the imaginary axis. This means for every RHP zero, there is a corresponding LHP zero equal in magnitude but opposite in sign. The symmetry of the s-plane results from ignoring the damping of the structure in the Euler-Bernoulli model and is confirmed by Spector and Flashner[20].

8.4.2 *Effects of Discretization*

When modeling a continuous system with a discrete model, one should check to make sure the discretization of the model does not affect the results. This is easily confirmed by studying a uniform beam. Using transfer matrices, a uniform beam can be modeled with one element or several elements. It has been confirmed in earlier publications that this has no effect.[8] For a tapered beam, the number of elements will be more critical because increasing the elements will decrease the discontinuities at each element interface. This should result in a better approximation of the tapered link. For nonuniform designs, the poles and zeros should converge to the actual values as the number of elements increases. This will be confirmed later in this section.

8.4.3 *Modeling of a Tapered Beam*

Another point to consider in the computer implementation of the RHP zeros problem is how well does the model represent the actual system. Although the model is limited to uniform elements, there are any number of combinations one can find to represent the system. This example examines two different methods for modeling a linearly tapered beam. As shown in Figure 8.4.1 the link is tapered along the length in the width dimension while the height was held constant. The taper is described by two dimensions: the width at the base, A, and the width at the tip, B. The degree of taper, R=A/B, is used to compare different designs.

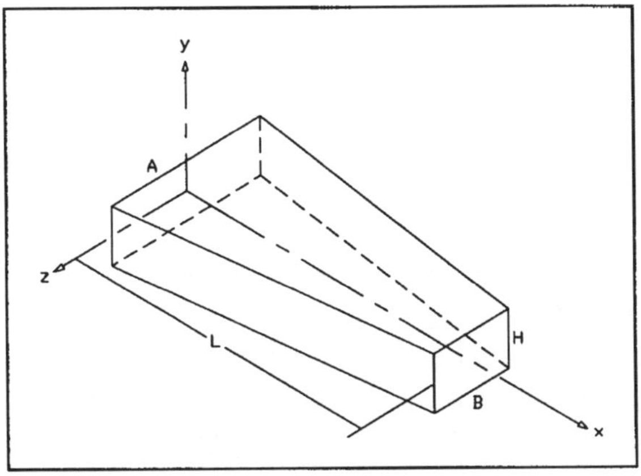

Figure 8.4.1: Tapered Link Diagram

Using Method 1 to model the tapered link, the beam is divided into elements of equal length. For a three element model with length L, each element will have length L/3. The height of each element is the same, while the width of each element changes linearly as a function of x. Figure 8.4.2 presents Method 1.

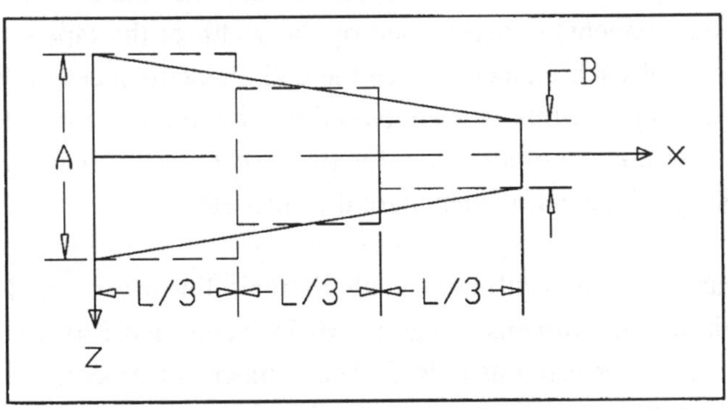

Figure 8.4.2: Modeling Method 1

Using Method 2 to model the tapered link, the beam is divided into elements so the first and last element have length one-half of the intermediate elements. For a three element model with length L, the first and last elements will have length L/4 and the middle element will have length L/2. Again the height of each element ss the same, while the width of each element changes linearly as a function of x. Figure 8.4.3 presents Method 2.

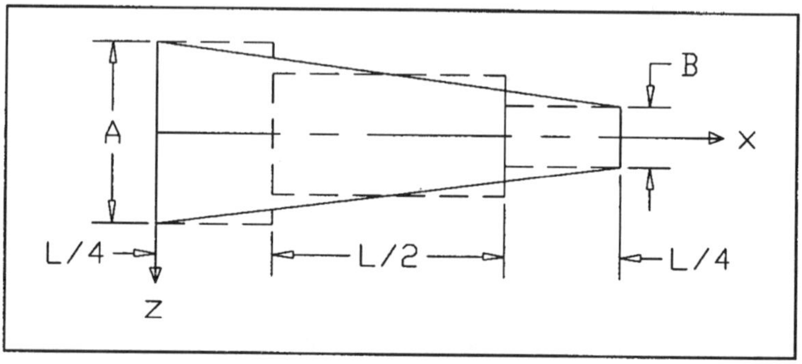

Figure 8.4.3: Modeling Method 2

Figures 8.4.2 and 8.4.3 illustrate the main difference between the two modeling methods. Method 2 compensates the elements at each end for meeting the specified end widths A and B. In both methods the width of intermediate elements is determined by the width of the tapered beam at the midpoint of each element. Since the end elements meet the specified A and B, the tapered link will not pass through the midpoint of these two elements. Method 2 compensates for this exception by making the end element lengths one half the length of the other elements.

To compare these two different methods for a linearly tapered beam, a beam with nominal dimensions and A=0.75 inches and B=0.25 inches is examined. This corresponds to R=3. The number of elements is increased with each method until the zeros and poles converge. Figure 8.4.4 presents the results from Method 1 where all elements are of equal length, and Figure 8.4.5 presents the results from Method 2 where the end elements are half the length of all other elements. Although only two methods are considered in this research, there are many different ways to discretize a nonuniform link.

The two methods are evaluated based on an error function. When the tapered beam is modeled with 80 elements, both methods converge to nearly identical values for the poles and zeros. These values, when NE=80, are taken to be the "correct' values and other cases are compared to this case. The error, e, is defined for the zeros as:

$$e = \left| \frac{z_{80,i} - z_{NE,i}}{z_{80,i}} \right| \tag{8.4.2}$$

where i refers to the ith zero

A similar definition is used for the poles. As the figures show, Method 2 provides better results for the same number of elements. For a targeted error of less than 1%, Method 2 reaches the target with NE=10 while Method 1 requires NE=20 to reach the target. Thus, compensating the end elements does provide a better model of a linearly tapered beam, and this method is used in the following examples unless specified otherwise

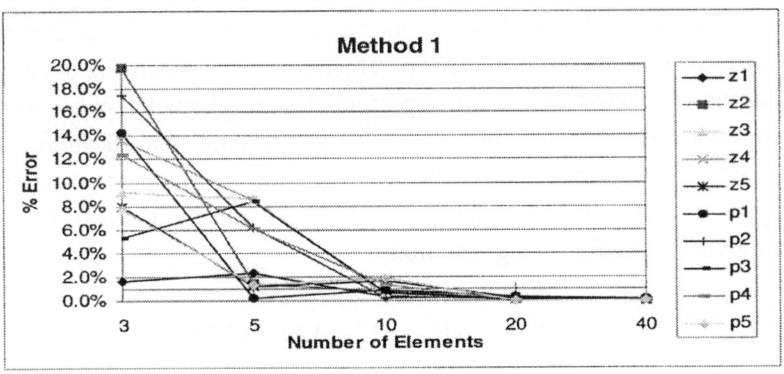

Figure 8.4.4: Percentage Error for Method 1 for Poles (pl...p5) and Zeros (zl...z5)

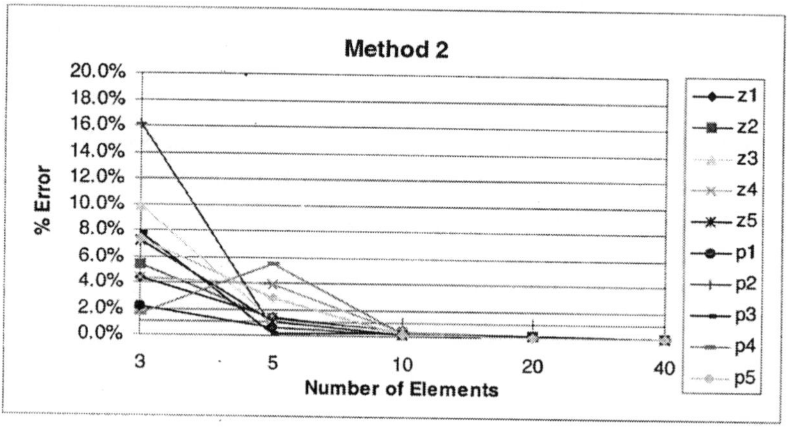

Figure 8.4.5: Percentage Error for Method 2 for Poles (p1...p5) and Zeros (z1...z5)

8.4.4 *Linear Taper Designs*

When comparing different link designs to evaluate pole/zero location as a function of link shape, it is necessary to keep some parameters constant to aid in the evaluation. For a single-link manipulator rotating in the horizontal plane, the link's mass moment of inertia about its axis of rotation, I_y, is of importance. This parameter directly affects the dynamic equations of motion and is an important design parameter in terms of motor selection. In the following studies, several link designs are evaluated for a given value of I_y. A tapered link's moment of inertia about its axis of rotation in terms of the links parameters: L, A, B, H, and ρ is given by the following equation:

$$I_y = \frac{\rho H}{48}(A^3 + A^2B + AB^2 + B^3 + 4AL^2 + 12BL^2) \qquad (8.4.3)$$

For a given tapered link design, one can use Eq. 8.4.3 to determine I_y. Knowing I_y, one can change the value of A and solve Eq. 8.4.3 for B. Since the equation is cubic in B, a commercial package like *Mathematica* can be used to solve for B. Following this method, a group of tapered link designs are generated for given values of I_y. Figures 8.4.6 shows pole/zero maps for selected values of R for I_y =764.05.

Several patterns are evident by examining the graph. First as a general rule, both the poles and zeros increase (move away from the origin) as the taper on the beam is increased. Increasing the taper effectively moves more of the link's mass closer to the base. Increasing the magnitude of the poles is often desirable to push them out of the system bandwidth and increase system response time. The ordering of poles and zeros is the second pattern recognized. In a minimum phase system, the poles and zeros will both lie on the imaginary axis in complex conjugate pairs and in an alternating order. This means, along the imaginary axis, the poles and zero are found in the order pl,zl,p2,z2, etc. or vice versa. Previous research 20 has found this alternating order of poles and zeros does not hold for nonminimum phase systems. It is not obvious in Figure 8.4.6, but the pole/zero order does not alternate. The actual order is: z1, pl, p2, z2, p3, z3, p4, p5, z4... p2 jumps in front of z2, and the same occurs for p5. This reordering of poles and zeros can be critical as accurate knowledge of the pole/zero order is important for control system design.

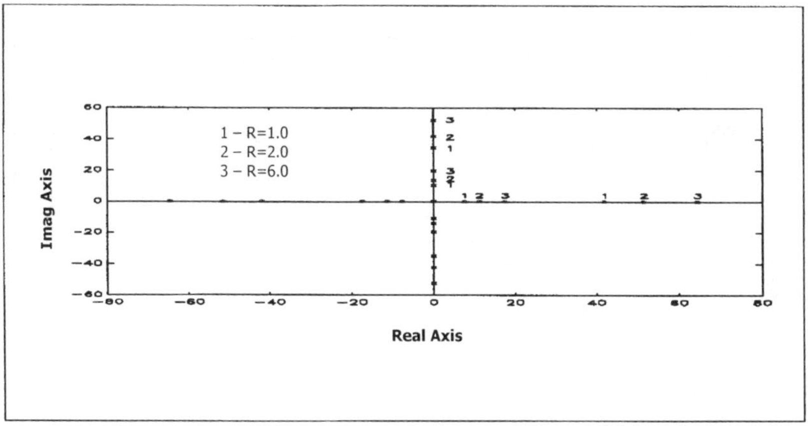

Figure 8.4.6: Pole/Zero Map of Selected Designs For I_y=764.05

8.4.4.1 *Designs With Constant I_y*

Important information is learned from examining the relationship between the taper ratio, R, and the values of the normalized zeros. The first normalized zero is of most importance. Figure 8.4.7 shows this relationship for data with Iy=764.05 and Iy=1528.1. Both curves are fitted with a third order polynomial. Even though the coefficients are different for each polynomial fit, the curves are nearly identical.

This illustrates an important relationship in the design of tapered links. For a given ratio R, the normalized zero will always remain the same. The designer can choose the location of the first pole and zero, determine the normalized zero, and then using Figure 8.4.8 find the appropriate taper ratio R. Of course there are constraints on this process. A ratio less than one corresponds to a taper with B greater than A, which is usually undesirable. At the other end, R is limited by the value of H. If A is larger than the value of H, the link will be wider at the base than it is tall, and the assumption that the link is stiff in the vertical plane will no longer be valid. Although the designer can choose the pole/zero relationship, the values of normalized zeros are limited to approximately 0.72-0.82 (according to Figure 8.4.7).

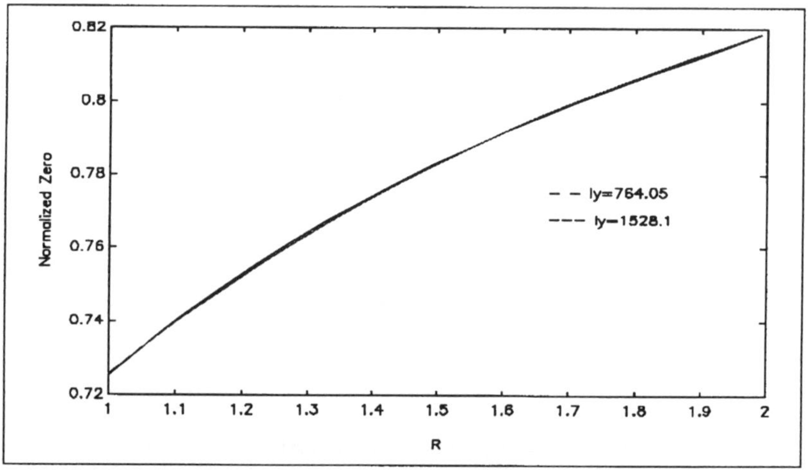

Figure 8.4.7: Comparison of Polynomial Curve Fits *8.4.4.2*

8.4.4.2 *Designs With Constant Poles and Zeros*

The previous example demonstrates how the designer can choose the pole/zero relationship and then determine the appropriate taper design. Once the taper is chosen, the designer can change the link height to independently adjust the value of I_y. Since the adjustment of H is out of the plane of motion, it has no effect on the location of poles and zeros. Combining this with the results from the previous example, the designer can effectively choose the location of poles and zeros and independently adjust the links moment of inertia about its axis of rotation to meet the needs of the particular system.

8.5 Conclusions

This algorithm is developed as a tool to locate the poles and zeros of a single-link manipulator modeled as a pinned-free Euler-Bernoulli beam. The algorithm uses transfer matrix theory to allow for variable cross-sections granting the designer new freedom in analysis of nonuniform link designs. The results are shown to be very accurate when system pole location is compared to analytic solutions for uniform beams. Several results from previous research are confirmed.

First, the reordering of poles and zeros is confirmed for nonminimum phase systems. Accurate knowledge of pole/zero order is critical for proper control system design. Even with very few elements in the model, this algorithm predicts the proper order of poles and zeros

Second, the examples presented suggest the nonminimum phase characteristics can not be eliminated by changing the structural design of the link. The system will be nonminimum phase above a finite frequency dictated by the location of the first nonminimum phase zero. It is possible that this frequency is out of the operating range and not of concern to the designer.

This algorithm is set up specifically for pinned-free boundary conditions of the model and determines pole and zero location based on a user determined frequency range. Linearly tapered beams have been presented in this chapter, but any type of nonuniform beam can be analyzed with this method. Slight modifications would also allow for different boundary conditions.

The design procedure for tapered beams allows the designer to choose the first pole and zero subject to certain physical constraints. These physical constraints only allow for approximately 25% variation in R. This zero to pole ratio defines a particular taper ratio according to the collected data. Keeping the ratio the same, the size of the taper can be changed to get the proper magnitude of the pole and zero. With the pole and zero placed, the height of the beam can be changed to adjust the link's moment of inertia about its axis of rotation. This procedure can be used to design tapered links to meet the particular requirements of the system.

The material presented is for a single-link manipulator modeled with pinned-free boundary conditions. This is a simplified model, but it is necessary to show transfer matrices yield good results for this case before progressing to more complicated problems. Now that transfer matrices have proven useful to solve for zero location, future work exists to extend these results.

First, a program could be developed so the user could input the desired boundary conditions which best represent the system. This could include hub inertia or end-point mass. Second, the program could be extended to multi-link designs to predict pole and zero location for different configurations. Transfer matrices have been derived for rotary joints and many other elements. The DSAP package developed by Book, et. al.[6] handles multi-link models and would be a good reference. Finally, the results for tapered link designs could be applied to the inverse dynamic algorithm developed by Kwon and Book[10]. This method requires mode shapes for the assumed modes and uses pinned-pinned boundary conditions as presented in Section 8.3.2.6.

References

1. Asada, H., Park, J.-H., and Rai, S., *"A Control-Configured Flexible Arm: Integrated Structure/Control Design,"* Proceedings of the 1991 IEEE International Conference on Robotics and Automation, Sacramento, California, April, 1991, pp. 2356-2362.
2. Bayo, E., "A Finite Element Approach to Control the End-Point Motion of a *Single-Link Flexible Robot*," Journal of Robotic Systems, Vol. 4, No. 1, 1987, pp. 63-75.
3. Beer, Ferdinand, and Johnson, Russell, Jr., *Vector Mechanics for Engineers, Statics and Dynamics,* Third Edition, McGraw-Hill, New York, 1977.
4. Book, W. J., *Design and Control of Flexible Manipulator Arms,* Ph.D. Thesis, Massachusetts Institute of Technology, April, 1974.
5. Book, W. J., and Kwon, D.-S., *"Contact Control for Advanced Applications of Light Weight Arms,"* Symposium on Control of Robots and Manufacturing, Arlington, Texas, 1990.
6. Book, W. J., Majette, M., and Ma, K., *The Distributed Systems Analysis Package (DSAP) and Its Application to Modeling Flexible Manipulators,* NASA Contract NAS 9-13809, Subcontract No. 551, School of Mechanical Engineering, Georgia Institute of Technology, 1979.
7. Churchill, R. V., and Brown, J. W., *Complex Variables and Applications,* Fifth Edition, McGraw-Hill Publishing Company, New York, 1990.
8. Girvin, D.L. Numerical Analysis of Right-Half Plane Zeros for a Single-Link Manipulator, Master's Thesis, Georgia Institute of Technology, Mechanical Engineering, March, 1992.
9. Kwon, D.-S., *An Inverse Dynamic Tracking Control for Bracing A Flexible Manipulator,* Ph.D. Dissertation, Georgia Institute of Technology, Mechanical Engineering, June, 1991.
10. Kwon, D.-S., and Book, W. J., *"An Inverse Dynamics Method Yielding Flexible Manipulator State Trajectories,"* Proceedings of the American Control Conference, June, 1990, pp. 186-193.
11. Majette, *M., Modal State Variable Control of a Linear Distributed Mechanical System Modeled with the Transfer Matrix Method,* Master's Thesis, Georgia Institute of Technology, Mechanical Engineering, June, 1985.
12. Meirovitch, *L., Elements of Vibrational Analysis,* McGraw-Hill, New York, 1986.
13. Misra, Pradee, *"On The Control of Non-Minimum Phase Systems,"* Proceedings of the 1989 American Control Conference, 1989, pp. 1295-1296.

14. Miu, D.K., *"Physical Interpretation of Transfer Function Zeros for Simple Control Systems with Mechanical Flexibilities,"* Journal of Dynamic Systems, Measurement, and Control, Vol. 113, September, 1991, pp. 419-424.
15. Nebot, E. M., Lee, G. K. F. and Brubaker, T. A., *"Experiments on a Single Link Flexible Manipulator,"* Proceedings from the USA-Japan Symposium on Flexible Automation Crossing Bridges: Advances in Flexible Automation and Robotics, 1988, pp. 391-398.
16. Park, J.-H. and Asada, H., *"Design and Analysis of Flexible Arms for Minimum-Phase Endpoint Control,"* Proceedings of the American Control Conference, 1990, pp. 1220-1225.
17. Park, J.-H. and Asada, H., "Design and Control of Minimum-Phase Flexible Arms with Torque Transmission Mechanisms," Proceedings of the 1990 IEEE International Conference on Robotics and Automation, 1990, pp. 1790-1795.
18. Pestel and Leckie, *Matrix Methods in Elastomechanics,* McGraw-Hill, New York, 1963.
19. Rao, Singiresu S., *Mechanical Vibrations,* Addison-Wesley Publishing Company, Reading, Massachusetts, 1986.
20. Spector, V. A. and Flashner, H., *"Modeling and Design Implications of Noncollocated Control in Flexible Systems,"* Journal of Dynamic Systems, Measurement, and Control, Vol. 112, June, 1990, pp. 186-193.
21. Spector, V. A. and Flashner, H., *"Sensitivity of Structural Models for Noncollocated Control Systems,"* Journal of Dynamic Systems, Measurement, and Control, Vol. 111, December, 1989, pp. 646-655.

Chapter 9

Optimum Shape Design of Flexible Manipulators with Tip Loads

The problem of optimum shape design of flexible manipulators is addressed using two different approaches. One approach employs an iterative method to solve the unconstrained analytical formulation of the optimal equations. In order to accommodate various constraints on link design, an optimization approach is developed which uses mathematical programming to solve the segmentized formulation. The goal of each approach is to find that unique shape which maximizes the fundamental frequency of vibration, since this frequency is the governing factor of manipulator speed. The requirement of multiple tip loads is formulated as a minimax problem. Also, a multi-link design is implemented which utilizes single-link solutions. A sensitivity analysis of design parameters is conducted to reveal the robustness of optimum designs.

9.1 Introduction

Research on flexible manipulators has increased dramatically in recent years, motivated by the need for higher speed, low weight, and better energy consumption. These issues are most relevant in the design of new-generation industrial robotic manipulators and in the field of robotic space applications. Many techniques have been applied to overcome tip oscillations caused by manipulator flexibility. Here, the control problem is addressed from a design perspective. The goal is to build a high-performance flexible manipulator such that the effect of its oscillation is reduced during normal operation. This design should also improve the performance of control algorithms. An important direction in flexible link research is the search for structural designs which optimize manipulator mass and speed. The design problems are:

1. Given a fixed mass for the link, find a shape that maximizes its speed.

2. Given a fixed speed for the link, find a shape that minimizes its mass.

How can these objectives be achieved? We need a design variable that relates directly to the maximum speed of the manipulator and is a function of shape, as well as material properties. The only design variable that fits these requirements is the fundamental frequency of vibration. Thus, Problem 1 is equivalent to finding a shape that maximizes the fundamental frequency of vibration of the link. A large fundamental frequency is desired since it generates a large bandwidth for the flexible manipulator, and this allows for fast motion and stable endpoint control. Problem 2 is known as the minimum weight problem and is dual to Problem 1. This problem is very important in that a significant reduction in energy costs can be achieved by using the optimally designed link. Also for space applications, where flexible manipulators have been generally recognized as an ideal tool for material handling and space-structure construction, minimum-weight manipulators are of special interest due to the strict weight constraints imposed by the loading capacity of space vehicles [2].

What is the difference between a flexible and rigid manipulator? A rigid manipulator, in theory, has an infinitely large fundamental frequency of vibration and thus can move at any desired speed without significant tip deflection. The drawbacks to this design are a large link mass and a correspondingly low energy efficiency. A flexible manipulator, on the other hand, has a much lower mass but experiences significant tip deflection under normal operating conditions, and the oscillations become severe when the manipulator speed approaches its fundamental frequency. The amount of deflection depends upon the following factors:

- Rotational velocity of the link
- Total length of the link
- Material properties of the link
- Shape of the link
- Mass, placement, and moment of tip load

In general, only one of these factors is not a specified design requirement and that is the link's shape function. So, the objective is to develop a flexible manipulator which is capable of high-speed motion with less energy consumption by increasing the fundamental frequency of vibration through optimum shape design.

It is shown that the fundamental frequency of a flexible manipulator can be increased substantially through the optimum tapering of its cross section. For example, numerical analysis indicates that, depending on the values of hub inertia and tip load parameters, an increment of up to 625% in the fundamental frequency can be achieved for manipulators with links of geometrically similar cross sections [34].

The first attempts to optimize a flexible beam using the fundamental frequency of vibration were by engineers working in the field of structural optimization. Niordson first showed that the optimum design of a simply supported beam will have a fundamental frequency of vibration that is 6.6% larger than that of the corresponding uniform beam [19]. This work was extended by others to include specific design requirements [6], [30], [26]. In the early 1970s, Karihaloo and Niordson showed that an increase of up to 678% in the fundamental frequency of vibration was possible for vibrating cantilevers with a tip mass [16]. This research was devoted to beams that were rigidly attached to an inertial frame-the simply supported beams at both ends and the cantilever at only one end. But, flexible robotic links must rotate about an axis at the link's base. Also, the base of the flexible link usually must be rigidly attached to a rotating hub for mechanical support. Thus, the results from structural optimization cannot be carried over into the field of robotics. This is an important point and one that is often overlooked.

Early work on flexible manipulators was carried out by Cannon and Schmitz [9], Book [5], Nguyen *et al* [20], Sakawa *et al* [29], and many others. Khorrami and Ozgoner [17] presented a set of *integro-partial* equations and used a perturbation method for control purposes. Besides link flexibility, Yang and Donath also considered the flexibility of the joint [46]. A complete study of different dynamic equations under various boundary conditions for flexible manipulators has been conducted by Bellezze, Lanari, and Ulivi [3]. Wang and Wen [41] derived the dynamic equations for flexible manipulators subject to large angular deformations and clarified some issues related to the specification of boundary conditions. A comprehensive study on the influence of rotatory inertia, shear deformation, and tip load on the vibration behavior of one-link flexible manipulators has been conducted by Wang and Guang [35]. Recently, Asada, Park, and Rai [1] have addressed the optimal design problem using a finite element computational model. As part of their effort toward a control-configured flexible arm, they have tried to increase the fundamental frequency of vibration of a flexible manipulator through optimum tapering of a beam of varying rectangular cross section. A theoretical increment of 43% in the fundamental frequency of the manipulator has been obtained.

In Section 2, a derivation of the Euler-Bernoulli and vibration equations is given along with details of the design model. Section 3 includes a derivation of the optimality equations. Then, a successive iteration scheme is developed, followed by an extensive analysis of the design model dynamics and numerical results. In Section 4, the segmentized formulation of optimum shape design is given, and a heuristic dynamic programming algorithm is presented along with numerical results. Since the tip load greatly influences the optimum frequency of a flexible arm and it is unavoidable for the arm to handle multiple tip loads, a minimax design method is proposed for the optimum shape in Section 5. Also, a new optimum formulation and numerical results are given for a two-link flexible manipulator, since a single-link arm is of little practical use. Section 6 is a look into the sensitivity of the two design formulations with respect to parameter changes. The chapter concludes with a discussion of contributions, results, and directions for future research.

9.2 Euler-Bernoulli Equations

9.2.1 *Design model*

The system model of the flexible manipulator is composed of a hub rigidly attached to a flexible link of length L with a tip load (see Fig. 9.1). The hub has a rotational inertia given by I_H. The tip load is modeled by its mass, *MP,* and moment of inertia, I_P. A constant torque, τ, is applied to the hub, which is located to the left of $x_1 = 0$. The angular displacement of the link axis, x_1, is given by θ. The transverse displacement of the link's neutral axis from the x_1-axis is given by ω. The tip load center of mass, which also includes the end-effector mass, is located at (a_c, b_c) in the (x_2, y_2) frame. Three coordinate systems are employed in the subsequent analytical derivation: inertial (x_o, y_o), link (x_1, y_1), and tip load *(x2, y2).*

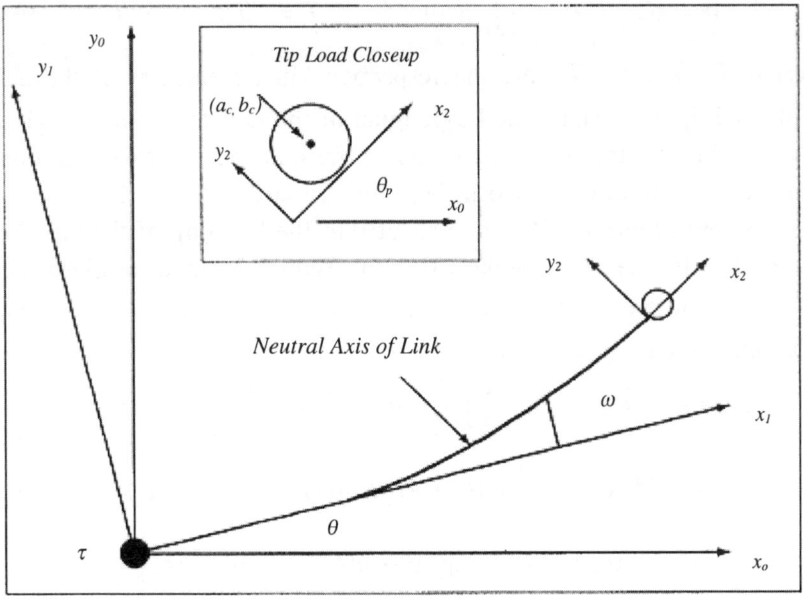

Figure 9.1: Design Model of Flexible Manipulator.

The dynamics of the flexible link will be modeled using the Euler-Bernoulli beam theory. Rotatory inertia, as well as shear deformation, will be ignored since these factors have been shown to have only a negligible effect on the fundamental frequency of vibration [35]. No rotatory inertia implies that the velocity of every point on a given cross section has the same velocity as the point of intersection with the neutral axis. The normal plane assumption is also used, which means that each transverse section remains plane and normal to the neutral axis after deformation.

9.2.2 *Variational derivation*

The basic equations of motion for the flexible manipulator are derived using variational calculus [27, 15]. From Hamilton's Principle,

$$\delta \int_{t_0}^{t_f} [T - P + W] dt = 0 \tag{9.1}$$

The work, potential energy, and kinetic energy are given by,

$$W = \tau\theta, \qquad P = \frac{1}{2}\int_0^L EI\omega''^2 dx_1, \quad T = T_H + T_L + T_P, \tag{9.2}$$

where T_H, T_L, and T_P are the respective kinetic energies of the hub, link, and tip load. Here, the Lagrangian of the isolated system is given by T-P. Since a torque, τ, is applied to the hub, external work is done on the system and that work, W, must be added to the Lagrangian. The potential energy, P, is due strictly to the bending of the link. For more details, see Timoshenko [31]. E is Young's Modulus of Elasticity, I is the second-area cross sectional moment of inertia, and $()' = \partial()/\partial x_1$. The kinetic energy components are given by,

$$T_H = \frac{1}{2}I_H\dot{\theta}^2, \quad T_L = \frac{1}{2}\int_0^L \left[\left(x_1\dot{\theta} + \dot{\omega}\right)^2 + \left(\omega\dot{\theta}\right)^2 \right] \lambda(x_1) dx_1 \tag{9.3}$$

$$\begin{aligned} T_P = {} & \frac{1}{2}M_p L^2\dot{\theta}^2 + \frac{1}{2}I_p\dot{\theta}_p^2 + M_p L\dot{\theta}\dot{\omega}(L) + \frac{1}{2}M_p\dot{\omega}(L)^2 \\ & + M_p\dot{\theta}_p \cos(\theta_p - \theta)\left[\left(L\dot{\theta} + \dot{\omega}(L)\right)a_c + \omega(L)\dot{\theta}b_c\right] \\ & + M_p\dot{\theta}_p \sin(\theta_p - \theta)\left[\omega(L)\dot{\theta}a_c - \left(L\dot{\theta} + \dot{\omega}(L)\right)b_c\right] \end{aligned} \tag{9.4}$$

where $\lambda(x_1)$ is the linear mass density of the link, θ_p is the angular displacement of the x_2-axis from the x_0-axis (see closeup in Fig. 9.1), and a dot indicates differentiation with respect to time. Here, $\omega(L)$ and $\dot{\omega}(L)$ are evaluated at the link's endpoint.

After *extended* variational calculus, the Nonlinear Euler-Bernoulli Equations can be arrived at,

$$I_H\ddot{\theta}+\frac{\partial}{\partial t}(f_1+f_2)=\tau \tag{9.5}$$

$$[D\omega''(x_1)]''+\lambda(x_1)\left[x_1\ddot{\theta}+\ddot{\omega}-\omega\dot{\theta}^2\right]=0 \tag{9.6}$$

with boundary conditions,

$$\omega\,(0)=0, \quad \omega'(0)=0 \tag{9.7}$$

$$D(L)\omega''(L)-M_p f_3+\frac{\partial}{\partial t}f_4=0 \tag{9.8}$$

$$[D(L)\omega''(L)]'-M_p\frac{\partial}{\partial t}f_5+M_p f_6=0 \tag{9.9}$$

Where,

$$D(x_1)=EI(x_1) \qquad \text{[bending rigidity]}$$

$$f_1=\int_0^L\lambda(x_1)\left[x_1\dot{\omega}+\left(x_1^2+\omega^2\right)\dot{\theta}\right]dx_1, \qquad f_2=f_4+f_7$$

$$f_3=\dot{\theta}_p\left(G_y\cos\psi-G_y\sin\psi\right), f_4=I_p\dot{\theta}_p+M_p\left(G_x\cos\psi+G_y\sin\psi\right)$$

$$f_5=L\dot{\theta}+\dot{\omega}(L)+\dot{\theta}_p\left(a_c\cos\psi-b_c\sin\psi\right)$$

$$f_6=\dot{\theta}^2\omega(L)+\dot{\theta}_p\dot{\theta}\left(b_c\cos+a_c\sin\right)$$

$$f_7=M_p\left\{L^2\dot{\theta}+L\dot{\omega}(L)+\dot{\theta}\omega^2(L)+\dot{\theta}_p\left[\cos\psi\left(a_cL+b_c\omega(L)\right)+\sin\psi\left(a_c\omega(L)-b_cL\right)\right]\right\},$$

$$\psi=\omega'(L), \qquad G_x=a_c\left(L\dot{\theta}+\dot{\omega}(L)\right)+b_c\dot{\theta}\omega(L),$$

$$G_y=a_c\dot{\theta}\omega(L)-b_c\left(L\dot{\theta}+\dot{\omega}(L)\right)$$

Due to the intractable nature of these equations, a linearization is in order. This will not only reduce computation time drastically, but will also allow for easier interpretation of results. One inherent assumption in the linearization process is that the link will undergo only small deformations; that is, ω will be small. If the link is very flexible and large displacements from the neutral axis are expected, then higher order terms in the above model must be considered.

To linearize this model, we drop all deformations and cross terms above first order. A further simplification can be achieved by making a change of variables as,

$$\upsilon(x_1,t)=\omega(x_1,t)+x_1\theta(t) \tag{9.10}$$

where v is the total displacement of the link. From this point on, all dynamic variables will be assumed to be a function of time. Then, the Linearized Euler-Bernoulli Model becomes,

$$[D(x_1)\upsilon''(x_1)]''+\lambda(x_1)\ddot{\upsilon}(x_1)=0 \tag{9.11}$$

$$I_H\ddot{\theta}-D(0)\upsilon''(0)=\tau \tag{9.12}$$

with boundary conditions,

$$\upsilon(0)=0, \qquad \upsilon'(0)=\theta \tag{9.13}$$

$$D(L)\upsilon''(L)+I_p\ddot{\upsilon}'(L)+M_pa_c\ddot{\upsilon}(L)=0 \tag{9.14}$$

$$[D(L)\upsilon''(L)]'-M_p[\ddot{\upsilon}(L)+a_c\ddot{\upsilon}'(L)]=0 \tag{9.15}$$

Eq. (9.11) is the well-known dynamics equation which all flexible links must satisfy. Eq. (9.12) is the torque relationship. Eq. (9.13) is the boundary condition at the hub, while Eqs. (9.14, 9.15) are the boundary conditions at the tip load. These are the linearized equations of motion for the system, upon which all subsequent derivations will be based. Note that the endpoint boundary conditions are independent of b_c, the transverse placement of the tip load center of mass. This makes sense, mechanically, since small deformations were assumed in the linearization process.

9.2.3 *Vibration equations*

Suppose the link undergoes harmonic vibrations in its plane of motion. Then, the total displacement can be expressed as,

$$\upsilon(x_1,t)=\upsilon(x_1)\sin(\omega t) \qquad \theta=\theta_0\sin(\omega t), \tag{9.16}$$

where ω is the frequency of vibration. The governing equations of harmonic vibration can then be written as,

$$[D(x_1)\upsilon''(x_1,t)]'' - \rho\omega^2 A(x_1)\upsilon(x_1) = 0 \tag{9.17}$$

$$I_H\upsilon'(0)\omega^2 + D(0)\upsilon(0)'' = 0 \tag{9.18}$$

with boundary conditions,

$$\upsilon(0) = 0 \tag{9.19}$$

$$D(L)\upsilon''(L) - I_p\omega^2\upsilon'(L) - M_p a_c\omega^2\upsilon(L) = 0 \tag{9.20}$$

$$[D(L)\upsilon''(L)]' + M_p\omega^2\upsilon(L) + M_p a_c\omega^2\upsilon'(L) = 0 \tag{9.21}$$

Here, the second boundary condition from Eq. (9.13) has been substituted into the torque equation. The substitution, $\lambda(x_1) = \rho A(x_1)$, has been used, where ρ is the volumetric mass density and $A(x_1)$ *is* the cross sectional area function. Now, we change to a variable, $\xi = x_1/L$ which indicates the position along the link axis in dimensionless coordinates. Then,

$$\frac{\partial}{\partial x_1} = \frac{1}{L}\frac{\partial}{\partial \xi}, \quad \frac{\partial^2}{\partial x_1^2} = \frac{1}{L^2}\frac{\partial^2}{\partial \xi^2}, \quad 0 \le \xi \le 1.$$

The equations of vibration become,

$$[D(\xi)\upsilon''(\xi)]'' - \rho L^4\omega^2 A(\xi)\upsilon(\xi) = 0 \tag{9.22}$$

$$I_H L\omega^2\upsilon'(0) + D(0)\upsilon(0)'' = 0 \tag{9.23}$$

with boundary conditions,

$$\upsilon(0) = 0 \tag{9.24}$$

$$D(1)\upsilon''(1) - I_p L\omega^2\upsilon'(1) - M_p a_c L^2\omega^2\upsilon(1) = 0 \tag{9.25}$$

$$[D(1)\upsilon''(1)]' + M_p L^3\omega^2\upsilon(1) + M_p a_c L^2\omega^2\upsilon'(1) = 0 \tag{9.26}$$

Multiplying Eq. (9.22) by $\upsilon(\xi)$, integrating by parts, and applying one boundary condition at each integration, the following equation can be derived,

$$-2M_p a_c L^2\omega^2\upsilon'(1)\upsilon(1) - M_p L^3\omega^2\upsilon(1)^2 - I_p L\omega^2\upsilon'(1)^2$$

$$-I_H L\omega^2 \upsilon'(0)^2 + \int_0^1 D(\xi)\upsilon''(\xi)^2 d\xi - \rho L^4 \omega^2 \int_0^1 A(\xi)\upsilon(\xi)^2 d\xi = 0$$

Solving this for ω^2 gives,

$$\omega^2 = \frac{\int_0^1 D(\xi)[\upsilon''(\xi)]^2 d\xi}{\rho L^4 \int_0^1 A(\xi)\upsilon^2(\xi)d\xi + LI_H\upsilon'(0)^2 + G_p} \tag{9.27}$$

where,

$$G_p = M_p L^3 \upsilon(1)^2 + 2a_c M_p L^2 \upsilon(1)\upsilon'(1) + I_p L\upsilon'(1)^2.$$

Eq. (9.27) *is* the basic equation of vibration which our system model must satisfy. Note that G_p *is* due only to the tip load, and if a tip load is absent, then $G_p = 0$.

9.3 Analytical Solution

9.3.1 *Optimality equations*

Now, the optimality equations will be derived using variational calculus, similar to that of Haftka, *et al* [14]. In order to simplify computations, the following momentarea relationship will be used,

$$I(\xi) = \gamma_p A^p(\xi) \qquad p = 1,2,3 \tag{9.28}$$

where, $p = 1 \Rightarrow$ Rectangular cross sections of uniform height
$p = 2 \Rightarrow$ Geometrically similar cross sections
$p = 3 \Rightarrow$ Rectangular cross sections of uniform width

This relation was first used by Brach [6] in the optimization of vibrating cantilevers, and is very useful in comparing different designs. Here, A *is* the cross sectional area of the link and γ_p is a constant. It should be noted that for $p = 2$, γ_p *will* depend on the shape of the cross section. For example, the geometrically similar cross section can be circular, square, or rectangular, each having its own unique moment.

Now, the nondimensional shape function is defined as,

$$\alpha(\xi)=\frac{\rho L}{M}A(\xi) \tag{9.29}$$

where M is the total link mass. The frequency equation, Eq. (9.27), then becomes,

$$\omega^2=\frac{\gamma_p EM^{p-1}}{\rho^p L^{p+3}}\left\{\frac{\int_0^1 \alpha^p(\xi)[\upsilon''(\xi)]^2 d\xi}{\int_0^1 \alpha(\xi)\upsilon^2(\xi)d\xi+M^{-1}L^{-3}\left[LI_H\upsilon'(0)^2+G_p\right]}\right\} \tag{9.30}$$

Converting parameters into their dimensionless form,

$$\lambda=\left[\frac{\rho^p L^{p+3}}{\gamma_p EM^{p-1}}\right]\omega^2 \qquad \text{[dimensionless eigenvalue]}$$

$$\eta=\frac{I_H}{ML^2} \qquad \text{[relative hub moment]}$$

$$\mu=\frac{M_p}{M} \qquad \text{[relative tip load mass]}$$

$$\zeta=\frac{a_c}{L} \qquad \text{[relative tip load placement]}$$

$$\kappa=\frac{I_p}{ML^2} \qquad \text{[relative tip load moment]}$$

Eq.(9.30) now becomes,

$$\lambda=\frac{\int_0^1 \alpha^p(\xi)[\upsilon''(\xi)]^2 d\xi}{\int_0^1 \alpha(\xi)\upsilon^2(\xi)d\xi+\eta\upsilon'(0)^2+H(1)} \tag{9.31}$$

where,

$$H(1)=2\zeta\mu\upsilon(1)\upsilon'(1)+\kappa\upsilon'(1)^2+\mu\upsilon(1)^2$$

The fundamental frequency eigenvalue, λ_f, is obtained by minimizing Eq. (9.31) over all admissible displacements,

$$\lambda_f=\min_{\upsilon(\xi)}\lambda \tag{9.32}$$

The maximum fundamental frequency eigenvalue, $\lambda_{f(\max)}$, is obtained by maximizing Eq. (9.32) over all admissible shape functions,

$$\lambda_{f(\max)} = \max_{\alpha(\xi)} \lambda_f \tag{9.33}$$

where the shape function must satisfy,

$$\int_0^1 \alpha(\xi)d\xi = 1 \quad \text{[constant volume constraint]} \tag{9.34}$$

Without this constraint, optimization would be meaningless since the volume, and likewise the mass, could be increased indefinitely making $\lambda_{f(man)}$ as large as desired. The optimum shape design problem now becomes:

Find a shape function, α_{opt} that satisfies the above constraint and maximizes the fundamental frequency eigenvalue, $\lambda_{f(\max)}$

In order to derive the optimization equations, we convert to an unconstrained problem by defining,

$$L_\upsilon = \lambda_{f(\max)} - \sigma\left\{\int_0^1 \alpha(\xi)d\xi - 1\right\} \tag{9.35}$$

which is a functional of α and υ. Here, σ is a Lagrange multiplier. Setting the total variation of L_υ to zero generates the following equations,

$$\left[\alpha^p \upsilon''\right]'' - \lambda\alpha\upsilon = 0 \qquad \text{[Dynamics Equation]} \tag{9.36}$$

$$p\alpha^{p-1}\upsilon''^2 - \lambda\upsilon^2 = \lambda\sigma \quad \text{[Optimality Equation]} \tag{9.37}$$

with boundary conditions,

$$\upsilon(0) = 0, \qquad \alpha^p(0)\upsilon''(0) + \lambda\eta\upsilon'(0) = 0 \tag{9.38}$$

$$\alpha^p(1)\upsilon''(1) - \kappa\lambda\upsilon'(1) - \zeta\mu\lambda\upsilon(1) = 0 \tag{9.39}$$

$$\left[\alpha^p(1)\upsilon''(1)\right]' + \zeta\mu\lambda\upsilon'(1) + \mu\lambda\upsilon(1) = 0 \tag{9.40}$$

It is easy to see that this solution is not unique. If $(\upsilon,\sigma,\alpha,\lambda)$ is a valid solution, then $(c\upsilon, c^2\sigma, \alpha, \lambda)$ is also a valid solution for any nonzero constant c. We can utilize this nonuniqueness in a constructive way by choosing $c = 1/\sqrt{\sigma}$. Then, the solution can be expressed as $\left(\frac{\upsilon}{\sqrt{\sigma}}, 1, \alpha, \lambda\right)$. Thus, we only need to find $\upsilon/\sqrt{\sigma}$ instead of υ and σ separately. And since the displacement function υ is not unique, we can set $\sigma = 1$, thereby simplifying the problem considerably.

Now, the optimality condition, Eq. (9.37), becomes,

$$p\alpha^{p-1}\upsilon''^2 - \lambda\upsilon^2 = \lambda \tag{9.41}$$

Eq. (9.41), along with Eqs. (9.36, 9.38, 9.39, 9.40), constitute the optimality equations upon which the following iterative solution will be based. Note that for p = 1, the shape function α disappears, and special techniques must be employed to find the solution [36].

The most important case is $p = 2$, as it gives the greatest increase in frequency in comparison with the other possible cross sectional shapes [34]. Thus, only geometrically similar cross sections will be considered in the following solutions due to the above reason and the fact that the $p = 1$ and $p = 3$ shapes and frequencies are mostly a rescaling of the $p = 2$ case.

9.3.2 *Iterative solution*

The optimality equations derived above constitute a nonlinear eigenvalue problem, which in general has no closed form solution. Thus, an iterative scheme is developed to find the numerical solution for a specified design vector, $D = [\eta, \mu, \zeta, \kappa]$ Formal integration of the dynamics equation, Eq. (9.36), yields, upon applying one boundary condition at each integration,

$$\upsilon''(\xi) = \frac{s\left[\upsilon^2(\xi)+1\right]^{\frac{p}{p+1}}}{\left|p\beta g(\xi) + p(1-\xi)^2\int_0^1 Q(x,\xi)dx\right|^{\frac{p-1}{p+1}}} \tag{9.42}$$

where s is the sign of the denominator and,

$$Q(x,\xi) = x\phi_\upsilon[\xi + x(1-\xi)]\upsilon[\xi + x(1-\xi)] \tag{9.43}$$

$$g(\xi) = \kappa\upsilon'(1) + \zeta\mu\upsilon(1) + \mu(1-\xi)(\zeta\upsilon'(1) + \upsilon(1)) \tag{9.44}$$

$$\alpha(\xi)=\frac{\phi_\upsilon}{\beta},\ \phi_\upsilon(\xi)=\left[\frac{\upsilon^2(\xi)+1}{\upsilon''^2(\xi)}\right]^{\frac{1}{p-1}},\ \beta=\left(\frac{p}{\lambda}\right)^{\frac{1}{p-1}}=\int_0^1\phi_\upsilon(\xi)d\xi \quad (9.45)$$

with boundary conditions, Eq. (9.38), expressed as,

$$\upsilon(0)=0, \qquad \upsilon'(0)=-\frac{\upsilon''(0)}{p\eta\beta|\upsilon''(0)|^{\frac{2p}{p-1}}} \quad (9.46)$$

Standard integration gives v as,

$$\upsilon(\xi)=\upsilon'(0)\xi+\xi^2\int_0^1(1-x)\upsilon''(x\xi)dx. \quad (9.47)$$

Here, the integral relations given below have been utilized, since single integration is much less computationally intensive than double integration.

$$\int_0^\xi\int_0^x G(s)dsdx=\xi^2\int_0^1(1-x)G(x\xi)dx, \quad (9.48)$$

$$\int_\xi^1\int_x^1 G(s)dsdx=(1-\xi)^2\int_0^1 xG[\xi+x(1-\xi)]dx. \quad (9.49)$$

The successive iteration scheme shown below is a variation of the one developed by Karihaloo and Niordson [16] for cantilever optimization.

1. Specify a design vector $D=[\eta,\mu,\zeta,\kappa]$
2. Select an initial β_0 and $\upsilon_0''(\xi)$.
3. Update $\upsilon_i(\xi)$ by using $\upsilon_i''(\xi)$ and β_i.
4. Update $\upsilon_i''(\xi)$ by using $\upsilon_{i+1}(\xi)$ and $\upsilon_i''(\xi)$.
5. Update β_i by using $\upsilon_{i+1}(\xi)$ and $\upsilon_{i+1}''(\xi)$
6. If a specified tolerance is not met, go to Step 3. Here,

$$\text{tolerance}=\frac{|\beta_{i+1}-\beta_i|}{|\beta_{i+1}|}+\frac{\|\upsilon_{i+1}-\upsilon_i\|}{\|\upsilon_{i+1}\|}+\frac{\|\upsilon_{i+1}''-\upsilon_i''\|}{\|\upsilon_{i+1}''\|}.$$

7. Using Eq. (9.45), $\lambda_{f(\max)}=\frac{p}{\beta^{p-1}}$ and $\alpha_{opt}=\frac{\phi_\upsilon}{\beta}$.

When the specified accuracy is met (typically .001), the shape function, α_{opt}, and the fundamental frequency eigenvalue, $\lambda_{f(\max)}$, are computed.

The actual optimum cross sectional area of the link can then be computed using Eq. (9.29). The optimum fundamental frequency of vibration, ω_{opt}, is found from the eigenvalue relation below Eq. (9.30). Cubic splines were used to approximate all functions and a recursive Simpson's formula was used for numerical integration.

In the introduction, the basic dual problems were given as:

1. Given a link mass M_o, find α_{opt} that gives $\omega_{\max}$.
2. Given a frequency ω_0, find α_{opt} that gives $M_{\min}$.

Problem 1 is based on finding a shape, α_{opt}, which gives $\lambda_{f(\max)}$ under the constraint of constant volume, V_0. Suppose there is a smaller mass, M_1, and a smaller volume, V_1, which generates the frequency λ_1. Now, the density, ρ, is given by $\frac{M_1}{V_1}$. Thus, $M_1 = \rho V_1$. Substitution of $V_1 = \int_0^L A(x_1)dx_1$ gives $M_1 = \rho \int_0^L A(x_1)dx_1$ Thus, if M_O decreases to M_1, then $A(x_1)$ must also decrease. Now, the bending rigidity, *EI,* is a function of *A* by Eq. (9.28). So, a decrease in *A will* cause a decrease in the bending rigidity, which will have the unavoidable effect of decreasing λ_f. Thus, it should be obvious that a smaller mass means a smaller fundamental frequency of vibration for links of the same length and density. All of this leads to the fact that M_O must necessarily be the minimum mass, $M_{\min}$ This implies that each optimum shape, α_{opt}, and frequency, $\lambda_{opt} = \lambda_{f(\max)}$ correspond to one unique mass, M_{opt}, which is minimum for that particular frequency. And since $\lambda_{f(\max)}$ is directly proportional to $\omega_{\max}^2$ via the relation below Eq. (9.30),

$$\omega_{\max} = \omega_0 \quad \Leftrightarrow \quad M_0 = M_{\min} \tag{9.50}$$

Note that this is true even if the optimum shape, α_{opt}, is not unique for that particular λ_{opt}. To date, the mathematical proof of the uniqueness of α_{opt} for each λ_{opt} has not been *successfully* demonstrated, although all of the numerical results points to this conclusion.

Utilizing this knowledge, relation (9.50) can be combined with the above iterative solution to Problem 1 into a simple algorithm which solves Problem 2. Here, λ_{opt} is computed using the above iterative solution. The algorithm is:

1. Set ω, ρ, L, E, and $\Xi = \lfloor I_H, M_p, a_c, I_p \rfloor$.
2. Initialize link mass, M_k.
3. Compute design vector $D = \lfloor \eta, \mu, \zeta, \kappa \rfloor$, where η, μ, and κ are functions of M_k.
4. Get λ_k, where $\lambda_k = \dfrac{C_M}{M_k}$ and $C_M = \dfrac{\rho^2 L^5 \omega^2}{\gamma_2 E}$ using the relation below Eq. (9.30).
5. Using *D*, find λ_{opt}.
6. Set error = $\left| \lambda_{opt} - \lambda_k \right|$.
7. If tolerance is not met, change mass and go to Step 3.

The minimum mass design results at the intersection of the λ_{opt} and λ_k plots versus mass. For a separate solution to Problem 2 based upon the optimality equations, see Wang and Russell [36].

9.3.3 *Numerical results and discussion*

Again, we have only considered the p = 2 case, and in particular, links with circular cross sections. The advantage of circular cross sections is that the frequency of vibration will be the same, regardless of the direction of movement in a 3-dimensional space frame. For example, a rectangular link will have different dynamical characteristics if the motion of the link is diagonal to a side as opposed to perpendicular. Circular cross sections also give a larger increment in the fundamental frequency of vibration compared with other possible shapes.

First, a general overview of the dynamical relations between $\lambda_{f(\max)}$ and the four design parameters, η, μ, ζ , and κ, will be given. It is easy to see that an increase in tip load mass, μ, will generate a larger bending moment in the link which manifests as a larger tip deflection during motion. Similarly, when the tip load moment, κ, is increased, a larger bending moment will be induced in the link, particularly towards the end of the link near the tip load. This should also increase the tip deflection during motion. Now, an increase in tip deflection implies a corresponding decrease in the fundamental frequency of vibration. So, increasing μ or κ should cause $\lambda_{f(\max)}$ to decrease. From mechanics, it should be obvious that the further away the tip load center of mass is from the endpoint of the link, the larger the induced bending moment will be, especially near the end of the link. As with the other parameters, we should expect a larger ζ to generate a smaller $\lambda_{f(\max)}$. Although, if ζ is small (on the order of 0.05), the effect should only be slight.

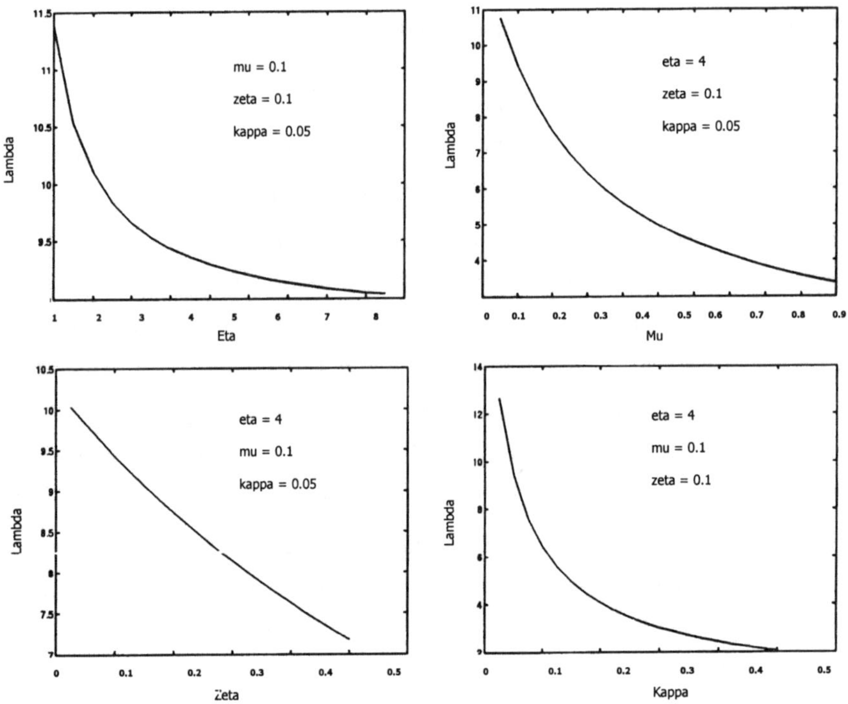

Figure 9.2: λ_{opt} vs. $D = [\eta, \mu, \zeta, \kappa]$.

Now, consider the effect of a change in hub inertia, η, on $\lambda_{f(\max)}$ large hub inertia will give the link a very stable platform. On first examination, this stability might seem to reduce vibrations and tip deflection, but in fact it has just the opposite effect. The reason for this is that a large hub inertia means that the base of the link is very stable. This very stability exacerbates tip deflection since the vibrations cannot be damped out by hub rotational vibrations. Conversely, when the hub is small and easy to rotate, it effectively damps out much of the link's vibrational modes. Thus, we should expect a smaller hub inertia to give a larger fundamental frequency of vibration. The numerical results given in Fig. 9.2 reflect the validity of these intuitive observations.

The following analysis is the result of extensive numerical simulations of different design vectors, *D,* using the above successive iteration scheme. The convergence of this method is very quick for large hub inertias *(η> 4).* Usually only 5 to 8 iterations at 10 seconds per iteration on *Matlab)* are required. For small hub inertias (η < 1), the required number of iterations increases dramatically, and if a large tip load is also present, it may not converge at all. These facts, along with the dynamics analysis given above, serve to point out that the hub inertia is very important to the optimum shape design. This is substantiated in the numerical results, as evidenced in Fig. 9.2. We also see that λ_{opt} is relatively insensitive to variations in normal tip load placements (ζ< 0.1). A thorough sensitivity analysis of each design parameter will be presented in Section 9.6.

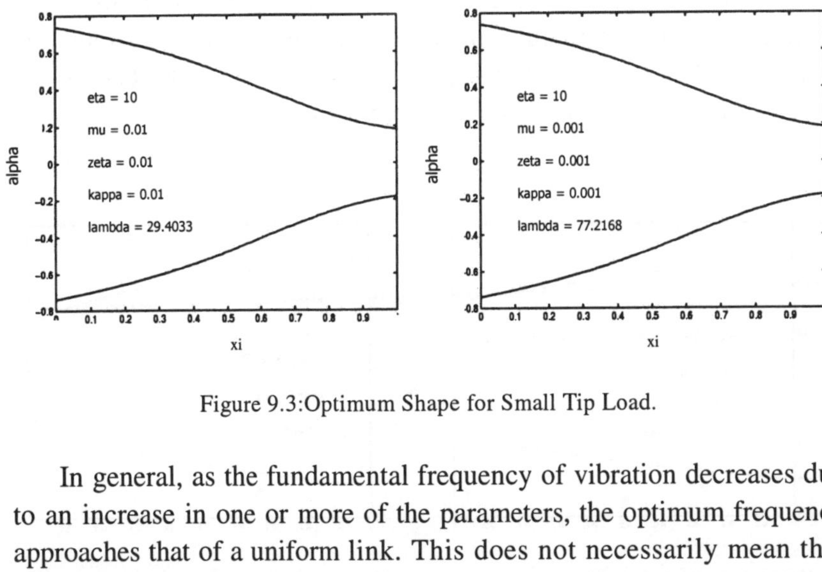

Figure 9.3:Optimum Shape for Small Tip Load.

In general, as the fundamental frequency of vibration decreases due to an increase in one or more of the parameters, the optimum frequency approaches that of a uniform link. This does not necessarily mean that the optimum shape becomes uniform. An example of this is μ, which generates a large amount of tapering near the tip load of the optimum shape for large parameter values (see Fig. 9.4). Also note the almost uniform shape generated by κ = 1.0. The reason for this can be explained from dynamics principles. When the manipulator slows down or speeds up, the large tip load moment will generate a large bending force near the end of the link. This would increase flexibility and thus lower the frequency. To compensate, the optimization scheme puts more mass here.

In Table 9.1, the gain in λ_f of the optimized link versus the uniform link is given. Note that although λ_{opt} decreases with η, the gain in λ_f actually increases. Thus, a smaller hub will give a smaller advantage in frequency to the optimized link. When κ = 0.5, there is very little gain in frequency since the shape becomes practically uniform. This serves to point out an important fact of optimum shape design, and that is, if the manipulator will be handling tip loads with relatively large moments, then it will most likely be more economical to use a less expensive uniform link. The largest gain, both here and in general, will be realized when the tip load parameters are small ($\mu = \kappa \leq 0.05$). As this is the case with many robotic applications, there will be many possibilities for optimum shape design in commercial and industrial situations. For example, when $\mu = \zeta = \kappa = 0.001$, the gain in ω over that of the corresponding uniform link is 625% (see Fig. 9.3).

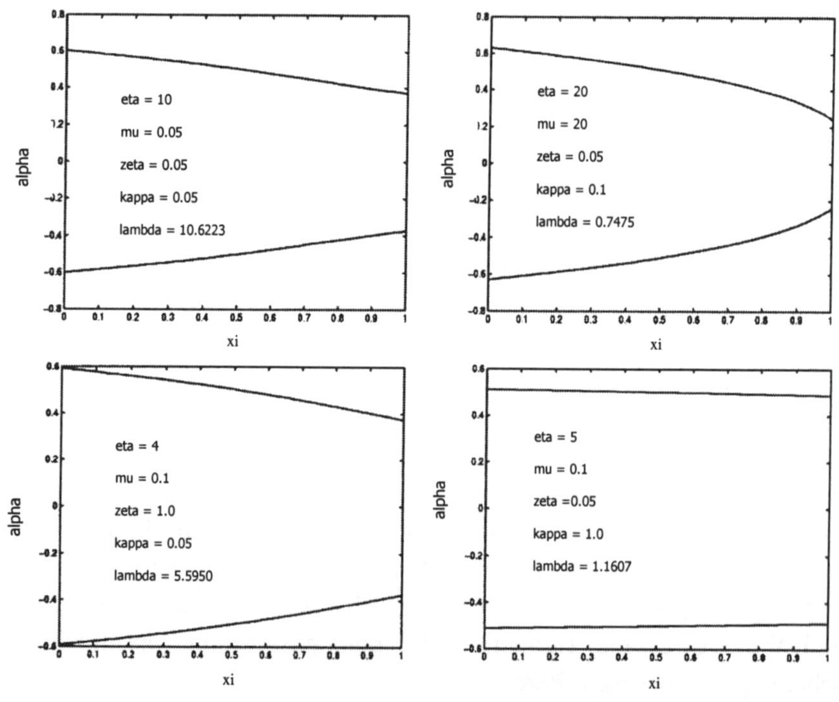

Figure 9.4:Optimum Shapes for large η, μ, ζ , and κ.

Table 9.1: Ratio of $\lambda_{opt} / \lambda_u$

	$(\mu,\zeta,\kappa)=$ 0.10,0.1,0.05)	$(\mu,\zeta,\kappa)=$ 0.05,0.1,0.05)	$(\mu,\zeta,\kappa)=$ 0.10,0.1,0.50)	$(\mu,\zeta,\kappa)=$ 0.50,0.1,0.10)
η=1.2	3.6704	3.8694	1.0843	1.8172
η=2	4.0219	4.5408	1.1602	1.9821
η=4	4.4335	5.0578	1.1959	2.0775
η=8	4.6478	5.335	1.2066	2.0961

9.4 Optimization Approach

9.4.1 *Segmentized formulation*

Much of the work being done on the design of optimum controllers for flexible manipulators is also based upon the fundamental frequency of vibration. Currently, this frequency has to be approximated using modal shape functions developed for uniform links. But since the links are nonuniform, there is an inherent inaccuracy in the design.

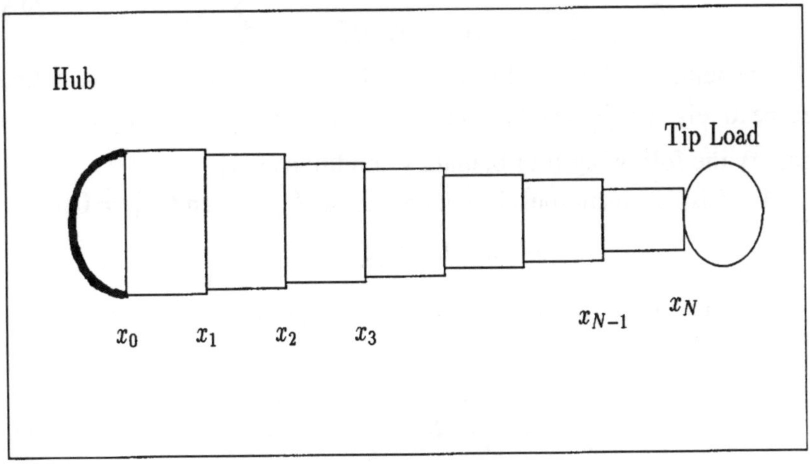

Figure 9.5: Design Model of Segmentized Link

An efficient method to find the fundamental frequency of vibration to any required degree of accuracy would result in a considerable improvement of optimum controllers. The analytical solution of Section 3 is for the unconstrained design. But, we would like a model to which any practical constraint could be applied. Also, we would like to have a more formal optimization method which allows the utilization of mathematical programming techniques. For these reasons, an innovative design method has been developed based on a segmentized solution of the vibration equations.

The flexible link is approximated by N discrete segments, each having a constant cross sectional area. A model of this design is shown in Fig. 9.5. The link, of length *L*, is rigidly attached to a hub of inertia I_H. Here, x is the link axis, and ω is the transverse displacement of the link's neutral axis from the *x*-axis.

Define local coordinates by:

$$\xi_i = \frac{x - x_{i-1}}{\Delta_i}, \qquad \Delta_i = x_i - x_{i-1}, \qquad x_{i-1} \le x \le x_i. \tag{9.51}$$

where,

$$\frac{\partial}{\partial x} = \frac{1}{\Delta_i}\frac{\partial}{\partial \xi_i}, \qquad \frac{\partial}{\partial x^2} = \frac{1}{\Delta_i^2}\frac{\partial^2}{\partial \xi_i^2}, \qquad \sum_{i=1}^{N} \Delta_i = L. \tag{9.52}$$

The bending rigidity and linear mass density of segment i are defined, respectively, as $\beta_i = (EI)_i$ and $\gamma_i = (\rho A)_i$. Now each segment must satisfy the following four boundary conditions [31]:

1. Displacement must be continuous at $\xi_{i-1} = 1$ and $\xi_i = 0$.

$$\upsilon_{i-1}(\xi_{i-1}) = \upsilon_i(\xi_i) \tag{9.53}$$

2. The first derivative must be continuous at $\xi_{i-1} = 1$ and $\xi_i = 0$.

$$\frac{1}{\Delta_{i-1}}\frac{\partial \upsilon_{i-1}(\xi_{i-1})}{\partial \xi_{i-1}} = \frac{1}{\Delta_i}\frac{\partial \upsilon_i(\xi_i)}{\partial \xi_i} \tag{9.54}$$

3. Bending moments must be equal at $\xi_{i-1} = 1$ and $\xi_i = 0$.

$$\frac{\beta_{i-1}}{\Delta_{i-1}^2}\frac{\partial^2 \upsilon_{i-1}(\xi_{i-1})}{\partial \xi_{i-1}^2}=\frac{\beta_i}{\Delta_i^2}\frac{\partial^2 \upsilon_i(\xi_i)}{\partial \xi_i^2} \tag{9.55}$$

4. Shearing forces must be equal at $\xi_{i-1}=1$ and $\xi_i=0$.

$$\frac{\beta_{i-1}}{\Delta_{i-1}^3}\frac{\partial^3 \upsilon_{i-1}(\xi_{i-1})}{\partial \xi_{i-1}^3}=\frac{\beta_i}{\Delta_i^3}\frac{\partial^3 \upsilon_i(\xi_i)}{\partial \xi_i^3} \tag{9.56}$$

Using the dynamics equation of vibration, Eq. (9.22),

$$\frac{\partial^4 \upsilon_i(\xi_i)}{\partial \xi_i^4}-k_i^4\upsilon_i(\xi_i)=0, \quad \text{where} \quad k_i^4=\frac{\gamma_i\Delta_i^4}{\beta_i}\omega^2. \tag{9.57}$$

The solution is given by,

$$\upsilon_i^{(j)}(\xi_i)=\phi_{ij}C_i \tag{9.58}$$

where,

$$C_i=[C_{0i} \quad C_{1i} \quad C_{2i} \quad C_{3i}]^T \qquad [C_{mn} = \text{Constant}]$$

$$\phi_{i0}=[\sin(k_i\xi_i) \quad \sinh(k_i\xi_i) \quad \cos(k_i\xi_i) \quad \cosh(k_i\xi_i)]$$

$$\phi_{i1}=\phi_{i0}' \qquad \phi_{i2}=\phi_{i1}' \qquad \phi_{i3}=\phi_{i2}' \qquad \phi_{i4}=k_i^4\phi_{i0}.$$

The torque equation and boundary conditions, Eqs.(9.23-9.26), become:

$$[I_H\omega^2\Delta_1\phi_{11}(0)+\beta_1k_1\phi_{12}(0)]C_1=0, \qquad \phi_{10}(0)C_1=0 \tag{9.59}$$

$$[\beta_Nk_N^2\phi_{N2}(k_N)-I_p\omega^2\Delta_Nk_N\phi_{N1}(k_N)-a_cM_p\omega^2\Delta_N^2\phi_{N0}(k_N)]C_N=0 \tag{9.60}$$

$$[\beta_Nk_N^3\phi_{N3}(k_N)+a_cM_p\omega^2k_N\Delta_N^2\phi_{N1}(k_N)+M_p\omega^2\Delta_N^3\phi_{N0}(k_N)]C_N=0 \tag{9.61}$$

Reformulating the interfacial relations, Eqs. (9.53-9.56), into transfer matrices,

$$C_i=\psi^{-1}M_{i-1}C_{i-1} \tag{9.62}$$

where,

$$\psi = \begin{bmatrix} 0 & 0 & 1 & 1 \\ 1 & 1 & 0 & 0 \\ 0 & 0 & -1 & 1 \\ -1 & 1 & 0 & 0 \end{bmatrix}, \quad M_j = \begin{bmatrix} sn_j & sh_j & cs_j & ch_j \\ \tau_j cs_j & \tau_j ch_j & -\tau_j sn_j & \tau_j sh_j \\ -\gamma_j sn_j & \gamma_j sh_j & -\gamma_j cs_j & \gamma_j ch_j \\ -\theta_j cs_j & \theta_j ch_j & \theta_j sn_j & \theta_j sh_j \end{bmatrix} \tag{9.63}$$

$$sn_j = \sin(k_j \xi_j), \qquad sh_j = \sinh(k_j \xi_j), \qquad cs_j = \cos(k_j \xi_j),$$
$$ch_j = \cosh(k_j \xi_j)$$

$$\tau_j = \frac{k_j}{k_{j+1}}, \qquad \delta_j = \frac{\Delta_{j-1}}{\Delta_j}, \qquad T_j = \frac{\beta_j}{\beta_{j-1}}, \qquad \gamma_j = \frac{\tau_j^2}{\tau_{j+1}\delta_{j+1}^2},$$

$$\theta_j = \frac{\tau_j^3}{\tau_{j+1}\delta_{j+1}^3}.$$

Relating the interior coefficient vectors, Eq. (9.62),

$$C_i = \phi_{i-1} C_{i-1}, \qquad \phi_i = \psi^{-1} M_i, \tag{9.64}$$

the endpoint coefficient vector relation can now be expressed as,

$$C_N = \phi C_1, \quad \text{where} \quad \phi = \prod_{N-1}^{1} \phi_i. \tag{9.65}$$

Substitution of *CN* into the boundary equations, Eqs. (60, 61), gives,

$$QC_1 = 0 \tag{9.66}$$

where,

$$Q = \begin{bmatrix} 0 & 0 & 1 & 1 \\ I_H \Delta_1 \omega^2 & I_H \Delta_1 \omega^2 & -\beta_1 k_1 & \beta_1 k_1 \\ & R_1 \phi & & \\ & R_2 \phi & & \end{bmatrix} \tag{9.67}$$

Here, R_1 and R_2 are the matrix coefficients of C_N in Eq. (9.60, 9.61), respectively. Requiring the determinant of Q to be zero gives the frequency of the link. The maximum speed design problem becomes,

$$\max \omega(X_\beta, X_\gamma; \Xi) \quad \text{subject to:} \tag{9.68}$$

$$\Delta X_\gamma^T = M_T, \qquad X_\gamma^L \le X_\gamma \le X_\gamma^U, \qquad X_\beta^L \le X_\beta \le X_\beta^U$$

where $X_\beta = [\beta_1, \ \beta_2, \ \cdots \ \beta_N]$, $X_\gamma = [\gamma_1, \ \gamma_2, \ \cdots \ \gamma_N]$, $\Xi = [I_H, \ M_p, \ a_c, \ I_p]$, $\Delta = [\Delta_1, \ \Delta_2, \ \cdots \ \Delta_N]$, and M_T is the total link mass.

9.4.2 *Segmentized optimization solutions*

The solution to the segmentized formulation can be achieved through the use of a number of mathematical programming methods. Here, two particular methods are presented: a dynamic programming (DP) algorithm and an adaptive random search algorithm (IHR).

9.4.2.1 Dynamic programming solution

The DP algorithm has been developed explicitly for this problem, considering the previous optimum results in Section 9.3. Knowledge of the optimum shapes generated by the iterative solution can simplify the algorithm considerably. Perhaps the most useful observation is that the optimum shape is monotonically decreasing towards the tip load. Of course, the constant volume constraint must always be satisfied.

These heuristics have been successfully integrated into a DP algorithm where the stages are represented by the number of segments in the current design, and the states at each stage are all admissible combinations of segment areas [48]. The general algorithm is presented below,

1. Set the tolerance of ω_{opt}.

2. Set the link parameters to the desired values.

3. Set the number of links, N, to 1.

4. Find $\omega_{opt}(1)$.

5. Set $N = 2N$.

6. Find $\omega_{opt}(N)$.

7. If tolerance is met, stop. Else, go to Step 5.

Here, Step 4 finds the uniform link frequency corresponding to the prescribed parameter values using equal segment lengths. Step 6 is obviously the crucial element in the algorithm. First, consider a 2-segment link. Since we know that in the optimum design, the area of segment 1 must be greater than or equal to that of segment 2, we begin by incrementing the area of segment 1 and decrementing the area of segment 2 by the same amount. This way the total volume will stay constant as required. So, the areas are changed slowly until a peak in ω is reached. For a 2-segment design, this will be optimal. If the tolerance of ω_{opt} has not been reached, the link will be divided into 4 segments. Then, segments 3 and 4 are optimized, after which segments 1 and 2 are optimized. One of the drawbacks to the above algorithm is that the number of links must double on each iteration of Steps 5 through 7, since an even number of equal segments cannot be split into an odd number of equal segments.

Here, a constant ρ and E are assumed. Also, the moment-area relation, Eq. (9.28), for circular cross sections has been used. In Step 4, area $A = M_T / (\rho L)$. In succeeding steps, A is kept constant by the balanced increments described above. This will ensure a constant volume. The above algorithm works very well because the fundamental frequency of vibration, ω, is a smooth function of area increments, as can be seen in Fig. 6. The figure on the left results from splitting a 2-segment link, while the figure on the right is due to the splitting of segments 7 and 8 in an 8-segment design.

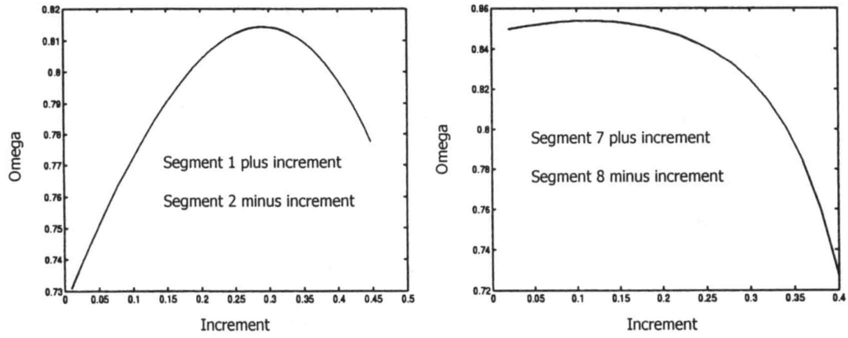

Figure 9.6: ω vs. Segment Increments

9.4.2.2 Adaptive random search solution

Now, the IHR algorithm will be considered. This algorithm is a slight modification of the Improving Hit and Run algorithm recently developed by Zabinsky, *et al* [49]. The modification of the algorithm was necessary to ensure a constant volume during the optimization process. The modified IHR algorithm is:

1 Set design vector $\Xi = [I_H, M_P, a_c, I_P]$ and number of segments, *N*. Set step multiplier MUL to 1 and FACTOR to 8.
2 Calculate ω_0 (frequency for uniform link).
3 Set minimum and maximum area constraints, A_{min} and A_{max}. Set $j = 1$.
4 Randomly select *N/2* of the *N* segments and mark them with a 1 in vector D_j of length *N*. Mark the remaining *N/2* segments with a -1. If *N* is odd, one randomly selected area will remain constant and will be marked by a 0.
5 Get *N/2* samples from a *N(0,1)* normal distribution and place them in each position of D_j where there is a 1. Place the negatives of this same sample in each position of D_j*;* where there is a-1. This arrangement will ensure a constant volume during optimization. Here, D_j is called the direction vector.

6 Generate a step size, S_j, uniformly from L_j, the set of feasible step sizes in the direction D_j, where,

$$L_j = \{S_j \in \Re : A_{min} < A_j + S_j D_j < A_{max}\}$$

If $L_j = 0$, go to Step 4.

7 Set $S_j = S_j * \text{MUL}$ while $|\text{MUL}| < 1$.

8 Update the area vector, A_j, *if* the frequency is improving:

$$A_{j+1} = \left\{ \begin{array}{ll} A_j + S_j D_j & \text{if } \omega(A_j + S_j D_j) > \omega_j \\ \hline A_j & \text{otherwise} \end{array} \right\}$$

Set $\omega_{j+1} = \omega(A_{j+1})$.

9 If $\omega_{j+1} > \omega_j$, set $\text{TRIM} = \dfrac{|\omega_{j+1} - \omega_j|}{\omega_j} * \text{FACTOR}$ and

$\text{MUL} = \text{MUL} - \text{MUL} * \text{TRIM}$.

10 If the stopping criteria is met, stop. Otherwise, increment j and go to Step 4.

Notice that the area function, A_j, is not updated unless the frequency is improving. Also, the step size, S_j, is adapted to the proportional increase in frequency when ω is improving through the use of MUL which is set in Step 9. Thus, as the segmentized shape approaches its optimum, the segment areas will change less and less. This is highly desirable since it eliminates the high fluctuations in the random direction vector, D_j, as the design approaches optimality.

9.4.3 *Numerical results and riscussion*

In Fig. 9.8, the optimum shapes and frequencies are given for a 1-,2-,4-, and 8-segment design. The segmentized shapes were produced by the above DP algorithm with design vector, $\Xi = [10,10,0.05,0.1]$. The corresponding optimum shape from Section 9.3 has been overlaid for comparison purposes. This particular example was chosen because of its highly nonlinear shape.

Here, with only 8 segments, the segmentized frequency is within 1% of optimum! Generally, most designs can be approximated very closely with only a few segments $(N < 10)$.

One obvious goal is to have the analytic design of Section 3 and the optimization design presented above agree with each other, i.e. the same shape should produce a similar frequency. To accomplish this comparison, the optimum shape, α_{opt}, and frequency, λ_{opt}, were computed using the iterative solution of Section 3 for two different designs as specified in Fig. 7. Then, α_{opt} was converted to an area function using Eq. (9.29), which was then used to approximate the area of each segment at its midpoint. The segmentized frequency was then computed for N = 1 to 25. The results can be seen in Fig. 9.7. Here, the first frequency, N = 1, is for the uniform link corresponding to the chosen design parameters. It is also apparent here that the segmentized frequency rapidly approaches that of the analytic optimum.

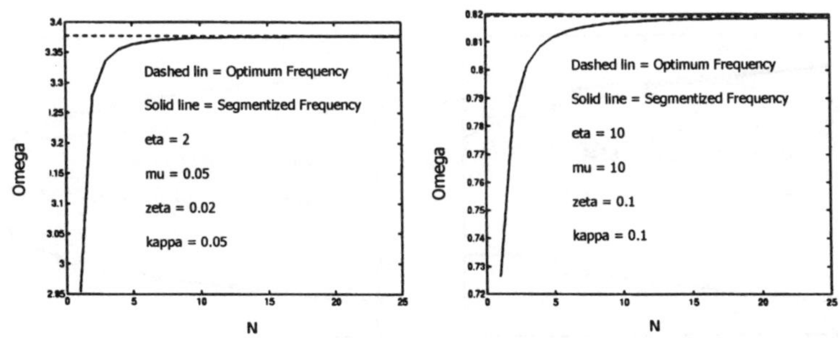

Figure 9.7: ω_{opt} vs. N

The IHR results are also very impressive. Obviously, the number of required iterations will be somewhat higher than that of the DP algorithm, depending on the number of segments in the design. One advantage of the IHR algorithm is that it is a truly impartial optimization process. That is, the segment area increments are chosen randomly in each iteration. Thus, this gives us yet one more valuable tool in comparing the merit of the other optimization processes.

The IHR algorithm has been tested on a 2-, 4-, and 8-segment design, and the results are very good. The average number of iterations needed to achieve a 1 % accuracy compared with the DP solution of Fig. 8 is only 6 for a 2-segment design. For a 4-segment design, the required number of iterations is 27. And for an 8-segment design, the required number of iterations is 138. Fig. 9.9 gives the results of the IHR solution using the same design vector as the DP solution shown in Fig. 9.8.

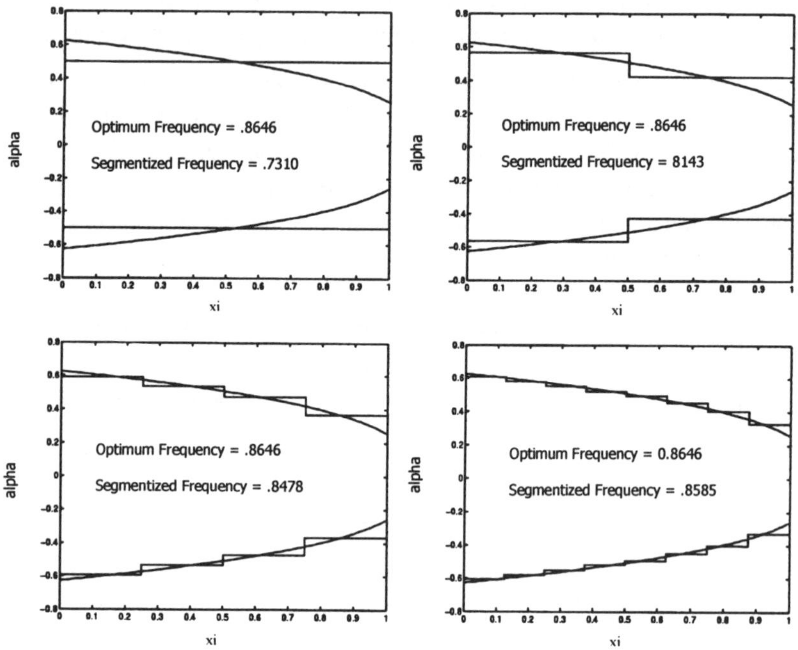

Figure 9.8: DP Optimum Shapes for $N = 1, 2, 4, 8$

Both the DP and IHR algorithms lend themselves readily to the incorporation of constraints into the design. Two common constraints are maximum and minimum area. Indeed, these are required in the IHR algorithm to compute the step size. One important constrained design is that of the circular tunnel cross section. This shape gives a very high rigidity factor with a smaller total mass. An added bonus is that the arm is hollow so that control cables and even mechanical drives can be located here inside the link. This is also the most commonly used design in the new composite links which promise to be exceptionally useful. For optimization results of the circular tunnel cross section and composite design implementations, see Russell [28].

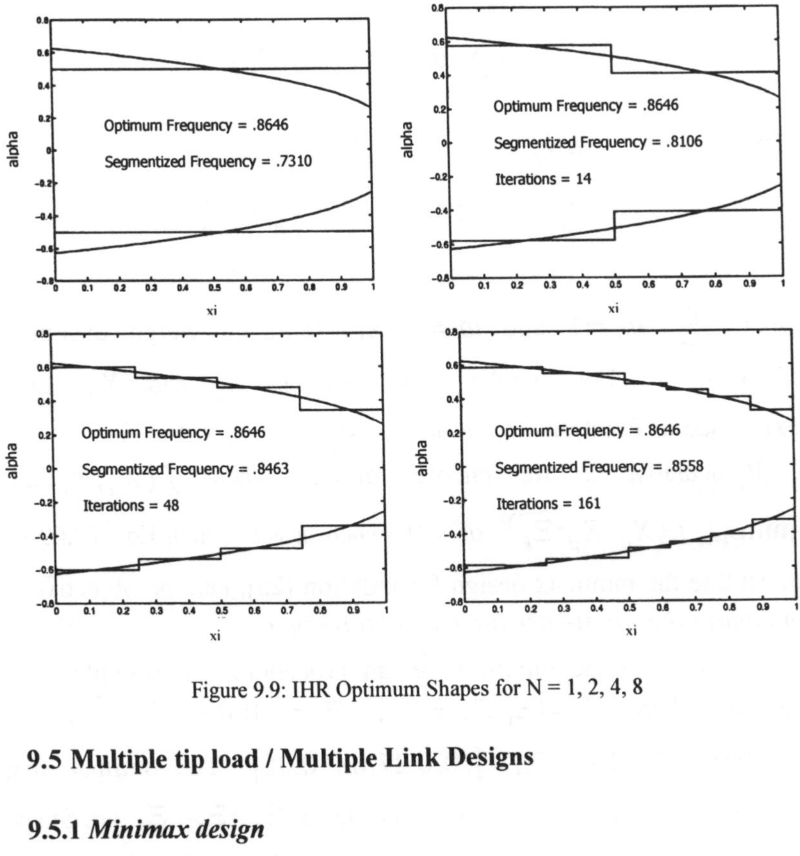

Figure 9.9: IHR Optimum Shapes for N = 1, 2, 4, 8

9.5 Multiple tip load / Multiple Link Designs

9.5.1 *Minimax design*

In most cases, a flexible manipulator has to perform tasks in different situations, particularly with different tip loads. This is reflected by different values of vector Ξ. Results from Section 9.3 indicate that the fundamental frequency of vibration varies dramatically with a change in tip load parameters. This naturally leads to the question of what is an "optimum" design for a manipulator dealing with multiple task situations. Consider the case of a finite number of task situations, i.e., $\Xi \in \{\Xi_1, \ldots, \Xi_n\}$. Ξ is defined below Eq. (9.68).

Using the segmentized formulation of Section 4, the maximum speed design problem for multiple tip loads can be formulated as,

$$\max_{Z} \sum_{k=1}^{n} \omega_k f\left(X\gamma, X_\beta, \Xi_k\right) \tag{9.69}$$

subject to $\Delta X_\gamma{}^T = \gamma_s, \qquad Z^L \le X_\beta \le X_\beta^U$

$$X_\gamma^L \le X_\gamma \le X_\gamma^U, \; X_\beta^L \le X_\beta \le X_\beta^U \tag{9.70}$$

where $\sum_{k=1}^{n} \omega_k = 1$, and $\omega_i = 1$ represents the weight assigned to situation i. For circular cross sections, X_γ and X_β can be expressed as functions of an independent design vector, Z.

Particularly, if we choose $\omega_i = 1$ when $f\left(X_\alpha, X_\beta; \Xi_k\right) = \min_{1 \le k \le n} f\left(X_\alpha, X_\beta; \Xi_k\right)$ $\omega_i = 0$ when $k \ne i$, then Eqs. (9.69-9.70) constitute the minimax design formulation [25], i.e., the objective is to *maximize the worst-case fundamental frequency.*

To illustrate the minimax design procedure, an example will be solved where $\Xi = [\Xi_1, \Xi_2, \Xi_3]$, $\Xi_1 = [10, 0.1, 0.1, 0.1]$, $\Xi_2 = [10, 0.01, 0.1, 0.15]$, $\Xi_3 = [10, 0.27, 0.2, 0.05]$. The optimal shapes for the 2-segment design corresponding to Ξ_1, Ξ_2, Ξ_3 are Shape 1, Shape 2, and Shape 3. These are shown in Fig. 9.10. $Shape_M$ is the minimax design shape. This example is particularly illustrative because the three shape frequencies can be plotted versus the segment increment (see Fig. (9.11). The minimax design is specified by the intersection of the ω_2 and ω_3 plots.

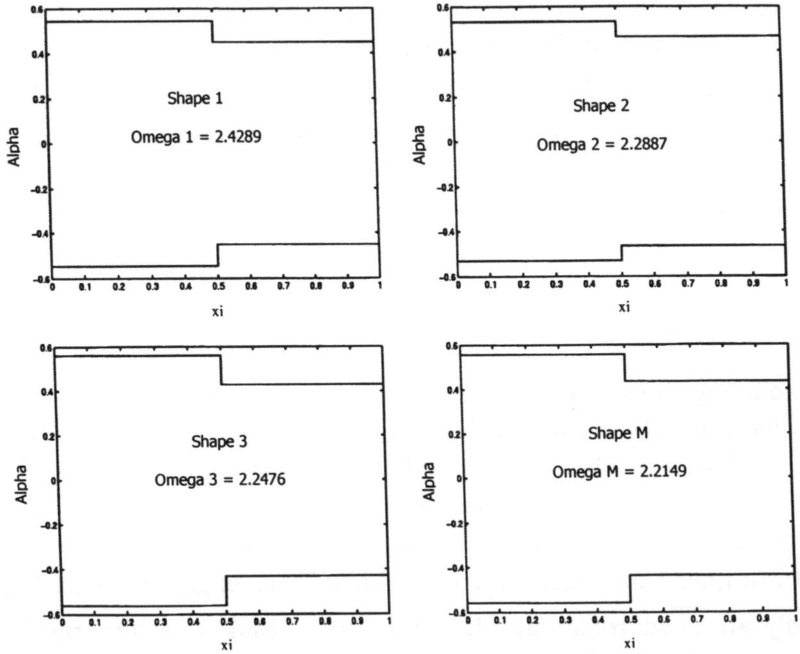

Figure 9.10: Optimum Shapes for Minimax Design

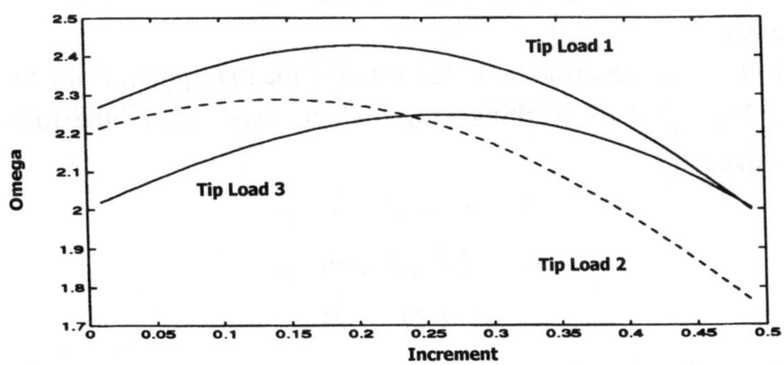

Figure 9.11: Omega vs. Increment.

Note that the frequency of the minimax design will always be less than or equal to the maximum of the shape with the minimum frequency. Equality only holds when the frequencies of the other shapes lie above this lowest frequency. When using the minimax design in practice, the average tip load will be specified along with upper and lower bounds on the usable range. These values will be used to select the weights, ω_i, with the weight corresponding to the average expected tip load being the largest. Once the minimax design vectors and weights have been specified, either the DP or IHR algorithm can be used to find the solution to Eqs. (9.69, 9.70) by searching over all combinations of shapes and design vectors.

9.5.2 *Two-link design*

Since a single-link manipulator has few practical applications, a two-link flexible manipulator will now be considered. The optimization model will be based on the single-link solutions presented in Section 9.3. Fig. 9.12 gives a view of the design model which is composed of two flexible links, Link 1 and Link 2, having respective masses M_1 and M_2. Link 1 has length L_1, while Link 2 has length L_2. Link 1 is rigidly attached to a hub, H_1, at its base. Link 2 is also rigidly attached to a hub, H_2, at its base, and this same hub is attached to an axis at the endpoint of Link 1. The working tip load is positioned at the endpoint of Link 2 with the normal tip load parameters given previously.

The general specifications for a two-link manipulator are its total mass, M_T, and its working range which must satisfy the following requirements,

$$L_1 + L_2 = MaxRange$$

$$|L_1 - L_2| \leq MinRange$$

$$M_1 + M_2 = M_T$$

Note, that although the total mass and working range are given as specifications, there is a choice of the mass and length distribution between the two links. Thus, both the mass and length distribution become design variables in the optimization process. Of course, it is easy to place constraints on these variables in this optimization process.

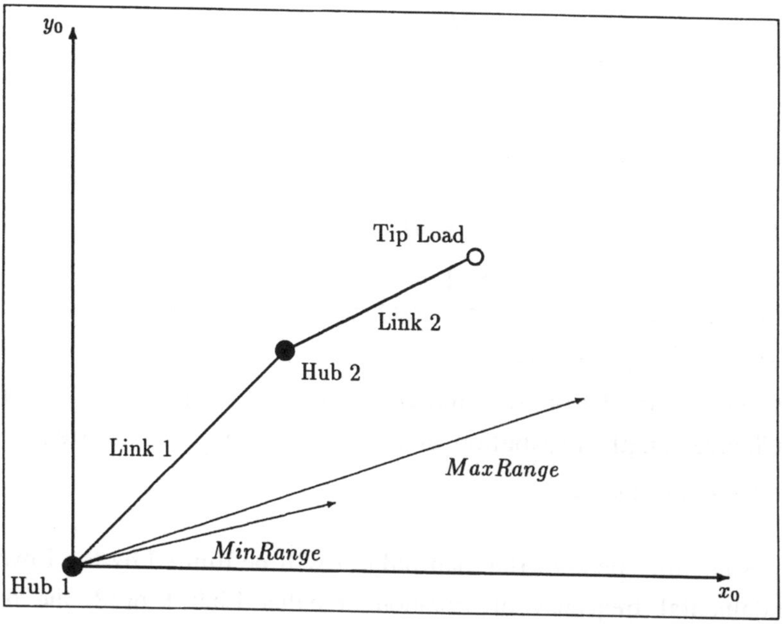

Figure 9.12: Two-Link Design Model of Flexible Manipulator.

The tip load and hub parameters for Link 1 are specified by,

$$\eta_1 = \frac{I_{H1}}{M_1 L_1^2},\ \mu_1 = \frac{M_2 + M_P + M_{H2}}{M_1},\ \zeta_1 = \frac{L_2}{2L_1}, \tag{9.71}$$

$$k_1 = \frac{1}{12} M_2 L_2^2 + \left(\frac{L_2}{2}\right)^2 M_P + \left(\frac{L_2}{2}\right)^2 M_{H2} \tag{9.72}$$

where M_{H2} is the mass of Hub 2. The moment, k_1, is approximated by the moment of inertia of Link 2, including Hub 2 and the tip load, about its center of mass which was assumed to be at $\frac{L_2}{2}$. The center of mass could be more accurately calculated numerically by using M_2 and the current shape function, α, for Link 2. But, this would drastically increase the computation time which is already large. The tip load and hub parameters for Link 2 are specified by,

$$\eta_2 = \frac{I_{H2}}{M_1 L_2^2}, \quad \mu_2 = \frac{M_P}{M_2}, \quad \zeta_2 = \frac{a_c}{L_2}, \quad k_2 = \frac{I_P}{M_2 L_2^2} \tag{9.73}$$

The optimization scheme is:

1 Initialize $L_1 = L_2$ and $M_1 = M_2$.
2 Optimize Link 2 using its tip load and Hub 2.
3 Optimize Link 1 using Hub 1 and Link 2 as its tip load.
4 Error=$|Freq1 - Freq2|$.
5 Change mass distribution and go to Step 2 if tolerance is not met.
6 Change length distribution and reset $M_1 = M_2$. Go to Step 2 if tolerance is not met.

Since the speed of the manipulator will be limited by the lowest fundamental frequency of vibration of either Link 1 or 2, the best possible system frequency occurs when $\omega_1 = \omega_2$. As can be expected from the previous analysis in Section 9.3, the large values of μ_1, ζ_1, and k_1 will require more mass to placed in Link 1 when $L1 = L2$. We could also expect Link 1 to be shorter in length than Link 2. Both a shorter length and a higher mass would increase the fundamental frequency of vibration of Link 1.

In the following results, it should be remembered that the analytical optimum of Section 9.3 is based on small deformations. Thus, this optimization scheme yields only an approximation of the actual governing system fundamental frequency of vibration. The amount of variation will depend upon material characteristics, as well as the average relative angular position of Link 2 relative to link 1. Fig. 9.13 gives the optimum design for a specified total mass and working range. The following values were used in the optimization process: *MaxRange* = 1 *m*, Minrange = 3.m, M_T = 10 kg, I_{H1} = 20 kg, I_{H2} = 4 kg, $E = 6.06\ E10 N/m^2$, density = 70 kg/m^3, and $\Xi = [20, 0.5, 0.02, 0.5]$.

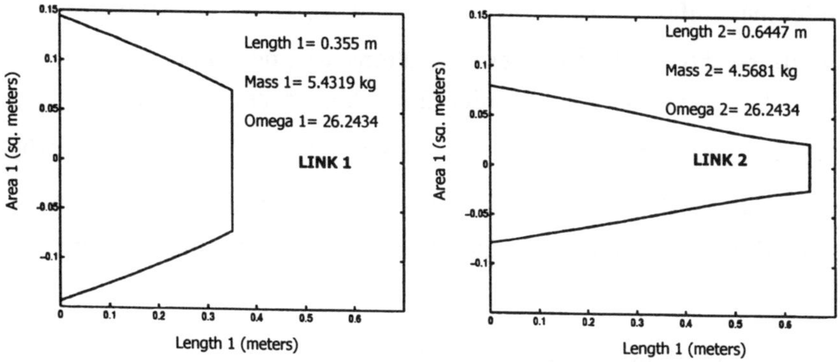

Figure 9.13: Optimum Shape for Two-Link Design.

9.6 Sensitivity Analysis

A variation in the hub inertia, or more likely, in the tip load will induce a corresponding change in the frequency of the link. For the purpose of link design, it is extremely important to know how sensitive the optimal frequency is with respect to changes in the design vector, D. For the optimal fundamental frequency λ_{opt}, from Eq. (9.33), a variation of λ_{opt} due to small changes of η, μ, ζ and k can expressed as,

$$\frac{\delta\lambda_{opt}}{\lambda_{opt}} = -S_\eta \frac{\delta\eta}{\eta} - S_\mu \frac{\delta\mu}{\mu} - S_{\mu\zeta} \frac{\delta(\mu\zeta)}{\mu\zeta} - S_k \frac{\delta k}{k}, \tag{9.74}$$

where sensitivity indices can be calculated according to,

$$\left(S_\eta, S_\mu, S_{\mu\zeta}, S_k\right) = \frac{\left(\eta\upsilon'(0)^2, \mu\upsilon(1)^2, 2\zeta\mu\upsilon(1)\upsilon'(1), k\upsilon'(1)^2\right)}{\int_0^1 \alpha\upsilon^2 d\xi + \eta\upsilon'(0)^2 + \mu\upsilon(1)^2 + 2\zeta\mu\upsilon(1)\upsilon'(1) + k\upsilon'(1)^2}$$

(9.75)

Note that although $\delta\eta$, $\delta\mu$, $\delta\zeta$, and δ_k generate corresponding variations $\delta\upsilon$ and $\delta\alpha$, these variations will not affect the value of λ_f when small terms of order 2 or higher are ignored, since λ_f arrives at one of its stationary values at υ_{opt} and α_{opt}. This is why we do not need to consider the variations of υ and α. It should be pointed out that sensitivity indices can also be formally defined as follows,

$$S_x = -\frac{\%\text{ change in }\lambda}{\%\text{ change in }x} = -\frac{\delta\lambda/\lambda}{\delta x/x} = -\frac{x}{\lambda}\frac{\delta\lambda}{\delta x}, \quad x=\eta,\mu,\mu\zeta,k \tag{9.76}$$

Clearly, from Eq. (9.75), $0 \le S_x \le 1$; therefore, *the vibration frequency always decreases as system parameters increase in value,* a conclusion which has been verified by numerical analysis (see Fig. 9.2). Using Eqs. (9.74,9.75), one can find the actual numerical values of sensitivity indices for any particular optimum shape design. In this section, numerical analysis is used to investigate the sensitivity of an optimum design to variations in $D=[\eta,\mu,\zeta,k]$.

In Fig. 9.14, the frequency response to variations in particular design parameters is given. The optimum design vector is $D=[4,0.1,0.1,0.05]$ which generates an optimum frequency, $\lambda_{out}=9.4331$. Note that the frequency is particularly sensitive to η and k. It is somewhat less sensitive to μ and almost completely insensitive to ζ.

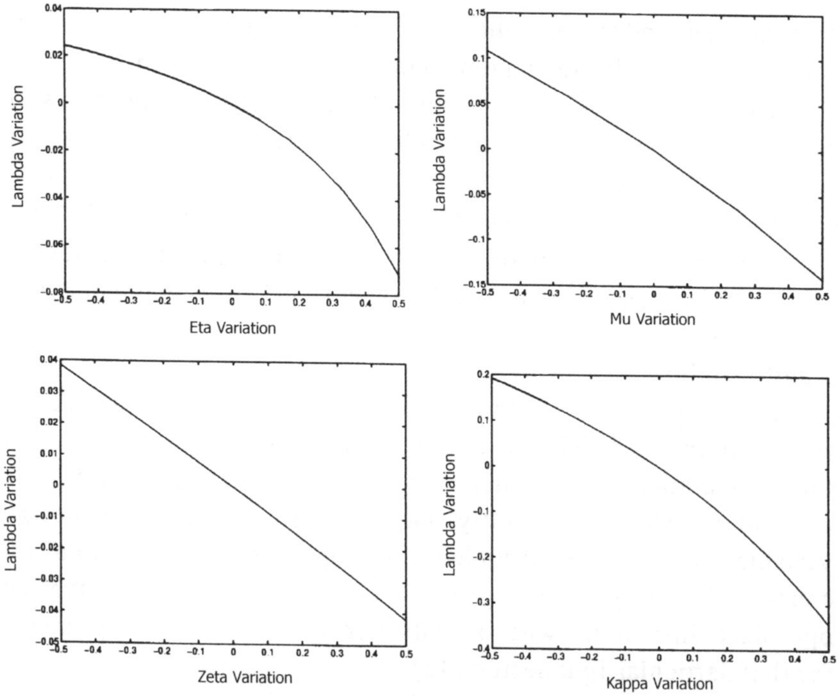

Figure 9.14: $\delta\lambda/\lambda_{out}$ Vs. $\delta X/X_{out}$

9.7 Conclusion

In this chapter, the problem of optimum shape design of flexible manipulators has been fully examined from two major directions: an analytical approach and an optimization approach. Both approaches utilized the Euler-Bernoulli equations of motion which were derived using the variational method. These equations were then used to derive the equations of vibration which govern the vibrational characteristics of all flexible manipulators.

In the analytical approach, the optimality equations were used to develop an iterative scheme which generated the optimum shape and frequency for a specified design vector. This is the solution to Problem 1 of Section 9.1. Here, only circular cross sections were considered due to their high frequency and varied applications. The two basic design problems as defined in Section 1 were shown to be dual to each other, and these duality requirements were then converted into an algorithm to solve the minimum mass problem (Problem 2).

Then, the numerical results of the analytic solution were presented and analyzed from an intuitive mechanical viewpoint. An increase in any design parameter was shown to generate a decrease in the fundamental frequency of vibration with η and k being most important. An increase in the fundamental frequency of vibration is most pronounced for small tip loads and large hub inertias, where a gain of over 600% is common in comparison with a uniform link.

In the optimization approach, a revolutionary new segmentized formulation was developed as a tool for optimum shape design, but this formulation is also ideal for use in optimizing flexible link controllers. More importantly, any required constraint on design variables can be easily included in the optimization algorithms. Of course, the truly *optimum flexible manipulator* will have an optimized link along with a controller which has been optimized for that particular link design [42].

To solve the segmentized formulation, two innovative algorithms were developed. The DP algorithm was shown to be both quick and accurate, requiring only a few segments. An adaptive random search algorithm (IHR) was modified to maintain a constant volume throughout the optimization process. An additional adaptive element was incorporated which increased the rate of convergence. The IHR results were comparable to the DP results, and both methods always converged to the analytical optimum of Section 3 as expected.

A minimax design method was presented for the application in which a manipulator is required to handle a wide variety of tip loads. This problem can be solved easily with either the DP or IHR algorithm. Also, a two-link optimum design was formulated and implemented, which produced the expected results of a shorter Link 1, in comparison with Link 2. Finally, a sensitivity analysis of the analytical solution of Section 3 was presented. Here, η and k were shown to have a pronounced effect upon the fundamental frequency of vibration. The tip load mass was also important, but the placement of its center of mass was relatively unimportant.

Thus, a complete picture of the optimum design of flexible manipulators has been presented. The next step in this research should be to incorporate a control algorithm into the optimum design process. This should be relatively easy since both the shape and control optimization processes can use the same segmentized algorithm (either DP or IHR). An exciting new shape in flexible link design is that of the variable hollow tunnel cross section which promises to reduce mass while increasing the fundamental frequency of vibration [28]. When this design is implemented with new composite materials, a truly superior flexible manipulator will be realized.

References

1. H. Asada, J.-H. Park, and S. Rai, "A Control-Configured Flexible Arm: Integrated Structure/Control Design," *IEEE Conf. on Robotics and Automation* (Sacramento, CA, 1991), Vol.3, pp. 2356-2362.
2. S. Bailey, *A Common Lunar Lander for the Space Exploration Initiative,* (NASA Johnson Space Center, Houston, TX, 1991).
3. F. Bellezze, L. Lanari, and G. Ulivi, "Exact Modeling of the Flexible Slewing Link," *Proceedings of IEEE Int. Conference on Robotics and Automation* (Cincinnati, OH, 1989), Vol.2, pp. 734-739.
4. W.K. Belvin and K.C. Park, "Structural Tailoring and Feedback Control Synthesis: An Interdisciplinary Approach," *Journal of Guidance and Control, Vol.* 13, No. 3 (1990), pp. 424-429.
5. W. J. Book, "Recursive Lagrangian Dynamics of Flexible Manipulator Arms," *International Journal of Robotics Research,* Vol.3, No.3 (1984), pp. 87-101.
6. R. M. Brach, "On the Extremal Fundamental Frequencies of Vibrating Beams," *International Journal of Solids Structures, Vol.* 4 (1968), pp. 667-674.
7. R. M. Brach, "On Optimal Design of Vibrating Structures", *Journal of Optimization Theory and Applications,* (1973), pp. 662-667.
8. H. Bremer, "On the Dynamics of Flexible Manipulators," *IEEE 1987 International Conference on Robotics and Automation, Vol.* 3, (1987).
9. R. H. Cannon, Jr. and E. Schmitz, "Precise Control of Flexible Manipulators," *Robotics Research: The First Intl. Symposium,* MIT Press, Cambridge, MA (1984), pp. 841-861.

10. A. De Luca and B. Siciliano, "Regulation of Flexible Arms Under Gravity," *IEEE Transactions on Robotics and Automation,* Vol.9, No.4 (1993), pp. 463-467.
11. S.D. Eppinger and W. Seering, "Modeling Robot Flexibility for Endpoint Force Control," *IEEE 1988 International Conference on Robotics and Automation, Vol.*1 (1988), pp. 165-170.
12. Frank, Paul M., *Introduction to System Sensitivity Theory,* (Academic Press, NY, 1978).
13. D. L. Girven and W. J. Book, "Numerical Analysis of Nonimimum Phase Zero for Nonuniform Link Design," in *Dynamics of Flexible Multibody Systems: Theory and Experiment,* ASME Winter Annual Meeting (Anaheim, CA, 1992), pp. 5-16.
14. R.T. Haftka, Z. Gurdal, and M.P. Kamat, *Elements of Structural Optimization* (Kluwer Academic Publishers, Boston, 1990).
15. R. Karel, Variational Methods in Mathematics, Science, and Engineering, (D. Reidel Publishing Co., Dordrecht, Holland, 1980).
16. B.L. Karihaloo and F.I. Niordson, "Optimum Design of Vibrating Cantilevers," *Journal of Optimization Theory and Applications,* Vol.11, No.6 (1973), pp. 638654.
17. F. Khorrami and Umit Ozgoner, "Perturbation Method in Control of Flexible Link Manipulators," *IEEE 1988 International Conference on Robotics and Automation* (Philadephia, PA, 1988), Vol. 1, pp. 310-315.
18. K.B. Lim and J.L. Junkins, "Robustness Optimization of Structural and Controller Parameters," *Journal of Guidance and Control,* Vol. 12, No. 1 (1989), pp. 89-96.
19. F. I. Niordson, "On the Optimal Design of a Vibrating Beam," *Quarterly of Applied Mathematics, Vol.* 23, No. 1 (1965), pp. 47-53.
20. P. K. Nguyen, R. Ravindran, R. Carr, D. M. Gossain, and K. H. Doetsch, "Structural Flexibility of the Shuttle Remote Manipulator System Mechanical Arm," *SPAR AeroSpace Ltd., Tech. Info. AIAA* (1982).
21. C.M. Oakley and R. H. Cannon, Jr., "End-Point Control of a Two-Link Manipulator with a Very Flexible Forearm: Issues and Experiments," *Proc. 1989 American Control Conference* (Pittsburgh, PA, 1989), pp. 1381-1388.
22. J.-H. Park and H. Asada, "Design and Control of Minimum-Phase Flexible Arms with Torque Transmission Mechanisms," *IEEE 1990 International Conference on Robotics and Automation, Vol.* 3 (1990), pp. 1790-1795.
23. J.-H. Park and H. Asada, "Integrated Structure/Control Design of Two-Link Nonrigid Robot Arm for High Speed Positioning," *IEEE Conf. on Robotics and Automation* (Nice, France, 1992), pp. 2356-2362.
24. A. Pil and H. Asada, "A Computer Integrated Development System for High Speed Robot Design Incorporating Structural Reconfiguration," *IEEE Conf. on Robotics and Automation* (Atlanta, GA, 1993), pp. 677-683.
25. E. P. Polak, "Basics of Minimax Algorithms," *Proceedings of the Summer School on Nonsmooth Optimization and Related Topics,* Erice, Italy (1988).

26. W. Prager and J. E. Taylor, "Problems of Optimal Structural Design," *ASME Trans. Journal of Applied Mechanics* (1968), pp. 102-106.
27. J.N. Reddy, *Energy and Variational Methods in Applied Mechanics,* (John Wiley and Sons, New York, 1984).
28. J. L. Russell, *Optimization Models for Flexible Manipulators* (1995), Ph. D. Dissertation, Department of Systems and Industrial Engineering, University of Arizona, Tucson.
29. T. Sakawa, F. Matsuno, and S. Fukushima, "Modeling and Feedback Control of a Flexible Arm," *Journal of Robotic Systems, Vol.* 2 (1985), pp. 453-472.
30. C. Y. Sheu, "Elastic Minimum-Weight Design for Specified Fundamental Frequency," *International Journal of Solids Structures, Vol.* 4 (1968), pp. 953-958.
31. S. Timoshenko and G. H. MacCullough, *Elements of Strength of Materials, third edition (D.* Van Nostrand Company, Inc., New York, 1949).
32. S. Timoshenko, D. H. Young, and W. Weaver, Jr., *Vibration Problem in engineering, third edition,* (D. Van Nostrand Company, Inc., New York, 1957).
33. Fei-Yue Wang, "Optimum Design of Vibrating Cantilevers: A Classical Problem Revisited," *J. of Optimization Theory and Applications, Vol.* 84, No. 3 (1995).
34. Fei-Yue Wang, "On the Extremal Fundamental Frequencies of One-Link Flexible Manipulators," *International Journal of Robotics Research, Vol.* 13, No. 2 (1994), pp. 162-170.
35. Fei-Yue Wang and G. G. Guang, "Influence of Rotatory Inertia, Shear Deformation and Loading on Vibration Behaviors of Flexible Manipulators," *Journal of Sound and Vibrations, Vol.* 171, No. 4 (1994), pp. 433-452.
36. Fei-Yue Wang and J. L. Russell, "Optimum Shape Construction of Flexible Manipulators with Tip Loads," *Proceedings of IEEE Intl. Conf. on Decision and Control* (Tucson, Arizona,1992), Vol. 1, pp. 311-316.
37. Fei-Yue Wang, "Optimum Design of Flexible Manipulators: Integration of Control and Construction," Dept. of Systems and Industrial Engineering, University of Arizona, SIE working paper #02-92 (1992).
38. Fei-Yue Wang and J. L. Russell, "Minimum-Weight Robot Arm for a Specified Fundamental Frequency," *Proceedings of the IEEE Conference on Robotics and Automation,* (Atlanta, Georgia, 1993), Vol. 1, pp. 490-495.
39. Fei-Yue Wang and J. L. Russell, "Optimum Shape Construction of Flexible Manipulators with Total Weight Constraint," *IEEE Transactions on Systems, Man, and Cybernetics, Vol.* 25, No. 3 (1995).
40. Fei-Yue Wang and J. L. Russell, "A New Approach to Optimum Flexible Link Design," *Proceedings of IEEE Int. Conf. on Robotics and Automation* (Nagoya, Japan, 1995).
41. Fei-Yue Wang and J. T. Wen, "Nonlinear Dynamical Model and Control for a Flexible Beam," CIRSSE Report #75, Rensselaer Polytechnic Inst., (Troy, New York, 1990).

42. Fei-Yue Wang, "Optimum Design of Flexible Robot Arms: A Mechatronic Approach," *Proc. of the First Chinese World Congress on Intelligent Control and Intelligent Automation* (Beijing, China, 1993), pp. 1160-1165.
43. Fei-Yue Wang, "Finding the Maximum Bandwidth of a Flexible Arm," *Proc. of 32nd IEEE Conf. on Decision and Control,* (San Antonio, TX, 1993), Vol. 1, pp. 619-625.
44. Fei-Yue Wang, "Optimization Formulations for Mechatronic Design of Flexible Robotic Manipulators," RAL Technical Report #45, SIE Dept, University of Arizona, Tucson, Arizona (1994).
45. E. J. Wiest and E. Polak, "On the Rate of Convergence of Two Minimax Algorithms," *Journal of Optimization Theory and Applications, Vol.* 71, No. 1 (1991), pp. 1-30.
46. G. B. Yang and M. Donath, "Dynamic Model of a One-Link Manipulator with Both Structural and Joint Flexibility," *IEEE 1988 International Conference on Robotics and Automation, Vol.* 6, No. 3 (1988), pp. 476-481.
47. K. Zuo and D. Wang, "Closed-loop Shaped Input Control of a Class of Manipulators with a Single Flexible Link," *IEEE International Conference on Robotics and Automation,* (Nice, France), Vol. 6, No. 3, pp. 476-481.
48. W.L. Winston, *Introduction to Mathematical Programming,* (PWS-Kent Publishing Co., Boston, 1991).
49. Z.B. Zabinsky, R.L. Smith, J.F. McDonald, H.E. Romeigh, and D.E. Kaufman, "Improving Hit-and-Run for Global Optimization," *Journal of Global Optimization, Vol.* 3 (1993), pp. 171-192.

Chapter 10

Mechatronic Design of Flexible Manipulators

The construction of lightweight manipulators with a larger speed range is one of the major goals in the design of well-behaving industrial robotic arms. Their use will lead to higher productivity and less energy consumption than is common with heavier, rigid arms. However, due to the flexibility involved with link deformation and the complexity of distributed parameter systems, modeling and control of flexible manipulators still remain a major challenge to robotic research. A compromise between modeling costs and control efficiency for real-time implementation is inevitable. The interdependency of subsystems results in a local optimal performance in the traditional design scheme. An important research topic in flexible manipulator design is the pursuit of better system performance while avoiding model-intensive or control-intensive work. This problem can be solved using the proposed mechatronic design approach. It treats the mechanical, electrical and control components of a flexible manipulator concurrently. The result is an global optimal design with an explicit link shape and controller parameters which result in the control problem and modeling accuracy no longer being critical for obtaining desired performance while the coupling effects among subsystems being taken into account automatically. Dynamics of flexible manipulators with shear force and rotatory inertia are derived, and state-space equations with the integration of DC motor dynamics are developed as a theoretical base for mechatronic designs. Case studies based on LQR formula and H_∞ are considered. The beam shape and controller parameters are obtained using an adaptive iterative algorithm with the accommodation of various geometrical constraints. Also, different output feedback strategies are investigated to evaluate the impacts of various controller structures.

10.1 Introduction

A manipulator becomes flexible when its link deformation cannot be ignored in the analysis of its performance. A manipulator which has large dimensions, is lightweight, fast, or handles a heavy load exhibits flexibility. Flexible manipulators utilize less energy due to their lightweight, higher productivity achieved through fast motion. They are safer to operate due to good damping behavior and the less pronounced interconnections between the different segments for multiple-link manipulators. These manipulators are found in space explorations (NASA Mar Exploration Mission STS95, 1998) because of the constraints on arm length, weight and 'gravity-loss'; in mining applications (Robotics Excavators) because of their heavy payload; in construction applications (Robotics Crane Systems) because of the length and heavy tip load; and where dexterous manipulators are required such as in medical operations or chip placement performed pick-and-place manipulators in electronic assembly manufacture. Even through, the NASA Remote Manipulator System has very low natural frequencies and consequently has to move slowly (0.5 deg/sec) in order to avoid vibrations because of its beam mass (450 kilograms) and its heavy payload (27; 200 kilograms).

In last decades, researches on flexible manipulators have increased dramatically. From the perspective of control, recent researches were found in Doyle, Francis, Gutierrez, Jnifence, Zabinsky and references therein. These studies did control-intensive work to improve beam performance. But they ignored the impact applied from the beam mechanic shape. The resulting system was only local optimal. In the other hand, a better beam shape to move faster but for less vibration is another topic. One of the first papers on optimal shape for flexible manipulators is Fujita by Karihaloo and Niordson. Extensive studies in optimal shape design have been conducted recently by Wang's group. Wang (1991) simplified substantially the scheme. In 1991, this group examined the external fundamental frequencies and developed an iterative scheme, producing the first one-link optimum manipulator shape. It was shown that by the proper selection of hub, the optimal link could improve the first natural frequency by 600%. A minimum weight design of flexible manipulators was developed by Wang and Russell (1992, 1993). In 1995, they also investigated a new approach, the segmentized scheme of optimal design, which treats the flexible beam as a collection of small lengths of rigid beam constrained by each other's interfacial conditions. The new technique converts the optimal shape design problem into a matrix determinant problem. In 1996, a new computing method for optimum mass and rigidity distribution was formulated by Wang, Zhou and Lever (1996) for a flexible manipulator with a tip-load. The robustness with respect to design specification and appropriate constraints was considered. Practical issues were addressed also. These studies focused on open-loop design. They only concerned the beam mechanic construction, that is proper design the flexible beam shape can make it suffering less vibration. But in reality, all manipulators must be in closed-loop to get high performance. The couplings between the controller and construction were not considered at larger in these designs.

10.1.1 *Why is Mechatronic Design*

Modeling accuracy and control efficiency of flexible manipulators: Flexible manipulators are distributed systems described by partial-differential equations, therefore, their dynamic behavior is of infinite degrees of freedom. From the perspective of control theory, it is impossible to design an infinite dimensional controller. The controlled plant must have a finite dimension, thereby requiring less significant terms in the model to be omitted. This causes model uncertainties since those terms generally are time-variable and system dependent. The boundary conditions set by tip-load, hub inertia, friction, rotatory inertia, and shear force also impact beam dynamics, which make the model implementation more complicated for the purpose of real-time control. But the model must be accurate enough to take control into account. The efficiency of modeling and the precision of control of flexible manipulators are contradictory and a compromise between them for real-time implementation is inevitable.

Interrelation and interdependency of subsystems of manipulator systems:
The complexity of manipulator systems is also due to the interrelation and interdependency of their subsystems. A manipulator system consists of different subsystems, which are:

- a kinematic system, mainly related to the beam structure,

- a control system,

- a driver system, or actuator,

- a measuring system, or sensors.

Some subsystems exert influence over others whereas some subsystems are only influenced by others. In traditional design, a manipulator link is designed first, followed by a driver system, measuring system, and then a control system as Fig.10.1. As a result, this traditional design scheme leads to a locally optimal solution, since these coupling effects have not been considered in the design process.

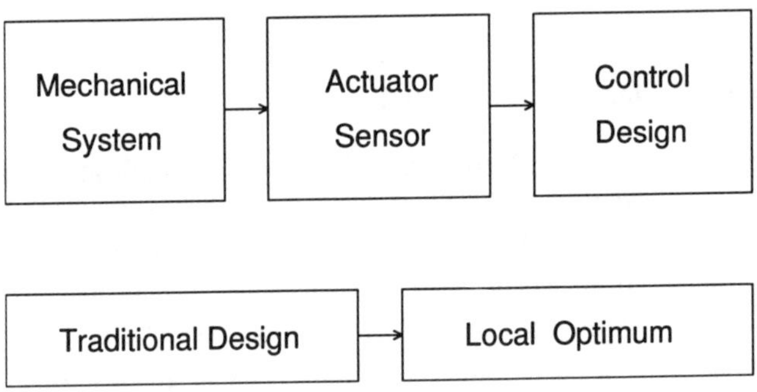

Fig. 10.1 Traditional Design Procedure

If a flexible manipulator system can be designed taking the interrelationships between subsystems into account while avoiding control-intensive or model-intensive work, the control and modeling problems will no longer be critical, thus improving productivity and reducing energy consumption. Mechatronic Design Method (MDM) fits these requirements exactly.

10.1.2 *What is Mechatronic Design*

For applications in manipulator systems, the MDM is *a system optimization through the integration of actuator, controller, system dynamics and sensor specifications as Fig. 10.2*. Comparing with traditional design, following features can be identified for mechatronic design.

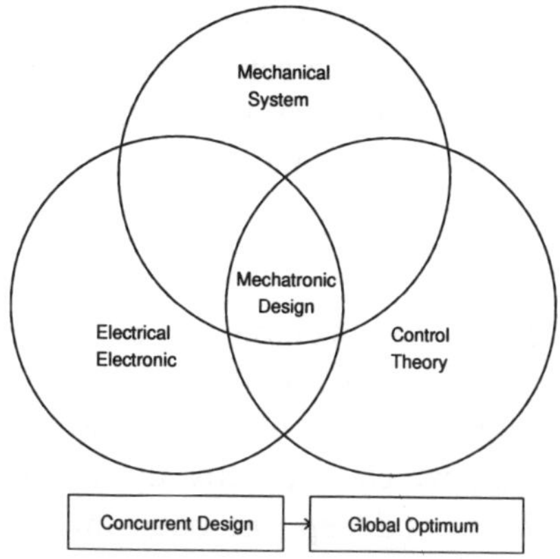

Fig. 10.2 Mechatronic Design Procedure

Different methodology: MDM treats the mechanical, electrical, and control components of a flexible manipulator concurrently, instead of sequentially from the very beginning of the design process. The coupling effects automatically are taken into account. This in turn results in global optimal objective functions in the system.

Different direction: MDM focuses on system overall design instead of modeling or control. That is, the objective is to design a better manipulator system so its control problem and dynamics accuracy will not be very critical for its performance. In other words, the result from this mechatronic design is better than what is obtained from model-intensive work or control-intensive work.

10.1.3 *How is Mechatronic Design*

Let *(A, B, C)* represent a flexible manipulator system, which is the mixture of beam shape, sensor specification, actuator selection and u the controller. J is performance index. Assume that Ω is the space of all feasible manipulator system and Λ is the feasible control space from the actuator. In traditional design,

$$J = \inf_{u \in \Lambda} J(u; A, B, C), \tag{10.1}$$

the objective is to find the optimal control for a system given by *(A, B, C)* only. But the mechatronic design is further to find the J^* by choosing different *(A, B, C)*, that is, choosing different beam shape, different sensor and actuator. Then the objective is to minimize the performance index J with respect to link construction, actuator selection, sensor specification, and control design, i.e.,

$$J^* = \inf_{(A,B,C)\in\Omega, u\in\Lambda} J(u;A,B,C) = \inf_{(A,B,C)\in\Omega, u\in\Lambda} [\inf J(u;A,B,C)] \quad (10.2)$$

Accordingly, the global optimization can be carried in two steps. The first step (the inner $\inf_{u\in\Lambda}$ loop) is to find the optimal control structure and its associated optimal value of the performance index based on a given plant *(A, B, C)*, which is exactly the traditional optimal control design problem. The second step (the outer $\inf_{(A,B,C)\in\Omega}$ loop) is to find the optimal feasible plant that will further minimize the performance index obtained by the optimal controller. In this chapter, assuming DC motor and sensors are given, the outer loop is only searching the beam shape distribution. The model of flexible manipulators based on modal shapes from system energy balance equation is derived in Section 10.2. The mechatronic design algorithm is schemed as in Section 10.3 and Section 10.4. Results are presented, followed by the conclusion in the end.

10.2 Dynamics of Flexible Manipulators

The dynamics of a flexible manipulator system are described by an infinite-dimensional mathematical model, since the model consists of partial differential equations. But to design a finite-dimensional controller, a finite-dimensional system model is needed. To achieve this goal, a finite dimensional approximation needs to be used to model a flexible manipulator, that is, to retain a finite number of modes, and to drop off the other, less significant modes based on the requirements of the controller. The N-mode expansion for the beam displacement $\omega(x,t)$ can be given by,

$$\omega(x,t) = \sum_{i=1}^{N} \varphi_i(x) q_i(t) \quad (10.3)$$

The separability in this case refers to describing the displacement as a series of terms which are products of two separate functions, each of that is a function of a single variable, a spatial variable x, and time t, respectively. φ_i is the ith modal shape, or eigenfunction. q_i is the corresponded generalized modal coordinate describing the flexible deformation.

The scheme in developing a mathematical model is to use the Lagrangian method or Hamiltonian's Principle to the total kinetic energy, total potential energy and virtual work done by the torque actuated to the joint. This method will not introduce extra errors into system and will be used to obtain the state-space model for a flexible manipulator suggested in this chapter.

10.2.1 *State-Space Equations of Flexible Manipulators*

The hub kinetics is written as,

$$T_h = \frac{1}{2} I_h [\dot{\theta}^2 + \dot{\omega}^2(0,t)'] \tag{10.4}$$

where $\omega = \partial\omega / \partial x = \omega'$. After applying Eq.10.3 into Eq.10.4,

$$T_h = \frac{1}{2} I_h \dot{\theta}^2 + \frac{1}{2} I_h \sum_{i=1}^{n} \sum_{j=1}^{n} \dot{q}_i \dot{q}_j \varphi_i'(0) \varphi_j'(0) \tag{10.5}$$

In the same way, the kinetics of the tip load is,

$$T_p = \frac{1}{2} M_p (\dot{\theta} L + \dot{w}(L,t))^2 + \frac{1}{2} J_p (\dot{\theta} + \dot{w}(L,t')^2 + a_c M (\dot{\theta} + \dot{w}(L,t')(\dot{\theta} L + \dot{w}(L,t) \tag{10.6}$$

which is the same as,

$$T_p = \frac{1}{2} \Gamma_1 \dot{\theta}^2 + \dot{\theta} \sum_{i=1}^{N} \dot{q} \Gamma_2(i) + \frac{1}{2} \sum_{i=1}^{n} \sum_{j=1}^{n} \dot{q}_i \dot{q}_j \Gamma_3(i,j) \tag{10.7}$$

where,

$$\begin{aligned}
\Gamma_1 &= (M_p L^2 + J_p + 2M_p L a_c), \\
\Gamma_2(i) &= (J_p + M_p L a_c) \psi_i'(L) + (L + a_c) M_p \psi_i(L). \\
\Gamma_3(i,j) &= J_p \psi_i'(L) \psi_j'(L) + M_p \psi_i(L) \psi_j(L) + 2a_c M_p \psi_i'(L) \psi_j(L)
\end{aligned} \tag{10.8}$$

The kinetic energy of the link,

$$T_L = \frac{1}{2}\int_0^L [(x\dot{\theta} + \dot{w})^2 + S(\dot{w}' + \dot{\theta})^2]\rho dx \tag{10.9}$$

is equal to,

$$T_L = \frac{1}{2}\Lambda_1\dot{\theta}^2 + \dot{\theta}\sum_{i=1}^{N}\dot{q}\Lambda_2(i) + \frac{1}{2}\sum_{i=1}^{n}\sum_{j=1}^{n}\dot{q}_i\dot{q}_j\Lambda_3(i,j) \tag{10.10}$$

where

$$\Lambda_1 = \int_0^L (x^2 + S)\rho dx,$$

$$\Lambda_2(i) = \int_0^L [x\psi_i(x) + S\psi_i^{'}(x)]\rho dx, \tag{10.11}$$

$$\Lambda_3(i,j) = \int_0^L [\psi_i(x)\psi_j(x) + S\psi_i^{'}(x)\psi_j^{'}(x)]\rho dx.$$

S is the function of cross section area of the link. The total potential energy is

$$P = \frac{1}{2}\int_0^L EI(x)w^2(x)^{''}dx = \frac{1}{2}\sum_{i=1}^{n}\sum_{j=1}^{n} q_i\, q_j k(i,j) \tag{10.12}$$

where $k(i,j) = \int_0^L EI(x)\psi_i^{''}(x)\psi_j^{''}(x)dx$

The generalized virtual work is derived as,

$$W = \tau v'(0,t) = \tau\theta + \tau w^{'}(0,t) = \tau\theta + \tau\sum_{i=1}^{N}\psi_i^{'}(0)q_i(t). \tag{10.13}$$

In order to apply the Hamiltonian Principle, let us substitute Eqs. 10.5, 10.7, 10.9, 10.11, 10.12 for the following formula and group it in terms of θ, δqi, and their derivatives,

$$\Xi = T_L + T_h + T_p - P + W = \frac{1}{2}\Omega_1\dot{\theta}^2 + \dot{\theta}\sum_{i=1}^{N}\dot{q}_i\Omega_2(i)$$

$$+\frac{1}{2}\sum_{i=1}^{N}\sum_{j=1}^{N}q_i q_j \Omega_3(i,j)+\tau\theta++\tau\sum_{i=1}^{N}\dot{\psi}_i(0)q-\frac{1}{2}\sum_{i=1}^{N}\sum_{j=1}^{N}q_i q_j k(i,j) \quad (10.14)$$

where

$$\begin{aligned}\Omega_1 &= I_h+\Gamma_1+\Lambda_1,\\ \Omega_2 &= \Gamma_2(i)+\Lambda_2(i)\\ \Omega_3(i,j) &= I_h\psi_i'(0)\psi_j'(0)+\Gamma_3(i,j)+\Lambda_3(i,j)\end{aligned} \quad (10.15)$$

After applying the Hamiltonian method, we obtain,

$$\delta\Xi=\Omega_1\dot{\theta}\,\delta\dot{\theta}+\sum_{i=1}^{N}\Omega\ \ (i)\dot{q}\delta\dot{\theta}+\dot{\theta}\sum_{i=1}^{N}\Omega_2(i)\delta\dot{q}_i$$

$$+\sum_{i=1}^{N}\sum_{j=1}^{N}\dot{q}_j\Omega_3(i,j)\delta\dot{q}_i+\tau\delta\theta+\tau\sum_{i=1}^{N}\psi_i'(0)\delta q_i-\sum_{i=1}^{N}\sum_{j=1}^{N}q_j k(i,j)\delta q_i. \quad (10.16)$$

since,

$$\int_{t_0}^{t_f}\dot{\theta}\delta\dot{\theta}dt=\dot{\theta}\delta\theta\,\Big|_{t_0}^{t_f}-\int_{t_0}^{t_f}\ddot{\theta}\delta\theta dt=-\int_{t_0}^{t_f}\ddot{\theta}\delta\theta dt.\quad \int_{t_0}^{t_f}\dot{q}_i\delta\dot{\theta}dt=-\int_{t_0}^{t_f}\ddot{q}_i\delta\theta dt, \quad (10.17)$$

$$\int_{t_0}^{t_f}\dot{\theta}\delta\dot{q}_i dt=-\int_{t_0}^{t_f}\ddot{\theta}\delta q_i dt,\quad \int_{t_0}^{t_f}\dot{q}_i\delta\dot{q}_j dt=-\int_{t_0}^{t_f}\ddot{q}_i\delta q_i dt.$$

The Hamilton Extended Principle results in,

$$\int_{t_0}^{t_f}\delta\Xi dt=\int_{t_0}^{t_f}\{[-\Omega_1\ddot{\theta}-\sum_{i=1}^{N}\ddot{q}_i\Omega_2(i)+\tau]\delta\theta$$

$$-[\ddot{\theta}\sum_{i=1}^{N}\Omega_2(i)+\sum_{i=1}^{N}\sum_{j=1}^{N}\ddot{q}_j\Omega_3(i,j)+\sum_{i=1}^{N}\sum_{j=1}^{N}q_j k(i,j)-\tau\sum_{i=1}^{N}\psi_i'(0)]\delta q_i\}dt=0. \quad (10.18)$$

These coefficients of $\delta\theta$, δq_i, $1\le i\le N$ must be zero, that is,

$$\Omega_1\ddot{\theta}+\sum_{i=1}^{N}\ddot{q}_i\Omega_2(i)-\tau=0, \quad (10.19)$$

$$-\ddot{\theta}\Omega_2(i)-\sum_{j=1}^{N}\ddot{q}_j\Omega_3(i,j)-\sum_{j=1}^{N}q_j k(i,j)+\tau\psi_i'(0)=0,\qquad 0\le i\le N \quad (10.20)$$

or in matrix form, they are,

$$M\ddot{x}+Kx=b\tau, \quad (10.21)$$

where,

$$x=[\theta,q_1,q_2,q_3,\ldots,q_N]^T_{(N+1,1)},\quad b=[1,\psi_1'(0),\psi_2'(0),\ldots,\psi_N'(0)]^T_{(N+1,1)}, \tag{10.22}$$

$$M=\begin{bmatrix} \Omega_1 & \Omega_2(1) & \Omega_2(2) & \cdot & \Omega_2(N) \\ \Omega_2(1) & \Omega_3(1,1) & \Omega_3(1,2) & \cdot & \Omega_3(1,N) \\ \cdot & \cdot & \cdot & \cdot & \cdot \\ \Omega_2(N) & \Omega_3(N,1) & \Omega_3(N,2) & \cdot & \Omega_3(N,N) \end{bmatrix},$$

$$K=\begin{bmatrix} 0 & 0 & 0 & \cdot & 0 \\ 0 & k(1,1) & k(1,2) & \cdot & k(1,N) \\ \cdot & \cdot & \cdot & \cdot & \cdot \\ 0 & k(N,1) & k(N,2) & \cdot & k(N,N) \end{bmatrix}.$$

As has been shown, $\Omega_3(i,j)=\Omega_3(j,i),\quad k(i,j)=k(j,i)$, thus the matrices M, K are symmetric and named mass and rigidity matrices respectively.

Next, the actuator dynamics need to be incorporated into the link system. It is assumed that the arm is driven by a permanent magnet DC-motor. Therefore, the actuator dynamics can be described as,

$$-J_m\ddot{\theta}-(B_m+\frac{K_bK_m}{R})\dot{\theta}+\frac{K_m}{R}v_c=\tau, \tag{10.23}$$

where J_m is the actuator inertia, B_m the friction coefficient, K_m the torque constant, K_b the back emf constant, R the armature resistance, and θ, v_c the hub rotation and armature voltage, respectively. In general, all motor circuit parameters can be considered as design variables. The overall state variable is defined as,

$$q=\begin{bmatrix} x \\ \dot{x} \end{bmatrix} \tag{10.24}$$

Combining Eq.10.21 with Eq.10.22, we present the overall system state space equations as,

$$\dot{q}=Aq+Bu,\qquad u=v_c, \tag{10.25}$$

where

$$A = \begin{bmatrix} 0 & I \\ -\overline{M}^{-1} K & -\overline{M}^{-1} \overline{B} \end{bmatrix}, \quad B = \begin{bmatrix} 0 \\ \overline{M}^{-1} \overline{D} \end{bmatrix}$$

$$\overline{M} = (M + J_m b \underline{e}_1)^{-1}, \quad \overline{B} = (B_m + \frac{K_b K_m}{R}) b \underline{e}_1, \tag{10.26}$$

$$\overline{D} = b \frac{K_m}{R}, \quad \underline{e}_1 = [1 \quad 0 \quad \dots \quad 0]_{1\times n}.$$

A is the function of beam construction, thus any changes in beam mechanical shape will result in a different A, which provides the basis for simultaneously optimal construction and control based on the mechatronic formulation discussed early.

10.2.2 *Output Specifications*

As can be seen from the system state-space equations, the state vector consists of θ, the generalized modal coordinates q_i and their first order derivatives. So all controllers based on the state feedback are indirect, which means that the states need to be predicted. This will trade off with computation time and bring one more inaccuracy into a closed-loop, cutoff closed-loop bandwidth. This presents hindrances to real-time processing and high motion speed. Output feedback takes precedence over state space feedback as far as the mechatronic approach and objectives of this chapter are concerned.

Tip deflection output, which is,

$$y = w(L,t) = \sum_{i=1}^{N} \psi_i(L) q_i = [0 \quad \psi_1(L) \quad \psi_2(L) \quad \dots \quad \psi_N(L) \quad \overline{0}] q, \tag{10.27}$$

where ψ_i is the i th eigenfunction, $\overline{0}_{1\times(n+1)}$ is zero vector. A CCD camera clipped on the hub can be used to measure the output.

The tip position output is,

$$y = v(L,t) = L\theta + \sum_{i=1}^{N} \psi_i(L) q_i = [L \quad \psi_1(L) \quad \psi_2(L) \quad \dots \quad \psi_N(L) \quad \overline{0}] q, \tag{10.28}$$

The hub tangent angle output is,

$$y = \Theta = \theta + \sum_{i=1}^{N} \psi_i'(0) q_i = [L \quad \psi_1'(0) \quad \psi_2'(0) \quad \dots \quad \psi_N'(0) \quad \overline{0}] q, \tag{10.29}$$

A potentiometer may be used for measuring Θ.

In this mechatronic design, output feedback may be one of above outputs, or the mixture, such as,

$$y = Cq = \begin{bmatrix} w(L,t) \\ v'(L,t) \end{bmatrix} q,$$
$$C = \begin{bmatrix} 0 & \psi_1(L) & \psi_2(L) & \dots & \psi_N(L) & \bar{0} \\ 1 & \psi_1'(0) & \psi_2'(0) & \dots & \psi_N'(0) & \bar{0} \end{bmatrix}. \tag{10.30}$$

The corresponding state-variable equations are,

$$\dot{q} = Aq + bu, \qquad y = Cq. \tag{10.31}$$

These output feedbacks will be used in control design in the later chapters to show the improvement of the suggested mechatronic approach and to make comparison between them.

10.3 Case I: Mechatronic Design Based on LQR Formula with IHR Programming

As stated, echatronic design is a global optimization of the overall system. For a flexible manipulator, the overall system is the integration of link dynamics, DC motor equation, measuring sensors, and the selected controller. The optimization process will result in an optimal link geometric distribution, an optimal controller structure subject to the performance requirement. To demonstrate this approach and also due to the computation complexity, the following restrictions are applied here. A rectangular beam is considered and divided into N segments equally along with the beam spatial coordinate. Each segment is uniform. In each segment, the width is the only variable to be optimized for the link geometrics.

It was pointed out that output feedback instead of state space feedback is a feasible choice. It is also predictable that the system optimum performance index may not matter in the selection of the controller to a large degree, since the resulting index is the result of the overall system optimization. Due to this factor, the linear quadratic regulator (*LQR*) is admissible as the selected controller.

Mechatronic design based on the (*LQR*) formula is discussed here. The *LQR* feedback is outlined in [4] as an inner loop followed by an Adaptive Iterative algorithm (*IHR*) as an outer loop searching for the beam width distribution. The mechatronic design procedure is addressed through detailing the integration of the inner loop with the outer loop.

10.3.1 *Inner Loop Optimization*

For a linear output feedback in the form of

$$u = \Re y, \tag{10.32}$$

where $\Re$ is a matrix of constant feedback coefficients to be determined. If the quadratic performance index (*PI*) is

$$J(u) = \frac{1}{2}\int_{t_0}^{\infty}(q^T(t)Qq(t) + u^T(t)Ru(t))dt, \tag{10.33}$$

where the Q , R are symmetric positive semidefinite weighting matrices. After substituting the feedback controller Eq.10.32 into Eq.10.33, also assuming that the system is asymptotically stable, so the *q(t)* vanishes with time,

$$\lim_{t\to\infty} q^T(t)Pq(t) = 0, \tag{10.34}$$

where *P* can be solved by Lyapunov equation,

$$A_c^T P + PA_c + C^T \Re^T R \Re C + Q = 0, \quad (A - B\Re C) = A_c. \tag{10.35}$$

Then,

$$J = \frac{1}{2}q^T(0)Pq(0) = \frac{1}{2}tr(PX), \tag{10.36}$$

where the $n \times n$ symmetric matrix *X* is defined by $q(0)q^T(0)$.

Following the substitution process in [4], two additional equations re described as,

$$\begin{aligned} &A_c S + SA_c^T + X = 0, \\ &\Re = R^{-1}B^T PSC^T(CSC^T)^{-1}, \end{aligned} \tag{10.37}$$

where $X = q(0)q^T(0)$, *S* is a symmetric matrix of Lagrange multipliers.

It is now clear that the problem of selecting $\Re$ to minimize *J* subject to the dynamical constraint of Eq.10.31 on the states is equivalent to the algebraic problem of selecting $\Re$ to minimize Eq. 10.37 subject to the constraint of Eq.10.35 on the auxiliary matrix *P*. To solve this modified problem, we use the Lagrange multiplier method.

The equations for *P*, *S* and $\Re$ are coupled among these nonlinear matrix equations, therefore, some trial-and-error iterative design methods have to be used to find the $\Re$. The process to find the feedback matrix $\Re$ is described as following for a given set of *A, B, C*.

LQR -- **Inner Loop Optimization:**

1. Initialize:
   $i = 0$, select an initial $\Re_0$ so that A_c is asymptotically stable; Set system initial state values $X = q(0)q^T(0)$, the Stop criteria $\wp$.
2. *i*th iteration:
   Set $Ai = A - B\Re_i C$.
   Using *lyap.m* in the Control Toolbox of Matlab to solve the equations $A_i^T P + PA_i + C1 = 0$, where $C1 = C^T \Re R \Re_i C + Q$ and $A_i S + SA_i^T = 0$ for P_i and S_i. Set $J = \frac{1}{2} tr(P_i X)$.
3. Updating:
   Gain updating direction: $\Delta\Re = R^{-1} B^T P_i S_i C^T (CS_i C^T)^{-1} - \Re_i$.
   Update gain: $\Re_{i+1} = \Re_i + \alpha \Delta \Re$,
   where α is chosen so that $A - B\Re_{i+1} C$ is asymptotically stable.
   Check the eigenvalues of new *A*, if unstable, go to Step 5.
4. Criteria:
   $$J_{i+1} = \frac{1}{2} tr(P_{i+1} X) \le J_i.$$
   If $(J_{i+1} - J_i) \le \wp$, the given criteria, go to 5. Otherwise set *i*=*i*+1 and go to 2
5. Stop:
   Set $\Re = \Re_{i+1}$, $J = J_{i+1}$.
   If index =0, go to step 2.

This process will converge to *J*, the minimum performance index for the set of *A, B, C* by searching proper feedback matrix $\Re$. In the next section, the outer loop searching the optimal beam shape will be discussed.

10.3.2 *Outer Loop Algorithm*

It was established that the components in the matrices of the state equations are related to a given link shape, that is, the matrices *A, B, C* are the functions of beam geometric distribution. Here a procedure to find out the link geometrics based on the Adaptive Iterative Hit and Run (*IHR*) algorithm [12] is conducted. Some restrictions shall be made clear from the perspective of the mechatronic design. First the beam weight holds constant, which, in terms of beam volume, means that the whole volume is constant. The only variable for the beam geometric is the width of the cross section for each segment, however, the sum of these width variables must remain a constant in order to keep the total volume constant based on *N* segment solution. Another consideration, realistically speaking, is that minimum stress on the beam is required, which in term of beam width, is its minimum width. These restrictions are applied to modify the *IHR* algorithm. This iterative procedure of the modified *IHR* can be specified as:

IHR -- **Outer Loop Optimization:**

1. Set uniform area A_0, minimum and maximum area constraints, A_{min} and A_{max} according to the beam strength requirement. Set j=1. Set beam material parameters.
2. Calculate the uniform beam *PI* as the starting value.
3. Initialize loop vectors, loop factors, direction vector *D* and stop criteria.
4. Select changing positions: randomly select $N/2$ of the *N* segments and mark them with 1's in vector D_j of length *N*. Mark the remaining $N/2$ segments with -1's. If *N* is odd, one randomly selected area will remain constant and will be marked by a 0.
5. Set direction vector: get $N/2$ samples from a *N*(0,1) normal distribution and place them in each position of D_j where there is a 1. Place the negatives of this same sample in each position of D_j where there is a -1. This arrangement will ensure a constant

volume during optimization. Here, D_j is called the direction vector.

6. Implement direction: generate a step size, S, uniformly from L_j the set of feasible step sizes in the direction D_j, where, $L_j = \{S \in \Re : A_{\min} < A_j + SD_j < A_{\max}\}$. If $L_j = 0$, go to Step 4.
7. Set $S = S * MUL$, while $0 < MUL < 1$.
8. Update the area vector, A_j,

   $A_{j+1} = A_j + SD_j$.

10.3.3 *Integrated Optimization Process*

Mechatronic design is to chooses the best plant and to finds out the associated controller for the performance requirement. It is an overall optimization process. This can be done by the integration of the inner loop optimization with the outer loop optimization as following.

Integrated Optimization

1. Set design vector $\Xi = \lfloor I_H, M_P, a_c, J_p L, B, E \rfloor$, and number of segments, N. Set step multiplier MUL to 1 and FACTOR to an appropriate constant.
2. Set weighting matrices Q and R.
3. Load precalculated coefficients for Modal Shapes.
4. Calculate A_0 (uniform shape radius or width), set initial $index = 0$, call ***LQR*** to get the corresponding J_0.
5. Update the area vector, A_j,

   1). call ***IHR***, to find the A_j

   2). based on Mode Shape coefficients and current link geometric, calculate mass matrix M, potential matrix K, vector B, C, and state equations after integrated with motor dynamics.

   3). call ***LQR*** to find J, If the J is improving:
   Set $J_{J+1} = J(A_{J+1})$, if J is improved, store it in an array and save this A_j. Update all loop variables and factors.

6. If all the differences between two J's in the array is smaller than stopping criteria, stop. Otherwise, go to Step 5.
7. Stop: Output optimal feedback matrix $\Re^*$, optimal performance index J^*, and optimal flexible link structure A^*.

10.3.4 *Results and Discussion*

In order to verify the mechatronic method presented in this chapter, a rectangular aluminum flexible link is used for simulation. The mechatronic algorithm is intended to find the beam geometric shape, or the width distribution so that the *PI* reaches the minimum.

To meet the minimum stress requirement pointed out earlier, the maximum width H_{max} and the minimum width H_{min} are set at twice and a quarter of the uniform width separately. For the *IHR* algorithm, all other criteria were set to 0.000001.

To fully test this mechatronic design algorithm, the different output feedback, combined with various state weighting matrix *Q*, and the number of segments *N*, but *R=I* (identity) were tried. Three sets of feedback were considered. They were the hub tangent angle $\{\Theta(0,t)\}$ feedback (Eq.10.29), the hub tangent angle with the tip deflection $\{\Theta(0,t),\omega(L,t)\}$ (Eq.10.27, Eq.10.29), and the hub tangent angle with the tip position $\{\Theta(0,t),v(L,t)\}$ (Eq.10.25, Eq.10.27). These sets of feedback have very clear physical meanings and are detectable. The initial state set is $q_1(0)=\left[\sqrt{10},0,\ldots 0\right]^T$. The scenario ({*feedback list*}, *number of segments*, *Q*) was used, meaning that the feedback strategy is as {*feedback list*}, that the beam is divided into *number of segments*, and that the value *Q* in LQR is *Q*. These notations are applied in the following figures and tables.

Note that it was necessary to show the *PI*'s with different number of segments. To do that, a performance index was set with $q=q_1(0)$, $Q=100\times I$, and with various $n=4,6,12$, and one of these three sets of feedback. For uniform and associated optimal shape, performance indexes were obtained with different feedback and different number of segments as Table 10.1. Column one is the type of feedback, while the column 2, column 3 and column 4 are for the different number of segments respectively. The performance indexes of the optimal shape are improved over that of the associated uniform shape. For example, when $n=4$ and $\{\Theta(0,t),\omega\}$feedback is applied, the *PI* for uniform shape is 3789.143, while the result of mechatronic design is smaller, which is 3591.341. But the performance indexes don't depend much on the number of segments. The associated feedback constants are listed in Table 10.2.

Table 10.1: Performance Index with $Q=100\times I$.

Feedback List	$n=4$	$n=6$	$n=12$
$\{\Theta(0,t)\}$Feedback-uniform shape	4665.504	4665.504	4665.504
$\{\Theta(0,t)\}$Feedback-optimal shape	4647.230	4646.571	4643.038
$\{\Theta(0,t), w\}$Feedback-uniform shape	3789.143	3789.143	3789.143
$\{\Theta(0,t), w\}$Feedback-optimal shape	3591.341	3585.945	3580.064
$\{\Theta(0,t), v\}$Feedback-uniform shape	4342.569	4342.569	4342.569
$\{\Theta(0,t), v\}$Feedback-optimal shape	4180.548	4178.569	4176.437

Table 10.2: Optimal Feedback Constants with $Q=100\times I$.

Feedback List	$n=4$	$n=6$	$n=12$
$\{\Theta(0,t)\}$Feedback-uniform	3.674	3.674	3.674
$\{\Theta(0,t)\}$Feedback-optimal	3.686	3.677	3.675
$\{\Theta(0,t), w\}$Feedback-uniform	(2.7599 –422.398)	(2.7599 –422.398)	(2.7599 –422.398)
$\{\Theta(0,t), w\}$Feedback-optimal	(2.498 –337.071)	(2.539 –380.396)	(2.423 –378.554)
$\{\Theta(0,t), v\}$Feedback-uniform	(2502.705 – 2499.521)	(2502.705 – 2499.521)	(2502.705 – 2499.521)
$\{\Theta(0,t), v\}$Feedback-optimal	(1768.423 – 1765.672)	(1732.346 – 1747.358)	(1746.276 – 1764.428)

The related optimal shapes are illustrated as following. In all figures, the solid outline is the optimal shape, and the dot one is for the uniform shape. The horizontal axis is the link spatial coordinate, the vertical one is the beam width.

Fig. 10.3 – 10.5 are for $\Theta(0,t)$ feedback (hub tangent angle) with $Q = 100 \times I$ and $n = 4,6,12$ separately. All three optimal shapes have one common feature with a large size at the hub end, a small size at the tip end.

Fig. 10.6 – 10.8 are for $(\Theta(0,t), \omega(L,t))$ feedback (hub tangent angle, tip deflection) with $Q = 100 \times I$ and $n = 4,6,12$ separately. In the same behavior, the best performance indexes are almost the same for different number of segments. A common feature for these three optimal shapes is a large beam size in the middle, and a relatively small size at both ends.

Fig. 10.9 – 10.11 are for $(\Theta(0,t), v(L,t))$ feedback (hub tangent angle, tip position) with $Q = 100 \times I$ and $n = 4,6,12$ respectively. As usual, the best performance indexes are almost the same for different numbers of segments. A common feature for these three optimal shapes is a relatively small beam dimension in the middle, and a large dimension at the ends.

As shown in these figures, the geometric shapes of these optimal shape are pretty much relied on the type of the feedback, not much relied on the number of segments. Next, this mechatronic approach was applied to different weighting matrix Q with these types of feedback. But the number of segments fixed as 4 since the number of segments plays a less important role here. Q are $50 \times I$ and $10 \times I$ separately. The performance indexes are listed in Table 10.3, where the column is for different type of feedback, and the row is for the value of Q. The associated feedback constants are listed in Table 10.4.

Table 10.3: Performance Index with Different $Q = 100 \times I$, $50 \times I, 10 \times I$.

Beam Shape	$\{\Theta(0,t)\}$	$\{\Theta(0,t), w\}$	$\{\Theta(0,t), v\}$
$Q = 100 \times I$ -uniform shape	4665.504	3781.547	4342.763
$Q = 100 \times I$ -optimal shape	4647.230	3591.538	4180.546
$Q = 50 \times I$ -uniform shape	2611.373	2174.499	2421.595
$Q = 50 \times I$ -optimal shape	2540.470	2098.374	2388.850
$Q = 10 \times I$ -uniform shape	679.856	630.266	670.482
$Q = 10 \times I$ -optimal shape	663.834	621.442	638.356

Table 10.4: Optimal Feedback Constants with Different $Q = 100 \times I$, $50 \times I$, $10 \times I$.

Beam Shape	$\{\Theta(0,t)\}$	$\{\Theta(0,t), w\}$	$\{\Theta(0,t), v\}$
$Q = 100 \times I$ -uniform shape	3.674	(2.7599-422.398)	(2502.705-2499.521)
$Q = 100 \times I$ -optimal shape	3.686	(2.498-337.071)	(1768.423-1765.672)
$Q = 50 \times I$ -uniform shape	(2.7599-422.398)	(2.279-395.458)	(2.7599-422.398)
$Q = 50 \times I$ -optimal shape	(2.759-422.398)	(2.096-335.370)	(2.423-378.554)
$Q = 10 \times I$ -uniform shape	1.659	(1.604-167.576)	(1138.098-1194.508)
$Q = 10 \times I$ -optimal shape	1.653	(1.386-166.142)	(1045.568-1065.754)

Fig. 10.12 – 10.13 are for $\Theta(0,t)$ feedback (hub tangent angle) with different weighting matrix Q. As usual, the optimal shapes have relatively large dimensions at the hub side.

Fig. 10.14 – 10.15 are for $(\Theta(0,\mathrm{t}), \omega(\mathrm{L,t}))$ feedback (hub tangent angle, tip deflection) with a different weighting matrix Q. The optimal shapes have a relatively large dimension at the middle.

Fig. 10.16 – 10.17 are for $(\Theta(0,t), v(L,t))$ feedback (hub tangent angle, tip position) with different weighting matrix Q. The optimal shapes have relatively small dimensions in the middle.

Fig. 10.18 – 10.19 are the hub tangent angle responses to initial state $q_1(0)$ for $\Theta(0,t)$ (hub tangent angle) and $(\Theta(0,t), w(L,t))$ (hub tangent angle, tip deflection) feedback separately.

Fig. 10.20 – 10.21 are the hub tangent angle responses to step input for $\Theta(0,t)$ (hub tangent angle) and $(\Theta(0,t), w(L,t))$ (hub tangent angle, tip deflection) feedback separately.

Fig. 10.22 – 10.23 are the tip deflection responses to initial $q_1(0)$ for $\Theta(0,t)$ (hub tangent angle) and ($\Theta(0,t), w(L,t)$) (hub tangent angle, tip deflection) feedback separately.

Fig. 10.24 – 10.25 are the tip deflection step input responses for $\Theta(0,t)$ (hub tangent angle) and ($\Theta(0,t), w(L,t)$) (hub tangent angle, tip deflection) feedback separately.

These results show that the performance index does not depend much on the number of segments, because according to this mechatronic algorithm, the final shape converges to its optimal shape. A system with two output feedbacks has a relatively lower performance index than one with only one output feedback, since the former has a higher degree of freedom to work with. The optimal shape for $\Theta(0,t)$ feedback is always a larger at the hub side, and smaller at the tip end. The optimal shape to ($\Theta(0,t), w(L,t)$) has smaller dimensions at the ends, while the optimal shape to { $\Theta(0,t), v(L,t)$ } is smaller in the middle. For optimal shapes, the initial responses have less vibration and fast convergence. The step input response of hub tangent angle for the optimal shape has a smaller rising time and the tip deflection converges much faster than those of uniform shapes.

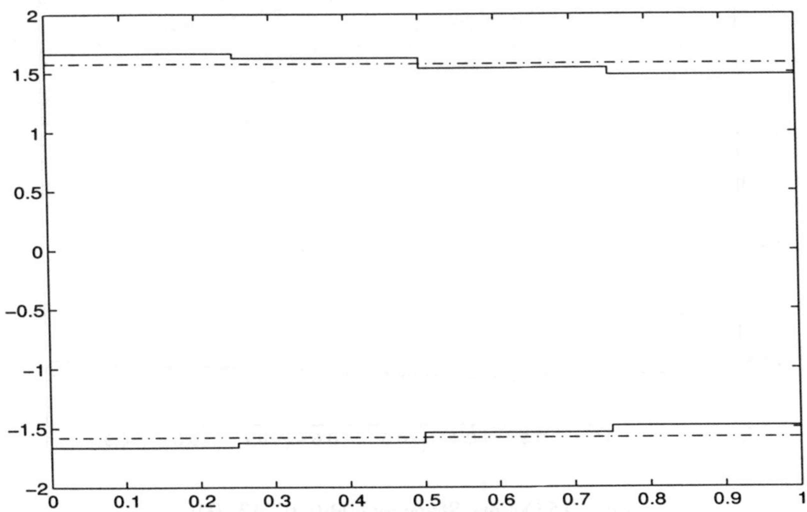

Fig. 10.3 Optimal Shape for ({$\Theta(0, t)$}, 4,100)

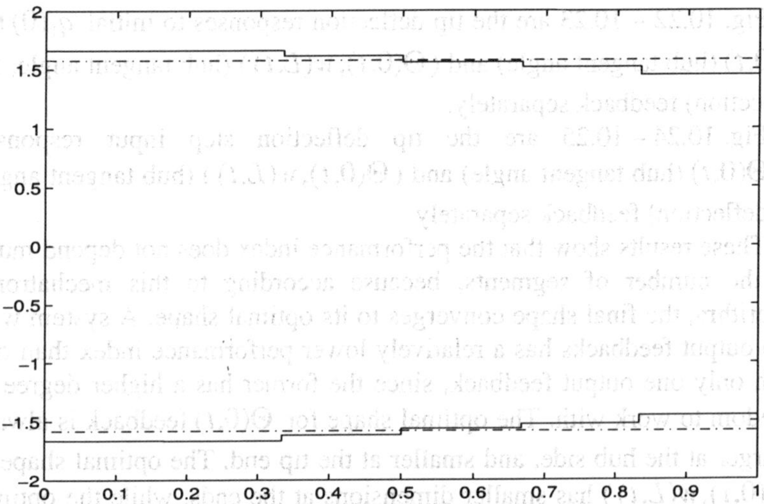

Fig. 10.4 Optimal Shape for ({Θ(0, *t*)}, 6,100)

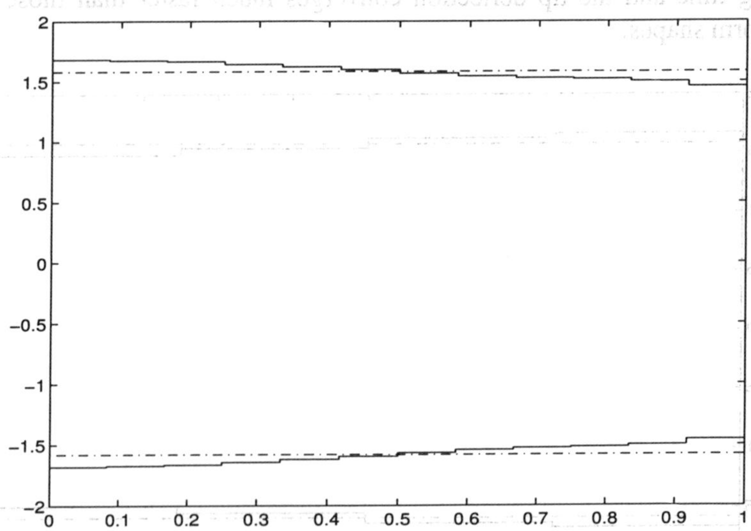

Fig. 10.5 Optimal Shape for ({Θ(0, *t*)}, 12,100)

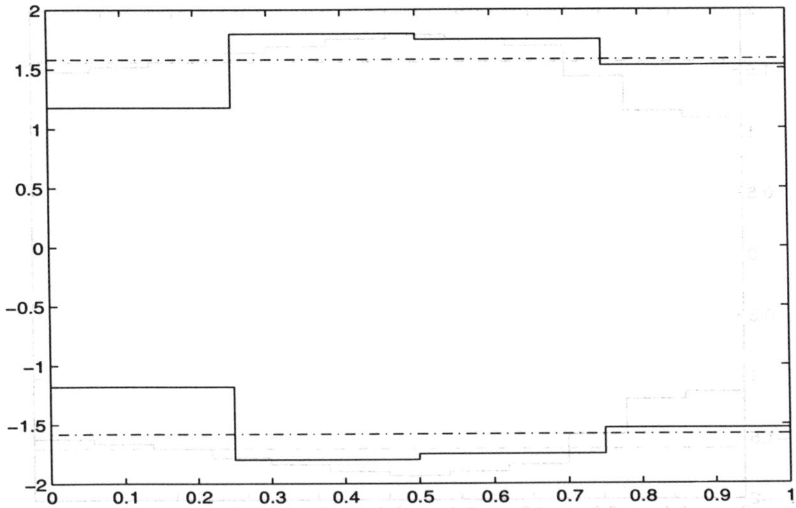

Fig. 10.6 Optimal Shape for ($\{\Theta(0,t), w\}$,4,100)

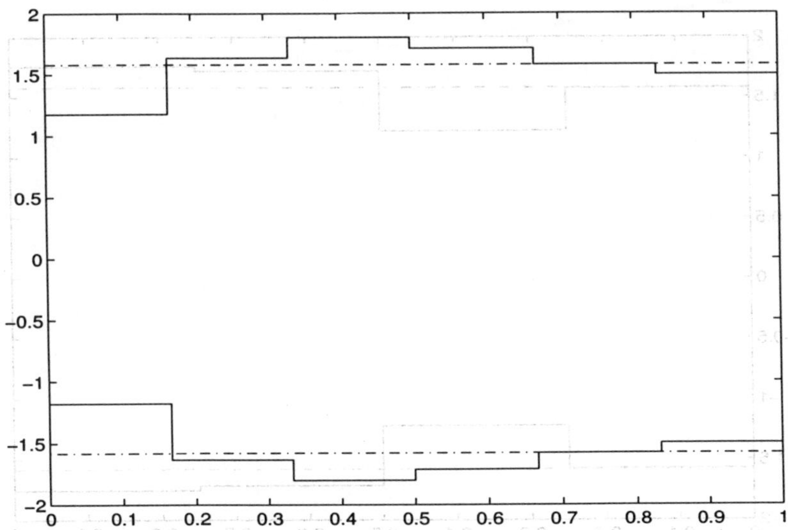

Fig. 10.7 Optimal Shape for ($\{\Theta(0,t), w\}$,6,100)

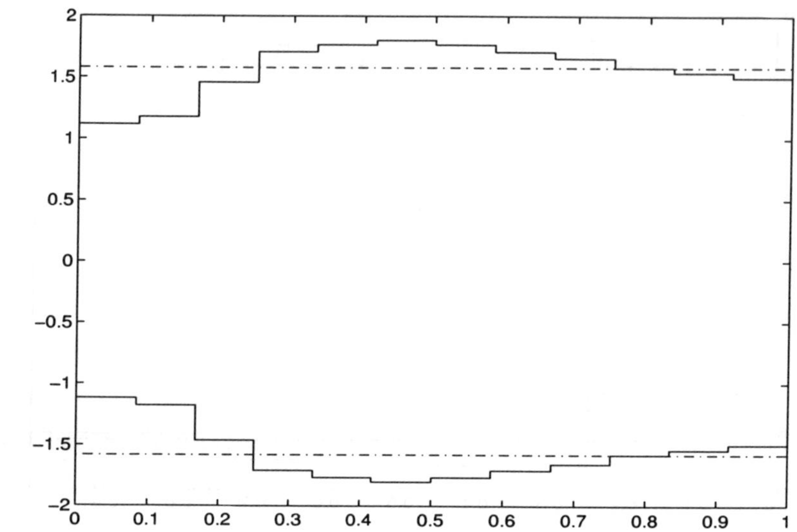

Fig. 10.8 Optimal Shape for ($\{\Theta(0,t), w\}$,12,100)

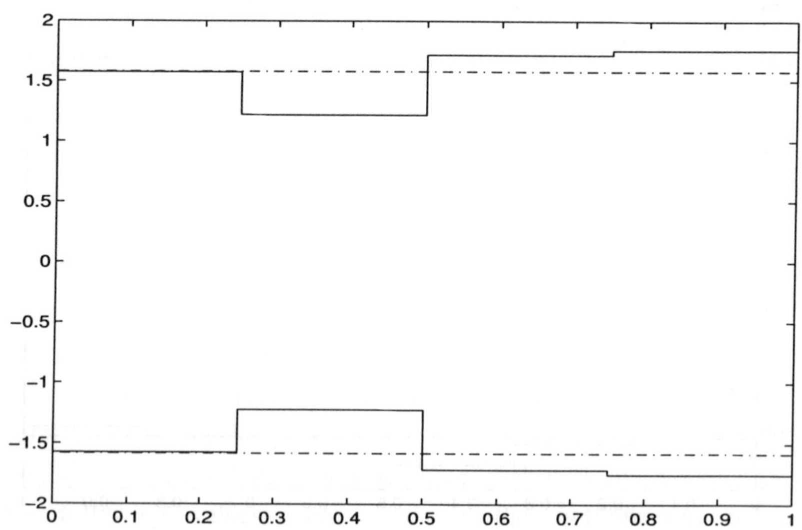

Fig. 10.9 Optimal Shape for ($\{\Theta(0,t), v\}$,4,100)

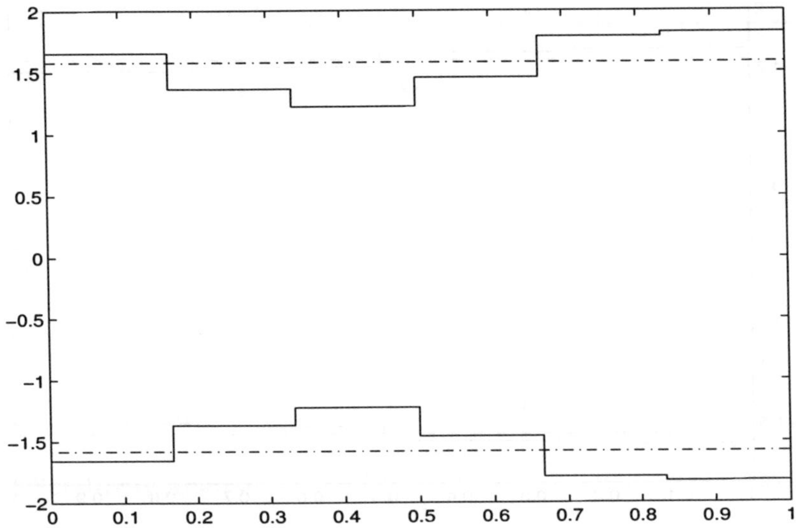

Fig. 10.10 Optimal Shape for ($\{\Theta(0,t), v\}$,6,100)

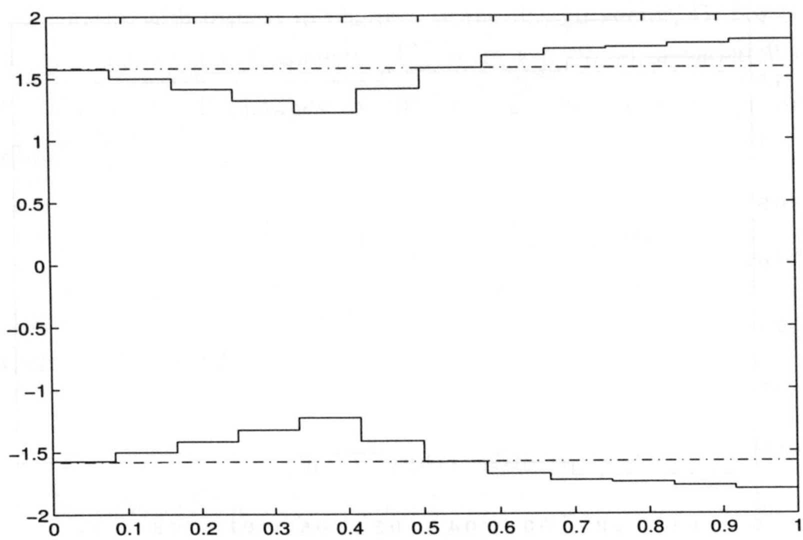

Fig. 10.11 Optimal Shape for ($\{\Theta(0,t), v\}$,12,100)

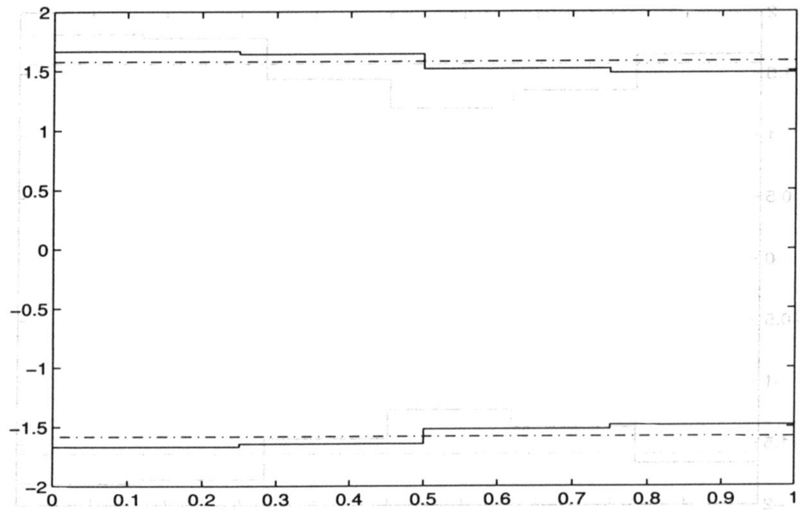

Fig. 10.12 Optimal Shape for ($\{\Theta(0,t)\},4,50$)

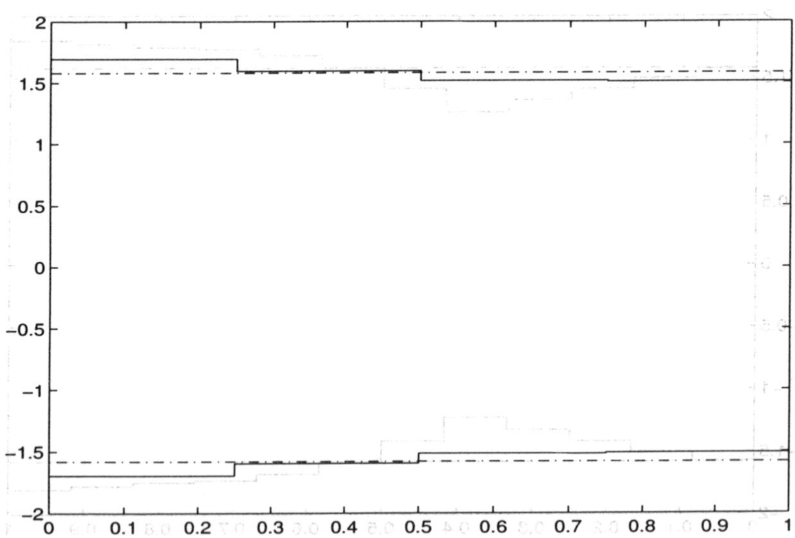

Fig. 10.13 Optimal Shape for ($\{\Theta(0,t)\},4,10$)

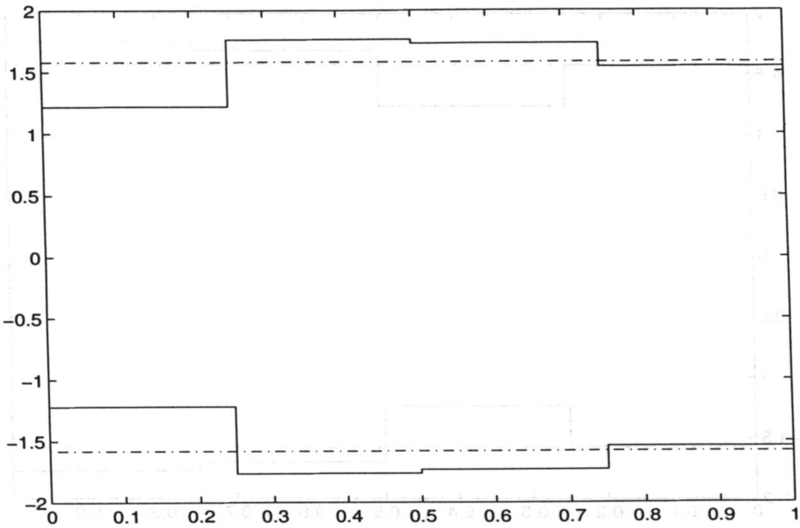

Fig. 10.14 Optimal Shape for $(\{\Theta(0,t), w\}, 4, 50)$

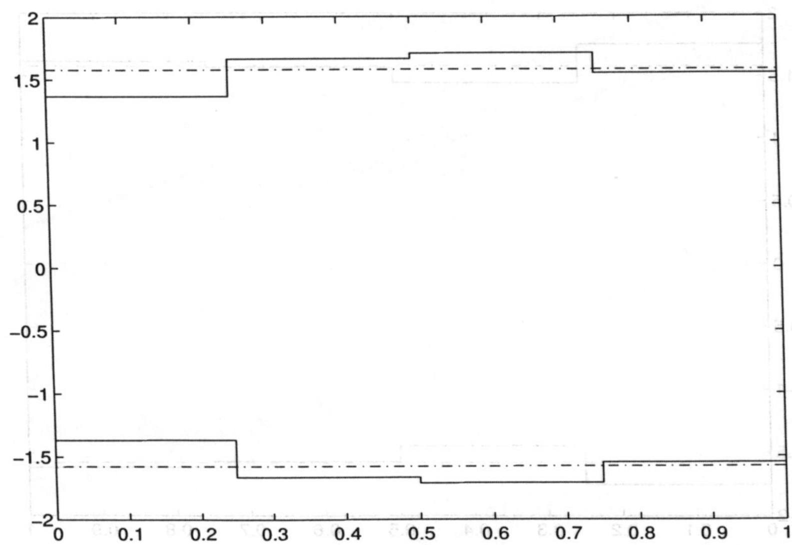

Fig. 10.15 Optimal Shape for $(\{\Theta(0,t), w\}, 4, 10)$

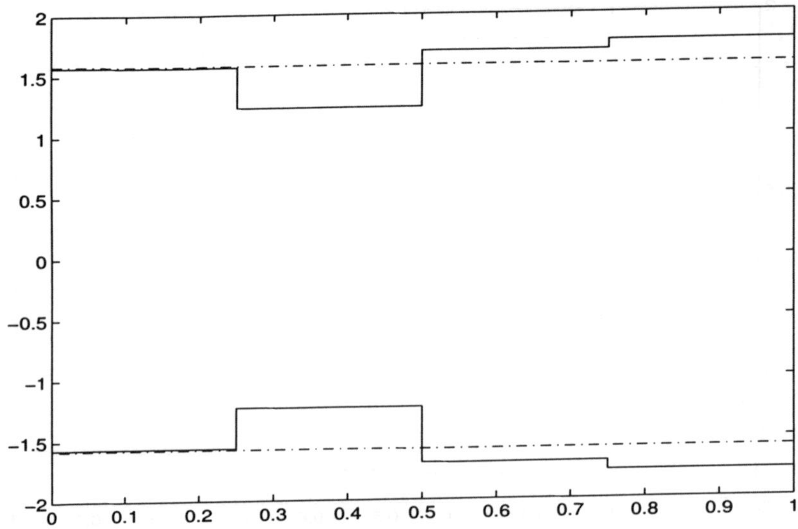

Fig. 10.16 Optimal Shape for ($\{\Theta(0,t), v\}$,4,50)

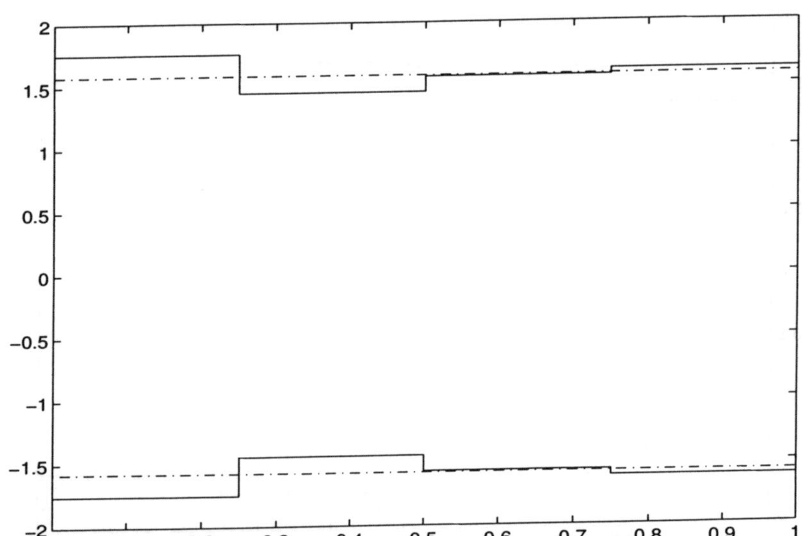

Fig. 10.17 Optimal Shape for ($\{\Theta(0,t), v\}$,4,10)

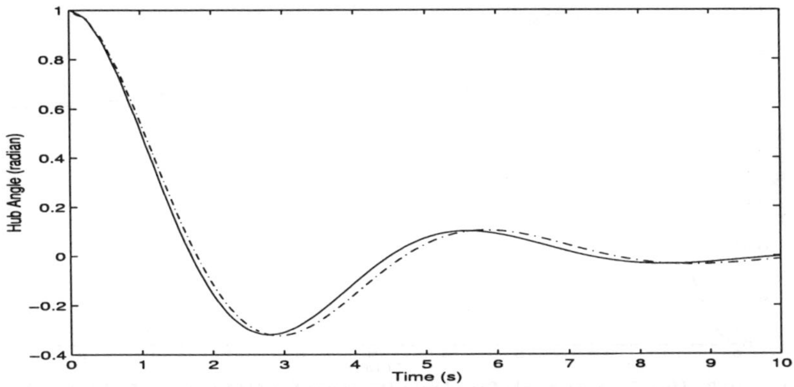

Fig 10.18 Hub Tangent Angle Initial Response for $(\{\Theta(0,t)\},4,100)$

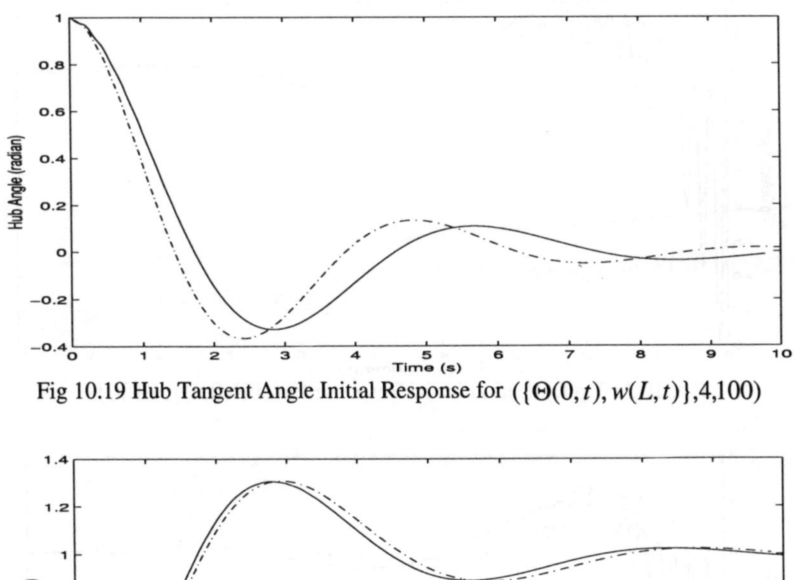

Fig 10.19 Hub Tangent Angle Initial Response for $(\{\Theta(0,t), w(L,t)\},4,100)$

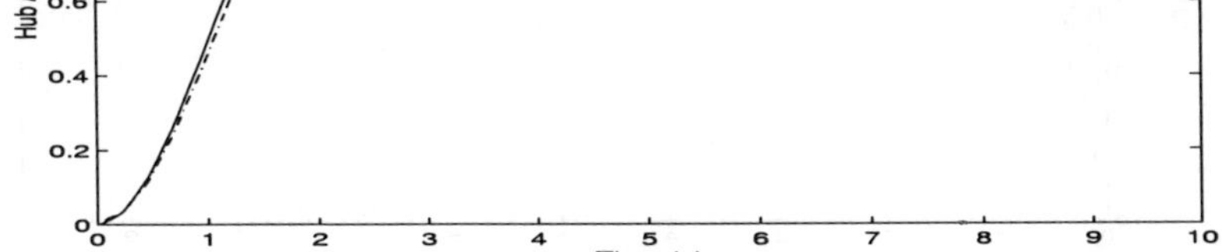

Fig. 10.20 Hub Tangent Angle Step Input Response for $(\{\Theta(0,t)\},4,100)$

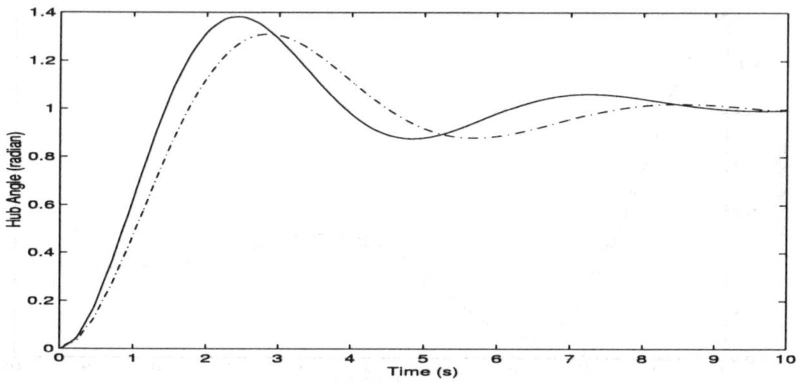

Fig. 10.21 Hub Tangent Angle Step Input Response for $(\{\Theta(0,t), w(L,t)\},4,100)$

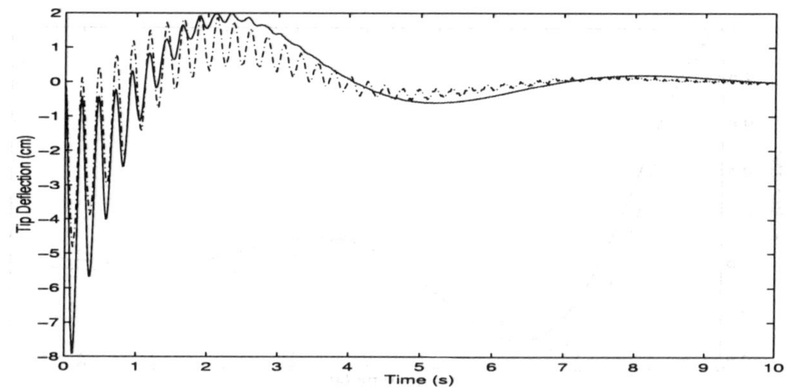

Fig. 10.22 Tip Deflection Initial Response for $(\{\Theta(0,t)\},4,100)$

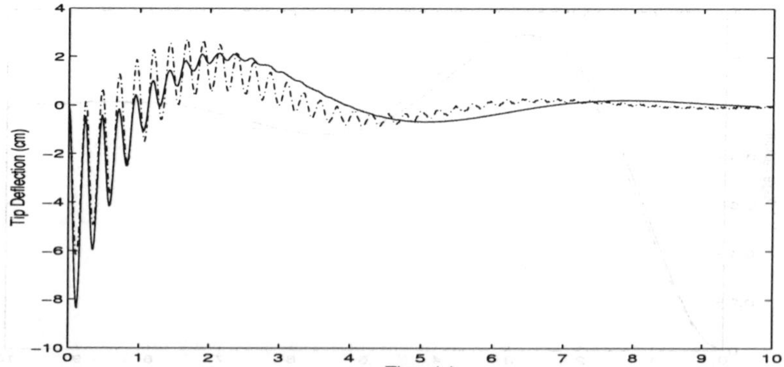

Fig. 10.23 Tip Deflection Initial Response for $(\{\Theta(0,t), w(L,t)\},4,100)$

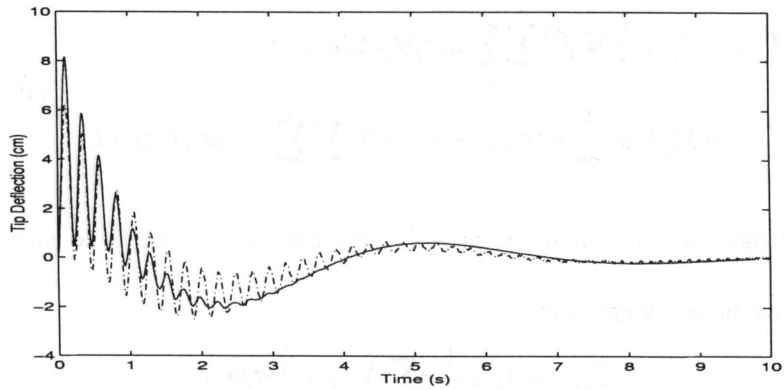

Fig. 10.24 Tip Deflection Step Input Response for $(\{\Theta(0,t)\},4,100)$

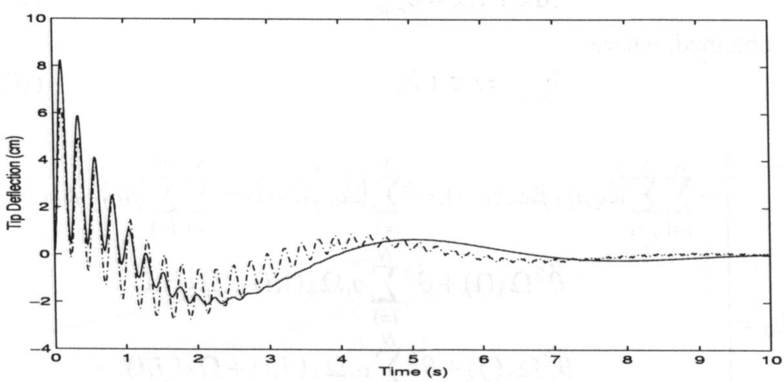

Fig. 10.25 Tip Deflection Step Input Response for $(\{\Theta(0,t), w(L,t)\},4,100)$

10.4 Case II: Mechatronic Design of Flexible Manipulators-based on H_∞ Controller with IHR Algorithm

10.4.1 *Generalized Plant of a Flexible Beam for* H_∞

In Eq.10.6, the omitted high order terms in the kinetics of the tip load are considered to be $\frac{1}{2}M_p\dot{\theta}^2 w(L,t)^2 + M_p(\dot{\theta} + \dot{w}(L,t))w(L,t)\dot{\theta}b_c$,

then Eq.10.7 is,

$$T_{pnew} = T_p + \frac{1}{2} M_p \dot{\theta}^2 \sum_{i=1}^{N} \sum_{j=1}^{N} q_i q_j \psi_i(L) \psi_j(L)$$
$$+ M_p b_c \dot{\theta}^2 \sum_{i=1}^{N} q_i \psi_i(L) + M_p b_c \dot{\theta}^2 \sum_{i=1}^{N} \sum_{j=1}^{N} q_i \dot{q}_j \psi_i(L) \psi_j(L) \qquad (10.39)$$

In the same way, if the term $\frac{1}{2}\int_0^L (w\dot{\theta})^2 \rho dx$ in Eq.10.8 is considered, then the link kinetics is,

$$T_{bnew} = T_b + \frac{1}{2} \dot{\theta} \sum_{i=1}^{N} \sum_{j=1}^{N} q_i q_j \int_0^L \psi_i \psi_j \rho dx \qquad (10.40)$$

After going through all derivative processes,

$$M\ddot{x} + Kx = b_{new} \qquad (10.41)$$

is obtained, where

$$b_{new} = b\tau + b_0 \qquad (10.42)$$

$$b_0 = \begin{bmatrix} -\sum_{i=1}^{N}\sum_{j=1}^{N}\left(\dot{\theta} q_i q_j\right)\Omega_4(i,j) - 2\sum_{i=1}^{N}\left(\dot{\theta} q_i\right)\Omega_5(i) - \sum_{i=1}^{N}\sum_{j=1}^{N}\left(q_i \dot{q}_j\right)\Omega_6(i,j) \\ \dot{\theta}^2 \Omega_5(1) + \dot{\theta}^2 \sum_{i=1}^{N} q_i \Omega_4(1,i) + \Omega_6(1,i) \\ \dot{\theta}^2 \Omega_5(j) + \dot{\theta}^2 \sum_{i=1}^{N} q_i \Omega_4(j,i) + \Omega_6(j,i) \\ \dot{\theta}^2 \Omega_5(N) + \dot{\theta}^2 \sum_{i=1}^{N} q_i \Omega_4(N,i) + \Omega_6(N,i) \end{bmatrix} \qquad (10.43)$$

$$\Omega_4(i,j) = M_p \psi_i(L) \psi_j(L) + \int_0^L \psi_i \psi_j \rho dx$$
$$\Omega_5(i) = M_p b_c \psi_i(L)$$
$$\Omega_6(i,j) = M_p b_c \psi_i(L) \psi_j(L)$$

Then Eq.10.31 will be,

$$\dot{q} = Aq + Bu + v_0 \qquad (10.44)$$

where

$$v_0 = \begin{bmatrix} 0 \\ \overline{M}^{-1} b_0 \end{bmatrix} \quad (10.45)$$

In this model, v_0 is a time-varying term, acting like a system disturbance but originating from model uncertainties. To describe these uncertainties, a frequency weighting function in practice in form of

$$v_0(s) = W_1(s)v(s) \quad (10.46)$$

is introduced, where

$$W_1(s) = C_{w1}(sI - A_{w_1})^{-1} B_{w_1} + D_{w_1} \quad (10.47)$$

$W_1(s)$ will be specified later.

If the fact that measured output y is affected by measuring noise and higher frequency model uncertainties is considered, then the system state-space model in Eq.10.25 turns into,

$$\dot{q} = A_g q + Bu + D_g v_0 \quad (10.48)$$

$$y = Cq + w_0 \quad (10.49)$$

Where $q = [\theta \quad q_1 \quad q_2 \quad \ldots \quad \dot{q}_1 \quad \dot{q}_1 \quad \ldots]'_{n \times n}$, D_g is an identity matrix with the same dimension of q

Following the same pattern, to describe w_0,

$$w_0(s) = W_2(s)w(s) \quad (10.50)$$

is introduced, where $w_2(s) = C_{w2}(sI - A_{w2})^{-1} B_{w2} + D_{w2}$. These coefficients will be specified in next section.

Naturally beam system stability is the main concern, therefore, the regulated variables are chosen as,

$$z_g = q = H_g x \quad (10.51)$$

where H_g is an identity matrix. The error vector' is

$$z = \begin{bmatrix} z_1 \\ z_2 \end{bmatrix} = \begin{bmatrix} \Theta z_g \\ ru \end{bmatrix} \quad (10.52)$$

where $\Theta = diag(\Theta_1 \quad \Theta_2 \quad \ldots \quad \Theta_n)$ is the weighting matrix on variable z_g, and Υ is the scalar factor. All states and the control variables are in the 'error vector', and the states are expected to die out to stabilize the closed-loop under the regulation of K, while the control variable will also become zero thus saving energy consumption.

Let the state variable vector be as

$$x = \begin{bmatrix} q \\ x_{w1} \\ x_{w2} \end{bmatrix}, d = \begin{bmatrix} v \\ w \end{bmatrix} \tag{10.53}$$

where x_{w_1} and x_{w_2} are the states of the frequency weighting function $w_1(s)$ and $w_2(s)$ respectively.

After some manipulations, the state-space representation by Doyle and Glover [1] [4] referred to is

$$A = \begin{bmatrix} A_g & D_g C_{w1} & 0 \\ 0 & A_{w1} & 0 \\ 0 & 0 & A_{w2} \end{bmatrix}, B_1 = \begin{bmatrix} D_g D_{w1} & 0 \\ B_{w1} & 0 \\ 0 & B_{w2} \end{bmatrix}, B_2 = \begin{bmatrix} B \\ 0 \\ 0 \end{bmatrix} \tag{10.54}$$

$$C_1 = \begin{bmatrix} \Theta H_g & 0 & 0 \\ 0 & 0 & 0 \end{bmatrix}, D_{11} = \begin{bmatrix} 0 & 0 \\ 0 & 0 \end{bmatrix}, D_{12} = \begin{bmatrix} 0 \\ rI \end{bmatrix},$$

$$C_2 = \begin{bmatrix} C & 0 & C_{w2} \end{bmatrix}, D_{21} = \begin{bmatrix} 0 & D_{w2} \end{bmatrix}, D_{22} = \begin{bmatrix} 0 \end{bmatrix},$$

After substituting $u(s) = -F(s)y(s)$, the transfer functions

$$T_{z_1 v} = \Theta H_g \phi (I + BFC\phi)^{-1} D_g W_1 \tag{10.55}$$

$$T_{z_2 v} = -rFC\phi (I + BFC\phi)^{-1} D_g W_1 \tag{10.56}$$

$$T_{z_1 w} = -\Theta H_g \phi BF (I + G_g F)^{-1} W_2 \tag{10.57}$$

$$T_{z_2 w} = -rF (I + G_g F)^{-1} W_2 \tag{10.58}$$

are obtained, where

$$\phi = (sI - A_g)^{-1}, G_g = C(sI - A_g)^{-1} B \tag{10.59}$$

From above equations, the H_∞ control problem now is to design $F(s)$ to internally stabilize, the system but satisfy

$$\left\| \begin{matrix} T_{z1v} & T_{z1w} \\ T_{z2v} & T_{z12} \end{matrix} \right\|_\infty \le r \tag{10.60}$$

One such controller is

$$k(s) := \begin{bmatrix} A_\infty & -Z_\infty L_\infty \\ F_\infty & 0 \end{bmatrix} \tag{10.61}$$

where

$$A_\infty = A + r^{-2} B_1 B_1' X_\infty + B_2 F_\infty + Z_\infty L_\infty G_2 \tag{10.62}$$

$$F_\infty = -B_2' X_\infty$$

$$L_\infty = -Y_\infty C_2'$$

$$Z_\infty = (T - r^{-2} Y_\infty X_\infty)^{-1}$$

and X_∞, Y_∞ are the solutions of two Hamiltonian matrices are involved in the H_∞ control problem,

$$H_\infty := \begin{bmatrix} A & r^{-2} B_1 B_1' - B_2 B_2' \\ -C_1 C_1 & -A \end{bmatrix} \tag{10.63}$$

$$J_\infty := \begin{bmatrix} A' & r^{-2} C_1 C_1' - C_2 C_2' \\ -B_1 B_1' & -A \end{bmatrix} \tag{10.64}$$

The purpose of the H_∞ controller here is to find out the minimum r thus attenuating z to maximum degree while still holding the necessary and sufficient stable conditions discussed in [1] [4]. The implementation algorithm is presented in the following section.

10.4.2 H_∞ *Controller Design*

To design the H_∞ controller discussed in the last section, those weighting functions and factors need to be specified. Because v_0 is in relatively low frequency range for low frequency model uncertainties, $W_1(s)$ with the 'cutting frequency' around 250 Hz is chosen. So $W_1(s)$ is as,

$$W_1(s) = \frac{9.79\times10^3}{(1+s/(1.72\pi))(1+s/(2.0\pi))} \qquad (10.65)$$

$W_2(s)$ assures robustness for high frequency noise.

$$W_1(s) = \frac{5.21\times10^{-7}(1+s/(0.02\pi))(1+s/(0.1\pi))(1+s/(200.0\pi))}{(1+s/(6.0\pi))(1+s/(20.0\pi))(1+s/(160.0\pi))} \qquad (10.66)$$

The parameters Θ, and Υ are adjusted by the cost,

$$J(u,(v,w)) = \int_{-\infty}^{\infty}[z_g^T Q z_g + u^T Ru - (v^T v + w^T w)]dt \qquad (10.67)$$

where $Q = \Theta^T\Theta$ and $R = r^2$. The H_∞ can be interpreted to reduce this cost. Based on the observations in Fujita, et al [3], and our simulations, the weighting matrix Θ for our beam system with ten states is

$$\Theta = diag(60.0 \quad 80.36 \quad 15.91 \quad 10.7 \quad 5.1 \quad 15.1 \quad 2.1 \quad 6.2 \quad 5.2 \quad 5.1) \qquad (10.68)$$

and the scalar factor $r = 5\times10^{-5}$. The output feedback used here is as in Eq.10.29.

In order to search the minimum value r, and since it is not known in which region such an index r constitutes a controller, an adequate number of random r's (usually $r \le 1$) are tested. All the r's to which such a controller exists are saved, and the minimum r represents the cost function for its beam shape. The loop in the H_∞ controller is,

H_∞ loop

1. For a given beam construction distribution, calculate beam system dynamics A_g, B, C.
2. Calculate generalized plant matrices,
   Get the state-space matrices for frequency weighting functions
   $A_{w_1}, B_{w_1}, C_{w_1}, D_{w_1}; A_{w_2}, B_{w_2}, C_{w_2}, D_{w_2}$;
   Get weighting factor Θ and scalar factor Υ;
   Generate $A, B_1, B_2, C_1, C_2, D_{11}, D_{12}, D_{21}, D_{22}$.
3. Search for minimum index r loop for a beam shape distribution,
   A) randomly pick a r;

B) check X_∞, Y_∞, ρ conditions;

C) decide to save this r in a vector or discard it;

D) choose next r which is smaller than the last valid r, go to B);

E) after a certain number of iterations, single out the minimum r from the vector to represent this shape. Save this r and its shape.

4. Check stop criteria. If the differences between the three smallest r's in this r vector meet the criteria, stop here. Otherwise, go back to Step 3 for another routine;
5. Calculate F_∞, L_∞, Z_∞, A_∞ to get the controller $K(s)$ based on the minimum r and A_g, B, C.

10.4.3 *Simulation Results*

There are two blocks in the searching loop, the H_∞ controller block and *IHR* algorithm block. The former is used to find the best controller which has a minimum index r based on the input of beam geometric shape as specified in last chapter. The latter is used to generate a feasible beam shape distribution. This loop stops until the index vector meets criteria as described in the last chapter.

For a given beam shape, one such admissible H_∞ controller associated with an index r is shown as Eq.10.51, which conforms to all the required conditions. The beam consists of 4 segments, and the number of degrees of its mechanic dynamics is set to 10 (the first four modal shapes are used here).

For Θ feedback controller as Eq.10.23, Fig. 10.26 is one of the best shapes, having a minimum index of $r = 0.832$.

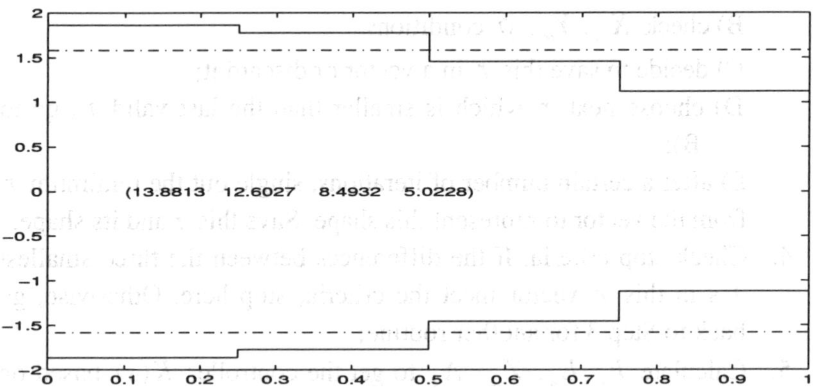

Fig. 10.26 Optimal Shapes for H_∞ Controller

The admissible H_∞ controller associated with the shape and this r is

$$-F(s) = 2.158\times10^3 \frac{(s+50.2655)(s+31.4159)(s+6.1133+18.8706i)}{(s+50.2577)(s+31.4219)(s+5.7432+18.5868i)}$$
$$\frac{(s+6.1133-18.8706i)(s+18.8496)(s+17.4371)(s+13.1358+11.0675i)}{(s+5.7432-18.5868i)(s+13.2079+11.0351i)(s+13.2079-11.0351i)(s+17.1713)}$$
$$\frac{(s+13.1358-11.0675i)(s+0.2660+9.2176i)(s+0.2660-9.2176i)}{(s+0.4437+9.9505i)(s+0.4437-9.9505i)(s+3.9752)(s+0.0376+5.0332i)}$$
$$\frac{(s+0.4664+3.9814i)(s+0.4664-3.9814i)(s+0.1299)(s+3.9720)}{(s+0.0376-5.0332i)(s+0.0658+2.4910i)(s+0.0658-2.4910i)(s+018.8489)} \qquad (10.69)$$

This controller is 15 degrees, ten of them having been contributed by the beam dynamics, two from the frequency weighting function $w_1(s)$, the rest originating from frequency weighting function $w_2(s)$.

This controller is tested with the same initial state as in Section 10. 3, $q_1(0) = [\sqrt{10},0,\ldots0]^T$. As usual, the dash curve is for uniform beam, the solid one is the response of the optimal beam shape. For Θ feedback, the initial response of the hub tangent angle is in Fig. 10.27. It has a small overshoot, but nevertheless steadily converges to the equilibrium point with several pulses. The optimal shape has faster convergency than the uniform. Fig. 10.28 is the associated initial responses of the tip deflection. It has a significant first deflective pulse and larger than that of uniform beam, but this deceases promptly, almost at the same speed as that of the hub tangent angle.

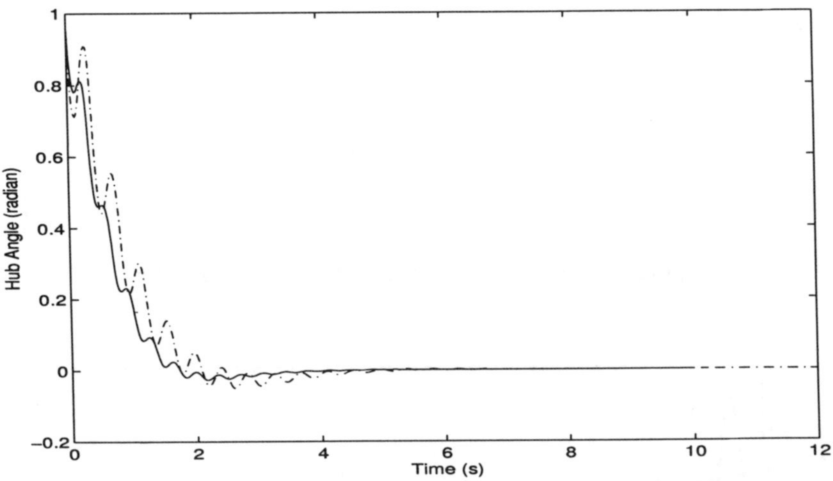

Fig. 10.27 Hub Tangent Angle Initial Response of $\Theta(H_\infty)$ Feedback

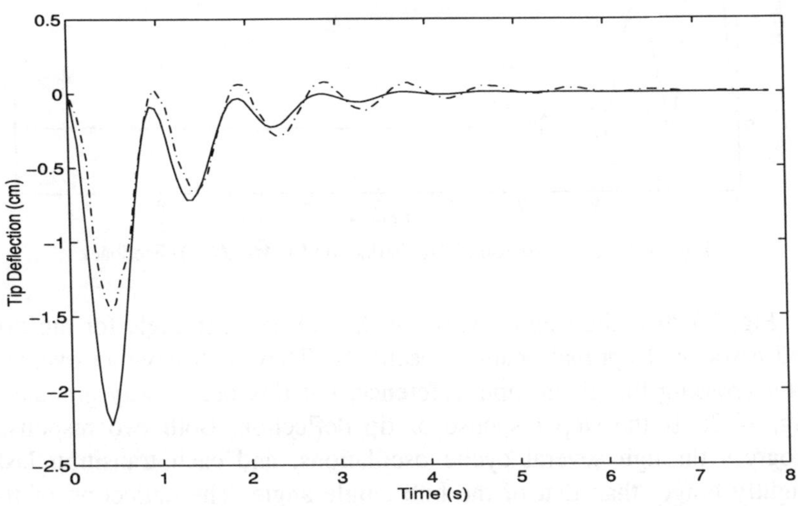

Fig. 10.28 Tip Deflection Initial Response of $\Theta(H_\infty)$ Feedback

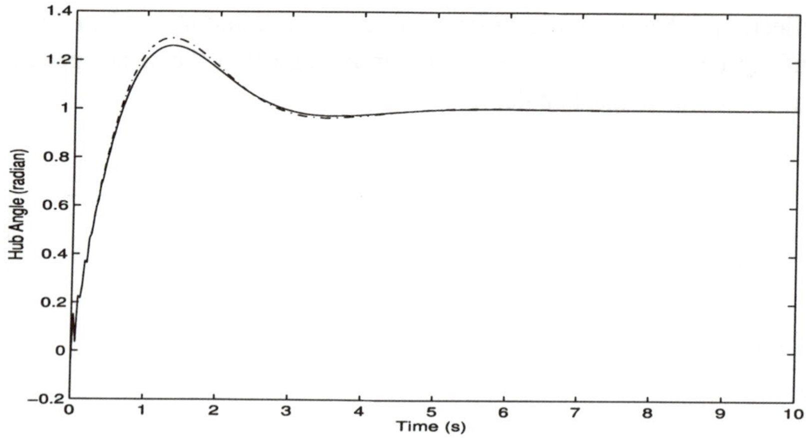

Fig 10.29 Step Response of Hub Angle for $\Theta(H_{\infty})$ Feedback

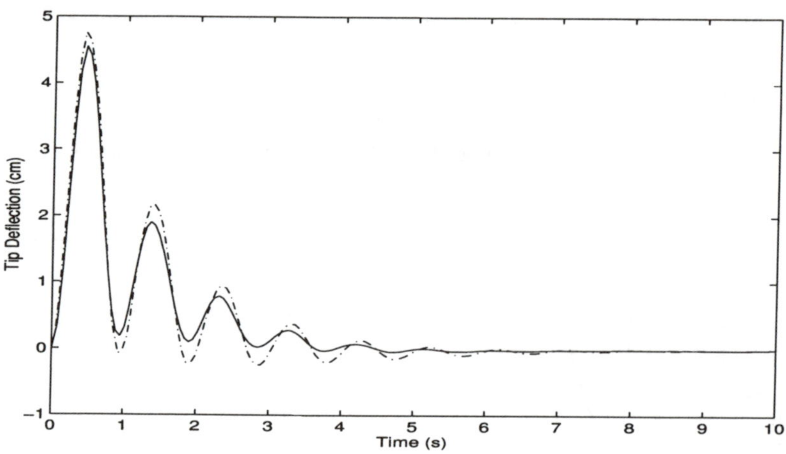

Fig.10.30 Step Response of Tip Deflection for $\Theta(H_{\infty})$) Feedback

Fig, 10.29 is the step response of the hub tangent angle for uniform and associated optimal beam respectively. They both have an overshot after crossing the given input reference, but this fades away gradually. Fig, 10.30 is the step response of tip deflection. Both two responses progress through several cyclic oscillations, and each transition lasts slightly longer than that of the hub tangle angle. The deflection of the optimal beam is smaller, and converges faster. The step-type disturbance is added as disturbance v for the stable test, as derived early, v is assumed to describe the system uncertainties. The results are as in Fig 10.31 and Fig 10.32. They have a notable deflective pulse, but recover to the equilibrium point promptly.

For tip deflection w feedback as Eq.10.21, the optimal shape is as Fig. 10.33 with r=0.952. The initial responses of the hub tangent angle is in Fig. 10.34. The response of the optimal beam has smaller vibration then the uniform beam, but both have higher frequency vibration then the responses of Θ feedback. Fig. 10.35 is the associated initial responses of the tip deflection. The optimal shape has a bigger first pulse, but decays faster for the whole transient.

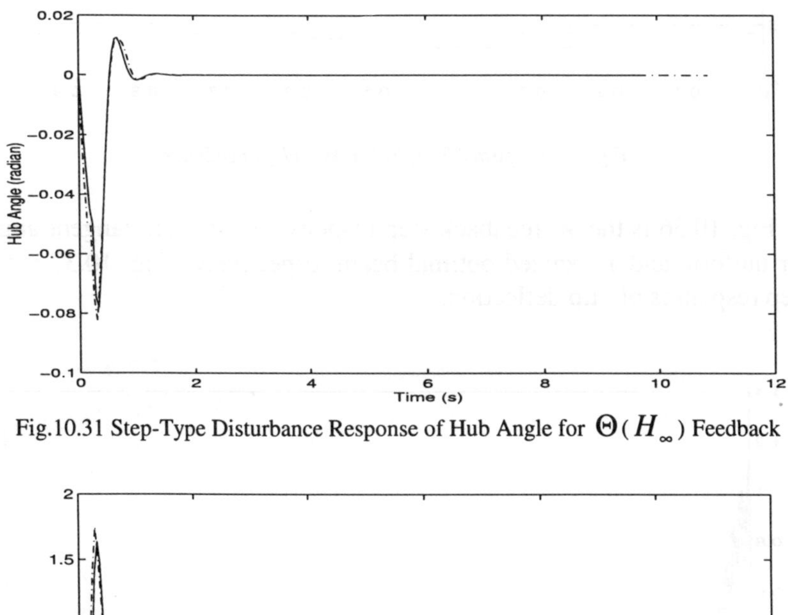

Fig.10.31 Step-Type Disturbance Response of Hub Angle for $\Theta(H_{\infty})$ Feedback

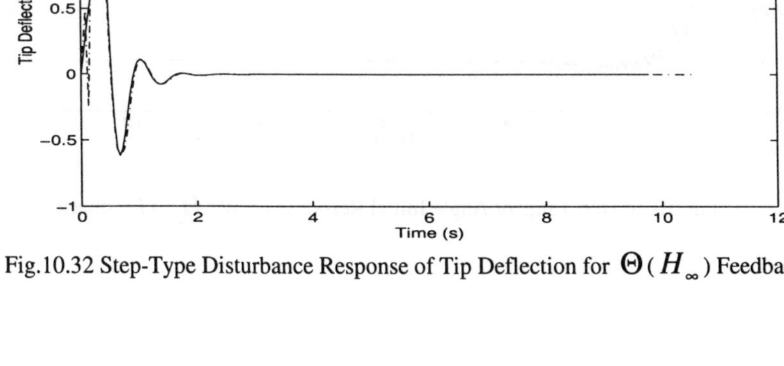

Fig.10.32 Step-Type Disturbance Response of Tip Deflection for $\Theta(H_{\infty})$ Feedback

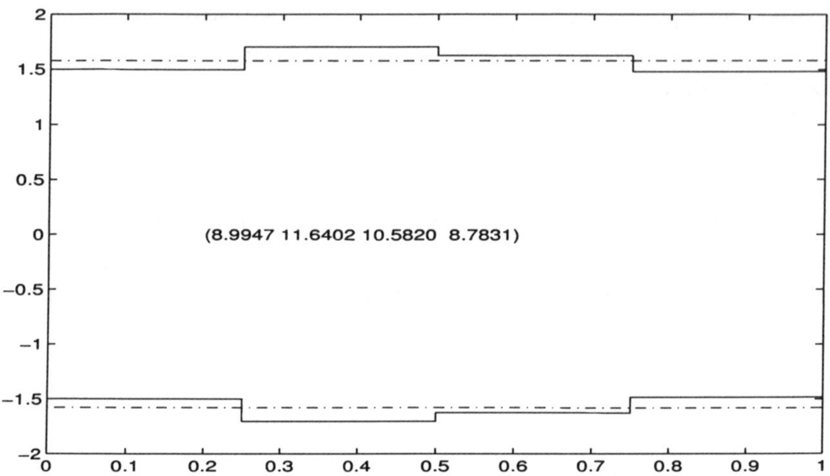

Fig.10.33 Optimal Shapes for $w\,(H_\infty)$ Feedback

Fig. 10.36 is the w feedback step responses of the hub tangent angle for uniform and associated optimal beam respectively. Fig. 10.37 is the step responses of tip deflection.

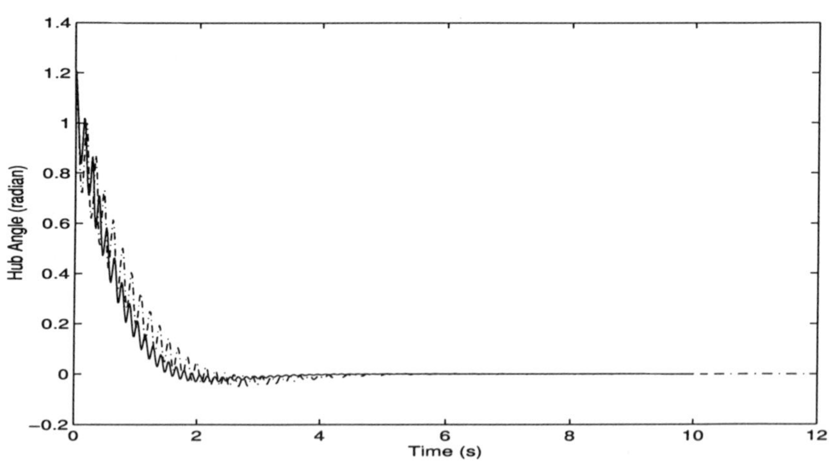

Fig. 10.34 Hub Tangent Angle Initial Response of $w\,(H_\infty)$ Feedback

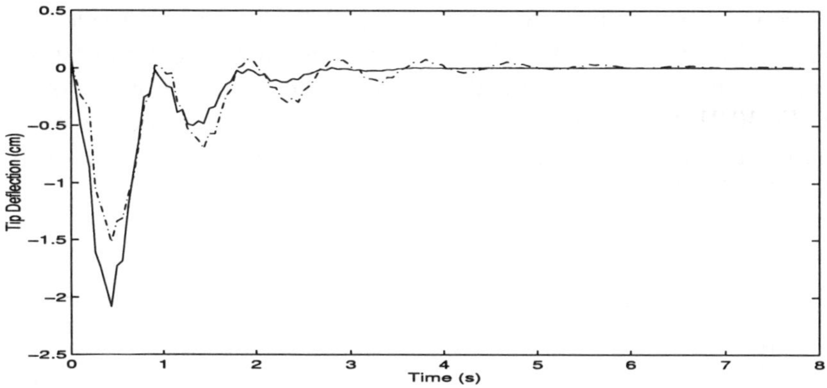

Fig. 10.35 Tip Deflection Initial Response of $w\,(H_\infty)$ Feedback

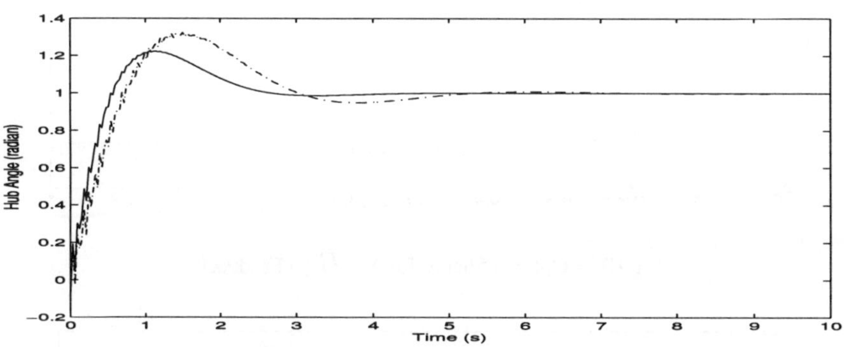

Fig. 10.36 Step Response of Hub Angle for $w\,(H_\infty)$ Feedback

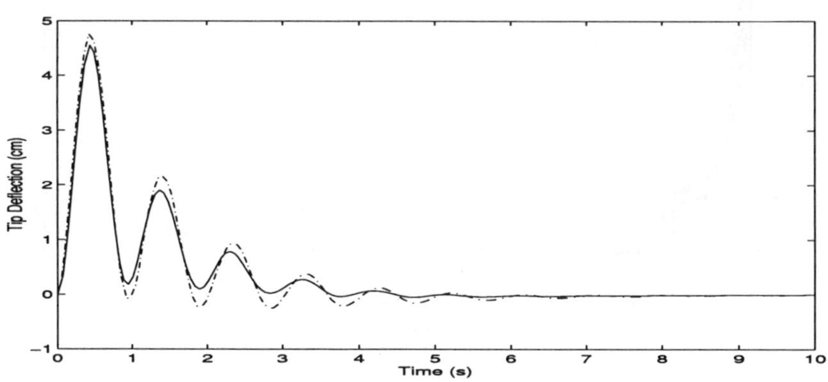

Fig. 10.37 Step Response of Tip Deflection for $w\,(H_\infty)$ Feedback

For the tip position v feedback as Eq.10.22, the optimal shape is as Fig.10.38 with a $r = 0.8162$. The initial response of the hub tangent angle is in Fig.10.39. Fig.10.30 is the associated initial responses of the tip deflection. Fig. 10.41 is tip position feedback step responses of the hub tangent angle for uniform and associated optimal beam respectively. Fig. 10.42 is the step responses of tip deflection.

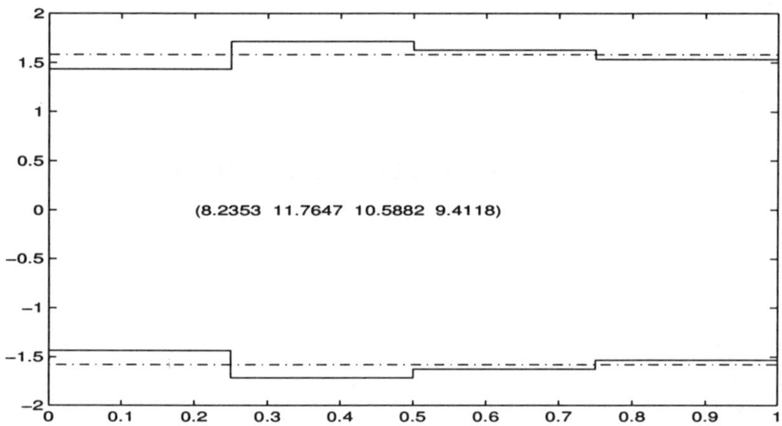

Fig.10.38 Optimal Shapes for v (H_∞) Feedback

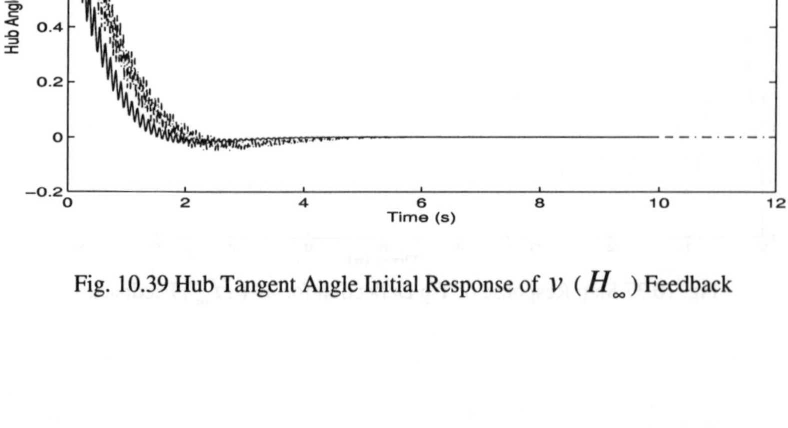

Fig. 10.39 Hub Tangent Angle Initial Response of v (H_∞) Feedback

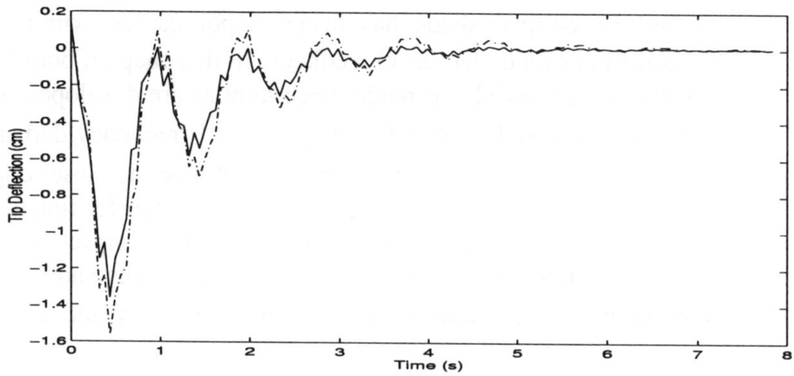

Fig. 10.40 Tip Deflection Initial Response of ν (H_∞) Feedback

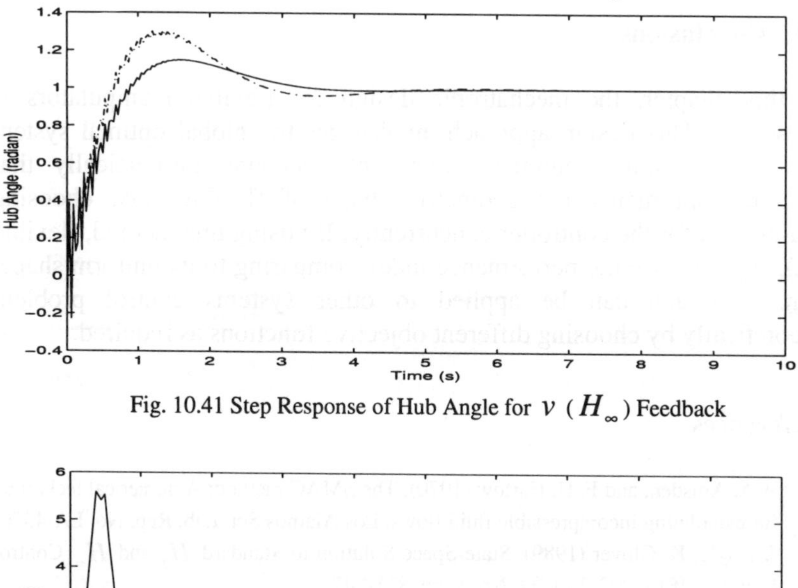

Fig. 10.41 Step Response of Hub Angle for ν (H_∞) Feedback

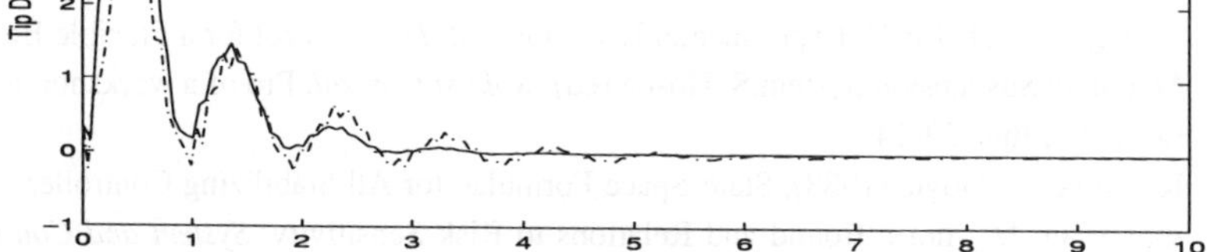

Fig. 10.42 Step Response of Tip Deflection for ν (H_∞) Feedback

From here, the optimal shape has faster responses less vibration cycles than uniform beam under the Θ feedback. In this chapter, both the generalized state-space model and model uncertainties are developed for H_∞ controller to completely avoid the complexity of frequency domain implementation. The design goal is to internally stabilize the close-loop, whiling minimizing the maximum singular value of the transfer function from characterized error inputs to error outputs, so that impacts from disturbances and system uncertainties vanish. All the results show that the H_∞ controller here is stable against system uncertainty and works robustly.

10.5 Conclusions

In this chapter, the mechatronic design for flexible manipulators is presented. This design approach emphasizes the global optimal system design by taking coupling effects into account automatically that includes optimizing the geometric shape of flexible link, choosing parameters for the controller concurrently. By using this method, flexible link can improve the performance index comparing to its uniform shape. This approach can be applied to other systems control problem theoretically by choosing different objective functions as required.

References

1. A.A. Amsden, and F. H. Harlow (1970). The SMAC method: A numerical technique for calculating incompressible fluid flows, Los Alamos Sci. Lab. Rep. No. LA-4370.
2. J. Doyle, K. Glover (1989), State-Space Solution to Standard H_2 and H_∞ Control Problem, IEEE AC, Vol. 34, No. 8, pp. 831-847.
3. B.Francis, A Course in H_∞ Control Theory, Springer-Verlag, Berlin-New York, 1998.
4. M. Fujita, et al (1991), Experimental Evaluation of H_∞ Control for a Flexible Beam Magnetic Suspension System,S. Hosoe (ed), *Robust Control*, Pro of a workshop held in Toyko, June 23-24.
5. K. Glover, J. Doyle (1988), State-Space Formulae for All Stabilizing Controller that satisfy an H_∞-norm Bound and Relations to Risk Sensitivity, *System and Control Letter*, Vol. 11, pp. 167-172.

6. L. B. Gutierrez, F. L. Lewis, J. A. Lowe (1998), Implementation of a Neural Network tracking Controller for a Single Flexible link: Comparison with PD and PID Controller, IEEE Trans. on Industrial Electronics, Vol. 45, No. 2,
7. A.Jnifene, A. Fahim (1997), A Computed Torque/Time Delay Approach to the End-Point Control of a One-Link Flexible Manipulator, *Dynamics and Control*, Vol. 7, pp. 177-189.
8. B. L. Karihaloo and F. I. Niordson (1973), Optimum Design of Vibrating Cantilevers, Journal of Optimization Theory and Applications, Vol. 11, No. 6, pp. 638-654.
9. F. L. Lewis, V. L. Syrmos (1995), Optimal *Control*, John Wiley and Sons, Inc., Second Edition.
10. K. Lezn, H. ozbay, et al (1991), Frequency Domain Analysis and Robust Control Design for an Ideal Flexible Beam, *Automatica*, Vol. 27, pp. 947-961.
11. A.Tchernychev, A. Sideris, J. Yu (1997), Constrained H_∞ Control of an Experimental Flexible Link, *Trans. of the ASME*, June.
12. Fei-Yue Wang (1994), On the Extermal Fundamental Frequencies of One-Link Flexible Manipulators, *International Journal of Robotics Research*, Vol. 13, No. 2, pp. 162-170.
13. Fei-Yue Wang (1995), Optimum Design of Vibrating Cantilevers: A Classical Problem Revisited, *J. of Optimization Theory and Applications*, Vol. 84, No. 3.
14. Fei-Yue Wang and J. L. Russell (1992), Optimum Shape Construction of Flexible Manipulators with Tip Loads, *Proceedings of IEEE Intl.\ Conf.\ on Decision and Control* (Tucson, Arizona), Vol. 1, pp. 311-316.
15. Fei-Yue Wang and J. L. Russell (1993), Minimum-Weight Robot Arm for a Specified Fundamental Frequency, *Proceedings of the IEEE Conference on Robotics and Automation*, (Atlanta, Georgia), Vol. 1, pp. 490-495.
16. Fei-Yue Wang and J. T. Wen (1990), *Nonlinear Dynamical Model and Control for a Flexible Beam*, CIRSSE Report \#75, Rensselaer Polytechnic Inst., (Troy, New York).
17. Z. B. Zabinsky, R. L. Smith, J. F. McDonald, H. Edwin (1993), Improving Hit-and-Run for Global Optimization, *Journal of Global Optimization*, Vol. 3, pp. 171-192.
18. Pixuan Zhou, Fei-Yue Wang (1999), Mechatronic-Based Integrated Construction and Control of Flexible Manipulators, *14th IFAC World Congress*, Beijing, P. R. China, 5 - 9, July.
19. Pixuan Zhou, Fei-Yue Wang, Weinong, Chen, Paul L. Lever (2001), Optimal Construction and Control of Flexible Manipulators: a Case Study Based on LQR Output Feedback, *Mechatronics,* Vol. 11 (2001) 59-77.
20. P. Zhou, M. Willians, F. Y. Wang (1996), On the Closed-Loop Design of Flexible Robotic links, *IEEE Conf. on Systems, Man, and Cybernetics*, China.
21. G. Zhu, T. Lee, S. Ge (1997), Tip Tracking Control of a Single-Link Flexible Robot: A Backstepping Approach, in *Dynamics and Control*, 7, 341-360.

Chapter 11

A Comprehensive Study of Dynamic Behaviors of Flexible Robotic Links: Modeling and Analysis

This chapter presents a dynamic analysis of lightweight flexible robotic manipulators that includes the effects of rotary inertia and shear deformation. Dynamic models of flexible manipulators have been formulated using both the Euler-Bernoulli and Timoshenko beam theories. Based on a complete analysis of natural frequencies and modal shape functions of vibration, analytic expressions of step responses and general solutions of flexible manipulators have been obtained. Using the method of modal expansion, a finite dimensional nonlinear dynamic model has been derived. Explicit solutions of the asymptotic behavior of high order modal frequencies and vibration modes are given and verified with the numerical results. Both asymptotic analysis and numerical results indicate that the effect of shear deformation is significant for high frequency vibrations or shorter manipulators. Simulations indicate that the effects of both rotary inertia and shear deformation are significant and should be included in the high precision and high performance control design of flexible manipulators.

11.1 Introduction

The link flexibility of a robotic manipulator must be considered in modeling and control when the manipulator is of large dimension or lightweight. Large manipulators play important roles in many applications, such as construction automation, environmental applications, and space engineering. Lightweight arms have great potential as design of high-performance industrial robotic manipulators since it allows high speed operation and low energy consumption. However, due to the complexity of the link deformation which is a distributed parameter system accurate modeling and high performance control of flexible manipulators poses a major challenge in practical design.

The concept of using flexible robotic manipulators was proposed as early as 1970's by W. J. Book. Earlier models of arms are based on simple assumed mode method. Some method to improve the accuracy of assumed mode method can be found in Gu and Tongue [39]. Over the last decade, a significant effort has been made in modeling and control of one-link flexible manipulators, which are essential for the understanding of general multiple-link flexible manipulators. The most commonly used deformation model in the current robotic literature uses the Euler-Bernoulli beam theory. Based on this theory, various dynamic equations have been formulated for one-link flexible manipulators [1, 3, 15, 18, 25, 33]. The early works were conducted by Cannon and Schmitz [3] and Sakawa, Matsuno and Fukushima [18]. Khorrami and Ozguner, modeled the link by a set of integral-partial differential equations and used perturbation method to design control. Besides the link flexibility, Yang and Donath [35] also considered the flexibility of the joints. A comprehensive study on different dynamic equations under various boundary conditions for flexible manipulators were carried out by Bellezza, Lanari, and Ulivi [1]. Wang and Wen [25] derived the dynamic equations for flexible manipulators subject to large rotation during the deformation and clarified some issues related to the specification of boundary conditions. Several investigations had also been made from the perspective of mechanism where model complexity was not a concern [2, 13, 19]. Issues related to control design have been studied by De Luca and Scukuabo [10, 11], Tzes and Yurkovich [22], Wang and Vidyasagar [23, 24], Kelemen and Bagchi [14], and Cetinkunt and Yu [4]. Various problems of optimal manipulator design based on the Euler-Bernoulli model have been investigate by Wang and Russell [26, 27, 28, 29, 30].

Most efforts in the modeling and control of flexible manipulators have been on the understanding of a single flexible link. Due to the complexity involved with link deformation, however, establishing the exact dynamic model even for one link flexible manipulators is unrealistic, and certain simplifying approximations about the link deformation have to be made. In addition to small deformation, the most commonly used assumption in the robot literature is that it can be satisfactorily modeled by the Euler-Bernoulli beam theory. Under such an assumption, the effects of both rotary inertia and shear deformation of a flexible link are neglected. However, it is well known in mechanics, particularly in beam and plate theory, that both of them have significant effects on dynamic behaviors of flexible structures [7, 9 20, 12, 31, 32]. For example, it has been found that for shorter beams or thicker plates, the effects of rotary inertia or shear deformation could cause a magnitude of difference in the calculation of high order vibration frequencies. In [33] a similar conclusion has also been drawn by studying on influences of rotary inertia and shear deformation on vibration frequencies of flexible manipulators.

The purpose of this chapter is to conduct a comprehensive study on the dynamic effects of rotary inertia and shear deformation on one-link flexible manipulators. Section 11.2 presents basic equations for kinematic analysis and energy calculation. Based on Euler-Bernoulli and Timoshenko beam models, nonlinear dynamic equations and their linearized versions are given in section 11.3 and 11.4, respectively. Section 11.5 discusses influences of rotary inertia and shear deformation on vibrations, Furthermore, analytical expression of step responses and general solutions of flexible arms are derived in Section 11.6. Section 11.7 concludes the chapter.

11.2 Problem Description and Energy Calculations

11.2.1 *Problem Statement*

Consider a flexible manipulator carrying a tip load. It consists of a flexible beam attached on a rigid hub in the horizontal plane. The coordinate systems are shown in Fig. 11.1: $(x_{b,}\, y_b)$ is a fixed base coordinate system, (x, y) is the local coordinate system attached to the hub, and the Fig. 11.2 shows (a, b) is the local coordinate system attached to the tip load. It is assumed that initially the neutral longitudinal axis (x-axis) of the beam and a-axis of the tip load coincides with the x_b-axis.

The motion of the manipulator system is described by the rotation angle $\theta(t)$ of the hub, lateral displacement $w(x,t)$ and rotation

$\varphi(x,t)$ of the cross section beam from the deformation position. We will use a prime to indicate the derivative with respect to x, and a dot for derivative with respect to time. For example, $w' = \dfrac{\partial w}{\partial x}$, $\dot{w} = \dfrac{\partial w}{\partial t}$. Both Euler-Bernoulli and Timoshenko beam theories are investigated in this chapter. In the Euler-Bernoulli theory, the so-called normal plane assumption is used, i.e., it is assumed that any transverse section of the beam remains a plane and normal to the longitudinal axis of the beam after deformation [21]. In this case $\varphi = \partial w / \partial x = w'$ and the shear deformation of the beam is neglected. The Timoshenko theory takes the effect of shear deformation into account, replacing the normal plane assumption by the more accurate one plane assumption. The plane assumption states that any transverse section of the beam remains a plane but may not be perpendicular to the longitudinal axis of the beam after deformation [21]. As a result, beam rotation φ remains independent of w.

In the following, the equation of motion for the flexible manipulators will be derived by using the Hamilton's Principle [17],

$$\int_{t_0}^{t_f} (\delta T + \delta W - \delta P) dt = 0$$

Where T, P and W are the kinetic energy, potential energy of the system, and the work done by external forces respectively.

The total kinetic energy is calculated as follows,

$$T = \frac{1}{2} I_H \dot{\theta}^2 + T_b + T_p \tag{11.2.1}$$

Where I_H is the rotational inertia of the hub, T_b and T_p are the kinetic energies of the beam and tip load, respectively.

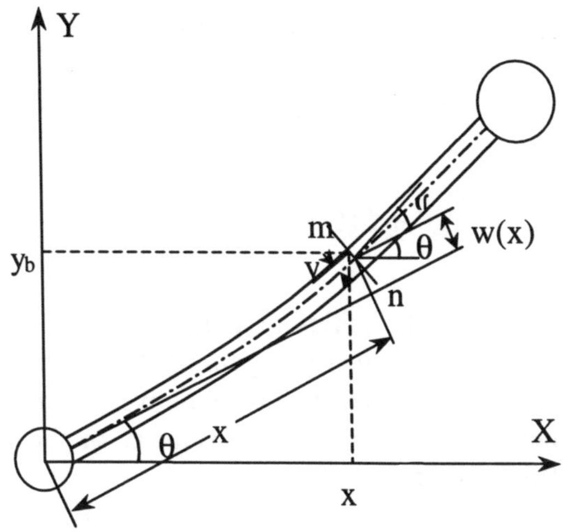

Fig. 11.1 A flexible beam fixed on a rigid hub in the horizontal plane

11.2.2 *Kinetic Energy of the Beam*

$$T_b = \int_0^L \iint_{sb} 1/2\rho_v(\dot{x}_b^{\,2} + \dot{y}_b^{\,2})ds_b \cdot dx \qquad (11.2.2)$$

let

$A = \iint_{sb} ds_b$; area of beam section

$y_c = \iint yds_b = 0$; coordinate of center

$I = \iint_{sb} y^2 ds_b$; moment of inertia

$\rho = \rho_v A$; mass density per unit length

sb is the area of the cross section;

ρv is the mass density per unit volume.

Let a point on the beam has coordinates (x, y) before deformation. It's position in base coordinate system is

$$x_b = x\cos\theta - w(x)\sin\theta - y\sin[\varphi(x)+\theta]$$

$$y_b = x\sin\theta + w(x)\cos\theta - y\cos[\varphi(x)+\theta]$$

$$T_b = \frac{1}{2}\int_0^L \rho[(x\dot{\theta} + \dot{w})^2 + (w\dot{\theta})^2 + S(\dot{\theta} + \dot{\varphi})^2]dx$$

Let $S = \dfrac{I}{A}$ and $\Delta_x = x\dot{\theta} + \dot{w}(x,t)$, $\Delta_y = \dot{\theta}w(x,t)$ (11.2.3)

Then $T_b = \dfrac{1}{2}\int_0^L \rho[\Delta_x{}^2 + \Delta_y{}^2 + S(\dot{\theta} + \dot{\varphi}^2)]dx$ (11.2.4)

Note that parameter S is the radius of gyration of the cross-section, reflecting the effect of rotary inertia. Ignoring the rotary inertia (set $S = 0$) implies that one assumes that velocity of every point on the same cross section is identical and equal to the velocity of the point at the neural axis on that cross section.

11.2.3 *Kinetic Energy of the Tip Mass*

The base coordinates of the point on the beam with $x = L$ after motion is,

$$x_c = L\cos\theta - w(L)\sin\theta$$
$$y_c = L\sin\theta + w(L)\cos\theta$$

the kinetic energy of the tip mass (which represents the payload) is,

$$T_p = \frac{1}{2}\iint_{sp} \rho_p(\dot{x}_p{}^2 + \dot{y}_p{}^2)ds_p \tag{11.2.5}$$

where, $ds_p = dadb$ is the area occupied by the tip load. ρ_p is the mass density per unit area tip mass.

The base coordinate of a point (a, b) on the tip load is,

$$x_p = L\cos\theta - w(L)\sin\theta + a\cos[\theta + \varphi(L)] - b\sin[\theta + \varphi(L)]$$
$$y_p = L\sin\theta + w(L)\cos\theta + a\sin[\theta + \varphi(L)] + b\cos[\theta + \varphi(L)]$$

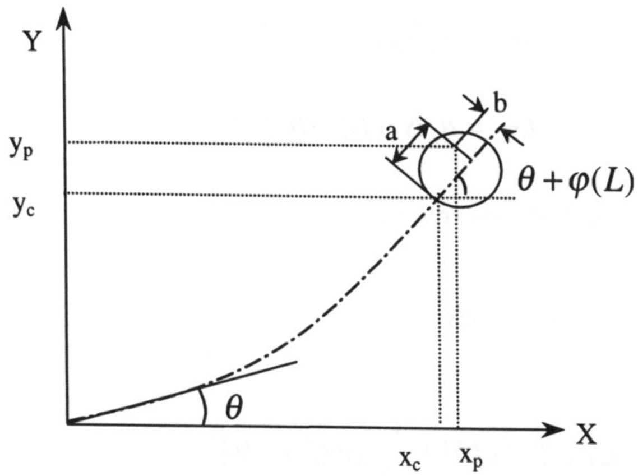

Fig.11.2 The relation between local and base coordinate for tip

Where L is the length of the beam.
Let

$M_p = \iint_{sp} \rho_p ds_p$ mass of tip

$J_p = \frac{1}{2}\iint_{sp} \rho_p (a^2+b^2) ds_p$ moment of inertia of the tip mass w. r. t the tip of the beam

$a_c = \iint_{sp} \rho_p\, a ds_p / M_p$ center of mass position in (a, b) coordinates

$b_c = \iint_{sp} \rho_p\, b ds_p / M_p$ center of mass position in (a, b) coordinates

Therefore,

$$T_p = \frac{1}{2}\iint_{sp} \rho_p (\dot{x}_p{}^2 + \dot{y}_p{}^2) ds_p$$

$$= \frac{M_p}{2}[\Delta_x{}^2(L)+\Delta_y{}^2(L)] + \frac{J_p}{2}[\dot{\varphi}(L)+\dot{\theta}]^2 + M_p[\dot{\varphi}(L)+\dot{\theta}]$$

$$+\{[a_c\Delta_x(L)+b_c\Delta_y(L)]\cos\varphi(L)+[a_c\Delta_y(L)-b_c\Delta_x(L)]\sin\varphi(L)\}$$

Let

$$G_x = a_c \Delta_x(L) + b_c \Delta_y(L),$$

$$G_y = a_c \Delta_y(L) - b_c \Delta_x(L).$$

We obtain the tip kinetic energy as,

$$T_p = \frac{1}{2}\iint_{sp} \rho_p (\dot{x}_p{}^2 + \dot{y}_p{}^2) ds_p$$

$$= \frac{M_p}{2}[\Delta_x{}^2(L) + \Delta_y{}^2(L)] + \frac{J_p}{2}[\dot{\varphi}(L) + \dot{\theta}]^2$$

$$+ M_p[\dot{\varphi}(L) + \dot{\theta}](G_x \cos\varphi + G_y \sin\varphi) \qquad (11.2.6)$$

If one assumes that tip load is a point mass M_p then $J_p = 0, G_x = G_y = 0$.

11.2.4 *Total Potential Energy*

The relation between the bending moment and deformation is,

$$M(x,t) = EI(x)\frac{\partial \varphi(x,t)}{\partial x}$$

and the relation between the shearing force and deformation is given by,

$$Q(x,t) = kGA(x)(\varphi - w')$$

From beam theory [21], the total potential energy can be found as,

$$P = \frac{1}{2}\int_0^L [D\varphi'^2 + C(\varphi - w')^2] ds \qquad (11.2.7)$$

where D and C are bending rigidity and shear rigidity of the beam. For beam of uniform cross section, $D = EI$ and $C = kGA$, where E is Young's modulus, and G the shear modulus, and k the shape factor [21, 8].

11.2.5 *The Work Done by External Force*

The work done by external forces to the manipulator system is,

$$\delta W = \tau \delta \theta \tag{11.2.8}$$

where τ is the torque applied at the hub.

11.3 Derivation of Equations of Motion

Substitute Eqs. (11.2.1-11.2.6) into the Hamilton's Principle, we can get the equation of motion equations of flexible arms. This section presents the nonlinear equations which are obtained without ignoring any terms during the entire derivation process. After a long tedious process of variation calculation and manipulation the final results are summarized at the end.

11.3.1 *Euler-Bernoulli Model Derivation*

For Euler-Bernoulli Model, there are three variables, $\theta, w(x), and \varphi(x)$. However, φ is related to w.

$$\varphi = \frac{\partial w}{\partial x}, \ \dot{\varphi} = \frac{d}{dt}\left(\frac{\partial w}{\partial x}\right) = \dot{w}'.$$

The resulting motion equations are found to be,

$$(Dw'')'' + \rho(x\ddot{\theta} + \ddot{w} - S\ddot{w}'' - \dot{\theta}^2 w) = 0, \tag{11.3.1}$$

$$I_H \ddot{\theta} + \frac{\partial}{\partial t}\int_0^L \rho[x\dot{w} + (x^2 + w^2)\dot{\theta} + S(\dot{\theta} + \dot{w}')]dx + H_\theta = \tau, \tag{11.3.2}$$

with the corresponding boundary conditions,

$$x=0,\quad w=0,\ w'=0; \tag{11.3.3}$$

$$x=L,\ Dw''+H_{\varphi}=0,\ (Dw'')'-\rho S(\ddot{w}'+\ddot{\theta})-H_{w}=0; \tag{11.3.4}$$

Where

$$F_x=\Delta_x(L)+(a_c\cos\varphi-b_c\sin\varphi)(\dot{\varphi}+\dot{\theta}),$$
$$F_y=\Delta_y(L)+(a_c\sin\varphi+b_c\cos\varphi)(\dot{\varphi}+\dot{\theta}),$$
$$F_{\varphi\theta}=J_p[\dot{\varphi}(l)+\dot{\theta}]+M_p(G_x\cos\varphi+G_y\sin\varphi),$$
$$H_\theta=\frac{\partial}{\partial t}\{F_{\varphi\theta}+M_p[LF_x+w(L)F_y]\},$$
$$H_w=M_p(\dot{F}_x-\dot{\theta}F_y),$$
$$H_\varphi=\dot{F}_{\varphi\theta}-M_pF_\varphi.$$

11.3.2 *Timoshenko Theory Model*

For Timoshenko Model, there are three independent variables, $\theta, w(x)$ and $\varphi(x)$.
φ is treated as an independent variable in the derivation, and the motion equations motion are found to be,

$$(D\varphi')'-c(\varphi-w')-\rho S(\ddot{\varphi}+\ddot{\theta})=0, \tag{11.3.5}$$

$$C(\varphi-w')+\rho(x\ddot{\theta}+\ddot{w}-\dot{\theta}^2w)=0, \tag{11.3.6}$$

$$I_H\ddot{\theta}+\frac{\partial}{\partial t}\int_0^L\rho[x\dot{w}+(x^2+w^2)\dot{\theta}+S(\dot{\theta}+\dot{w}')]dx+H_\theta=\tau, \tag{11.3.7}$$

with boundary conditions,

$$x=0,\quad w=0,\quad \varphi=0; \tag{11.3.8}$$

$$x=L,\quad D\varphi'+H_\varphi=0,\quad C(\varphi-w')-H_w=0. \tag{11.3.9}$$

where $H_{\theta,}H_{\varphi}$ and H_w are given by the same equations as in the Euler-Bernoulli model. Note that now we have three, instead of two, coupled motion equations.

11.4 Linearization

11.4.1 *Introduction*

In order to analyze fundamental vibration behaviors and the time response of the manipulator system, we need to linearize the nonlinear dynamic equations. The dynamic models after linearization are easier to analyze and easier for control design.

To linearize the dynamic models described in the previous section, we neglect all deformation terms which are higher than the first order. This is justified due to the small deformation assumption. All the second or higher order displacement and strain terms are ignored [21] for both the Euler-Bernoulli and Timoshenko beam theories. In addition, we ignore all the second or higher order cross terms. Note that these simplifications have been adopted in almost all the published works on modeling of flexible manipulators. As we can see later from the simulation results, nonlinear terms have noticeable effects on dynamic responses of flexible arms only when motion speeds are extremely high. After a tedious process of simplification, we arrive at the following relationships,

$$H_\theta = H_\varphi + LH_{W,} \quad H_w = m_p\dot{F}_x, \quad H_\varphi = \dot{F}_{\varphi\theta}$$
$$F_x = \Delta_x(x) + a_c(\dot{\varphi}+\dot{\theta})$$
$$F_{\varphi\theta} = J_p(\dot{\varphi}+\dot{\theta}) + a_c M_p \Delta_x(L)$$

Therefore:

$$\frac{\partial}{\partial t}\int_0^L \rho[s(\dot{w}+s\dot{\theta})+S(\dot{\theta}+\dot{\varphi})]ds + H_\theta = -D\varphi'(0) \qquad (11.4.1)$$

which enables us to eliminate the integral terms in dynamic equations.

To make the form of dynamic equations simpler, we further define the total deflection v and rotation α of the manipulator as,

$$v(x,t) = w(x,t) + x\theta(t), \qquad \alpha(x,t) = \varphi(x,t) + \theta(t) \tag{11.4.2}$$

In terms of the total deflection and rotation, the linearized dynamic equations of flexible manipulators can be summarized as follows.

11.4.2 *Euler-Bernoulli Model after Linearization*

Using Eq. (11.4.1), we have

$$(Dv'')'' + \rho(\ddot{v} - S\ddot{v}'') = 0 \tag{11.4.3}$$

$$I_H\ddot{\theta} - Dv''(0) = \tau \tag{11.4.4}$$

with boundary conditions,

$$x = 0, \quad v = 0, \quad v' = \theta; \tag{11.4.5}$$

$$x = L, Dv'' + J_p\ddot{v}' + a_cM_p\ddot{v} = 0, \ (Dv'')' - \rho S\ddot{v}' = M_p(\ddot{v} + a_c\ddot{v}'). \tag{11.4.6}$$

Note that in the dynamic models presented in [1], neither the rotary inertia nor the size of the tip load had been considered (i.e., S = 0 and $a_c = 0$ were assumed).

11.4.3 *Timoshenko Model after Linearization*

Similarly, we have

$$(D\alpha')' - C(\alpha - v') - \rho S\ddot{\alpha} = 0, \qquad [C(\alpha - v')]' + \rho\ddot{v} = 0, \tag{11.4.7}$$

$$I_H\ddot{\theta} - D\alpha'(0) = \tau, \tag{11.4.8}$$

with boundary conditions,

$$x = 0, \quad v = 0, \quad \alpha = \theta; \tag{11.4.9}$$

$$x = L, \quad D\alpha' + J_p\ddot{\alpha} + a_cM_p\ddot{v} = 0, \qquad C(\alpha - v') = M_p(\ddot{v} + a_c\ddot{\alpha}). \tag{11.4.10}$$

Note that when no tip load is present, the equations in this case are the same as those given in [33].

From boundary conditions (11.4.5) or (11.4.6), we find the equation (11.4.6) can be used to calculate hub rotation θ and thus to eliminate θ from dynamic equations. To compensate for the lost boundary condition, Eq. (11.4.4) or (11.4.8) can be considered as the new condition. As a matter of fact, Eq. (11.4.4) or (11.4.8) does represent the dynamic torque balance equation of the hub according to Euler-Bernoulli or Timoshenko beam model.

11.4.4 *Dimensionless Functions, Variables and Parameters*

The linearized equation of motion may be further simplified by introducing the following dimensionless variables and parameters,

$$z(\xi)=\frac{v}{L},\quad \xi=\frac{x}{L},\quad \delta=\frac{S}{L^2},\quad \sigma=\frac{CL^2}{D},\quad c^2=\frac{\rho L^2}{D},\qquad (11.4.11)$$

$$\mu=\frac{M_p}{\rho L},\qquad \eta=\frac{I_H}{\rho L^3},\qquad \kappa=\frac{J_p}{\rho L^3},\qquad \zeta=\frac{a_c}{L}.\qquad (11.4.12)$$

The resulting equations are:

Euler-Bernoulli Model with Dimensionless Variables

$$z'''' + c^2\ddot{z} - \delta c^2\ddot{z}'' = 0,\qquad (11.4.13)$$

boundary conditions,

$$z(0)=0,\qquad \eta c^2\ddot{z}'(0) - z''(0) = \frac{\tau L}{D};\qquad (11.4.14)$$

$$z''(1) + c^2[\kappa\ddot{z}'(1) + \zeta\mu\ddot{z}(1)] = 0,$$

$$z'''(1) - \delta c^2\ddot{z}'(1) - \mu c^2[\ddot{z}(1) + \zeta\ddot{z}'(1)] = 0\qquad (11.4.15)$$

Timoshenko Model with Dimensionless Variables

$$\alpha'' - \sigma(\alpha - z') - \delta c^2 \alpha = 0; \quad \sigma(\alpha - z')' + c^2 \ddot{z} = 0. \tag{11.4.16}$$

Boundary conditions,

$$z(0) = 0, \quad \eta c^2 \ddot{\alpha}(0) - \alpha'(0) = \frac{\tau L}{D}, \tag{11.4.17}$$

$$\alpha'(1) + c^2[\kappa \ddot{\alpha}(1) + \zeta \mu \ddot{z}(1)] = 0,$$
$$\sigma[\alpha(1) - z'(1)] - \mu c^2[\ddot{z}(1) + \zeta \ddot{\alpha}(1)] = 0. \tag{11.4.18}$$

A prime now indicates the differentiation with respect to the dimensionless coordinate ξ.

Note that if the shear deformation is very small, σ must be very large since the shear modulus G is large in this case. From Eq. (11.4.16), α must approach z' for all ξ in order to keep $\sigma(\alpha - z')$ of finite value. Actually, the first equation of Eq. (11.4.16) reveals that $\sigma(\alpha - z') \to \alpha'' - \delta c^2 \ddot{\alpha} \to z''' - \delta c^2 \ddot{z}'$ as $\sigma \to \infty$. Substitute this result into the second equation and boundary conditions, we can show easily that the Timoshenko model reduces to the Euler-Bernoulli model when $\sigma \to \infty$.

Sometime it is much easier to deal with a single higher order decoupled partial differential equation than several coupled low order equations. Using the technique developed in [34], the two coupled equations in (11.4.16) can be decoupled by the following transformation,

$$z = \phi', \quad \alpha = \phi'' - \frac{c^2}{\sigma} \ddot{\phi}, \tag{11.4.19}$$

where ϕ is a new function which has to satisfy the equation,

$$\phi'''' - (\frac{1}{\sigma} + \delta) c^2 \ddot{\phi}'' + c^2 \ddot{\phi} + \frac{\delta}{\sigma} c^4 \ddot{\ddot{\phi}} = 0. \tag{11.4.20}$$

Thus, we have replaced two coupled second order equations involving two variables by an equivalent fourth order equation with one variable.

11.5 Natural Frequencies and Model Shape Functions

Based on the results in the previous section, the harmonic vibration equation of one-link flexible manipulators can be obtained easily as follows.

Euler-Bernoulli Model

Mathematically, synchronous motions imply that the solution of Eqs. (11.4.13), (11.4.14) and (11.4.15) is separable in the spatial variable and time, and hence it has the form,

$$z(\xi,t) = z(\xi)e^{iwt} \tag{11.5.1}$$

where $z(\xi)$ depends on the spatial position alone and e^{iwt} depends on time alone.
Introducing Eq. (11.5.1) into Eqs. (11.4.13), (11.4.14) and (11.4.15), we can write,

$$z''''+\delta m^2 z'' - m^2 z = 0, \tag{11.5.2}$$

boundary conditions,

$$z(0)=0, \quad z''(0)+\eta m^2 z'(0) = 0; \tag{11.5.3}$$

$$z''(1) - m^2[\kappa z'(1) + \zeta\mu z(1)] = 0,$$

$$z'''(1) + m^2[(\delta+\mu\zeta)z'(1) + \mu z(1)] = 0. \tag{11.5.4}$$

where $m = c\omega$ is the dimensionless frequency, and ω is the circular vibration frequency.
The solution of Eq. (11.5.2) has exponential form

$$z(\xi) = Ae^{\lambda\xi} \tag{11.5.5}$$

Introducing Eq. (11.5.5) into Eq. (11.5.2) and dividing through by $Ae^{\lambda\xi}$, we obtain the characteristic equation,

$$\lambda^4 + \delta m^2 \lambda^2 - m^2 = 0 \tag{11.5.6}$$

and,

$$\lambda_{1,3} = \pm m\left(\frac{-\delta + \sqrt{\delta^2 + 4m^{-2}}}{2}\right)^{\frac{1}{2}}, \quad \lambda_{2,4} = \pm m\left(\frac{\delta + \sqrt{\delta^2 + 4m^{-2}}}{2}\right)^{\frac{1}{2}} i \cdot \tag{11.5.7}$$

Let $$\beta_1 = \left[\frac{\sqrt{\delta^2 + 4m^{-2}} + \delta}{2}\right]^{\frac{1}{2}}, \beta_2 = \left[\frac{\sqrt{\delta^2 + 4m^{-2}} - \delta}{2}\right]^{\frac{1}{2}}, \tag{11.5.8}$$

$$\lambda_j = \begin{cases} \pm m\beta_1 i \\ \pm m\beta_2 \end{cases}, \qquad j = 1,2,3,4\,. \tag{11.5.9}$$

$$k = \frac{m\lambda_1\lambda_2}{\lambda_1^{\ 2} + \lambda_2^{\ 2}} = \frac{1}{\sqrt{\delta^2 + 4m^{-2}}}\,.$$

The characteristic equation for determining the natural frequency can be found as,

$$\sum \delta(m) = \begin{bmatrix} z_{11}(m) & z_{12}(m) \\ z_{21}(m) & z_{22}(m) \end{bmatrix} = 0\,,$$

where

$$\begin{bmatrix} z_{11} & z_{12} \\ z_{21} & z_{22} \end{bmatrix} = \begin{bmatrix} Z''(1) - m^2(\kappa Z'(1) + \zeta\mu Z(1)) \\ Z'''(1) + m^2((\delta + \mu\zeta)Z'(1) + \mu Z(1) \end{bmatrix}$$

in which function $Z = (z_s, z_h)$ is defined as,

$$z_s(\xi) = \beta_2 \sin\lambda_1\xi + \eta k(\cos\lambda_1\xi - \cosh\lambda_2\xi)\,,$$
$$z_h(\xi) = \beta_1 \sin\lambda_2\xi + \eta k(\cos\lambda_1\xi - \cosh\lambda_2\xi)\,;$$

Fig. 11.3 presents ratios of the 1st, 2nd, and 3rd order natural frequencies calculated from the Euler-Bernoulli model without ($\delta = 0$)

and with $(\delta \neq 0)$ the rotary inertia taken into account, respectively. No tip load is assumed in these two figures. Clearly, the effect of the rotary inertia on vibration frequency is significant, especially for higher order vibration modes.

In the Fig. 11.3, H is the beam height and L is the beam length. m_2 and m_1 are the frequency of the models with and without rotary inertia respectively. Curve A is the first mode, B is the second and C is the third.

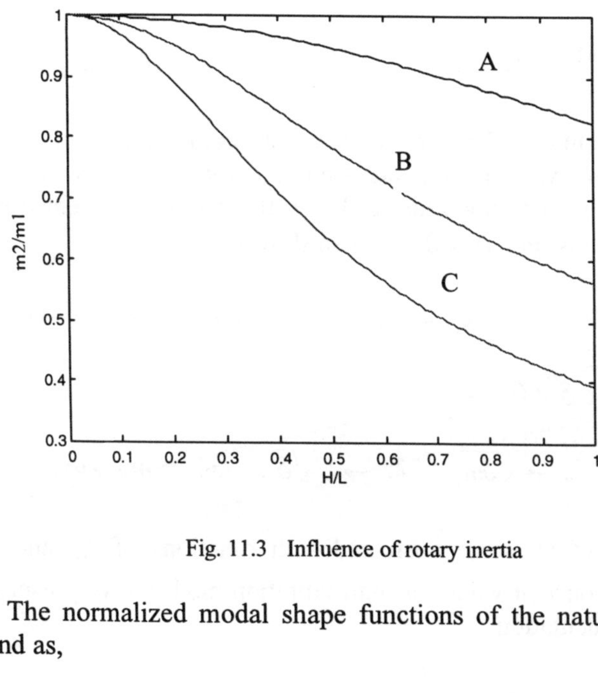

Fig. 11.3 Influence of rotary inertia

The normalized modal shape functions of the natural vibration can found as,

$$z(\xi;m) = z_s(\xi) - \frac{z_{11}(m)}{z_{12}(m)} z_h(\xi),$$

$$z_i(\xi) = \frac{z(\xi;m)}{\langle z(\xi;m_i) | z(\xi;m_i) \rangle_\delta}, \qquad i = 0,1,2,\ldots,$$

where $\langle * | * \rangle_\delta$ is the generalized inner product for the Euler-Bernoulli model,

$$\langle f | z \rangle_\delta \equiv \int_0^1 (fz + \delta f'z') d\xi + \eta f'(0) z'(0) + \mu f(1) z(1) + \mu\zeta [f'(1) z(1) + f(1) z'(1)] + \kappa f'(1) z'(1) .$$

Once can show the following orthogonal relationship between modal shape functions,

$$\langle z_i | z_j \rangle_\delta = \delta_{ij} , \quad \int_0^1 z_i'' z_j'' d\xi = m^2 \delta_{ij} , \quad \delta_{ij} = \begin{cases} 0 & i \neq j \\ 1 & i = j \end{cases} \tag{11.5.11}$$

For high vibration modes, the computation of exponential functions involved in cosh and sinh functions becomes very difficult. Therefore it is very important to know the asymptotic behavior of high order vibrations. For $\delta = 0$, one can show that,

$$\lambda_1 \to \sqrt{m} , \quad \lambda_2 \to \sqrt{m} , \quad \text{as} \quad m \to \infty .$$

However, for $\delta \neq 0$,

$$\lambda_1 \to \sqrt{\delta m} , \quad \lambda_2 \to 1/\sqrt{\delta} , \quad \text{as} \quad m \to \infty .$$

Therefore, some cosh and sinh functions of z_s and z_h will approach a constant value for high vibration modes, a very useful fact in numerical calculation.

Timoshenko Model

Mathematically, synchronous motions imply that the solution of Eqs. (11.4.15), (11.4.16) and (11.4.17) is separable in the spatial variable and time, and hence it has the form,

$$\phi(\xi, t) = \phi(\xi) e^{iwt} \tag{11.5.12}$$

where $\phi(\xi)$ depends on the spatial position alone and e^{iwt} depends on time alone.
Introducing Eq. (11.5.12) into Eqs. (11.4.15), (11.4.16) and (11.4.17), we can write,

$$\phi'''' - (\frac{1}{\sigma} + \delta)m^2\phi'' + m^2\phi + \frac{\delta}{\sigma}c^2m^2\phi = 0 \quad (11.5.13)$$

Boundary conditions,

$$\phi'(0) = 0, \quad \phi'''(0) + \frac{m^2}{\sigma}\phi'(0) + \eta m^2[\phi''(0) + \frac{m^2}{\sigma}\phi(0)] = 0, \quad (11.5.14)$$

$$\phi'''(1) - m^2\kappa\phi(1) + m^2(\frac{1}{\sigma} - \zeta\mu)\phi'(1) - \kappa m^4\phi(1) = 0,$$

$$\zeta\mu\phi''(1) + \mu\phi'(1) + (1 + \zeta\frac{\mu}{\sigma})\phi(1) = 0. \quad (11.5.15)$$

The solution of Eq. (11.5.13) has exponential form

$$\phi(\xi) = Ae^{\lambda\xi} \quad (11.5.16)$$

Introducing Eq. (11.5.16) into Eq. (11.5.13) and dividing through by $Ae^{\lambda\xi}$, we obtain the characteristic equation,

$$\lambda^4 + (\delta + \frac{1}{\sigma})m^2\lambda^2 - (1 + \frac{\delta c^2}{\sigma})m^2 = 0 \quad (11.5.17)$$

Which has the roots,

$$\lambda_{1,3} = \pm m\left(\frac{-\delta_s + \sqrt{\delta_s{}^2 + 4\delta_d m^{-2}}}{2}\right)^{\frac{1}{2}}, \quad \lambda_{2,4} = \pm m\left(\frac{\delta_s + \sqrt{\delta_s{}^2 + 4\delta_d m^{-2}}}{2}\right)^{\frac{1}{2}} i \quad (11.5.18)$$

Where $\delta_s = (\frac{1}{\sigma} + \delta)$, $\delta_d = (\frac{\delta}{\sigma}c^2 + 1)$.

Let $$\beta_1 = \left[\frac{\sqrt{\delta_d^{\ 2} + 4\delta_d m^{-2}} + \delta_s}{2}\right]^{\frac{1}{2}}, \beta_2 = \left[\frac{\sqrt{\delta_d^{\ 2} + 4\delta_d m^{-2}} - \delta_s}{2}\right]^{\frac{1}{2}}, \tag{11.5.19}$$

$$\lambda_j = \begin{cases} \pm m\beta_1 i \\ \pm m\beta_2 \end{cases}, \qquad j = 1,2,3,4. \tag{11.5.20}$$

$$k_i = \frac{m^3[m^2 + (-1)^i \sigma\lambda_i^{\ 2}]}{\sigma\lambda_1\lambda_2(\lambda_1^{\ 2} + \lambda_2^{\ 2})} = \frac{m[1 + (-1)^i \sigma\beta_i^{\ 2}]}{\sigma\beta_1\beta_2(\beta_1^{\ 2} + \beta_2^{\ 2})}.$$

The characteristic equation for determining the natural frequencies is

$$\sum\sigma(m) = \det\begin{bmatrix} \phi_{11}(m) & \phi_{12}(m) \\ \phi_{21}(m) & \phi_{22}(m) \end{bmatrix} = 0,$$

where

$$\begin{bmatrix} \phi_{11} & \phi_{12} \\ \phi_{21} & \phi_{22} \end{bmatrix} = \begin{bmatrix} \Phi'''(1) - m^2\kappa\Phi''(1) + m^2(1/\sigma - \zeta\mu)\Phi'(1) - \kappa m^4\,\Phi(1)/\sigma \\ \zeta\mu\Phi''(1) + \mu\Phi'(1) + (1 + \zeta\,\mu/\sigma)\Phi(1) \end{bmatrix}$$

in which $\Phi = (\phi_s, \phi_h)$ is defined as,

$$\phi_s(\xi) = \cos\lambda_1\xi + \eta k_1(\beta_2 \sin\lambda_1\xi - \beta_1 \sinh h\lambda_2\xi),$$
$$\phi_h(\xi) = \cosh\lambda_2\xi + \eta k_2(\beta_2 \sin\lambda_1\xi - \beta_1 \sinh h\lambda_2\xi);$$

Fig. 11.4 presents ratios of the 1st, 2nd, and 6th order frequencies calculated from the Euler-Bernoulli model $(\delta = 0)$ and Timoshenko $(\delta = 0, \sigma \neq 0)$, models, respectively. Tip load is set to zero. Like the rotary inertia, the effect of the shear deformation on vibration frequency is also significant, especially for higher order vibration modes. Figs. 6, 7 agree with the results given in [33].
In the Fig. 11.4, H is the beam height and L is the beam length, m_3 and m_1 are the frequency of the models with shear deformation and without

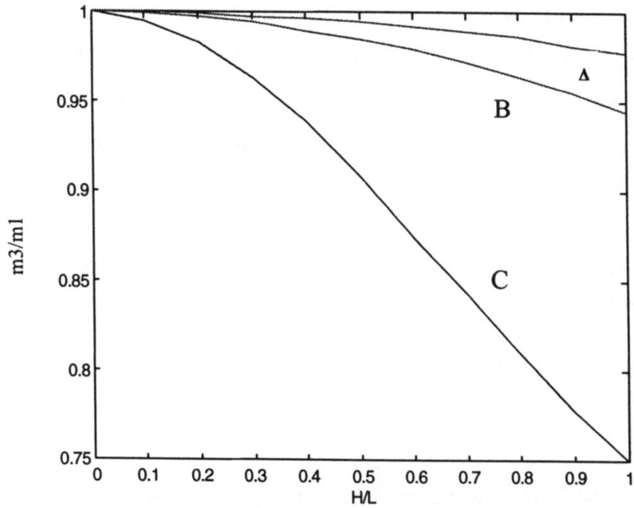

Fig. 11.4 Influence of shear deformation

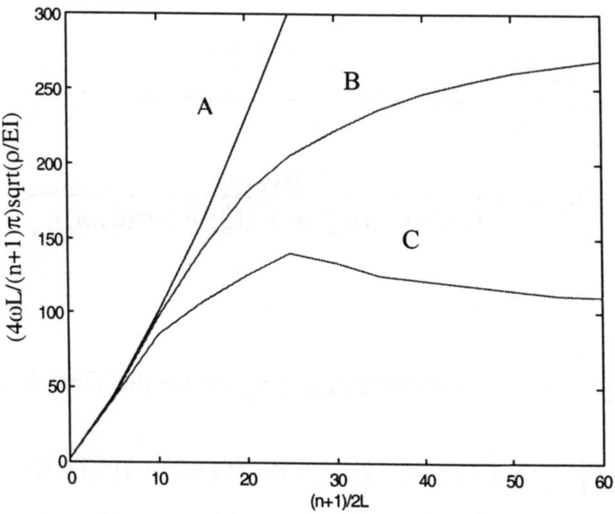

Fig. 11.5 Influence of rotary inertia and shear deformation

rotary inertia respectively. Curve A is the first mode, curve B is the second mode and curve C is the sixth mode.

In Figure 11.5, different modal frequencies calculated from Euler-Bernoulliith and without rotary inertia and Timoshenko models, respectively are illustrated, in terms of $(4\omega L/(n+1)\pi)\sqrt{\rho/EI}$ vs. $(n+1)/(2L)$. This indicates clearly that the effect of rotary inertia and shear deformation is profound for high order vibration modes. In the Figure 11.5, ω is the natural frequency, $c = L\sqrt{\rho/EI}$, n is the order of vibration mode.
Curve A is Euler-Bernoulli model without rotary inertia, curve B is Euler-Bernoulli model with rotary inertia, curve C is Timoshenko model.

The normalized modal shape functions of natural vibration can be found as,

$$\phi(\xi;m) = \phi_s(\xi) - \frac{\phi_{11}(m)}{\phi_{12}(m)}\phi_h(\xi), \quad z(\xi;m) = \phi'(\xi;m),$$

$$\alpha(\xi;m) = \phi''(\xi;m) + \frac{m^2}{\sigma}\phi(\xi;m),$$

$$z_{\sigma i}(\xi) = \frac{z(\xi;m)}{\langle z(\xi;m_i),\alpha(\xi;m_i) | z(\xi;m_i),\alpha(\xi;m_i)\rangle_\sigma},$$

$$\alpha_i(\xi) = \frac{\alpha(\xi;m)}{\langle z(\xi;m_i),\alpha(\xi;m_i) | z(\xi;m_i),\alpha(\xi;m_i)\rangle_\sigma},$$

$$i = 0,1,2,....,$$

where $\langle * | * \rangle_\delta$ is the generalized inner product for the Timoshenko model,

$$\langle f | z \rangle_\sigma \equiv \int_0^1 (fz + \delta g\alpha)d\xi + \eta g(0)\alpha(0) + \mu f(1)z(1)$$
$$+ \mu\zeta[g(1)z(1) + f(1)\alpha(1)] + \kappa g(1)\alpha(1) = 0.$$

Similar to the Euler-Bernoulli Model, the following orthogonal relationship is true,

$$\langle z_{\sigma i},\alpha_i \,|\, z_{\sigma j},\alpha_j\rangle_\sigma = \delta_{ij}, \qquad \int_0^1 \alpha_i'\alpha_j' + \sigma(\alpha - z_i')(\alpha - z_j')d\xi = m^2\delta_{ij}, \tag{11.5.22}$$

Note that $\langle f\,|\,z\rangle_\sigma = \langle f, f' \,|\, z, z'\rangle_\sigma$.

When $m^2 > \sigma/\delta$, β_2 becomes an imaginary number. In this case,

$$\beta_i = \left[\frac{\delta_s - (-1)^i\sqrt{\delta_d{}^2 + 4m^{-2}}}{2}\right]^{\frac{1}{2}}, \qquad k_i = \frac{m[1-\sigma\beta_i{}^2]}{\sigma\beta_1\beta_2(\beta_1{}^2+\beta_2{}^2)}, \qquad i = 1,2;$$

and cosh and sinh functions in ϕ_s and ϕ_h are replaced by cos and sin functions, respectively. Therefore, for large natural vibration frequencies, the computational problem associated with the exponential functions does not appear here.

11.6 Step Responses and General Solutions

In this section, we present analytic expressions of step and general responses for flexible manipulators. The results in section 11.5 are used extensively here. The detailed derivation of these responses is quite tedious and only the final results are given in this section.

For step input $\tau =$ constant, one can find step responses of system (4.13-4.15) and (4.16-4.18) as,

$$z(\xi,t) = z_{\delta p}(\xi,t) + \tau\sum_{i=0}^{\infty}[c_{i1}{}^{\delta}\cos\omega_i(t-t_0) + c_{i2}{}^{\delta}\sin\omega_i(t-t_0)]z_i(\xi) \tag{11.6.1}$$

$$z_\sigma(\xi,t) = z_{\sigma p}(\xi,t) + \tau\sum_{i=0}^{\infty}[c_{i1}{}^{\sigma}\cos\omega_i(t-t_0) + c_{i2}{}^{\sigma}\sin\omega_i(t-t_0)]z_{\sigma i}(\xi) \tag{11.6.2}$$

$$\alpha(\xi,t) = \alpha_{\sigma p}(\xi,t) + \tau\sum_{i=0}^{\infty}[c_{i1}{}^{\sigma}\cos\omega_i(t-t_0) + c_{i2}{}^{\sigma}\sin\omega_i(t-t_0)]\alpha_i(\xi) \tag{11.6.3}$$

where z and (z_σ, α) are for the Euler-Bernoulli (Timoshenko) and $z_{\delta p}, z_{\sigma p}$ and $\alpha_{\sigma p}$ are the corresponding particular solutions of these models,

$$(z_{\delta p}, z_{\sigma p}) = a\tau\left[\frac{\xi(t-t_0)^2}{2} - c^2(A_\delta, A_\sigma)\right]; \; (\alpha_{\sigma p} = a\tau\left[\frac{(t-t_0)^2}{2} - c^2 A_\alpha\right];$$

Where,

$$A_\delta(\xi) = \frac{\xi^5}{5!} - c_3\frac{\xi^3}{3!} + c_2\frac{\xi^2}{2!};$$

$$A_\sigma(\xi) = \frac{\xi^5}{5!} - (c_3 + \frac{1}{\sigma})\frac{\xi^3}{3!} + c_2\frac{\xi^2}{2!} + \frac{c_1}{\sigma}\xi;$$

$$A_\alpha(\xi) = \frac{\xi^4}{4!} - c_3\frac{\xi^2}{2!} + c_2\xi.$$

And,

$$c_1 = \frac{1}{2} + (1+\zeta)\mu, \quad c_2 = \frac{1}{3} + \delta + \kappa + 2\zeta\mu + \mu, \quad c_3 = c_1 + \delta,$$

$$a = \frac{1}{\rho L^3(\eta + c_2)}.$$

To determine the coefficients in (11.5.15-11.5.17), we consider the initial conditions,

$$z(\xi, t_0) = H_{\delta 0}(\xi), \quad \dot{z}(\xi, t_0) = \dot{H}_{\delta 0}(\xi),$$

$$z_\sigma(\xi, t_0) = H_{\sigma 0}(\xi), \quad \dot{z}_\sigma(\xi, t_0) = \dot{H}_{\sigma 0}(\xi),$$

$$\alpha(\xi, t_0) = A_{\sigma 0}(\xi), \quad \dot{\alpha}(\xi, t_0) = \dot{A}_{\sigma 0}(\xi).$$

Using orthogonal relationships (5.5) and (5.10), we found,

$$c_{i1}{}^\delta = ac^2\beta_i{}^\delta + \langle H_{\delta 0} | z_i \rangle_\delta, \quad c_{i2}{}^\delta = \frac{c}{m_i}\langle \dot{H}_{\delta 0} | z_i \rangle_\delta;$$

$$c_{i1}{}^{\sigma} = ac^2\beta_i{}^{\sigma} + \left\langle H_{\sigma 0}, A_{\sigma 0} \middle| z_{\sigma i}, \alpha_i \right\rangle_{\sigma},$$

$$c_{i2}{}^{\sigma} = \frac{c}{m_i}\left\langle \dot{H}_{\sigma 0}, \dot{A}_{\sigma 0} \middle| z_{\sigma i}, \alpha_i \right\rangle_{\sigma},$$

where

$$\beta_i{}^{\delta} = \left\langle A_{\delta} \middle| z_i \right\rangle_{\delta}, \quad \beta_i{}^{\sigma} = \left\langle A_{\sigma}, A_{\alpha} \middle| z_{\sigma i}, \alpha_i \right\rangle_{\sigma}.$$

For zero initial conditions, step responses become,

$$z(\xi,t) = z_{\delta p}(\xi,t) + \tau\sum_{i=0}^{\infty}\beta_i{}^{\delta}\cos\omega_i(t-t_0)z_i(\xi)$$

$$z_{\sigma}(\xi,t) = z_{\sigma p}(\xi,t) + \tau\sum_{i=0}^{\infty}\beta_i{}^{\sigma}\cos\omega_i(t-t_0)z_{\sigma i}(\xi)$$

$$\alpha(\xi,t) = \alpha_{\sigma p}(\xi,t) + \tau\sum_{i=0}^{\infty}\beta_i{}^{\sigma}\cos\omega_i(t-t_0)\alpha_i(\xi)$$

Therefore, after taking derivatives with respect to time, we find impulse responses of flexible manipulators as,

$$z_h(\xi,t) = a\xi(t-t_0) - \sum_{i=0}^{\infty}\beta_i{}^{\delta}\omega_i\sin\omega_i(t-t_0)z_i(\xi) \tag{11.6.4}$$

$$z_{h\sigma}(\xi,t) = a\xi(t-t_0) - \sum_{i=0}^{\infty}\beta_i{}^{\sigma}\omega_i\sin\omega_i(t-t_0)z_{\sigma i}(\xi) \tag{11.6.5}$$

$$\alpha_h(\xi,t) = a(t-t_0) - \sum_{i=0}^{\infty}\beta_i{}^{\sigma}\omega_i\sin\omega_i(t-t_0)\alpha_i(\xi) \tag{11.6.6}$$

Using the Duham's Theorem [17], we can find the response of a flexible manipulator under a general input $\tau = \tau(t)$. For the Euler-Bernoulli model, the response is,

$$z(\xi,t) = \int z_h(\xi,t)\tau(t-s)ds, \tag{11.6.7}$$

and for the Timoshenko model,

$$z_\sigma(\xi,t) = \int_{t_0}^{t} z_{\sigma h}(\xi,t)\tau(t-s)ds \,, \alpha(\xi,t) = \int_{t_0}^{t} \alpha_h(\xi,t)\tau(t-s)ds \tag{11.6.8}$$

The corresponding transfer functions can be determined easily as,

$$H_{z\tau}^{\ \delta}(\xi,s) = \frac{\Gamma(z(\xi,t)}{\Gamma(\tau(t)} = \frac{a\xi}{s^2} - \sum_{i=1}^{\infty} \frac{\beta_i^{\ \delta}\omega_i}{s^2+\omega_i^{\ 2}} z_i(\xi) \tag{11.6.9}$$

$$H_{z\tau}^{\ \sigma}(\xi,s) = \frac{\Gamma(z_\sigma(\xi,t)}{\Gamma(\tau(t)} = \frac{a\xi}{s^2} - \sum_{i=1}^{\infty} \frac{\beta_i^{\ \sigma}\omega_i}{s^2+\omega_i^{\ 2}} z_{\sigma i}(\xi) \tag{11.6.10}$$

$$H_{\alpha\tau}(\xi,s) = \frac{\Gamma(\alpha(\xi,t)}{\Gamma(\tau(t)} = \frac{a\xi}{s^2} - \sum_{i=1}^{\infty} \frac{\beta_i^{\ \sigma}\omega_i}{s^2+\omega_i^{\ 2}} \alpha_i(\xi) \tag{11.6.11}$$

These results are very useful in control design for flexible manipulators.

11.7 Conclusion

Based on the two dynamic models for one-link flexible manipulators, a comprehensive study on the dynamic effects of rotary inertia and shear deformation has been conducted in the chapter. Analytical expressions of vibration frequencies and shape functions, step responses, and general solutions of one-link flexible manipulators have been derived. To illustrate the effects of rotary inertia and shear deformation, various numerical results have been presented through the chapter. Results obtained in this chapter have shown clearly that both rotary inertia and shear deformation have significant influences on the dynamic behaviors and should be considered in control design.

References

1. F. Bellezze, L. Lanari and G. Ulivi, Exact Modeling of the Flexible Slewing Link, IEEE 1990 International Conference on Robotics and Automation, Vol.2, pp.734-739, 1990.

2. M. Boutaghou and A. G. Erdman, A Unified Approach for the Dynamics of Beams Undergoing Arbitrary Spatial Motion, Journal of Vibration and Acoustics-Transactions of the ASME, Vol.113, No.4, pp.494-507, 1991.
3. R. H. Cannon, Jr. and E. Schmitz, Precise Control of Flexible Manipulators, Robotics Research: The First Int. Symposium, MIT Press, Cambridge, MA, pp.841-861, 1984.
4. S. Cetinkunt and W. L. Yu, Closed Loop Behavior of a Feedback Controlled Flexible Arm: A Comparative Study, Int. J. of Robotics Res., Vol.10 No.3, pp.263-275, 1991.
5. Y. Chait, A Natural Modal Expansion for the Flexible Robot Arm Problem via a Self-Adjoint Formulation, IEEE Trans, on robotics and Automation Vol.6 No.5, pp.601-604, 1987.
6. R. W. Clough and J. Penzien, Dynamics of Structures, McGraw-Hill, New York, 1975.
7. G. R. Cowper, On the Accuracy of Timoshenko's Beam theory, Journal of the Engineering Mechanics Division, Proc. of the ASME, Vol.94, EM6, pp.1447-53, 1968.
8. G. R. Cowper, The Shear Coefficients in Timoshenko's Beam theory, Journal of Applied mechanics, Vol.33, pp.335-339, 1968.
9. R. M. Davies, A Critical Study of the Hopkins Pressure Bar, Philosophical Transactions of Royal Society, Serial A, 240, pp.454-461, 1948.
10. A. De Luca, P. Lucibello, and G. Ulivi, Inversion Techniques for Trajectory Control of Flexible Robot Arms, Journal of Robotic Systems, Vol.6, No.4, pp.325-344, 1989.
11. A. De luca, B. Siciliano, Trajectory Control of Non-linear One Link Flexible Arm, Int. Journal of Control, Vol.50, pp1699-1715, 1989.
12. T. C. Huang, The Effect of Rotary Inertia and Shear Deformations on the Frequency and Nominal Mode Equations of Uniform Beams with Simple End Conditions, Journal of Applied Mechanics, Vol.28 pp.579-589, 1961.
13. T. R. Kane, R. R. Ryan and A. K. Bamerjee, Dynamics of a Cantilever Beam Attached to a Moving Bar, Journal of Guidance and Control, Vol.10, pp.139-151, 1989.
14. M. Kelemen and A. Bagchi, Modeling and feedback Control of a Flexible Arm of a Robot for Prescribed Frequency-domain Tolerances, IFAC Journal Automatica, Vol.29, No.4, pp.899-909, 1993.
15. F. Khorrami and Umit Ozguner, Perturbation Method in Control of Flexible Link Msnipultors, IEEE Int. Conf. On Robotics and Automation, Vol.1, pp310-315, 1988.
16. K. A. Morris and M. Vidyasagar, A Comparison of Different Models for Beam Vibrations from the Standpoint of Control Design, J. of Dynamic Systems, Measurement, and Control, Vol.112, pp349-356, 1990.
17. A Pars, A treatise on Analytical Dynamics, OX BOX press, Woodbridge, CT, 1979.
18. T. Sakawa, F. Matsuno, and S. Fukushima, Modeling and Feedback Control of a Flexible Arm, Journal of Robotic Systems, Vol.2, pp.453-472, 1985.
19. J. C. Simo and L. Vu-Quoc, On the Dynamics of Flexible Beams Under Large Overall Motions, Part I and II, Journal of Applied Mechanics, Vol.53, pp.849-863, 1986.
20. S. Timoshenko, D. H. Young, and W. Weaver, Jr., Vibration Problems, in Engineering, New York: D. Van Nostrand Company, Inc., Third Edition, 1957.

21. S. Timoshenko and G. H. MacCullough, Elements of Strength of Materials, Third Edition. D. Van Nostrand Company, Inc., New York, 1949.
22. A. P. Tzes and S. Yurkovich, Application and Comparison of One- Line Identification Methods for Flexible Manipulator Control, International Journal of Robotics Research, Vol.10, No.5, pp.515-527, 1991.
23. D. Wang and M. Vidyasagar, Control of a Flexible Beam for Optimum Step Response, IEEE Conf. On Robotics and Automation (Rayleigh, NC), pp.1567-1572, 1987.
24. D. Wang and M. Vidyasagar, Passive Control of a Stiff Flexible Link, International Journal of Robotics Research, Vol.11, No.3, pp.572-578, 1992.
25. F. Y. Wang and John T. Wen, Nonlinear Dynamical Model and Control for a Flexible Beam, CIRSSE Report # 75, RPI, Tory, New York, 1990.
26. F. Y. Wang, Optimum Design of Vibrating Cantilevers: A Classical Problem Revisited, J. of Optimization Theory and Applications, Vol.84, No.3, 1995.
27. F. Y. Wang, On the Extreme Fundamental Frequencies of One-ink Flexible Manipulators, Int. Journal of Robotics Research, Vol.13, No.2, pp.162-170, 1994.
28. F. Y. Wang, Finding the Maximum Bandwidth of a Flexible Arm, Proc. of 32nd IEEE Conf. On Decision and Control (San Antonio, TX), Vol.1, pp.619-625, 1993.
29. F. Y. Wang and Jeffery L. Russell, Minimum-Weight Robot Arm for a Specified Fundamental Frequency, Proc. of the IEEE Conf. on Robotics and Automation (Atlanta, Georgia), Vol.1, pp.490-495, 1993.
30. F. Y. Wang and Jeffery L. Russell, Optimum Shape Construction of flexible Manipulators with Total Weight Constraint, IEEE Transactions on Systems, Man and Cybernetics, Vol.25, No.3, 1995.
31. F. Y. Wang and F. B. He, Free Vibration of Transversely Isotropic Plates, Journal of Vibration and Shock, Vol.5, No.1, pp1-11, 1986.
32. F. Y. Wang and Y. Zhou, On the Vibration Modes of Three-Dimensional Micropolar Elastic Plates, Journal of Sound and Vibration, Vol.146, No.1, pp.1-16, 1991.
33. F. Y. Wang and G.G. Guan, Influences of Rotary Inertia, shear Deformation and Tip load on Vibrations of Flexible Manipulators, Journal of Sound and Vibrations. Vol.171, No.4, pp.433-452. 1994.
34. F. Y. Wang, On the Solutions of Eringen's Micropolar Plate Equations and Other Approximate Equations, Int. J. Engr Sci., Vol.28, No.9, pp.919-935, 1990.
35. G. B. Yang and M. Donath, Dynamic Model of a One Link Manipulator with Both Structural and Joint Flexibility, IEEE Int. Conf. on Robotics and Automation, Vol.6, No.3, pp.454-461, 1989.
36. Leonard Meirovitch, Principles and Techniques of vibrations, Prentice-Hall, Inc. New Jersey 1997.
37. T. E. Albert, Book, W. J. and Dickerson, S. L., "Experiments in Augmenting Active Control of a Flexible Structure with Passive Damping", AIAA paper 86-0176, AIAA 24th Aerospace Sciences Meeting, Reno, Nevada, January 6-9, 1986.
38. Book, W. J., "Recursive Lagrangian Dynamics of Flexible Manipulator Arms", The International Journal of robotic Research, Vol.3, No.3, pp.87-101, 1984
39. K. Gu and B. H. Tongue, " A Method to Improve the Modal convergence for Structures with External Forcing", ASME Journal of Applied Mechanics, Vol.54, No.4, 1987.

40. Hastings, G. and Book, W. J., " Experiments in the Optimal Control of Flexible Manipulator", Proceedings of American Control Conference, Boston, MA, pp.728-729, 1985.

